TABLE OF ATOMIC MASSES (WEIGHTS) BASED ON CARBON-12

Name	Symbol	Atomic No.	Atomic Mass	Name	Symbol	Atomic No.	Atomic Mass
Actinium	Ac	89	(227)[a]	Molybdenum	Mo	42	95.94
Aluminum	Al	13	26.98154	Neodymium	Nd	60	144.24
Americium	Am	95	(243)[a]	Neon	Ne	10	20.179
Antimony	Sb	51	121.75	Neptunium	Np	93	237.0482[b]
Argon	Ar	18	39.948	Nickel	Ni	28	58.71
Arsenic	As	33	74.9216	Niobium	Nb	41	92.9064
Astatine	At	85	(210)[a]	Nitrogen	N	7	14.0067
Barium	Ba	56	137.34	Nobelium	No	102	(259)[a]
Berkelium	Bk	97	(247)[a]	Osmium	Os	76	190.2
Beryllium	Be	4	9.01218	Oxygen	O	8	15.9994
Bismuth	Bi	83	208.9804	Palladium	Pd	46	106.4
Boron	B	5	10.81	Phosphorus	P	15	30.97376
Bromine	Br	35	79.904	Platinum	Pt	78	195.09
Cadmium	Cd	48	112.40	Plutonium	Pu	94	(244)[a]
Calcium	Ca	20	40.08	Polonium	Po	84	(210)[a]
Californium	Cf	98	(251)[a]	Potassium	K	19	39.098
Carbon	C	6	12.011	Praseodymium	Pr	59	140.9077
Cerium	Ce	58	140.12	Promethium	Pm	61	(145)[a]
Cesium	Cs	55	132.9054	Protactinium	Pa	91	231.0359[b]
Chlorine	Cl	17	35.453	Radium	Ra	88	226.0254[b]
Chromium	Cr	24	51.996	Radon	Rn	86	(222)[a]
Cobalt	Co	27	58.9332	Rhenium	Re	75	186.2
Copper	Cu	29	63.546	Rhodium	Rh	45	102.9055
Curium	Cm	96	(247)[a]	Rubidium	Rb	37	85.4678
Dysprosium	Dy	66	162.50	Ruthenium	Ru	44	101.07
Einsteinium	Es	99	(252)[a]	Samarium	Sm	62	150.4
Erbium	Er	68	167.26	Scandium	Sc	21	44.9559
Europium	Eu	63	151.96	Selenium	Se	34	78.96
Fermium	Fm	100	(257)[a]	Silicon	Si	14	28.086
Fluorine	F	9	18.99840	Silver	Ag	47	107.868
Francium	Fr	87	(223)[a]	Sodium	Na	11	22.98977
Gadolinium	Gd	64	157.25	Strontium	Sr	38	87.62
Gallium	Ga	31	69.72	Sulfur	S	16	32.06
Germanium	Ge	32	72.59	Tantalum	Ta	73	180.9479
Gold	Au	79	196.9665	Technetium	Tc	43	98.9062[b]
Hafnium	Hf	72	178.49	Tellurium	Te	52	127.60
Helium	He	2	4.00260	Terbium	Tb	65	158.9254
Holmium	Ho	67	164.9304	Thallium	Tl	81	204.37
Hydrogen	H	1	1.0079	Thorium	Th	90	232.0381[b]
Indium	In	49	114.82	Thulium	Tm	69	168.9342
Iodine	I	53	126.9045	Tin	Sn	50	118.69
Iridium	Ir	77	192.22	Titanium	Ti	22	47.90
Iron	Fe	26	55.847	Tungsten	W	74	183.85
Krypton	Kr	36	83.80	Unnilhexium	Unh	106	(263)[a]
Lanthanum	La	57	138.9055	Unnilpentium	Unp	105	(262)[a]
Lawrencium	Lr	103	(260)[a]	Unnilquadium	Unq	104	(261)[a]
Lead	Pb	82	207.2	Uranium	U	92	238.029
Lithium	Li	3	6.941	Vanadium	V	23	50.9414
Lutetium	Lu	71	174.97	Xenon	Xe	54	131.30
Magnesium	Mg	12	24.305	Ytterbium	Yb	70	173.04
Manganese	Mn	25	54.9380	Yttrium	Y	39	88.9059
Mendelevium	Md	101	(258)[a]	Zinc	Zn	30	65.38
Mercury	Hg	80	200.59	Zirconium	Zr	40	91.22

[a]Mass number of most stable or best-known isotope

[b]Mass number of the isotope of longest half-life

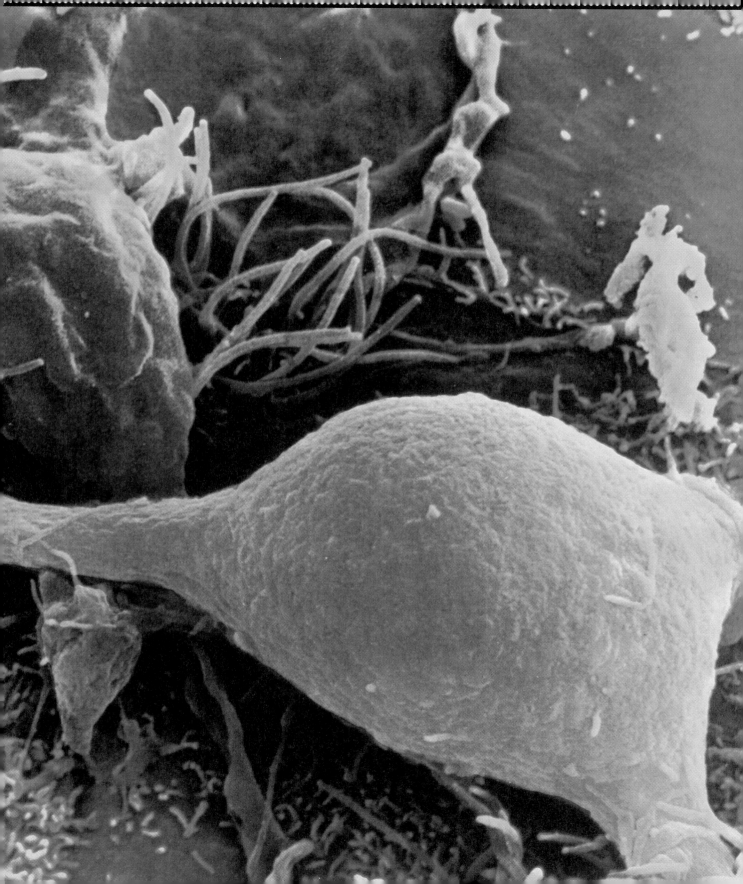

Foundations of College Chemistry

Brief Edition

Morris Hein
Mount San Antonio College

Susan Arena
University of Illinois, Urbana-Champaign

Brooks/Cole Publishing Company
I(T)P An International Thomson Publishing Company

Pacific Grove • Albany • Bonn • Boston • Cincinnati • Detroit • London • Madrid • Melbourne • Mexico City • New York
Paris • San Francisco • Singapore • Tokyo • Toronto • Washington

Sponsoring Editors: *Lisa J. Moller, Harvey C. Pantzis*
Marketing Team: *Connie Jirovsky, Jean Thompson, Tracey Armstrong*
Editorial Associate: *Beth Wilbur*
Production Service: *Julie Kranhold, Ex Libris*
Production Services Coordinator: *Jamie Sue Brooks*
Interior Design: *Nancy Benedict*
Cover Design: *Kelly Shoemaker*
Cover Photo: *Art Wolfe/Tony Stone Images*
Typesetting: *Monotype Composition Company, Inc.*
Cover Printing: *Color Dot Graphics, Inc.*
Printing and Binding: *Quebecor Printing/Hawkins*
Credits continue on page P-1.

For more information, contact:

BROOKS/COLE PUBLISHING COMPANY
511 Forest Lodge Road
Pacific Grove, CA 93950
USA

International Thomson Editores
Campos Eliseos 385, Piso 7
Col. Polanco
11560 México D. F. México

International Thomson Publishing Europe
Berkshire House 168-173
High Holborn
London WC1V 7AA
England

International Thomson Publishing GmbH
Königswinterer Strasse 418
53227 Bonn
Germany

Thomas Nelson Australia
102 Dodds Street
South Melbourne, 3205
Victoria, Australia

International Thomson Publishing Asia
21 Henderson Road
#05-10 Henderson Building
Singapore 0135

Nelson Canada
1120 Birchmount Road
Scarborough, Ontario
Canada M1K 5G4

International Thomson Publishing Japan
Hirakawacho Kyowa Building, 3F
2-2-1 Hirakawacho
Chiyoda-ku, Tokyo 102
Japan

Printed in the United States of America

10 9 8 7 6 5 4 3 2 1

Library of Congress Cataloging-in-Publication Data

Hein, Morris
    Foundations of college chemistry / Morris Hein, Susan Arena.—
Brief ed.
    p.   cm.
    Includes index.
    ISBN 0-534-33919–0
        1. Chemistry.   I. Arena, Susan [date]– . II. Title.
    QD33.H45   1996c
540—dc20   95-35611

The margin text (captions describing cover/interior photos):

♦ *Chemistry is at work in every aspect of our lives. Scientists can actually see matter at the molecular level by using electron microscopes, scanning electron microscopes, and atomic-force microscopes—the first page of this book shows a neuron photographed by an electron microscope.*

♦ *Petrochemical plants operate to convert crude oil into useful petroleum products by separating out fractions of the oil to be used for different purposes.*

♦ *The array of beautiful colors in many fabrics comes from synthetic aniline dyes, the first of which was discovered in 1856 by a student chemist who was attempting to make an entirely different substance.*

♦ *Under inspection, a computer chip creates a colorful and intricate maze. These chips are made from the metalloid silicon.*

♦ *A worker in a pharmaceutical plant monitors the complex and precise machinery used in pill production.*

♦ *Green leaves turn red and gold in autumn. As the days grow shorter and the nights grow chilly, the chlorophyll decomposes. The green fades away while other pigments remain, especially the reds and yellows.*

An education isn't how much you have committed to memory, or even how much you know. It's being able to differentiate between what you do know and what you don't. It's knowing where to go to find out what you need to know; and it's knowing how to use the information once you get it.

William Feather

Foundations of College Chemistry was originally intended for students who had never taken a chemistry course or who had a significant interruption in their studies but planned to continue on with the general chemistry sequence. Since its publication this book has helped define the preparatory chemistry course. The broader acceptance of previous editions indicates that the text has developed a much wider audience. In addition to preparatory chemistry, our text has found extensive use in one-semester general purpose courses, such as those for applied health fields, and in courses for nonscience majors—a diverse population of students. Our goal for this revision is to continue to make introductory chemistry accessible to all beginning students. The central focus is still the same as it has been from the first edition: How can we make this material interesting and understandable to students? And how can we teach them the problem-solving skills they will need?

In preparing the Ninth Edition we considered the comments and suggestions of students and instructors to design a revision that builds on the strengths of previous editions and presents chemistry as a vital, coherent, and interesting subject. We have especially tried to relate chemistry to the lives of our students as we develop the principles that form the foundation for the further study of chemistry.

Development of Problem-Solving Skills

We all want our students to develop real skills in solving problems. We believe that a key reason for the longevity of this text is the fact that the problem-solving approach we take works for students. A problem-solving approach (sometimes called a *dimensional analysis* approach) is a step-by-step process that allows students to use units and show the change from one unit to the next. Students can learn most easily from defining and demonstrating concepts and problems step by step. In this edition we continue to show many examples, beginning with simple substitutions, progressing to the use of algorithms, and moving toward more complex problems. The examples show how to incorporate fundamental mathematical skills, scientific notation, and significant figures by following the rules consistently. Painstaking care has been taken to show each step in the problem-solving process (*see pp. 107, 138*) and to give *alternative methods for solution* (ratio/proportion, algebraic, for example) where appropriate. These alternative methods give students flexibility in choosing the method that works best for them. In this edition we have used four significant figures for atomic and molar masses for consistency and for rounding off answers appropriately. We have been careful to follow the rules set down in providing answers, correctly rounded so that students who have difficulty with mathematics do not become confused.

Fostering Student Skills. *Attitude* plays a critical role in problem solving. We encourage students to learn that a systematic approach to solving problems is better than simple memorization. We begin to establish this attitude in Chapter 2. Through-out the book we encourage students to begin by writing down the facts or data

Dimensional analysis, or factor-label analysis, is explained in Section 2.8, p. 23. Beginning students are encouraged to use this approach until they become comfortable with the terms used in calculations.

Alternative methods for solution: See, for example, pp. 37–38 and 178–179.

Rounding off answers is presented on pages 14–16; additional hints are given where appropriate (see p. 31).

Problem-solving steps are printed in blue: see pages 149–150, 183–184, and 235 for examples.

Preface

Boxed rules and equations: see pages 58, 85, 202, and 363 for examples.

Practice Problems: see p. 113 for an example and p. 123 for answers.

Questions review key terms, concepts, figures, and tables see pp. 142–143; **Paired Exercises:** pp. 143–145 **Additional Exercises:** pp. 145–146.

given in a problem *(see p. 268)* and to think their way through the problem to an answer, which is then checked to see if it makes sense. Once we have laid the foundations of concepts, we highlight the steps in blue so students can locate them easily. Important rules and equations are highlighted in colored boxes for emphasis and ready reference.

Student Practice. *Practice problems* follow most of the examples in the text. Answers are provided at the end of each chapter for all of the practice problems. We have expanded and updated the number of end-of-chapter exercises. Each exercise set begins with a *Questions* section that helps students review key terms and concepts, as well as material presented in tables and figures. These are followed by a *Paired Exercises* section, where two similar exercises are presented side by side. These paired exercises cover concepts as well as numerical exercises. The section called *Additional Exercises* includes further practice on chapter concepts presented in a more random order. Challenging questions and exercises are denoted with an asterisk. Answers for *all* even-numbered questions and exercises appear in Appendix V.

Emphasis on Practical Aspects

We continue to emphasize the less theoretical aspects of chemistry early in the book, leaving the more abstract theory for later. This sequence seems especially appropriate in a course where students are encountering chemistry for the very first time. Atoms, molecules, and reactions are all an integral part of the chemical nature of matter. A sound understanding of these topics allows the student to develop a basic understanding of chemical properties and vocabulary.

Chapter 2 presents the basic mathematics and language of chemical calculations, including an explanation of the metric system and significant figures. Chapter 3 introduces the vocabulary of chemical substances, defining matter and the systems of naming and classifying elements. In Chapter 4 we present chemical properties—the ability of a substance to form new substances. Then, in Chapter 5, students encounter the history and terms of basic atomic theory.

We continue to present new material at a level appropriate for the beginning student by emphasizing nomenclature, composition of compounds, and reactions in Chapters 6 through 9 before moving into the details of modern atomic theory. The eighth edition chapter "The Periodic Table" has been integrated into the extensively revised Chapter 10, "Modern Atomic Theory," and into Chapter 11, "Chemical Bonds." Those instructors who feel it is essential to cover atomic theory and bonding early in the course can assign Chapters 10 and 11 immediately following Chapter 5. Students gain confidence in their own ability to identify and work with chemicals in the laboratory before tackling the abstract theories of matter. As practicing chemists we have little difficulty connecting theory and chemical properties. Students, especially those with no prior chemistry background, may not share this ability to connect the abstract and the practical. Finally, the entire text has been reexamined and the prose updated and rewritten to improve its clarity.

Learning Aids

To help the beginning student gain the confidence necessary to master technical material we have developed and enhanced a series of learning aids:

- Important **terms** are set off in boldface type where they are defined, and are printed in blue in the margin. These terms are also printed in boldface type in the index.

- Worked **examples** with steps included show students the *how* of problem solving before they are asked to tackle problems on their own.

- **Practice problems** permit immediate reinforcement of a skill shown in the example problems. Answers are provided at the end of the chapter to encourage students to check their problem solving immediately.

- **End of chapter exercises** have been significantly revised, with approximately 200 new exercises, many emphasizing concepts and applications. Numerous existing problems have been shortened to fewer parts.

- A list of **Concepts in Review** given at the end of each chapter guides students in determining the most important concepts in the chapter.

- A **Review of Mathematics** is provided in Appendix I.

- **Units of measurement** are shown in table format in Appendix III and in the endpapers. (*see p. A-12*)

- **Answers** to the even-numbered questions and exercises are given in Appendix V.

- **Each chapter opens** with a *color photograph* relating the chapter to our daily lives. A *chapter preview list* assists students in viewing the topics covered in the chapter, and the *introductory paragraph* further connects the chapter topic to everyday life.

- Each chapter contains at least one special **Chemistry in Action** section that shows the impact of chemistry in a variety of practical applications. These essays cover such topics as food additives, scuba diving, and gumballs. Other Chemistry in Action essays introduce experimental information on new chemical discoveries and applications.

terms: pp. 47, 241, 319

worked examples and practice problems: pp. 17, 113, 377

answers to practice problems: pp. 123, 396

end of chapter exercises: pp. 80–82, 142–146

Concepts in Review: pp. 40, 209

Chapter 1 begins with a garden photo as the metaphor for the diversity of the material world, which chemistry seeks to understand, explain, and utilize (pp. 1–2). Chapter 14 opens with an illustration of a surfer in the ocean, which is an aqueous solution.

Chemistry in Action: See p. 135, The Taste of Chemistry, and pp. 232–233, Goal! A Spherical Molecule

New to This Edition

A number of changes have been made in this edition, but the level of the material remains the same. The entire text was reviewed and reworked to improve the explanations and to provide students with greater assistance in solving problems. Specific new features and changes include the following:

Chapter 1 is completely rewritten to give students a better understanding of the scientific process by introducing the course with a narrative account of the discovery of nitinol, often called memory metal. Also included in this chapter is material on the benefits and risks of science in our high tech world.

A new design includes an updated art program in full color and numerous photographs to increase visual appeal and to highlight chemistry applications to the business of living. We also use color to identify and set off study aids.

- **Key terms** are listed alphabetically at the end of each chapter with section references to assist students in review of new vocabulary.

nitinol: pp. 3–5
risks and benefits: pp. 9–10

key terms: pp. 80, 209

Preface

paired exercises: pp. 118–119, 393–394

additional exercises: pp. 190–192, 395–396

marginal notations: 195, 263, 327

Chemistry in Action: see p. 76, Fast Energy or Fast Fat? and p. 308, How Sweet It Is!

steps for problem solving: pp. 149–150, 235

boxed statements and equations: pp. 58–59, 267

- A **glossary** is provided to help students review key terms. As a study aid, section numbers are given for each term to guide the student to the contextual definition. The glossary pages are color tinted to provide ready access.

- This edition features **paired exercises** at the end of most chapters. Two parallel exercises are given, side by side, so the student can use the same problem-solving skills with two sets of similar information. Answers to the even-numbered paired exercises are given in Appendix V.

- **Additional exercises** are provided at the end of most chapters. They are arranged in a more random order, to encourage students to review the chapter material.

- **Marginal notations** have been added to help students in understanding basic concepts and problem-solving techniques. These are printed in magenta ink to clearly distinguish them from text and vocabulary terms.

- Twelve new **Chemistry in Action** essays have been added, including such topical information as controlling graffiti and the fat content of fast food. Other Chemistry in Action sections have been revised and updated.

- **Steps for solving problems** are printed in blue for easy reference.

- **Important statements,** equations, and laws are boxed and highlighted for emphasis.

Available in Three Versions

For the convenience of instructors and to accommodate the various lengths of academic terms, three versions of this book are available. *Foundations of College Chemistry, 9th Edition* includes 20 chapters and is our main text. *Foundations of College Chemistry, 6th Alternate Edition,* provides a shorter paperbound version (17 chapters) of the same material without the nuclear, organic, and biochemistry chapters. *Foundations of College Chemistry, Brief Edition,* offers an even shorter paperbound edition (15 chapters), but without the chemical equilibrium and oxidation-reduction chapters.

A Complete Ancillary Package

The following comprehensive teaching package accompanies these books.

For the student:

Study Guide by Peter Scott of Linn-Benton Community College is a self-study guide for students. For each chapter it includes a self-evaluation section with various types of exercises, one or more "challenge problems," and answers and solutions to all the exercises. A recap section gives a concise summary of material learned in the chapter and a note about where the student will go from there.

Student Solutions Manual by Morris Hein and Susan Arena includes answers and solutions to all end-of-chapter questions and exercises.

Foundations of Chemistry in the Laboratory, 9th Edition, by Morris Hein, Leo R. Best, and Robert L. Miner includes 28 experiments for a laboratory program that may accompany the lecture course. Featuring updated information on waste disposal and emphasizing safe laboratory procedures, the lab manual also includes study aids and exercises.

A Basic Math Approach to Concepts of Chemistry, 6th edition, by Leo Michels is a self-paced paperbound workbook that has proven itself an excellent resource for students needing help with mathematical aspects of chemistry. Evaluation tests are provided for each unit and the test answers are given in the back of the book. A glossary is also included.

Brooks/Cole Exerciser (BCX) 2.0 by Laurel Technical Services is a text-specific software tutorial with exercises from the main text and the *Study Guide*. The program monitors student progress and generates reports. This software is available for DOS, DOS/Windows, and Macintosh platforms.

For the instructor:

Instructor's Manual for Foundations of College Chemistry, 9th edition, by Morris Hein and Susan Arena includes a copy of the test questions provided electronically in EXP-Test, Review Exercise Worksheets, answers to the test item questions, and answers to the Review Exercise Worksheets.

Instructor's Manual for Foundations of Chemistry in the Laboratory, 9th Edition, includes information on the management of the lab, evaluation of experiments, notes for individual experiments, a list of reagents needed, and answer keys to each experiment's report form and to all exercises.

EXP-Test, a computerized test generation system, is available for IBM PCs or Compatibles. A Macintosh version, *Chariot Microtest III,* is also available.

Transparencies in full color include illustrations from the text, enlarged for use in the classroom and lecture halls.

Acknowledgments

Books are the result of a collaborative effort of many talented and dedicated people. Among the friends and colleagues who have helped us, we appreciate the enthusiasm and energy of Connie Grosse, teacher at El Dorado High School in Southern California, who developed new exercises and problems for the revision. Our thanks to Don DeCoste, University of Illinois, for his long hours of reading and evaluating the manuscript for accuracy, and to Iraj Behbahani, Mt. San Antonio College, for checking the solutions to each and every question and exercise in this revision. These skillful people provided support and helpful criticism throughout the project. We are grateful for the many helpful comments from colleagues and students who, over the years, have made this book possible. We hope they will continue to share their ideas for change with us, either directly or through our publisher.

We are especially thankful for the support, friendship, and constant encouragement of our spouses, Edna and Steve, who have endured many lost weekends and been patient and understanding through the long hours of this process. Their optimism and good humor have given us a sense of balance and emotional stability.

No textbook can be completed without the untiring efforts of many publishing professionals. Special thanks to the talented staff at Brooks/Cole, especially Jamie Sue Brooks, Production Services Manager, and Beth Wilbur, Editorial Associate, who worked very hard to control the many aspects of this project. Much credit goes to Julie Kranhold of Ex Libris for her unfailing attention to detail and persistence in moving the book through the numerous stages of production. Julie is amazing

Preface

in her ability to track each change and translate our ideas for art and photos into the reality of this edition. Our special thanks also to Nancy Benedict for the colorful and appealing design of the book. We appreciate the guidance of Harvey Pantzis, Executive Editor, and his support at critical times during the revision process. Our heartfelt thanks also to a great group of ITP/Brooks/Cole book "reps" for their interest and enthusiasm as they talk with instructors across the country.

Our sincere appreciation goes to the following reviewers who were kind enough to read and give their professional comments: Kathleen Ashworth, Yakima Valley Community College; Ann Barber, Manatee Community College; Mark Bishop, Monterey Peninsula College; Eugene Boney, Ocean County College; John Chapin, St. Petersburg Junior College; William Hausler, Madison Area Technical College; Margaret Holzer, California State University, Northridge; James Jacobs, University of Rhode Island; William Nickels, Schoolcraft College; Jeffrey Schneider, State University of New York, Oswego; Donald Wink, University of Illinois, Chicago; and Donald Young, Ashville-Buncombe Community College.

Morris Hein
Susan Arena

Morris Hein is professor emeritus of chemistry at Mt. San Antonio College, where he regularly taught the preparatory chemistry course. His name has become synonymous with clarity, meticulous accuracy, and a step-by-step approach that students can follow. Over the years, more than two million students have learned chemistry using a text by Morris Hein. In addition to *Foundations of College Chemistry, Ninth Edition,* he is co-author of *College Chemistry: An Introduction to General, Organic, and Biochemistry, Fifth Edition,* and *Introduction to Organic and Biochemistry.* He has also written *Foundations of Chemistry in the Laboratory, Ninth Edition,* and *College Chemistry in the Laboratory, Fifth Edition.*

Susan Arena currently teaches general chemistry and is director of the Merit Program for Emerging Scholars at the University of Illinois, Urbana-Champaign. She collaborated with Morris Hein on the seventh edition of *Foundations of College Chemistry,* and became co-author in the eighth edition. She is also co-author of *College Chemistry: An Introduction to General, Organic, and Biochemistry, Fifth Edition,* and *Introduction to Organic and Biochemistry.*

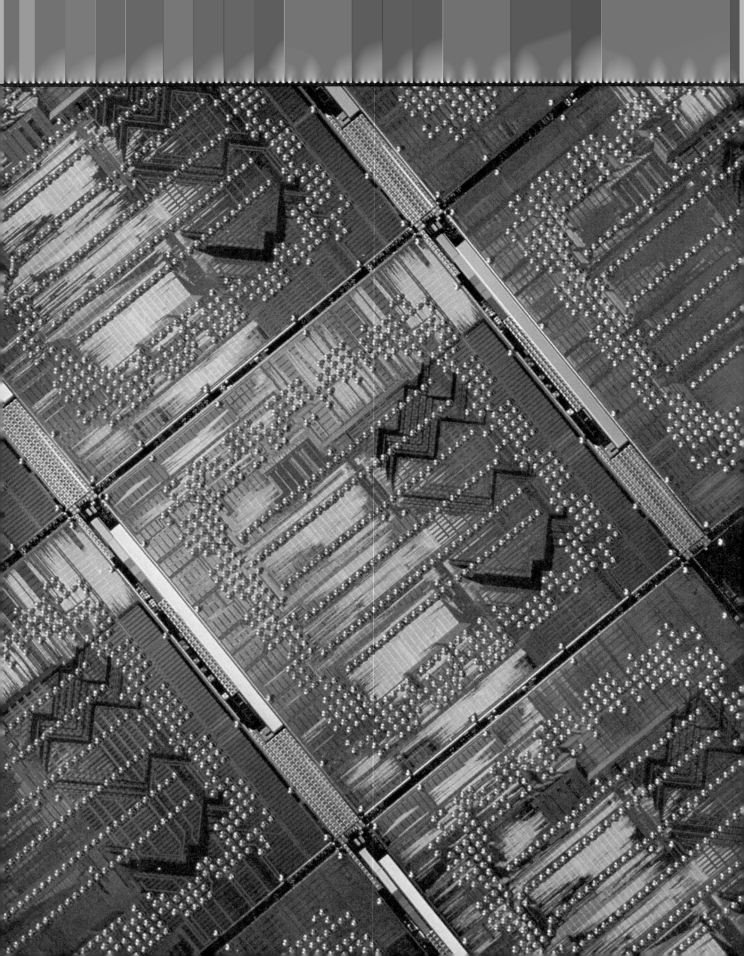

Chemistry in Action

Brief Contents

Detailed Contents

Detailed Contents

Detailed Contents

Have you ever strolled through a spring garden and been amazed at the diversity of colors in the flowers? Or perhaps you have curled up in front of a winter fire and become fascinated watching the flames. And think of those times when you have dropped a beverage container on a hard floor, and were relieved to find that it was plastic instead of glass. All of these phenomena are the result of chemistry—not in the laboratory but rather in our everyday lives. Chemical changes can bring us beautiful colors, warmth and light, or new and exciting products. Chemists seek to understand, explain, and utilize the diversity of materials we find around us.

1.1 Why Study Chemistry?

Chemistry is a subject that is fascinating to many people. Learning about the composition of the world around us can lead to interesting and useful inventions and new technology. More than likely you are taking this chemistry course because someone has decided that it is an important part of your career goals. The field of chemistry is central to an understanding of many fields including agriculture, astronomy, animal science, geology, medicine, applied health technology, fire science, biology, molecular biology, and materials science. Even if you are not planning to work in any of these fields, chemistry is used by each of us every day in our struggle to cope with our technological world. Learning about the benefits and risks associated with chemicals will help you to be an informed citizen, able to make intelligent choices concerning the world around you. Studying chemistry teaches you to solve problems and communicate with others in an organized and logical manner. These skills will be helpful in college and throughout your career.

1.2 The Nature of Chemistry

chemistry

Key words are highlighted in bold and color in the margin to alert you to new terms defined in the text.

◀ **Chapter Opening Photo:** Chemistry can help us to understand nature, its beauty, and its complexity. The colors of this spring garden result from a series of chemical reactions.

What is chemistry? A popular dictionary gives this definition: "**Chemistry** is the science of the composition, structure, properties, and reactions of matter, especially of atomic and molecular systems." Another, somewhat simpler definition is: "Chemistry is the science dealing with the composition of *matter* and the changes in composition that matter undergoes." Neither of these definitions is entirely adequate. Chemistry, along with the closely related science of physics, is a fundamental branch of knowledge. Chemistry is also closely related to biology, not only because living organisms are made of material substances but also because life itself is essentially a complicated system of interrelated chemical processes.

The scope of chemistry is extremely broad. It includes the whole universe and everything, animate and inanimate, in it. Chemistry is concerned not only with the composition and changes in the composition of matter, but also with the energy and energy changes associated with matter. Through chemistry we seek to learn and to understand the general principles that govern the behavior of all matter.

The field of chemistry is multi-layered and complex as you can see from the work of these two men. *Right:* This chemical operator is monitoring the process for production of a new medicine for postmenopausal osteoporosis at a plant in Northern Ireland. *Left:* This research chemist is developing a manufacturing process for hepatitis A vaccine.

The chemist, like other scientists, observes nature and attempts to understand its secrets: What makes a rose red? Why is sugar sweet? What is occurring when iron rusts? Why is carbon monoxide poisonous? Why do people wither with age? Problems such as these—some of which have been solved, some of which are still to be solved—are part of what we call chemistry.

A chemist may interpret natural phenomena, devise experiments that will reveal the composition and structure of complex substances, study methods for improving natural processes, or, sometimes, synthesize substances unknown in nature. Ultimately, the efforts of successful chemists advance the frontiers of knowledge and at the same time contribute to the well-being of humanity.

1.3   The Process of Chemistry

How does a chemist discover a new compound and study its behavior? How are these new substances developed into useful items for our everyday life? A combination of hard work, accident, and luck produced "memory metal," an advanced substance now used to make toys, braces for teeth, and to treat problems such as blood clots and reattachment of tendons. None of these applications was the target of the search for a new material. They all developed from the investigation of the properties of memory metal and the creativity of people associated with the work.

The discovery of "memory metal" began with William J. Buehler, a physical metallurgist at the Naval Ordnance Laboratory (NOL) in Maryland. He had the task of finding a substance composed of two or more metals that would be suitable for the nose cone of the Navy's underwater missile launchers. Buehler was experiencing difficulties at this particular time in his life and so he became particularly immersed in his work. He was tireless in his experimenting and eventually eliminated all but twelve compounds. He then began measuring impact resistance on them. His test was simple but effective: He made a button of the compound to be tested and hit it with a hammer. One of the substances, a 50%-50% mixture of nickel and titanium, produced a substance with greater impact resistance, elasticity, malleability, and resistance to fatigue than the others. Buehler named the substance *nitinol,* from *Ni*ckel *Ti*tanium *N*aval *O*rdnance *L*aboratory.

To begin testing, Buehler and his staff varied the percentages of nickel and titanium to determine the effect of composition on the properties of the compound. They made a series of bars in a furnace and, after cooling them, they smoothed them on a shop grinder. Buehler accidentally dropped one of the bars and noticed that it made a dull thud—much like that of a bar of lead. This aroused his curiosity and he began dropping other bars. To his surprise, he found that the cooled bars made a dull sound whereas the warm ones made a bell-like tone. Fascinated, he then began reheating and cooling the bars, discovering that the sound changed consistently back and forth from dull (cool) to bell-like (warm)—variances that indicated a change in the atomic structure of the metal.

During a project review at NOL, Buehler demonstrated the fatigue resistance of nitinol by bending a long strip of wire into accordion folds. He passed it around, letting the directors flex it and straighten it. One of them wanted to see what would happen when it was heated and held the pleated nitinol over a flame. To everyone's amazement it stretched out into a straight wire. Buehler recognized that this behavior related to the different sounds made when nitinol was heated or cooled.

Buehler recruited Frederick Wang, a crystallographer, to define the "memory" properties of the metal. Wang determined the atomic structural changes that give nitinol its unique characteristic of memory—changes that involve the rearrangement of the position of particles within the solid. Changes between solids and liquids or liquids and gases are well known (e.g., boiling water or melting ice), but the same sort of changes can occur between two solids. In the case of nitinol, a *parent shape* (the shape to which you want it to return) has to be defined. To fix the parent shape of nitinol it must be heated. No changes are apparent, but when it cools, the atoms in the solid rearrange into a slightly different structure. Thus, whenever the nitinol is heated the atoms rearrange themselves back to the structure necessary to produce the parent shape.

Buehler and Wang then began the process of moving nitinol from the experimental world of the laboratory to the commercial world of applications. By the late sixties nitinol was being used in pipe couplings in the aircraft industry.

In 1968, George Andreason, a dentist, began experimenting with nitinol in his metal-working shop where he made jewelry as a hobby. He developed a fine wire that could be molded to fit a patient's mouth (parent position), which when cooled could be bent to fit the misaligned teeth. When warmed to body temperature, it would exert a gentle constant pressure on the teeth. This was a major breakthrough in orthodontics, cutting the treatment time in half from that of steel braces.

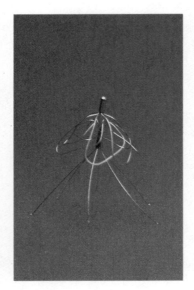

This nitinol filter can trap potentially fatal blood clots. When cooled below body temperature, it is collapsed into a straight bundle of wire. Then, with minor surgery, it is inserted into a large vein where it springs back into shape as it reaches body temperature.

◀ **William Buehler demonstrating the unique properties of nitinol at the Naval Ordnance Laboratory in 1969.**

Dr. Wang left NOL in 1980 to become a supplier of nitinol to the many and varied manufacturers who wanted this exciting new product for such things as eyeglass frames, which can take extreme abuse (e.g., bending them, sitting on them, or twisting them) and return to their original shape; antiscalding devices for showerheads and faucets, which automatically shut off if the water approaches 120°F; and toys such as blinking movie posters and tail-swishing dinosaurs.

The uses of nitinol are varied and growing. Today it has many applications in the fields of medicine, engineering, and safety; in housewares; and even in the lingerie business in underwire brassieres. Nitinol was the first of the "intelligent" materials—ones that respond to changes in the environment.

1.4   The Scientific Method

The nitinol story illustrates how chemists and other scientists work together to solve a specific problem. Chemists, metallurgists, physicists, engineers, and a wide variety of technicians were involved in the development of nitinol. As scientists conduct studies they ask many questions, and their questions often lead in directions that are not part of the original problem. The amazing developments from chemistry and technology usually result from the use of the scientific method. Although complete agreement is lacking on exactly what is meant by "using the scientific method," the general approach is as follows.

1. **Collect the facts or data** that are relevant to the problem or question at hand. This is usually done by planned experimentation. In the nitinol story, Buehler compared the properties of various substances until he selected nitinol. He also investigated the properties of nitinol, including its ability to produce different sounds when warm and cool, and its unique ability to return to its original shape. Analyze the data to find trends or regularities that are pertinent to the problem.

2. **Formulate a hypothesis** that will account for the data that have been accumulated and that can be tested by further experimentation. Wang and Buehler investigated the properties and proposed a hypothesis regarding the two phases of nitinol. Further tests supported their model.

3. **Plan and do additional experiments to test the hypothesis.** The nitinol group continued its testing process until the mechanism was well understood.

4. **Modify the hypothesis** as necessary so that it is compatible with all the pertinent data.

hypothesis

theory

scientific laws

Confusion sometimes arises regarding the exact meanings of the words *hypothesis, theory,* and *law*. A **hypothesis** is a tentative explanation of certain facts that provides a basis for further experimentation. A well-established hypothesis is often called a **theory**. Thus a theory is an explanation of the general principles of certain phenomena with considerable evidence or facts to support it. Hypotheses and theories explain natural phenomena whereas **scientific laws** are simple statements of natural phenomena to which no exceptions are known under the given conditions.

Although these four steps are a broad outline of the general procedure that is followed in most scientific work, they are not a recipe for doing chemistry or any other science (Figure 1.1). But chemistry is an experimental science, and much of its progress has been due to application of the scientific method through systematic research.

Many theories and laws are studied in chemistry, which make the study of any science easier because they summarize particular aspects of that science. Some of the theories advanced by great scientists in the past have since been substantially altered and modified. Such changes do not mean that the discoveries of the past are less significant than those of today. Modification of existing theories in the light of new experimental evidence is essential to the growth and evolution of scientific knowledge.

1.5   Relationship of Chemistry to Other Sciences and Industry

Besides being a science in its own right, chemistry is the servant of other sciences and industry. Chemical principles contribute to the study of physics, biology, agriculture, engineering, medicine, space research, oceanography, and many other sciences. Chemistry and physics are overlapping sciences, since both are based on the properties and behavior of matter. Biological processes are chemical in nature. The metabolism of food to provide energy to living organisms is a chemical process. Knowledge of molecular structure of proteins, hormones, enzymes, and the nucleic acids is assisting biologists in their investigations of the composition, development, and reproduction of living cells.

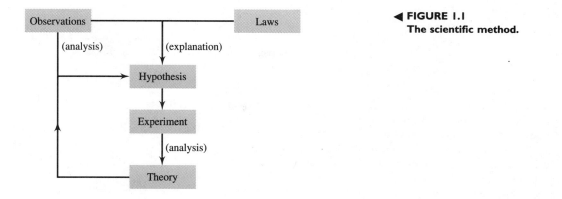

Chemistry is playing an important role in alleviating the growing shortage of food in the world. Agricultural production has been increased with the use of chemical fertilizers, pesticides, and improved varieties of seeds. Chemical refrigerants make possible the frozen food industry, which preserves large amounts of food that might otherwise spoil. Chemistry is also producing synthetic nutrients, but much remains to be done as the world population increases relative to the land available for cultivation. Expanding energy needs have brought about difficult environmental problems in the form of air and water pollution. Chemists and other scientists are working diligently to alleviate these problems.

Advances in medicine and chemotherapy, through the development of new drugs, have contributed to prolonged life and the relief of human suffering. More than 90% of the drugs and pharmaceuticals being used in the United States today have been developed commercially within the past 50 years. The plastics and polymer industry, unknown 60 years ago, has revolutionized the packaging and textile industries and is producing durable and useful construction materials. Energy derived from chemical processes is used for heating, lighting, and transportation. Virtually every industry is dependent on chemicals—for example, the petroleum, steel, rubber, pharmaceutical, electronic, transportation, cosmetic, space, polymer, garment, aircraft, and television industries.

People outside of science usually have the perception that science is an intensely logical field. They picture the white-coated chemist going from hypothesis to experiment and then to laws and theories without error or emotion. Quite often scientific discoveries are the result of trial and error. In the nitinol story the memory nature of the material was discovered because someone wanted to see what would happen if a flame was applied to compressed nitinol. Buehler's creativity and insight and his ability to relate this chance discovery to his experiments enabled him to make the connection. This is an excellent example of what is meant by serendipity in science. Buehler was searching for a material to use as a nose cone in missiles, but the results of his experiments were quite unexpected. He and his coworkers discovered a material that has applications in medicine, engineering, and everyday life.

The nitinol story also illustrates the fact that scientists do not work alone in their laboratories. Buehler, an engineer, was joined by Wang, a crystallographer, and many other chemists, engineers, dentists, and physicians in developing the applications of nitinol. Each of them had a contribution to make to the body of

Serendipity in Science

Discoveries in the world of chemistry are for the most part made by people who are applying the scientific method in their work. Occasionally, important discoveries are made by chance, or through serendipity. But even when serendipity is involved, a discovery is more likely to be made by someone with a good knowledge of the field. Louis Pasteur summed this up in a statement made long ago: "Chance favors the prepared mind." In chemistry serendipity often can lead to whole new fields and technology.

The synthetic dye industry began in 1856 when William Perkin, an 18-year-old student at the Royal College of Chemistry in London, was attempting to synthesize quinine, a drug used to treat malaria. He reacted two chemicals, aniline sulfate and potassium dichromate, and obtained a black paste. Perkin then extracted the paste with alcohol. Upon evaporating the alcohol, violet crystals appeared that, when dissolved in water, made a beautiful purple solution. He so enjoyed the color he began investigating the solution; he then determined the purple color had a strong affinity for silk. Perkin had discovered the first synthetic aniline dye. Recognizing the commercial possibilities, he immediately left school and, with his father and an older brother, went into the dye manufacturing business. His dye, known as mauve, quickly became a success and inspired other research throughout Europe. By 1870, cloth could be purchased in more and brighter synthetic colors than were ever available with natural dyes.

A second, more recent, account of chance events in chemistry also led to a multimillion dollar industry (see table). In 1965 James Schlatter was researching anti-ulcer drugs for the pharmaceutical firm G. D. Searle. In the course of his work he accidentally ingested a small amount of

The Discovery of Artificial Sweeteners		
Sweetener	Date	Discoverer
Saccharin	1878	I. Remsen and C. Fahlberg
Cyclamate	1937	M. Sveda
Aspartame	1965	J. Schlatter

his preparation and found, to his surprise, it had an extremely sweet taste. (*Note:* Tasting chemicals of any kind in the laboratory is not a safe procedure.) When purified, the sweet-tasting substance turned out to be aspartame, a molecule consisting of two amino acids joined together. Since only very small quantities are necessary to produce sweetness, it proved to be an excellent low calorie artificial sweetener. Today, under the trade names of "Equal" and "Nutrasweet," aspartame is one of the cornerstones of the artificial sweetener industry.

The wide array of beautiful colors seen in this fabric shop come from synthetic dyes, which have been available for more than 100 years.

knowledge surrounding the substance nitinol. Chemistry is a field in which teamwork and cooperation play a vital role in understanding complex systems.

1.6   Risks and Benefits

The nitinol story is only one example of the many problems confronting us today that rely on science for an answer. Virtually every day we read or hear about stories such as:

- developing an AIDS vaccine
- banning the use of herbicides and pesticides
- analyzing DNA to determine genetic disease, biological parents, or to place a criminal at the scene of a crime
- removing asbestos from public buildings
- removing lead from drinking water
- the danger of radon in our homes
- global warming
- the hole in the ozone layer
- health risks associated with coffee, margarine, saturated fats, and other foods
- burning of tropical rain forests and the effect on global ecology

DNA analysis is playing an ever-important role in such fields as genetics, disease control, and crime.

Which of these risks present true danger to us and which pose no great problem? All of these problems will be around for many years and new ones will be continually added to the list. Wherever we live and whatever our occupation, each of us is exposed to chemicals and chemical hazards every day. The question we must answer is: Do the risks outweigh the benefits?

Risk assessment is a process that brings together professionals from the fields of chemistry, biology, toxicology, and statistics in order to determine the risk associated with exposure to a certain chemical. Assessment of risk involves determining both the probability of exposure and the severity of that exposure. Once this is done an estimate can be made of the overall risk. Studies have shown how people perceive various risks. The perception of risk depends on some rather interesting factors. Voluntary risks, such as smoking or flying, are much more easily accepted than involuntary ones, such as herbicides on apples or asbestos in buildings. People also often conclude that anything "synthetic" is bad while anything "organic" is good. Risk assessment may provide information on the degree of risk but not on whether the chemical is "safe." Safety is a qualitative judgment based on many personal factors including beliefs, preferences, benefits, and costs.

Once a risk has been assessed the next step is to manage it. This involves ethics, economics, and equity as well as government and politics. For example, some things are perceived as low risk by scientists (such as asbestos in buildings) but are classified as high risk by the general public. This inconsistency may result in the expenditure of millions of dollars to rid the public of a perceived threat that is much lower than they believe.

Risk management involves value judgments that integrate social, economic, and political issues. These risks must be weighed against the benefits of new technol-

Asbestos, once a widely used building material, was banned by the EPA in 1986 because of its health hazards. Today, the removal of asbestos from our homes and our public buildings has become a big concern. ▶

ogy and products in order to make the decisions required at home, in our local communities, and around the world. We use both risk assessment and risk management to decide whether to buy a certain product (such as a pesticide), take a certain drug (such as a pain reliever), or eat certain foods (such as hot dogs). We must realize that all risks can never be eliminated. Our goal is to minimize unnecessary risks and to make responsible decisions regarding the risks in our lives and our environment.

The theories and models used in risk assessment are based on concepts learned in chemistry—they are based on assumptions and therefore contain uncertainties. By improving your understanding of the concepts of chemistry you will be better able to understand the capabilities and limitations of science. You can then intelligently question the process of risk assessment and make decisions that will lead to a better understanding of our world and our responsibilities to each other.

Key Terms

The terms listed here have all been defined within the chapter. The section number is given for each term.

These terms can also be found in the Glossary.

chemistry (1.2)	scientific laws (1.4)
hypothesis (1.4)	theory (1.4)

mass

weight

◀ **Chapter Opening Photo: Measuring instruments come in various shapes and sizes, all necessary to allow components to fit together, and permit us to quantify our world.**

Doing an experiment in chemistry is very much like cooking a meal in the kitchen. It is important to know the ingredients *and* the amounts of each in order to have a tasty product. Working on your car requires specific tools, in exact sizes. Buying new carpeting or draperies is an exercise in precise and accurate measurement for a good fit. A small difference in the concentration or amount of medication a pharmacist gives you may have significant effects on your well-being. In all of these cases, the ability to measure accurately and a strong foundation in the language and use of numbers provide the basis for success. In chemistry, we begin by learning the metric system and the proper units for measuring mass, length, volume, and temperature.

2.1 Mass and Weight

Chemistry is an experimental science. The results of experiments are usually determined by making measurements. In elementary experiments the quantities that are commonly measured are mass, length, volume, pressure, temperature, and time. Measurements of electrical and optical quantities may also be needed in more sophisticated experimental work.

Although mass and weight are often used interchangeably, the two words have quite different meanings. The **mass** of a body is defined as the amount of matter in that body. The mass of an object is a fixed and unvarying quantity that is independent of the object's location. The mass of an object can be measured on a balance by comparison with other known masses.

An everyday example of a balance is a child's seesaw, shown in Figure 2.1. If children of equal mass sit on opposite ends the seesaw balances (2.1a). If one child is heavier than the other the seesaw sinks on the side holding the heavier child (2.1b). To bring the seesaw back into balance additional mass must be added to the side holding the lighter child (2.1c). Another common example of a balance is shown in Figure 2.2. These balances were used by assayers during the gold rush to determine the mass of gold brought in by the prospectors. The gold was placed on one pan of the balance and known masses on the other side until the pans were level with each other. (Several modern balances are shown in Figure 2.5, page 28.)

The **weight** of a body is the measure of the earth's gravitational attraction for that body. Weight is measured on a device called a scale, which measures force against a spring. Unlike mass, weight varies in relation to the position of an object on earth or its distance from the earth.

Consider an astronaut of mass 70.0 kilograms (154 pounds) who is being shot into orbit. At the instant before blast-off the weight of the astronaut is also 70.0 kilograms. As the distance from the earth increases and the rocket turns into an orbiting course, the gravitational pull on the astronaut's body decreases until a state of weightlessness (zero weight) is attained. However, the mass of the astronaut's body has remained constant at 70.0 kilograms during the entire event.

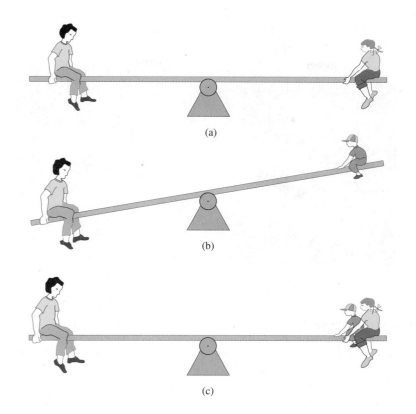

◀ **FIGURE 2.1**
We experience our first lessons in balancing from playing on a seesaw.

2.2 Measurement and Significant Figures

To understand certain aspects of chemistry it is necessary to set up and solve problems. Problem solving requires an understanding of the elementary mathematical operations used to manipulate numbers. Numerical values or data are obtained from measurements made in an experiment. A chemist may use these data to calculate the extent of the physical and chemical changes occurring in the substances that are being studied. By appropriate calculations the results of an experiment can be compared with those of other experiments and summarized in ways that are meaningful.

The result of a measurement is expressed by a numerical value together with a unit of that measurement. For example,

$$\overbrace{70.0 \text{ kilograms}}^{\text{numerical value}} = 154 \underbrace{\text{ pounds}}_{\text{unit}}$$

Numbers obtained from a measurement are never exact values. They always have some degree of uncertainty due to the limitations of the measuring instrument and the skill of the individual making the measurement. The numerical value recorded

FIGURE 2.2
An assayer's balance is used for weighing gold.

for a measurement should give some indication of its reliability (precision). To express maximum precision this number should contain all the digits that are known plus one digit that is estimated. This last estimated digit introduces some uncertainty. Because of this uncertainty every number that expresses a measurement can have only a limited number of digits. These digits, used to express a measured quantity, **significant figures**   are known as **significant figures**, or **significant digits**.

Suppose we measure temperature on a thermometer calibrated in degrees and observe that the mercury stops between 21 and 22 (see Figure 2.3a). We then know that the temperature is at least 21 degrees and is less than 22 degrees. To express the temperature with greater precision, we estimate that the mercury is about two-tenths the distance between 21 and 22. The temperature is, therefore, 21.2 degrees. The last digit (2) has some uncertainty because it is an estimated value. The recorded temperature, 21.2 degrees, is said to have three significant figures. If the mercury stopped exactly on the 22 (Figure 2.3b), the temperature would be recorded as 22.0 degrees. The zero is used to indicate that the temperature was estimated to a precision of one-tenth degree. Finally, look at Figure 2.3c. On this thermometer, the temperature is recorded as 22.11°C (four significant figures). Since the thermometer is calibrated to tenths of a degree, the first estimated digit is the hundredths.

Some numbers are exact and have an infinite number of significant figures. Exact numbers occur in simple counting operations; when you count 25 dollars, you have exactly 25 dollars. Defined numbers, such as 12 inches in 1 foot, 60 minutes in 1 hour, and 100 centimeters in 1 meter, are also considered to be exact numbers. Exact numbers have no uncertainty.

Evaluating Zero

In any measurement all nonzero numbers are significant. However, zeros may or may not be significant depending on their position in the number. Here are some rules for determining when zero is significant.

1. Zeros between nonzero digits are significant:
   205 has three significant figures

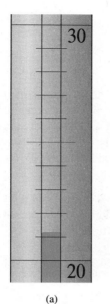

(a)

(b)

(c)

◀ **FIGURE 2.3**
Measuring temperature with various degrees of precision.

2.05 has three significant figures
61.09 has four significant figures

2. Zeros that precede the first nonzero digit are not significant. These zeros are used to locate a decimal point:

   0.0025 has two significant figures (2, 5)
   0.0108 has three significant figures (1, 0, 8)

3. Zeros at the end of a number that include a decimal point are significant:

   0.500 has three significant figures (5, 0, 0)
   25.160 has five significant figures
   3.00 has three significant figures
   20. has two significant figures

4. Zeros at the end of a number without a decimal point are not considered significant:

   1000 has one significant figure
   590 has two significant figures

One way of indicating that these zeros are significant is to write the number using a decimal point and a power of 10. Thus, if the value 1000 has been determined to four significant figures, it is written as 1.000×10^3. If 590 has only two significant figures, it is written as 5.9×10^2.

Rules for significant figures should be memorized for use throughout the text.

Practice 2.1

How many significant figures are in each of these numbers?

(a) 4.5 inches   (e) 25.0 grams
(b) 3.025 feet   (f) 12.20 liters
(c) 125.0 meters   (g) 100,000 people
(d) 0.001 mile   (h) 205 birds

Answers to Practice Exercises are found at the end of each chapter.

2.3   Rounding Off Numbers

rounding off numbers

In calculations we often obtain answers that have more digits than we are justified in using. It is necessary, therefore, to drop the nonsignificant digits in order to express the answer with the proper number of significant figures. When digits are dropped from a number, the value of the last digit retained is determined by a process known as **rounding off numbers**. Two rules will be used in this book for rounding off numbers:

Not all schools use the same rules for rounding. Check with your instructor for variations in these rules.

Rule 1   When the first digit after those you want to retain is 4 or less, that digit and all others to its right are dropped. The last digit retained is not changed. The following examples are rounded off to four digits:

74.693 = 74.69
— This digit is dropped.

1.00629 = 1.006
— These two digits are dropped.

Rule 2   When the first digit after those you want to retain is 5 or greater, that digit and all others to the right are dropped and the last digit retained is increased by one. These examples are rounded off to four digits:

1.026868 = 1.027
— These three digits are dropped.
— This digit is changed to 7.

18.02500 = 18.03
— These three digits are dropped.
— This digit is changed to 3.

12.899 = 12.90
— This digit is dropped.
— These two digits are changed to 90.

Practice 2.2

Round off these numbers to the number of significant digits indicated:
(a) 42.246 (four digits)   (d) 0.08965 (two digits)
(b) 88.015 (three digits)   (e) 225.3 (three digits)
(c) 0.08965 (three digits)   (f) 14.150 (three digits)

2.4   Scientific Notation of Numbers

The age of the earth has been estimated to be about 4,500,000,000 (4.5 billion) years. Because this is an estimated value, say to the nearest 0.1 billion years, we are justified in using only two significant figures to express it. Thus, we write it, using a power of 10, as 4.5×10^9 years.

Very large and very small numbers are often used in chemistry and can be simplified and conveniently written using a power of 10. Writing a number as a power of 10 is called **scientific notation**.

To write a number in scientific notation, move the decimal point in the original number so that it is located after the first nonzero digit. This new number is multiplied by 10 raised to the proper power (exponent). The power of 10 is equal to the number of places that the decimal point has been moved. If the decimal is moved to the left, the power of 10 will be a positive number. If the decimal is moved to the right, the power of 10 will be a negative number.

The scientific notation of a number is the number written as a factor between 1 and 10 multiplied by 10 raised to a power. For example,

$$2468 = 2.468 \times 10^3$$

number   scientific notation
of the number

Examples show you problem-solving techniques in a step-by-step form. Study each one and then try the Practice Exercises.

Write 5283 in scientific notation.

Example 2.1

5283.   Place the decimal between the 5 and the 2. Since the decimal was moved three places to the left the power of 10 will be 3, and the number 5.283 is multiplied by 10^3.

Solution

5.283×10^3

Write 4,500,000,000 in scientific notation (two significant figures).

Example 2.2

4 500 000 000.   Place the decimal between the 4 and the 5. Since the decimal was moved nine places to the left the power of 10 will be 9, and the number 4.5 is multiplied by 10^9.

Solution

4.5×10^9

Write 0.000123 in scientific notation.

Example 2.3

0.000123   Place the decimal between the 1 and the 2. Since the decimal was moved four places to the right the power of 10 will be -4, and the number 1.23 is multiplied by 10^{-4}.

Solution

1.23×10^{-4}

Practice 2.3

Write the following numbers in scientific notation:
(a) 1200 (four digits)   (c) 0.0468
(b) 6,600,000 (two digits)   (d) 0.00003

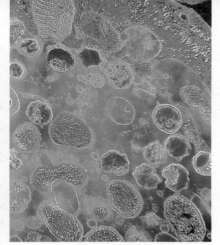

Left: Distances between the moon and stars are often so great they must be expressed in scientific notation. Right: These *Chlamydia* bacteria are so small, on the other hand, they need great magnification in order to be seen (18,000X).

2.5 Significant Figures in Calculations

The results of a calculation based on measurements cannot be more precise than the least precise measurement.

Multiplication or Division

In calculations involving multiplication or division, the answer must contain the same number of significant figures as in the measurement that has the least number of significant figures. Consider the following examples:

Use your calculator to check your work in the examples. Compare your results to be sure you understand the mathematics.

Example 2.4

$190.6 \times 2.3 = 438.38$

Solution

The value 438.38 was obtained with a calculator. The answer should have two significant figures, because 2.3, the number with the fewest significant figures, has only two significant figures.

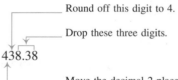

Round off this digit to 4.

Drop these three digits.

438.38

Move the decimal 2 places to the left to express in scientific notation.

The correct answer is 440 or 4.4×10^2.

Example 2.5

$$\frac{13.59 \times 6.3}{12} = 7.13475$$

Solution

The value 7.13475 was obtained with a calculator. The answer should contain two significant figures because 6.3 and 12 have only two significant figures.

Drop these four digits.

7.13475

This digit remains the same.

The correct answer is 7.1.

Practice 2.4

134 in. $\times$ 25 in. = ?

Practice 2.5

$$\frac{213 \text{ miles}}{4.20 \text{ hours}} = ?$$

Practice 2.6

$$\frac{2.2 \times 273}{760} = ?$$

Addition or Subtraction

The results of an addition or a subtraction must be expressed to the same precision as the least precise measurement. This means the result must be rounded to the same number of decimal places as the value with the fewest decimal places.

Add 125.17, 129, and 52.2.

Example 2.6

Solution

$$
\begin{array}{r}
125.17 \\
129. \\
\underline{52.2} \\
306.37 \quad (306)
\end{array}
$$

The number with the least precision is 129. Therefore the answer is rounded off to the nearest unit: 306.

Subtract 14.1 from 132.56.

Example 2.7

Solution

$$
\begin{array}{r}
132.56 \\
-14.1 \\
\hline
118.46 \quad (118.5)
\end{array}
$$

14.1 is the number with the least precision. Therefore, the answer is rounded off to the nearest tenth: 118.5.

Subtract 120 from 1587.

Example 2.8

Solution

$$
\begin{array}{r}
1587 \\
-120 \\
\hline
1467 \quad (1.47 \times 10^3)
\end{array}
$$

120 is the number with least precision. The zero is not considered significant; therefore, the answer must be rounded to the nearest ten: 1470 or 1.47×10^3.

Example 2.9   Add 5672 and 0.00063.

Solution
$$5672$$
$$+\ \ 0.00063$$
$$\overline{5672.00063}\quad (5672)$$

The number with least precision is 5672. Therefore, the answer is rounded off to the nearest unit: 5672.

Example 2.10 $\dfrac{1.039 - 1.020}{1.039} = 0.018286814$

Solution   The value 0.018286814 was obtained with a calculator. When the subtraction in the numerator is done,

$$1.039 - 1.020 = 0.019$$

the number of significant figures changes from four to two. Therefore, the answer should contain two significant figures after the division is carried out:

Drop these six digits.

0.018286814

This digit remains the same.

The correct answer is 0.018, or 1.8×10^{-2}.

Practice 2.7

How many significant figures should the answer contain in each of these calculations?

(a) 14.0×5.2 (e) $119.1 - 3.44$

(b) 0.1682×8.2 (f) $\dfrac{94.5}{1.2}$

(c) $\dfrac{160 \times 33}{4}$ (g) $1200 + 6.34$

(d) $8.2 + 0.125$ (h) $1.6 + 23 - 0.005$

If you need to brush up on your math skills refer to the "Mathematical Review" in Appendix I.

Additional material on mathematical operations is given in Appendix I, "Mathematical Review." Study any portions that are not familiar to you. You may need to do this at various times during the course when additional knowledge of mathematical operations arises.

2.6   The Metric System

metric system or SI

The **metric system**, or **International System** (**SI**, from *Système International*), is a decimal system of units for measurements of mass, length, time, and other physical quantities. It is built around a set of standard units and uses factors of 10 to express

TABLE 2.1 Prefixes and Numerical Values for SI Units*

Prefix	Symbol	Numerical value	Power of 10 equivalent
exa	E	1,000,000,000,000,000,000	10^{18}
peta	P	1,000,000,000,000,000	10^{15}
tera	T	1,000,000,000,000	10^{12}
giga	G	1,000,000,000	10^{9}
mega	M	1,000,000	10^{6}
kilo	k	1,000	10^{3}
hecto	h	100	10^{2}
deka	da	10	10^{1}
—	—	1	10^{0}
deci	d	0.1	10^{-1}
centi	c	0.01	10^{-2}
milli	m	0.001	10^{-3}
micro	μ	0.000001	10^{-6}
nano	n	0.000000001	10^{-9}
pico	p	0.000000000001	10^{-12}
femto	f	0.000000000000001	10^{-15}
atto	a	0.000000000000000001	10^{-18}

* The more commonly used prefixes are in color.

larger or smaller numbers of these units. To express quantities that are larger or smaller than the standard units, prefixes are added to the names of the units. These prefixes represent multiples of 10, making the metric system a decimal system of measurements. Table 2.1 shows the names, symbols, and numerical values of the prefixes. Some examples of the more commonly used prefixes are:

$$1\ kilo\text{meter} = 1000 \text{ meters}$$
$$1\ kilo\text{gram} = 1000 \text{ grams}$$
$$1\ milli\text{meter} = 0.001 \text{ meter}$$
$$1\ micro\text{second} = 0.000001 \text{ second}$$

The seven standard units in the International System, their abbreviations, and the quantities they measure are given in Table 2.2. Other units are derived from these units.

The metric system, or International System, is currently used by most of the countries in the world, not only in scientific and technical work but also in commerce and industry.

Most products today list both systems of measurement on their labels.

TABLE 2.2 International System's Standard Units of Measurement

Quantity	Name of unit	Abbreviation
Length	meter	m
Mass	kilogram	kg
Temperature	kelvin	K
Time	second	s
Amount of substance	mole	mol
Electric current	ampere	A
Luminous intensity	candela	cd

2.7   Measurement of Length

Standards for the measurement of length have an interesting history. The Old Testament mentions such units as the *cubit* (the distance from a man's elbow to the tip of his outstretched hand). In ancient Scotland the inch was once defined as a distance equal to the width of a man's thumb.

meter (m)

Reference standards of measurements have undergone continuous improvements in precision. The standard unit of length in the metric system is the **meter**. When the metric system was first introduced in the 1790s, the meter was defined as one ten-millionth of the distance from the equator to the North Pole, measured along the meridian passing through Dunkirk, France. In 1889 the meter was redefined as the distance between two engraved lines on a platinum–iridium alloy bar maintained at 0° Celsius. This international meter bar is stored in a vault at Sèvres near Paris. Duplicate meter bars have been made and are used as standards by many nations.

By the 1950s length could be measured with such precision that a new standard was needed. Accordingly, the length of the meter was redefined in 1960 and again in 1983. The latest definition is: A meter is the distance that light travels in a vacuum during 1/299,792,458 of a second.

A meter is 39.37 inches, a little longer than 1 yard. One meter contains 10 decimeters, 100 centimeters, or 1000 millimeters (see Figure 2.4). A kilometer contains 1000 meters. Table 2.3 shows the relationships of these units.

TABLE 2.3 Metric Units of Length

Unit	Abbreviation	Meter equivalent	Exponential equivalent
kilometer	km	1000 m	10^3 m
meter	m	1 m	10^0 m
decimeter	dm	0.1 m	10^{-1} m
centimeter	cm	0.01 m	10^{-2} m
millimeter	mm	0.001 m	10^{-3} m
micrometer	μm	0.000001 m	10^{-6} m
nanometer	nm	0.000000001 m	10^{-9} m
angstrom	Å	0.0000000001 m	10^{-10} m

FIGURE 2.4 ▶
Comparison of the metric and American systems of length measurement: 2.54 cm = 1 in.

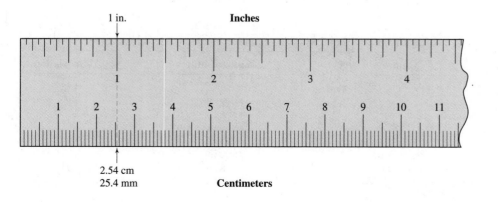

1 in.   Inches

2.54 cm
25.4 mm   Centimeters

The nanometer (10^{-9} m) is used extensively in expressing the wavelength of light as well as in atomic dimensions. See inside back cover for a complete table of common conversions. Other important relationships are:

See inside back cover for a table of conversion.

$$1 \text{ m} = 100 \text{ cm} = 1000 \text{ mm} = 10^6 \text{ } \mu\text{m} = 10^{10} \text{ Å}$$
$$1 \text{ cm} = 10 \text{ mm} = 0.01 \text{ m}$$
$$1 \text{ in.} = 2.54 \text{ cm}$$
$$1 \text{ mile} = 1.609 \text{ km}$$

2.8 Problem Solving

Many chemical principles are illustrated by mathematical concepts. Learning how to set up and solve numerical problems in a systematic fashion is *essential* in the study of chemistry. This skill, once acquired, is also very useful in other study areas. A calculator will save you much time in computation.

Usually a problem can be solved by several methods. But in all methods it is best, especially for beginners, to use a systematic, orderly approach. The *dimensional analysis,* or *factor-label, method* is stressed in this book because

1. It provides a systematic, straightforward way to set up problems.
2. It gives a clear understanding of the principles involved.
3. It trains you to organize and evaluate data.
4. It helps to identify errors because unwanted units are not eliminated if the setup of the problem is incorrect.

The basic steps for solving problems are

Steps for solving problems are highlighted in color for easy reference.

1. Read the problem very carefully to determine what is to be solved for, and write it down.
2. Tabulate the data given in the problem. Even in tabulating data it is important to label all factors and measurements with the proper units.
3. Determine which principles are involved and which unit relationships are needed to solve the problem. Sometimes it is necessary to refer to tables for needed data.
4. Set up the problem in a neat, organized, and logical fashion, making sure that unwanted units cancel. Use sample problems in the text as guides for making setups.
5. Proceed with the necessary mathematical operations. Make certain that the answer contains the proper number of significant figures.
6. Check the answer to see if it is reasonable.

Just a few more words about problem solving. Don't allow any formal method of problem solving to limit your use of common sense and intuition. If a problem is clear to you and its solution seems simpler by another method, by all means use it. But in the long run you should be able to solve many otherwise difficult problems by using the dimensional analysis method.

The dimensional analysis method of problem solving converts one unit to another unit by the use of conversion factors.

Important equations are
boxed or highlighted in color.

$$\text{unit}_1 \times \text{conversion factor} = \text{unit}_2$$

If you want to know how many millimeters are in 2.5 meters, you need to convert meters (m) to millimeters (mm). Therefore, you start by writing

$$\text{m} \times \text{conversion factor} = \text{mm}$$

This conversion factor must accomplish two things. It must cancel (or eliminate) meters, and it must introduce millimeters—the unit wanted in the answer. Such a conversion factor will be in fractional form and have meters in the denominator and millimeters in the numerator:

$$\cancel{m} \times \frac{\text{mm}}{\cancel{m}} = \text{mm}$$

We know that 1 m = 1000 mm. From this relationship we can write two conversion factors—1 m per 1000 mm and 1000 mm per 1 m:

$$\frac{1\ \text{m}}{1000\ \text{mm}} \quad \text{and} \quad \frac{1000\ \text{mm}}{1\ \text{m}}$$

Using the conversion factor 1000 mm/1 m, we can set up the calculation for the conversion of 2.5 m to millimeters,

$$2.5\ \cancel{m} \times \frac{1000\ \text{mm}}{1\ \cancel{m}} = 2500\ \text{mm} \quad \text{or} \quad 2.5 \times 10^3\ \text{mm}$$

(two significant figures)

Note that, in making this calculation, units are treated as numbers; meters in the numerator are canceled by meters in the denominator.

Now suppose you need to change 215 centimeters to meters. We start with

$$\text{cm} \times \text{conversion factor} = \text{m}$$

The conversion factor must have centimeters in the denominator and meters in the numerator:

$$\cancel{cm} \times \frac{\text{m}}{\cancel{cm}} = \text{m}$$

From the relationship 100 cm = 1 m, we can write a factor that will accomplish this conversion:

$$\frac{1\ \text{m}}{100\ \text{cm}}$$

Now set up the calculation using all the data given.

$$215\ \cancel{cm} \times \frac{1\ \text{m}}{100\ \cancel{cm}} = \frac{215\ \text{m}}{100} = 2.15\ \text{m}$$

Some problems may require a series of conversions to reach the correct units in the answer. For example, suppose we want to know the number of seconds in 1 day. We need to go from the unit of days to seconds in this manner:

day → hours → minutes → seconds

Units are emphasized in problems by using color and flow diagrams to help visualize the steps in the process.

This series requires three conversion factors, one for each step. We convert days to hours (hr), hours to minutes (min), and minutes to seconds (s). The conversions can be done individually or in a continuous sequence:

$$\text{day} \times \frac{\text{hr}}{\text{day}} \rightarrow \text{hr} \times \frac{\text{min}}{\text{hr}} \rightarrow \text{min} \times \frac{\text{s}}{\text{min}} = \text{s}$$

$$\text{day} \times \frac{\text{hr}}{\text{day}} \times \frac{\text{min}}{\text{hr}} \times \frac{\text{s}}{\text{min}} = \text{s}$$

Inserting the proper factors we calculate the number of seconds in 1 day to be

$$1\ \text{day} \times \frac{24\ \text{hr}}{1\ \text{day}} \times \frac{60\ \text{min}}{1\ \text{hr}} \times \frac{60\ \text{s}}{1\ \text{min}} = 86{,}400.\ \text{s}$$

All five digits in 86,400 are significant, since all the factors in the calculation are exact numbers.

The dimensional analysis, or factor-label, method used in the preceding work shows how unit conversion factors are derived and used in calculations. After you become more proficient with the terms, you can save steps by writing the factors directly in the calculation. The problems that follow give examples of the conversion from American to metric units.

Label all factors with the proper units.

How many centimeters are in 2.00 ft?

Example 2.11

The stepwise conversion of units from feet to centimeters may be done in this manner. Convert feet to inches; then convert inches to centimeters:

ft → in. → cm

The conversion factors needed are

$$\frac{12\ \text{in.}}{1\ \text{ft}} \quad \text{and} \quad \frac{2.54\ \text{cm}}{1\ \text{in.}}$$

$$2.00\ \text{ft} \times \frac{12\ \text{in.}}{1\ \text{ft}} = 24.0\ \text{in.}$$

$$24.0\ \text{in.} \times \frac{2.54\ \text{cm}}{1\ \text{in.}} = 61.0\ \text{cm}$$

Since 1 ft and 12 in. are exact numbers, the number of significant figures allowed in the answer is three, based on the number 2.00.

How many meters are in a 100.-yd football field?

Example 2.12

The stepwise conversion of units from yards to meters may be done in this manner, using the proper conversion factors:

$$yd \rightarrow ft \rightarrow in. \rightarrow cm \rightarrow m$$

$$100. \,\cancel{yd} \times \frac{3 \text{ ft}}{1 \,\cancel{yd}} = 300. \text{ ft} \qquad (3 \text{ ft/yd})$$

$$300. \,\cancel{ft} \times \frac{12 \text{ in.}}{1 \,\cancel{ft}} = 3600 \text{ in.} \qquad (12 \text{ in./ft})$$

$$3600 \,\cancel{in.} \times \frac{2.54 \text{ cm}}{1 \,\cancel{in.}} = 9144 \text{ cm} \qquad (2.54 \text{ cm/in.})$$

$$9144 \,\cancel{cm} \times \frac{1 \text{ m}}{100 \,\cancel{cm}} = 91.4 \text{ m} \qquad (1 \text{ m/100 cm}) \qquad \text{(three significant figures)}$$

Examples 2.11 and 2.12 may be solved using a linear expression and writing down conversion factors in succession. This method often saves one or two calculation steps and allows numerical values to be reduced to simpler terms, leading to simpler calculations. The single linear expressions for Examples 2.11 and 2.12 are

$$2.00 \,\cancel{ft} \times \frac{12 \,\cancel{in.}}{1 \,\cancel{ft}} \times \frac{2.54 \text{ cm}}{1 \,\cancel{in.}} = 61.0 \text{ cm}$$

$$100. \,\cancel{yd} \times \frac{3 \,\cancel{ft}}{1 \,\cancel{yd}} \times \frac{12 \,\cancel{in.}}{1 \,\cancel{ft}} \times \frac{2.54 \,\cancel{cm}}{1 \,\cancel{in.}} \times \frac{1 \text{ m}}{100 \,\cancel{cm}} = 91.4 \text{ m}$$

Using the units alone (Example 2.12), we see that the stepwise cancellation proceeds in succession until the desired unit is reached.

$$\cancel{yd} \times \frac{\cancel{ft}}{\cancel{yd}} \times \frac{\cancel{in.}}{\cancel{ft}} \times \frac{\cancel{cm}}{\cancel{in.}} \times \frac{m}{\cancel{cm}} = m$$

Practice 2.8

How many meters are in 10.5 miles?

Example 2.13   How many cubic centimeters (cm^3) are in a box that measures 2.20 in. by 4.00 in. by 6.00 in.?

Solution   First we need to determine the volume of the box in cubic inches ($in.^3$) by multiplying the length times the width times the height:

$$2.20 \text{ in.} \times 4.00 \text{ in.} \times 6.00 \text{ in.} = 52.8 \text{ in.}^3$$

Now we need to convert $in.^3$ to cm^3, which can be done by using the inches and centimeters relationship three times:

$$\cancel{in.}^3 \times \frac{\text{cm}}{\cancel{in.}} \times \frac{\text{cm}}{\cancel{in.}} \times \frac{\text{cm}}{\cancel{in.}} = cm^3$$

$$52.8 \,\cancel{in.}^3 \times \frac{2.54 \text{ cm}}{1 \,\cancel{in.}} \times \frac{2.54 \text{ cm}}{1 \,\cancel{in.}} \times \frac{2.54 \text{ cm}}{1 \,\cancel{in.}} = 865 \text{ cm}^3$$

A driver of a car is obeying the speed limit of 55 miles per hour. How fast is it traveling in kilometers per second?   **Example 2.14**

Two conversions are needed to solve this problem: **Solution**

$$\text{mi} \rightarrow \text{km}$$
$$\text{hr} \rightarrow \text{min} \rightarrow \text{s}$$

To convert mi → km,

$$\frac{55 \; \cancel{\text{mi}}}{\text{hr}} \times \frac{1.609 \text{ km}}{1 \; \cancel{\text{mi}}} = 88 \frac{\text{km}}{\text{hr}}$$

Next we must convert hr → min → s. Notice that hours is in the denominator of our quantity, so the conversion factor must have hours in the numerator:

$$\frac{88 \text{ km}}{\cancel{\text{hr}}} \times \frac{1 \; \cancel{\text{hr}}}{60 \; \cancel{\text{min}}} \times \frac{1 \; \cancel{\text{min}}}{60 \text{ s}} = 0.024 \frac{\text{km}}{\text{s}}$$

Practice 2.9

How many cubic meters are in a room measuring 8 ft × 10 ft × 12 ft?

<table>
<tr><td>**2.9**</td><td># Measurement of Mass</td></tr>
</table>

The gram is used as a unit of mass measurement, but it is a tiny amount of mass; for instance, a nickel has a mass of about 5 grams. Therefore the *standard unit* of mass in the SI system is the **kilogram** (equal to 1000 g). The amount of mass in a kilogram is defined by international agreement as exactly equal to the mass of a platinum–iridium cylinder (international prototype kilogram) kept in a vault at Sèvres, France. Comparing this unit of mass to 1 lb (16 oz), we find that 1 kg is equal to 2.205 lb. A pound is equal to 453.6 g (0.4536 kg). The same prefixes used in length measurement are used to indicate larger and smaller gram units (see Table 2.4).

kilogram

TABLE 2.4 Metric Units of Mass

Unit	Abbreviation	Gram equivalent	Exponential equivalent
kilogram	kg	1000 g	10^3 g
gram	g	1 g	10^0 g
decigram	dg	0.1 g	10^{-1} g
centigram	cg	0.01 g	10^{-2} g
milligram	mg	0.001 g	10^{-3} g
microgram	μg	0.000001 g	10^{-6} g

FIGURE 2.5 ▶
(a) A quadruple beam balance with a precision of 0.01 g; (b) a single pan, top-loading balance with a precision of 0.001 g (1 mg); (c) a digital electronic analytical balance with a precision of 0.0001 g; and (d) a digital electronic balance with a precision of 0.001 g.

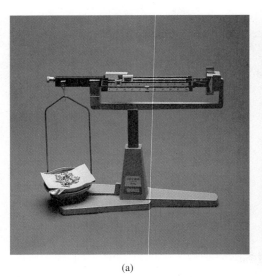

(a)

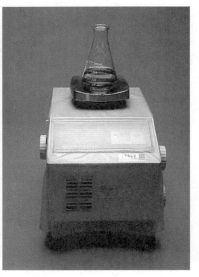

(b)

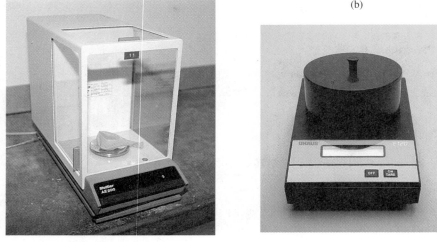

(c)   (d)

A balance is used to measure mass. Some balances will determine the mass of objects to the nearest microgram. The choice of balance depends on the precision required and the amount of material. Several balances are shown in Figure 2.5.

It is convenient to remember that

1 g = 1000 mg
1 kg = 1000 g
1 kg = 2.205 lb
1 lb = 453.6 g

To change grams to milligrams, multiply grams by the conversion factor 1000 mg/g. The setup for converting 25 g to milligrams is

$$25 \text{ g} \times \frac{1000 \text{ mg}}{1 \text{ g}} = 25{,}000 \text{ mg} \qquad (2.5 \times 10^4 \text{ mg})$$

Note that multiplying a number by 1000 is the same as multiplying the number by 10^3 and can be done simply by moving the decimal point three places to the right:

$$6.428 \times 1000 = 6428 \qquad (6.428)$$

To change milligrams to grams, multiply milligrams by the conversion factor 1 g/ 1000 mg. For example, to convert 155 mg to grams:

$$155 \text{ mg} \times \frac{1 \text{ g}}{1000 \text{ mg}} = 0.155 \text{ g}$$

Mass conversions from American to metric units are shown in Examples 2.15 and 2.16.

Example 2.15

A 1.50-lb package of baking soda costs 80 cents. How many grams of this substance are in this package?

Solution

We are solving for the number of grams equivalent to 1.50 lb. Since 1 lb = 453.6 g, the factor to convert pounds to grams is 453.6 g/lb:

$$1.50 \text{ lb} \times \frac{453.6 \text{ g}}{1 \text{ lb}} = 680. \text{ g}$$

Note: The cost of the baking soda has no bearing on the question asked in this problem.

Example 2.16

Suppose four ostrich feathers weigh 1.00 lb. Assuming that each feather is equal in mass, how many milligrams does a single feather weigh?

Solution

The unit conversion in this problem is from 1 lb/4 feathers to milligrams per feather. Since the unit *feathers* occurs in the denominator of both the starting unit and the desired unit, the unit conversions needed are

$$\text{lb} \rightarrow \text{g} \rightarrow \text{mg}$$

$$\frac{1.00 \text{ lb}}{4 \text{ feathers}} \times \frac{453.6 \text{ g}}{1 \text{ lb}} \times \frac{1000 \text{ mg}}{1 \text{ g}} = \frac{113,400}{1 \text{ feather}} \qquad (1.13 \times 10^5 \text{ mg/feather})$$

Practice 2.10

You are traveling in Europe and wake up one morning to find your mass is 75.0 kg. Determine the American equivalent to see whether you need to go on a diet before you return home.

Practice 2.11

A tennis ball has a mass of 65 g. Determine the American equivalent in pounds.

FIGURE 2.6 ▶
Calibrated glassware for
measuring the volume of liquids.

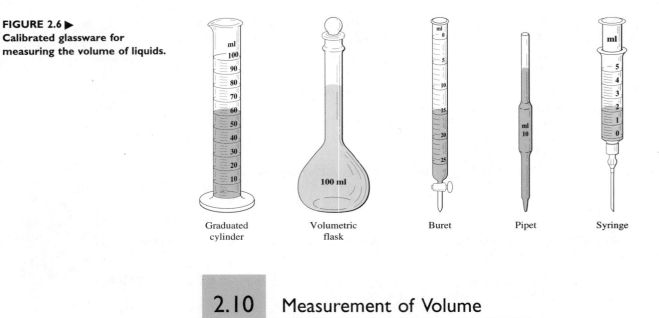

Graduated   Volumetric   Buret   Pipet   Syringe
cylinder   flask

2.10   Measurement of Volume

volume   **Volume**, as used here, is the amount of space occupied by matter. The SI unit of
volume is the *cubic meter* (m^3). However, the liter (pronounced *leeter* and abbreviated L) and the milliliter (abbreviated mL) are the standard units of volume used
liter   in most chemical laboratories. A **liter** is usually defined as 1 cubic decimeter
(1 kg) of water at 4°C.

The most common instruments or equipment for measuring liquids are the
graduated cylinder, volumetric flask, buret, pipet, and syringe, which are illustrated in
Figure 2.6. These pieces are usually made of glass and are available in various sizes.

It is convenient to remember that

$$1 \text{ L} = 1000 \text{ mL} = 1000 \text{ cm}^3$$
$$1 \text{ mL} = 1 \text{ cm}^3$$
$$1 \text{ L} = 1.057 \text{ qt}$$
$$946.1 \text{ mL} = 1 \text{ qt}$$

The volume of a cubic or rectangular container can be determined by multiplying
its length × width × height. Thus a 10-cm-square box has a volume of
10 cm × 10 cm × 10 cm = 1000 cm^3. Let's try some examples.

Example 2.17   How many milliliters are contained in 3.5 liters?

Solution   The conversion factor to change liters to milliliters is 1000 mL/L:

$$3.5 \text{ L} \times \frac{1000 \text{ mL}}{\text{L}} = 3500 \text{ mL} \qquad (3.5 \times 10^3 \text{ mL})$$

Liters may be changed to milliliters by moving the decimal point three places to
the right and changing the units to milliliters:

$$1.500 \text{ L} = 1500. \text{ mL}$$

How many cubic centimeters are in a cube that is 11.1 inches on a side?

Example 2.18

First change inches to centimeters. The conversion factor is 2.54 cm/in.:

Solution

$$11.1 \; \cancel{in.} \times \frac{2.54 \; cm}{1 \; \cancel{in.}} = 28.2 \; cm \; on \; a \; side$$

Then change to cubic volume (length × width × height):

28.2 cm × 28.2 cm × 28.2 cm = 22,426 cm^3 (2.24 × 10^4 cm^3)

Practice 2.12

A bottle of excellent chianti holds 750. mL. Determine the volume in quarts.

Practice 2.13

Milk is often purchased by the half gallon. Determine the number of liters necessary to equal this amount.

When doing problems with multiple steps you should round only at the end of the problem. We are rounding at the end of each step in example problems to illustrate the proper significant figures.

2.11   Measurement of Temperature

Heat is a form of energy associated with the motion of small particles of matter. The term *heat* refers to the quantity of energy within a system or to a quantity of energy added to or taken away from a system. *System* as used here simply refers to the entity that is being heated or cooled. Depending on the amount of heat energy present, a given system is said to be hot or cold. **Temperature** is a measure of the intensity of heat, or how hot a system is, regardless of its size. Heat always flows from a region of higher temperature to one of lower temperature. The SI unit of temperature is the kelvin. The common laboratory instrument for measuring temperature is a thermometer (see Figure 2.7).

heat

temperature

The temperature of a system can be expressed by several different scales. Three commonly used temperature scales are the Celsius scale (pronounced *sell-see-us*), the Kelvin (absolute) scale, and the Fahrenheit scale. The unit of temperature on the Celsius and Fahrenheit scales is called a *degree,* but the size of the Celsius and the Fahrenheit degree is not the same. The symbol for the Celsius and Fahrenheit degrees is °, and it is placed as a superscript after the number and before the symbol for the scale. Thus, 100°C means 100 *degrees Celsius.* The degree sign is not used with Kelvin temperatures.

degrees Celsius = °C
Kelvin (absolute) = K
degrees Fahrenheit = °F

On the Celsius scale the interval between the freezing and boiling temperatures of water is divided into 100 equal parts, or degrees. The freezing point of water is

FIGURE 2.7 ▶
Comparison of Celsius, Kelvin, and Fahrenheit temperature scales.

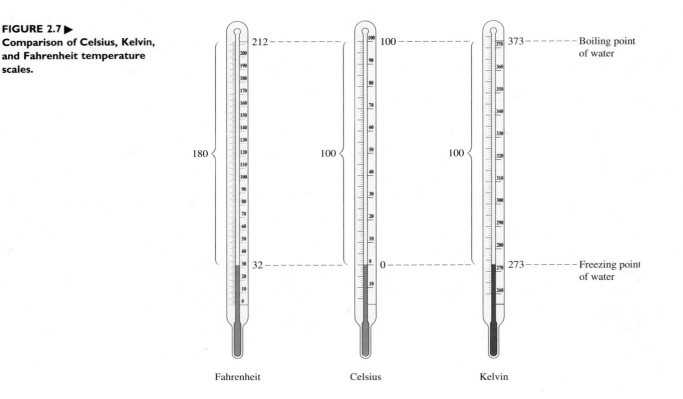

Fahrenheit   Celsius   Kelvin

assigned a temperature of 0°C and the boiling point of water a temperature of 100°C. The Kelvin temperature scale is known as the absolute temperature scale, because 0 K is the lowest temperature theoretically attainable. The Kelvin zero is 273.15 degrees below the Celsius zero. (A kelvin is equal in size to a Celsius degree.) The freezing point of water on the Kelvin scale is 273.15 K. The Fahrenheit scale has 180 degrees between the freezing and boiling temperatures of water. On this scale the freezing point of water is 32°F and the boiling point is 212°F.

$$0°C \cong 273 \text{ K} \cong 32°F$$

The three scales are compared in Figure 2.7. Although absolute zero (0 K) is the lower limit of temperature on these scales, temperature has no upper limit. (Temperatures of several million degrees are known to exist in the sun and in other stars.)

By examining Figure 2.7 we can see that there are 100 Celsius degrees and 100 kelvins between the freezing and boiling points of water, but there are 180 Fahrenheit degrees between these two temperatures. Hence, the size of a degree on the Celsius scale is the same as the size of 1 kelvin, but 1 Celsius degree corresponds to 1.8 degrees on the Fahrenheit scale.

$$\frac{180}{100} = 1.8$$

From these data, mathematical formulas have been derived to convert a temperature on one scale to the corresponding temperature on another scale:

$$K = °C + 273.15 \tag{1}$$

$$°F = (1.8 \times °C) + 32 \tag{2}$$

$$°C = \frac{°F - 32}{1.8} \tag{3}$$

The temperature at which table salt (sodium chloride) melts is 800.°C. What is this temperature on the Kelvin and Fahrenheit scales?

Example 2.19

We need to calculate K from °C, so we use formula (1). We also need to calculate °F from °C; for this calculation we use formula (2).

Solution

K = °C + 273.15

K = 800.°C + 273.15 = 1073 K

°F = (1.8 × °C) + 32

°F = (1.8 × 800.°C) + 32

°F = 1440 + 32 = 1472°F

800.°C = 1073 K = 1472°F

Remember, since the original measurement of 800.°C was to the units' place, the converted temperature is also to the units' place.

The temperature for December 1 in Honolulu, Hawaii, was 110.°F, a new record. Convert this temperature to °C.

Example 2.20

Formula (3) applies here.

Solution

$$°C = \frac{°F - 32}{1.8}$$

$$°C = \frac{110. - 32}{1.8} = \frac{78}{1.8} = 43°C$$

What temperature on the Fahrenheit scale corresponds to −8.0°C? (Notice the minus sign in this problem.)

Example 2.21

°F = (1.8 × °C) + 32

Solution

°F = [1.8 × (−8.0)] + 32 = −14.4 + 32

°F = 17.6°F

Since the original measurement is to the tenth the converted temperature is also to the tenth.

Temperatures used throughout this book are in degrees Celsius (°C) unless specified otherwise. The temperature after conversion should be expressed to the same precision as the original measurement.

Practice 2.14

Helium boils at 4 K. Convert this temperature to °C and then to °F.

Practice 2.15

"Normal" human body temperature is 98.6°F. Convert this to °C and K.

How Hot Is Hot?

How hot is it today? Does my son have a fever? What is the best temperature for storing mayonnaise? What is the optimum temperature for oil in my car? People have many reasons to determine the temperature of objects around them. In all cases an instrument for measuring the temperature is required. These instruments, called *thermometers,* come in a variety of shapes and sizes.

The Galileo thermometer is one of the more interesting and decorative room thermometers (see photo). These bulbs floating in the colored liquid are filled with liquids at different densities. As the temperature of the surrounding liquid (and the room) changes, the density of the globes change relative to the surrounding liquid, and they sink or float. Each globe is calibrated to sink at a particular temperature. The tag on the globe that is floating highest in the solution tells the temperature.

Traditional thermometers are calibrated tubes filled with a liquid that expands as it becomes warm. Many different liquids are used in thermometers, two of the most common being mercury and alcohol (usually dyed red or green). Mercury is the liquid of choice because it has a wide temperature range in the liquid phase and it is compact so the thermometer doesn't have to be very long. One problem with the mercury thermometer is that it could have toxic effects on the patient if were to break in the mouth. Also a mercury spill is tricky to clean up. Another drawback with this type of oral thermometer is that it takes approximately 3 minutes to stabilize and register accurately—a long time to keep your mouth closed if you have a severe cold or are congested.

Traditional home thermometers need to be shaken down in order to take a new temperature. Why do they require this

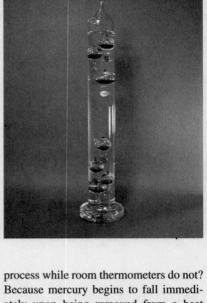

process while room thermometers do not? Because mercury begins to fall immediately upon being removed from a heat source, allowing no time for an accurate reading of the thermometer. To solve this problem a small kink is placed above the bulb in the tube (see diagram). The mercury is forced through this kink as it is being warmed, but it cannot fall back into the bulb as it cools. It remains in place until the thermometer is shaken, which forces the mercury through the kink and prepares the thermometer for its next use. Alcohol thermometers are more generally used because they are inexpensive, relatively safe, and work well at or near room temperature.

Today many hospitals, physicians' offices, and homes are equipped with electronic thermometers. These thermometers contain thermistors that are sensitive to temperature. A voltage reading is taken and associated with a number (calibrated into the thermometer). The temperature reading appears as a digital readout on the thermometer. These thermometers have several advantages, including the lack of toxic liquids, speed (they require much less time to register an accurate reading), convenience (they can be used with sterile cover to prevent the spread of infection), and size.

Engineers at Johns Hopkins Applied Physics Laboratory have built a battery-powered transmitting thermometer the size of an aspirin capsule that works after it is swallowed. It is capable of transmitting temperature measurements to within 0.01 degree until it passes out of the body (usually 1–2 days). This is useful in monitoring temperature patterns in the treatment of hypothermia, during which the body must be warmed at a slow, constant rate with continual temperature monitoring. The capsule can also be used to prevent hyperthermia in athletes, race drivers, or even in those taking routine treadmill tests.

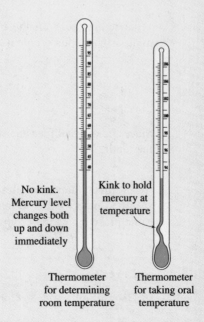

No kink. Mercury level changes both up and down immediately

Kink to hold mercury at temperature

Thermometer for determining room temperature

Thermometer for taking oral temperature

2.12 Density

Density *(d)* is the ratio of the mass of a substance to the volume occupied by that mass; it is the mass per unit of volume and is given by the equation

density

$$d = \frac{mass}{volume}$$

Density is a physical characteristic of a substance and may be used as an aid to its identification. When the density of a solid or a liquid is given, the mass is usually expressed in grams and the volume in milliliters or cubic centimeters.

$$d = \frac{mass}{volume} = \frac{g}{mL} \quad or \quad d = \frac{g}{cm^3}$$

Since the volume of a substance (especially liquids and gases) varies with temperature, it is important to state the temperature along with the density. For example, the volume of 1.0000 g of water at 4°C is 1.0000 mL, at 20°C it is 1.0018 mL, and at 80°C it is 1.0290 mL. Density therefore also varies with temperature.

The density of water at 4°C is 1.0000 g/mL, but at 80°C the density of water is 0.9718 g/mL.

$$d^{4°C} = \frac{1.0000 \text{ g}}{1.0000 \text{ mL}} = 1.0000 \text{ g/mL}$$

$$d^{80°C} = \frac{1.0000 \text{ g}}{1.0290 \text{ mL}} = 0.97182 \text{ g/mL}$$

The density of iron at 20°C is 7.86 g/mL.

$$d^{20°C} = \frac{7.86 \text{ g}}{1.00 \text{ mL}} = 7.86 \text{ g/mL}$$

The densities of a variety of materials are compared in Figure 2.8.

Densities for liquids and solids are usually represented in terms of grams per milliliter (g/mL) or grams per cubic centimeter (g/cm^3). The density of gases, however, is expressed in terms of grams per liter (g/L). Unless otherwise stated, gas densities are given for 0°C and 1 atmosphere pressure (discussed further in Chapter 13). Table 2.5 lists the densities of some common materials.

Suppose that water, Karo syrup, and vegetable oil are successively poured into a graduated cylinder. The result is a layered three-liquid system (Figure 2.9). Can we predict the order of the liquid layers? Yes, by looking up the densities in Table 2.5. Karo syrup has the greatest density (1.37 g/mL), and vegetable oil has the lowest density (0.91 g/mL). Karo syrup will be the bottom layer and vegetable oil will be the top layer. Water, with a density between the other two liquids, will form the middle layer. This information can also be determined by experiment. Vegetable oil, being less dense than water, will float when added to the beaker.

The density of air at 0°C is approximately 1.293 g/L. Gases with densities less than this value are said to be "lighter than air." A helium-filled balloon will rise rapidly in air because the density of helium is only 0.178 g/L.

FIGURE 2.8 ▶
(a) Comparison of the volumes of equal masses (10.0 g) of water, sulfur, and gold.
(b) Comparison of the masses of equal volumes (1.00 cm³) of water, sulfur, and gold. (Water is at 4°C; the two solids, at 20°C.)

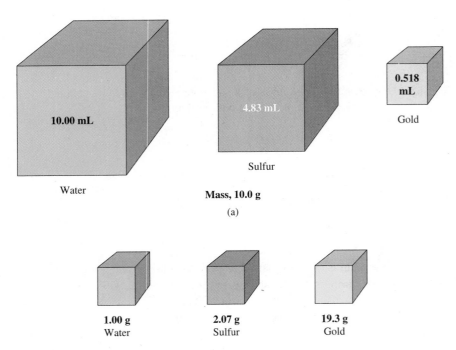

▲ FIGURE 2.9
Relative density of liquids. When three liquids are poured together, the liquid with the highest density will be the bottom layer. In the case of vegetable oil, water, and Karo syrup, vegetable oil is the top layer.

Vegetable oil

Water

Karo syrup

TABLE 2.5 Densities of Some Selected Materials*

Liquids and solids		Gases	
Substance	Density (g/mL at 20°C)	Substance	Density (g/mL at 0°C)
Wood (Douglas fir)	0.512	Hydrogen	0.090
Ethyl alcohol	0.789	Helium	0.178
Vegetable oil	0.91	Methane	0.714
Water (4°C)	**1.000**	Ammonia	0.771
Sugar	1.59	Neon	0.90
Glycerin	1.26	Carbon monoxide	1.25
Karo syrup	1.37	Nitrogen	1.251
Magnesium	1.74	**Air**	**1.293**
Sulfuric acid	1.84	Oxygen	1.429
Sulfur	2.07	Hydrogen chloride	1.63
Salt	2.16	Argon	1.78
Aluminum	2.70	Carbon dioxide	1.963
Silver	10.5	Chlorine	3.17
Lead	11.34		
Mercury	13.55		
Gold	19.3		

*For comparing densities the density of water is the reference for solids and liquids; air is the reference for gases.

Healthy Measurements

Obesity, a known health hazard, is among the top ten factors in cardiovascular disease. Simply measuring the weight of an individual is not a good indicator of leanness or obesity. A person may appear to be thin, yet have a high percentage of body fat. Someone else may appear "overweight" in comparison to published height-weight charts, but actually be especially lean as a result of a large percentage of muscle mass. To assess the body composition of an individual requires a measurement of the percent body fat.

Body fat is defined by health science professionals as the percentage of weight attributable to fat. It is the sum of the essential fat, surrounding and cushioning the internal organs, and the storage fat, which acts as a reservoir for energy in the body. A variety of techniques are currently in use to measure the body composition, including skin-fold tests, bioelectrical impedance, and hydrostatic weighing.

Hydrostatic weighing is considered to be one of the most accurate methods for determining body density. The individual is weighed in air, then seated on a chair suspended from a scale and lowered into a tank of warm water. After exhaling as much as possible the individual is submerged in the water and remains under the surface for 5–7 seconds. The underwater weight is recorded during this time. A series of calculations can then be made to determine percent body fat. The basis for these calculations lies in the variation in density of different tissues shown in the above table. A person with more bone and muscle mass will weigh more in water and have a higher body density.

As in all measurements errors in measured values result in variations in calculated results. A 100-g error in underwater weight results in a 1% body fat error. Various values for percent body fat calculations are shown in the table above.

Percent Body Fat

Classification	Male	Female
Lean	<8%	<15%
Healthy	8–15%	15–22%
Plump	16–19%	23–27%
Fat	20–24%	28–33%
Obese	>24%	>33%
Average college-age	15%	25%
Average middle-age	23%	32%
Distance runner	4–9%	6–15%
Tennis player	14–17%	19–22%

This person is being weighed hydrostatically to determine body composition.
▼

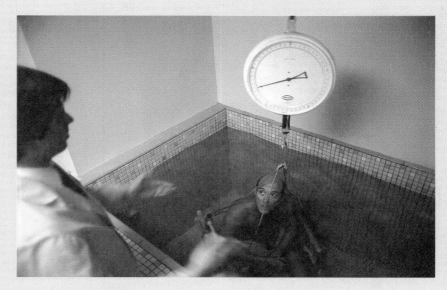

Density of Various Tissues in the Human Body

Bone	3.0 g/cm^3
Muscle	1.06 g/cm^3
Water	1.00 g/cm^3
Fat	0.9 g/cm^3

When an insoluble solid object is dropped into water, it will sink or float, depending on its density. If the object is less dense than water, it will float, displacing a *mass* of water equal to the mass of the object. If the object is more dense than water, it will sink, displacing a *volume* of water equal to the volume of the object. This information can be utilized to determine the volume (and density) of irregularly shaped objects.

specific gravity

The **specific gravity** (sp gr) of a substance is the ratio of the density of that substance to the density of another substance, usually water at 4°C. Specific gravity has no units because the density units cancel. The specific gravity tells us how many times as heavy a liquid, a solid, or a gas is as compared to the reference material. Since the density of water at 4°C is 1.00 g/mL, the specific gravity of a solid or liquid is the same as its density in g/mL without the units.

$$\text{sp gr} = \frac{\text{density of a liquid or solid}}{\text{density of water}}$$

Sample calculations of density problems follow.

Example 2.22   What is the density of a mineral if 427 g of the mineral occupy a volume of 35.0 mL?

Solution   We need to solve for density, so we start by writing the formula for calculating density:

$$d = \frac{\text{mass}}{\text{volume}}$$

Then we substitute the data given in the problem into the equation and solve:

$$\text{mass} = 427 \text{ g} \qquad \text{volume} = 35.0 \text{ mL}$$

$$d = \frac{\text{mass}}{\text{volume}} = \frac{427 \text{ g}}{35.0 \text{ mL}} = 12.2 \text{ g/mL}$$

Example 2.23   The density of gold is 19.3 g/mL. What is the mass of 25.0 mL of gold?

Solution   Two ways to solve this problem are: (a) Solve the density equation for mass, then substitute the density and volume data into the new equation and calculate. (b) Solve by dimensional analysis.

When alternative methods of solution are available, more than one is shown in the example. Choose the method you are most comfortable using to solve the problem.

Method 1 (a) Solve the density equation for mass:

$$d = \frac{\text{mass}}{\text{volume}} \qquad d \times \text{volume} = \text{mass}$$

(b) Substitute the data and calculate.

$$\text{mass} = \frac{19.3 \text{ g}}{\text{mL}} \times 25.0 \text{ mL} = 483 \text{ g}$$

Method 2 Dimensional analysis: Use density as a conversion factor, converting

$$mL \rightarrow g$$

The conversion of units is

$$mL \times \frac{g}{mL} = g$$

$$25.0 \; mL \times \frac{19.3 \; g}{mL} = 483 \; g$$

Calculate the volume (in mL) of 100. g of ethyl alcohol. **Example 2.24**

From Table 2.5 we see that the density of ethyl alcohol is 0.789 g/mL. This density **Solution**
also means that 1 mL of the alcohol has a mass of 0.789 g (1 mL/0.789 g).

> **Method 1:** This problem may be done by solving the density equation for
> volume and then substituting the data in the new equation.
>
> $$d = \frac{mass}{volume}$$
>
> $$volume = \frac{mass}{d}$$
>
> $$volume = \frac{100. \; g}{0.789 \; g/mL} = 127 \; mL$$

Method 2: Dimensional analysis. For a conversion factor, we can use either

$$\frac{g}{mL} \quad or \quad \frac{mL}{g}$$

In this case the conversion is from $g \rightarrow mL$, so we use mL/g. Substituting the data,

$$100. \; g \times \frac{1 \; mL}{0.789 \; g} = 127 \; mL \; of \; ethyl \; alcohol$$

The water level in a graduated cylinder stands at 20.0 mL before and at 26.2 mL **Example 2.25**
after a 16.74-g metal bolt is submerged in the water. (a) What is the volume of the
bolt? (b) What is the density of the bolt?

(a) The bolt will displace a volume of water equal to the volume of the bolt. Thus **Solution**
the increase in volume is the volume of the bolt.

$$
\begin{array}{rl}
26.2 \; mL = & volume \; of \; water \; plus \; bolt \\
-20.0 \; mL = & volume \; of \; water \\
\hline
6.2 \; mL = & volume \; of \; bolt
\end{array}
$$

(b) $d = \dfrac{mass \; of \; bolt}{volume \; of \; bolt} = \dfrac{16.74 \; g}{6.2 \; mL} = 2.7 \; g/mL$

Practice 2.16

Pure silver has a density of 10.5 g/mL. A ring sold as pure silver has a mass of 25.0 g. When placed in a graduated cylinder the water level rises 2.0 mL. Determine whether the ring is actually pure silver or if the customer should see the Better Business Bureau.

Practice 2.17

The water level in a metric measuring cup is 0.75 L before the addition of 150. g of shortening. The water level after submerging the shortening is 0.92 L. Determine the density of the shortening.

Concepts in Review

The major concepts of the chapter are listed in this section to help you review the chapter.

1. Differentiate between mass and weight. Indicate the instruments used to measure each.
2. Know the metric units of mass, length, and volume.
3. Know the numerical equivalent for the metric prefixes *deci, centi, milli, micro, nano, kilo,* and *mega.*
4. Express any number in scientific notation.
5. Express answers to calculations to the proper number of significant figures.
6. Set up and solve problems utilizing the method of dimensional analysis (factor-label method).
7. Convert measurements of mass, length, and volume from American units to metric units, and vice versa.
8. Make temperature conversions among Fahrenheit, Celsius, and Kelvin scales.
9. Differentiate between heat and temperature.
10. Calculate the density, mass, or volume of an object from the appropriate data.

Key Terms

The terms listed here have all been defined within the chapter. The section number is given for each term.

density (2.12)   meter (2.7)   specific gravity (2.12)
heat (2.11)   metric system (SI) (2.6)   temperature (2.11)
kilogram (2.9)   rounding off numbers (2.3)   volume (2.10)
liter (2.10)   scientific notation (2.4)   weight (2.1)
mass (2.1)   significant figures (2.2)

Questions

Questions refer to tables, figures, and key words and concepts defined within the chapter. A particularly challenging question or exercise is indicated with an asterisk.

1. Determine how many centimeters make up 1 km. (Table 2.3)

2. Determine the metric equivalent of 3 in. (Figure 2.4)

3. Why is the neck of a 100-mL volumetric flask narrower than the top of a 100-mL graduated cylinder? (Figure 2.6)

4. Describe the order of the following substances (top to bottom) if these three substances were placed in a 100-mL graduated cylinder: 25 mL glycerin, 25 mL mercury, and a cube of magnesium 2.0 cm on an edge. (Table 2.5)

5. Arrange these materials in order of increasing density: salt, vegetable oil, lead, and ethyl alcohol. (Table 2.5)

6. Ice floats in vegetable oil and sinks in ethyl alcohol. The density of ice must lie between what numerical values? (Table 2.5)

7. Distinguish between heat and temperature.

8. Distinguish between density and specific gravity.

9. Why is measuring the weight of a person a poor indication of leanness or obesity?

10. How is body density determined? What is the basis for this determination?

11. What are some of the important advantages of the metric system over the American system of weights and measurements?

12. State the rules used in this text for rounding off numbers.

13. Compare the number of degrees between the freezing point of water and its boiling point on the Fahrenheit, Kelvin, and Celsius temperature scales. (Figure 2.7)

14. Which of the following statements are correct? Rewrite the incorrect statements to make them correct:
    (a) The prefix *micro* indicates one-millionth of the unit expressed.
    (b) The length 10 cm is equal to 1000 mm.
    (c) The number 383.263 rounded to four significant figures becomes 383.3.
    (d) The number of significant figures in the number 29,004 is five.
    (e) The number 0.00723 contains three significant figures.
    (f) The sum of $32.276 + 2.134$ should contain four significant figures.
    (g) The product of 18.42 cm $\times$ 3.40 cm should contain three significant figures.
    (h) One microsecond is 10^{-6} second.
    (i) One thousand meters is a longer distance than 1000 yards.
    (j) One liter is a larger volume than 1 quart.
    (k) One centimeter is longer than 1 inch.
    (l) One cubic centimeter (cm^3) is equal to 1 milliliter.
    (m) The number 0.0002983 in exponential notation is 2.983×10^{-3}.
    (n) $(3.0 \times 10^4)(6.0 \times 10^6) = 1.8 \times 10^{11}$
    (o) Temperature is a form of energy.
    (p) The density of water at 4°C is 1.00 g/mL.
    (q) A pipet is a more accurate instrument for measuring 10.0 mL of water than is a graduated cylinder.

Paired Exercises

These exercises are paired. Each odd-numbered exercise is followed by a similar even-numbered exercise. Answers to the even-numbered exercises are given in Appendix V.

Metric Abbreviations

15. State the abbreviation for each of the following units:
    (a) gram
    (b) microgram
    (c) centimeter
    (d) micrometer
    (e) milliliter
    (f) deciliter

16. State the abbreviation for each of the following units:
    (a) milligram
    (b) kilogram
    (c) meter
    (d) nanometer
    (e) angstrom
    (f) microliter

Significant Figures, Rounding, Exponential Notation

17. For the following numbers, tell whether the zeros are significant:
  (a) 503
  (b) 0.007
  (c) 4200
  (d) 3.0030
  (e) 100.00
  (f) 8.00×10^2

18. Are the zeros significant in these numbers?
  (a) 63,000
  (b) 6.004
  (c) 0.00543
  (d) 8.3090
  (e) 60.
  (f) 5.0×10^{-4}

19. How many significant figures are in each of the following numbers?
  (a) 0.025
  (b) 22.4
  (c) 0.0404
  (d) 5.50×10^3

20. State the number of significant figures in each of the following numbers:
  (a) 40.0
  (b) 0.081
  (c) 129,042
  (d) 4.090×10^{-3}

21. Round each of the following numbers to three significant figures:
  (a) 93.246
  (b) 0.02857
  (c) 4.644
  (d) 34.250

22. Round each of the following numbers to three significant figures:
  (a) 8.8726
  (b) 21.25
  (c) 129.509
  (d) 1.995×10^6

23. Express each of the following numbers in exponential notation:
  (a) 2,900,000
  (b) 0.587
  (c) 0.00840
  (d) 0.0000055

24. Write each of the following numbers in exponential notation:
  (a) 0.0456
  (b) 4082.2
  (c) 40.30
  (d) 12,000,000

25. Solve the following problems, stating answers to the proper number of significant figures:
  (a) $12.62 + 1.5 + 0.25 =$
  (b) $(2.25 \times 10^3)\,(4.80 \times 10^4) =$
  (c) $\dfrac{452 \times 6.2}{14.3} =$
  (d) $0.0394 \times 12.8 =$
  (e) $\dfrac{0.4278}{59.6} =$
  (f) $10.4 + 3.75 \times (1.5 \times 10^4) =$

26. Evaluate each of the following expressions. State the answer to the proper number of significant figures:
  (a) $15.2 - 2.75 + 15.67$
  (b) 4.68×12.5
  (c) $\dfrac{182.6}{4.6}$
  (d) $1986 + 23.84 + 0.012$
  (e) $\dfrac{29.3}{284 \times 415}$
  (f) $(2.92 \times 10^{-3})\,(6.14 \times 10^5)$

27. Change these fractions into decimals. Express each answer to three significant figures:
  (a) $\dfrac{5}{6}$ **(b)** $\dfrac{3}{7}$ **(c)** $\dfrac{12}{16}$ **(d)** $\dfrac{9}{18}$

28. Change each of the following decimals to fractions in lowest terms:
  (a) 0.25 **(b)** 0.625 **(c)** 1.67 **(d)** 0.8888

29. Solve each of these equations for x:
  (a) $3.42x = 6.5$
  (b) $\dfrac{x}{12.3} = 7.05$
  (c) $\dfrac{0.525}{x} = 0.25$

30. Solve each equation for the variable:
  (a) $x = \dfrac{212 - 32}{1.8}$
  (b) $8.9\,\dfrac{g}{mL} = \dfrac{40.90\ g}{x}$
  (c) $72°F = 1.8x + 32$

Unit Conversions

31. Complete the following metric conversions using the correct number of significant figures:
    (a) 28.0 cm to m
    (b) 1000. m to km
    (c) 9.28 cm to mm
    (d) 10.68 g to mg
    (e) 6.8×10^4 mg to kg
    (f) 8.54 g to kg
    (g) 25.0 mL to L
    (h) 22.4 L to μL

32. Complete the following metric conversions using the correct number of significant figures:
    (a) 4.5 cm to Å
    (b) 12 nm to cm
    (c) 8.0 km to mm
    (d) 164 mg to g
    (e) 0.65 kg to mg
    (f) 5.5 kg to g
    (g) 0.468 L to mL
    (h) 9.0 μL to mL

33. Complete the following American/metric conversions using the correct number of significant figures:
    (a) 42.2 in. to cm
    (b) 0.64 mi to in.
    (c) 2.00 in.² to cm²
    (d) 42.8 kg to lb
    (e) 3.5 qt to mL
    (f) 20.0 gal to L

34. Make the following conversions using the correct number of significant figures:
    (a) 35.6 m to ft
    (b) 16.5 km to mi
    (c) 4.5 in.³ to mm³
    (d) 95 lb to g
    (e) 20.0 gal to L
    (f) 4.5×10^4 ft³ to m³

35. An automobile traveling at 55 miles per hour is moving at what speed in kilometers per hour?

36. A cyclist is traveling downhill at 55 km/hr. How fast is she moving in feet per second?

37. Carl Lewis, a sprinter in the 1988 Olympic Games, ran the 100.-m dash in 9.92 s. What was his speed in feet per second?

38. Al Unser, Jr., qualified for the pole position at the 1994 Indianapolis 500 at a speed of 229 mph. What was his speed in kilometers per second?

39. When the space probe *Galileo* reached Jupiter in 1995, it was traveling at an average speed of 27,000 miles per hour. What was its speed in kilometers per second?

40. The sun is approximately 93 million miles from the earth. How many seconds will it take for light from the sun to travel to the earth if the velocity of light is 3.00×10^8 m/s?

41. How many kilograms does a 176-lb man weigh?

42. The average mass of the heart of a human baby is about 1 oz. What is its mass in milligrams?

43. A regular aspirin tablet contains 5.0 grains of aspirin. How many grams of aspirin are in one tablet (1 grain = 1/7000. lb)?

44. An adult ruby-throated hummingbird has an average mass of 3.2 g, while an adult California condor may attain a weight of 21 lb. How many hummingbirds would it take to equal the mass of one condor?

45. A bag of pretzels has a mass of 283.5 g and costs $1.49. If a bag contains 18 pretzels determine the cost of a pound of pretzels?

46. The price of gold varies greatly and has been as high as $875 per ounce. What is the value of 250 g of gold at $350 per ounce? Gold is priced by troy ounces (14.58 troy ounces = 1 lb).

47. At 35¢/L how much will it cost to fill a 15.8-gal tank with gasoline?

48. How many liters of gasoline will be used to drive 525 miles in a car that averages 35 miles per gallon?

*49. Assuming that there are 20. drops in 1.0 mL, how many drops are in 1 gallon?

50. How many liters of oil are in a 42-gal barrel of oil?

*51. Calculate the number of milliliters of water in a cubic foot of water.

*52. Oil spreads in a thin layer on water called an "oil slick." How much area in m³ will 200 cm³ of oil cover if it forms a layer 0.5 nm thick?

53. A textbook is 27 cm long, 21 cm wide, and 4.4 cm thick. What is the volume in:
    (a) cubic centimeters?
    (b) liters?
    (c) cubic inches?

54. An aquarium measures 16 in. × 8 in. × 10 in. How many liters of water does it hold? How many gallons?

Temperature Conversions

55. Normal body temperature for humans is 98.6°F. What is this temperature on the Celsius scale?

56. Driving to the grocery store you notice the temperature is 45°C. Determine what this temperature is on the Fahrenheit scale and what season of the year it might be.

57. Make the following conversions and include an equation for each one:
(a) 162°F to °C
(b) 0.0°F to K
(c) −18°C to °F
(d) 212 K to °C

58. Make the following conversions and include an equation for each one:
(a) 32°C to °F
(b) −8.6°F to °C
(c) 273°C to K
(d) 100 K to °F

*59. At what temperature are the Fahrenheit and Celsius temperatures exactly equal?

*60. At what temperature are Fahrenheit and Celsius temperatures the same in value but opposite in sign?

Density

61. Calculate the density of a liquid if 50.00 mL of the liquid has a mass of 78.26 g.

62. A 12.8-mL sample of bromine has a mass of 39.9 g. What is the density of bromine?

63. When a 32.7-g piece of chromium metal was placed into a graduated cylinder containing 25.0 mL of water, the water level rose to 29.6 mL. Calculate the density of the chromium.

64. An empty graduated cylinder has a mass of 42.817 g. When filled with 50.0 mL of an unknown liquid it has a mass of 106.773 g. What is the density of the liquid?

65. Concentrated hydrochloric acid has a density of 1.19 g/mL. Calculate the mass of 250.0 mL of this acid.

66. What mass of mercury (density 13.6 g/mL) will occupy a volume of 25.0 mL?

Additional Exercises

These exercises are not paired or labeled by topic and provide additional practice on the concepts covered in this chapter.

67. One liter of homogenized whole milk has a mass of 1032 g. What is the density of the milk in grams per milliliter? In kilograms per liter?

68. The volume of blood plasma in adults is 3.1 L. Its density is 1.03 g/cm^3. Approximately how many pounds of blood plasma are there in your body?

*69. The dashed lane markers on an interstate highway are 2.5 ft long and 4.0 in. wide. One (1.0) qt of paint covers 43 ft^2. How many dashed lane markers can be painted with 15 gal of paint?

70. Will a hollow cube with sides of length 0.50 m hold 8.5 L of solution? Depending on your answer, how much additional solution would be required to fill the container or how many times would the container need to be filled to measure the 8.5 L?

*71. The accepted toxic dose of mercury is 300 μg/day. Dental offices sometimes contain as much as 180 μg of mercury per cubic meter of air. If a nurse working in the office ingests 2 × 10^4 L of air per day, is he or she at risk for mercury poisoning?

72. Which is the higher temperature, 4.5°F or −15°C?

*73. A flask containing 100. mL of alcohol ($d = 0.789$ g/mL) is placed on one pan of a two-pan balance. A larger container, with a mass of 11.0 g more than the empty flask, is placed on the other pan of the balance. What volume of turpentine ($d = 0.87$ g/mL) must be added to this container to bring the two pans into balance?

74. Suppose you have samples of two metals, A and B. Use the data below to determine which sample occupies the larger volume

	A	B
Mass	25 g	65 g
Density	10 g/mL	4 g/mL

*75. As a solid substance is heated its volume increases but its mass remains the same. Sketch a graph of density vs. temperature showing the trend you expect. Briefly explain.

76. A 35.0-mL sample of ethyl alcohol (density 0.789 g/mL) is added to a graduated cylinder that has a mass of 49.28 g. What will be the mass of the cylinder plus the alcohol?

77. You are given three cubes, A, B, and C; one is magnesium, one is aluminum, and the third is silver. All three cubes have the same mass, but cube A has a volume of 25.9 mL, cube B has a volume of 16.7 mL, and cube C has a volume of 4.29 mL. Identify cubes A, B, and C.

*78. A cube of aluminum has a mass of 500. g. What will be the mass of a cube of gold of the same dimensions?

79. A 25.0-mL sample of water at 90°C has a mass of 24.12 g. Calculate the density of water at this temperature.

80. The mass of an empty container is 88.25 g. The mass of the container when filled with a liquid (d = 1.25 g/mL) is 150.50 g. What is the volume of the container?

81. Which liquid will occupy the greater volume, 50 g of water or 50 g of ethyl alcohol? Explain.

82. A gold bullion dealer advertised a bar of pure gold for sale. The gold bar had a mass of 3300 g and measured 2.00 cm × 15.0 cm × 6.00 cm. Was the bar pure gold? Show evidence for your answer.

83. The largest nugget of gold on record was found in 1872 in New South Wales, Australia, and had a mass of 93.3 kg. Assuming the nugget is pure gold, what is its volume in cubic centimeters? What is it worth by today's standards if gold is \$345/oz? (14.58 troy oz = 1 lb.)

*84. Forgetful Freddie placed 25.0 mL of a liquid in a graduated cylinder with a mass of 89.450 g when empty. When Freddie placed a metal slug with a mass of 15.454 g into the cylinder, the volume rose to 30.7 mL. Freddie was asked to calculate the density of the liquid and of the metal slug from his data, but he forgot to obtain the mass of the liquid. He was told that if he found the mass of the cylinder containing the liquid and the slug, he would have enough data for the calculations. He did so and found its mass to be 125.934 g. Calculate the density of the liquid and of the metal slug.

Answers to Practice Exercises

2.1 (a) 2, (b) 4, (c) 4, (d) 1, (e) 3, (f) 4, (g) 1, (h) 3

2.2 (a) 42.25 (Rule 2), (b) 88.0 (Rule 1), (c) 0.0897 (Rule 2), (d) 0.090 (Rule 2), (e) 225 (Rule 1), (f) 14.2 (Rule 2)

2.3 (a) 1200 = 1.200×10^3 (left means positive exponent), (b) 6,600,000 = 6.6×10^6 (left means positive exponent), (c) 0.0468 = 4.68×10^{-2} (right means negative exponent), (d) 0.00003 = 3×10^{-5} (right means negative exponent)

2.4 3350 in.^{2} = 3.4×10^3 in.^{2}

2.5 50.7 mph

2.6 0.79

2.7 (a) 2, (b) 2, (c) 1, (d) 2, (e) 4, (f) 2, (g) 2, (h) 2

2.8 1.69×10^4 m

2.9 30 m^3 or 3×10^1 m^3

2.10 165 lb

2.11 0.14 lb

2.12 0.793 qt

2.13 1.89 L (the number of significant figures is arbitrary)

2.14 $-269°C$, $-452°F$

2.15 37.0°C, 310.2 K

2.16 The density is 13 g/mL; therefore the ring is *not* pure silver.

2.17 0.88 g/mL

Throughout our lives we seek to bring order into the chaos that surrounds us. To do this, we classify things according to their similarities. In the library, we find books grouped according to the subject, and then by author. Our local department store organizes its merchandise by the size and style of clothing, as well as by the type of customer. The ball park or theater classifies its seats by price and location. The biologist divides the living world into plants and animals; this broad classification is further simplified into various phyla and on to specific genera. In chemistry, this classification process begins with mixtures (such as air or vinegar) and then on to pure substances (such as water or mercury). This process continues and ultimately leads us to the fundamental building blocks of matter—the elements.

3.1   Matter Defined

The entire universe consists of matter and energy. Every day we come into contact with countless kinds of matter. Air, food, water, rocks, soil, glass, and this book are all different types of matter. Broadly defined, **matter** is *anything* that has mass and occupies space.

matter

Matter may be quite invisible. For example, if an apparently empty test tube is submerged mouth downward in a beaker of water, the water rises only slightly into the tube. The water cannot rise further because the tube is filled with invisible matter: air (see Figure 3.1).

To the eye, matter appears to be continuous and unbroken. However, it is actually discontinuous and is composed of discrete, tiny particles called *atoms*. The particulate nature of matter will become evident when we study atomic structure and the properties of gases.

3.2   Physical States of Matter

Matter exists in three physical states: solid, liquid, and gas. A **solid** has a definite shape and volume, with particles that cohere rigidly to one another. The shape of a solid can be independent of its container. For example, a crystal of sulfur has the same shape and volume whether it is placed in a beaker or simply laid on a glass plate.

solid

Most commonly occurring solids, such as salt, sugar, quartz, and metals, are *crystalline*. The particles that form crystalline materials exist in regular, repeating, three-dimensional, geometric patterns. Because their particles do not have any regular, internal geometric pattern, such solids as plastics, glass, and gels are called **amorphous** solids. (*Amorphous* means without shape or form.)

amorphous

liquid

A **liquid** has a definite volume but not a definite shape, with particles that cohere firmly but not rigidly. Although the particles are held together by strong attractive forces and are in close contact with one another, they are able to move freely. Particle mobility gives a liquid fluidity and causes it to take the shape of the container in which it is stored.

◀ **Chapter Opening Photo: Apples are classified by color, taste, and variety before they are shipped to consumers.**

FIGURE 3.1
An apparently empty test tube is submerged, mouth downward, in water. Only a small volume of water rises into the tube, which is actually filled with air.

TABLE 3.1 Common Materials in the Solid, Liquid and Gaseous States of Matter

Solids	Liquids	Gases
Aluminum	Alcohol	Acetylene
Copper	Blood	Air
Gold	Gasoline	Butane
Polyethylene	Honey	Carbon dioxide
Salt	Mercury	Chlorine
Sand	Oil	Helium
Steel	Vinegar	Methane
Sulfur	Water	Oxygen

TABLE 3.2 Physical Properties of Solids, Liquids, and Gases

State	Shape	Volume	Particles	Compressibility
Solid	Definite	Definite	Rigidly cohering; tightly packed	Very slight
Liquid	Indefinite	Definite	Mobile; cohering	Slight
Gas	Indefinite	Indefinite	Independent of each other and relatively far apart	High

gas   A **gas** has indefinite volume and no fixed shape, with particles that move independently of one another. Particles in the gaseous state have gained enough energy to overcome the attractive forces that held them together as liquids or solids. A gas presses continuously in all directions on the walls of any container. Because of this quality a gas completely fills a container. The particles of a gas are relatively far apart compared with those of solids and liquids. The actual volume of the gas particles is usually very small in comparison with the volume of the space occupied by the gas. A gas therefore may be compressed into a very small volume or expanded almost indefinitely. Liquids cannot be compressed to any great extent, and solids are even less compressible than liquids.

If a bottle of ammonia solution is opened in one corner of the laboratory, we can soon smell its familiar odor in all parts of the room. The ammonia gas escaping from the solution demonstrates that gaseous particles move freely and rapidly and tend to permeate the entire area into which they are released.

Although matter is discontinuous, attractive forces exist that hold the particles together and give matter its appearance of continuity. These attractive forces are strongest in solids, giving them rigidity; they are weaker in liquids but still strong enough to hold liquids to definite volumes. In gases the attractive forces are so weak that the particles of a gas are practically independent of one another. Table 3.1 lists a number of common materials that exist as solids, liquids, and gases. Table 3.2 summarizes comparative properties of solids, liquids, and gases.

◄ Water can exist as a solid (snow), a liquid (water), and a gas (steam) as shown here at Yellowstone National Park.

3.3   Substances and Mixtures

The term *matter* refers to all materials or material things that make up the universe. Many thousands of different and distinct kinds of matter or substances exist. A **substance** is a particular kind of matter with a definite, fixed composition. A substance, sometimes known as a *pure substance,* is either an element or a compound. Familiar examples of elements are copper, gold, and oxygen. Familiar compounds are salt, sugar, and water.

    We can classify a sample of matter as either *homogeneous* or *heterogeneous* by examining it. **Homogeneous** matter is uniform in appearance and has the same properties throughout. Matter consisting of two or more physically distinct phases is **heterogeneous.** A **phase** is a homogeneous part of a system separated from other parts by physical boundaries. A **system** is simply the body of matter under consideration. Whenever we have a system in which visible boundaries exist between the parts or components, that system has more than one phase and is heterogeneous. It does not matter whether these components are in the solid, liquid, or gaseous states.

    A pure substance may exist as different phases in a heterogeneous system. Ice floating in water, for example, is a two-phase system made up of solid water and liquid water. The water in each phase is homogeneous in composition, but because two phases are present, the system is heterogeneous.

    A **mixture** is a material containing two or more substances and can be either heterogeneous or homogeneous. Mixtures are variable in composition. If we add a

substance

homogeneous

heterogeneous
phase
system

mixture

FIGURE 3.2 ▶
Classification of matter. A pure substance is always homogeneous in composition, whereas a mixture always contains two or more substances and may be either homogeneous or heterogeneous.

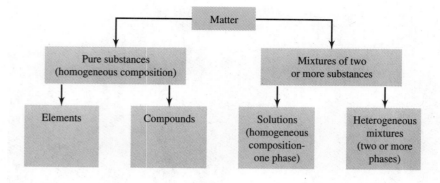

Flow charts can help you to visualize the connections between concepts.

spoonful of sugar to a glass of water, a heterogeneous mixture is formed immediately. The two phases are a solid (sugar) and a liquid (water). But upon stirring the sugar dissolves to form a homogeneous mixture or solution. Both substances are still present: All parts of the solution are sweet and wet. The proportions of sugar and water can be varied simply by adding more sugar and stirring to dissolve.

Many substances do not form homogeneous mixtures. If we mix sugar and fine white sand, a heterogeneous mixture is formed. Careful examination may be needed to decide that the mixture is heterogeneous because the two phases (sugar and sand) are both white solids. Ordinary matter exists mostly as mixtures. If we examine soil, granite, iron ore, or other naturally occurring mineral deposits, we find them to be heterogeneous mixtures. Air is a homogeneous mixture (solution) of several gases. Figure 3.2 illustrates the relationships of substances and mixtures.

3.4   Elements

All the words in the English dictionary are formed from an alphabet consisting of only 26 letters. All known substances on earth—and most probably in the universe, too—are formed from a sort of "chemical alphabet" consisting of 111 presently known elements. An **element** is a fundamental or elementary substance that cannot be broken down by chemical means to simpler substances. Elements are the building blocks of all substances. The elements are numbered in order of increasing complexity beginning with hydrogen, number 1. Of the first 92 elements, 88 are known to occur in nature. The other four—technetium (43), promethium (61), astatine (85), and francium (87)—either do not occur in nature or have only transitory existences during radioactive decay. With the exception of number 94, plutonium, elements above number 92 are not known to occur naturally but have been synthesized, usually in very small quantities, in laboratories. The discovery of trace amounts of element 94 (plutonium) in nature has been reported recently. The syntheses of elements 110 and 111 were reported in 1994. No elements other than those on the earth have been detected on other bodies in the universe.

element

Most substances can be decomposed into two or more simpler substances. Water can be decomposed into hydrogen and oxygen. Sugar can be decomposed into carbon, hydrogen, and oxygen. Table salt is easily decomposed into sodium and chlorine. An element, however, cannot be decomposed into simpler substances by ordinary chemical changes.

If we could take a small piece of an element, say copper, and divide it and subdivide it into smaller and smaller particles, we would finally come to a single unit of copper that we could no longer divide and still have copper. This smallest particle of an element that can exist, is called an **atom**, which is also the smallest unit of an element that can enter into a chemical reaction. Atoms are made up of still smaller subatomic particles. But these subatomic particles (described in Chapter 5) do not have the properties of elements.

atom

3.5 Distribution of Elements

Elements are distributed very unequally in nature, as shown in Figure 3.3. At normal room temperature two of the elements, bromine and mercury, are liquids. Eleven elements, hydrogen, nitrogen, oxygen, fluorine, chlorine, helium, neon, argon, krypton, xenon, and radon, are gases. All the other elements are solids.

Ten elements make up about 99% of the mass of the earth's crust, seawater, and atmosphere. Oxygen, the most abundant of these, constitutes about 50% of this mass. The distribution of the elements shown in Table 3.3 includes the earth's crust to a depth of about 10 miles, the oceans, fresh water, and the atmosphere but does not include the mantle and core of the earth, which are believed to consist of metallic iron and nickel. Because the atmosphere contains relatively little matter, its inclusion has almost no effect on the distribution shown in Table 3.3. But the inclusion of fresh and salt water does have an appreciable effect since water contains about 11.2% hydrogen. Nearly all of the 0.87% hydrogen shown in the table is from water.

The average distribution of the elements in the human body is shown in Figure 3.4. Note again the high percentage of oxygen.

FIGURE 3.3
Distribution of elements in nature.
▼

Galaxies

Earth

Humans

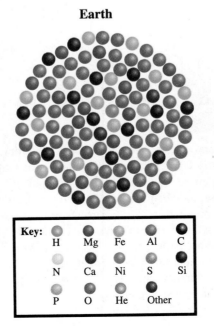

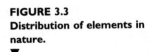

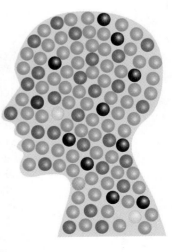

TABLE 3.3	Distribution of the Elements in the Earth's Crust, Seawater, and Atmosphere		
Element	**Mass percent**	**Element**	**Mass percent**
Oxygen	49.20	Chlorine	0.19
Silicon	25.67	Phosphorus	0.11
Aluminum	7.50	Manganese	0.09
Iron	4.71	Carbon	0.08
Calcium	3.39	Sulfur	0.06
Sodium	2.63	Barium	0.04
Potassium	2.40	Nitrogen	0.03
Magnesium	1.93	Fluorine	0.03
Hydrogen	0.87		
Titanium	0.58	All others	0.47

FIGURE 3.4 ▶
Average elemental composition of the human body.

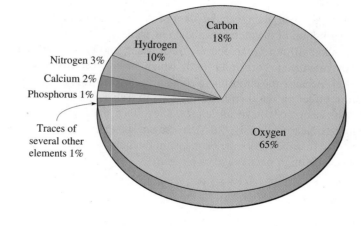

3.6   Names of the Elements

The names of the elements come to us from various sources. Many are derived from early Greek, Latin, or German words that generally described some property of the element. For example, iodine is taken from the Greek word *iodes,* meaning violetlike. Iodine, indeed, is violet in the vapor state. The name of the metal bismuth had its origin from the German words *weisse masse,* which means white mass. Miners called it *wismat*; it was later changed to *bismat,* and finally to bismuth. Some elements are named for the location of their discovery—for example, germanium, discovered in 1886 by a German chemist. Others are named in commemoration of famous scientists, such as einsteinium and curium, named for Albert Einstein and Marie Curie, respectively.

3.7   Symbols of the Elements

We all recognize Mr., N.Y., and Ave. as abbreviations for mister, New York, and avenue. In a like manner chemists have assigned an abbreviation to each element; these are called **symbols** of the elements. Fourteen of the elements have a single letter as their symbol, six have three-letter symbols, and the rest have two letters. A symbol stands for the element itself, for one atom of the element, and (as we shall see later) for a particular quantity of the element.

symbol

Controversy exists over the names for the elements with three-letter symbols. (See Chemistry in Action, p. 103.)

Rules governing symbols of elements are as follows:

1. Symbols are composed of one, two, or three letters.
2. If one letter is used, it is capitalized.
3. If two or three letters are used, only the first is capitalized.

Examples: Sulfur S   Barium Ba

The symbols and names of all the elements are given in the table on the inside front cover of this book. Table 3.4 lists the more commonly used symbols. If we examine this table carefully, we note that most of the symbols start with the same letter as the name of the element that is represented. A number of symbols, however, appear to have no connection with the names of the elements they represent (see Table 3.5). These symbols have been carried over from earlier names (usually in Latin) of the elements and are so firmly implanted in the literature that their use is continued today.

Special care must be taken in writing symbols. Begin each with a capital letter and use a lowercase second letter if needed. For example, consider Co, the symbol for the element cobalt. If through error CO (capital C and capital O) is written, the two elements carbon and oxygen (the *formula* for carbon monoxide) are represented

TABLE 3.4 Symbols of the Most Common Elements

Element	Symbol	Element	Symbol	Element	Symbol
Aluminum	Al	Fluorine	F	Phosphorus	P
Antimony	Sb	Gold	Au	Platinum	Pt
Argon	Ar	Helium	He	Potassium	K
Arsenic	As	Hydrogen	H	Radium	Ra
Barium	Ba	Iodine	I	Silicon	Si
Bismuth	Bi	Iron	Fe	Silver	Ag
Boron	B	Lead	Pb	Sodium	Na
Bromine	Br	Lithium	Li	Strontium	Sr
Cadmium	Cd	Magnesium	Mg	Sulfur	S
Calcium	Ca	Manganese	Mn	Tin	Sn
Carbon	C	Mercury	Hg	Titanium	Ti
Chlorine	Cl	Neon	Ne	Tungsten	W
Chromium	Cr	Nickel	Ni	Uranium	U
Cobalt	Co	Nitrogen	N	Zinc	Zn
Copper	Cu	Oxygen	O		

This tiny and powerful computer chip is made of silicon, a metalloid.

J. J. Berzelius (1779–1848) devised the chemical symbol system in use today.

TABLE 3.5 Symbols of the Elements Derived from Early Names*		
Present name	**Symbol**	**Former name**
Antimony	Sb	Stibium
Copper	Cu	Cuprum
Gold	Au	Aurum
Iron	Fe	Ferrum
Lead	Pb	Plumbum
Mercury	Hg	Hydrargyrum
Potassium	K	Kalium
Silver	Ag	Argentum
Sodium	Na	Natrium
Tin	Sn	Stannum
Tungsten	W	Wolfram

* These symbols are in use today even though they do not correspond to the current name of the element.

instead of the single element cobalt. Another example of the need for care in writing symbols is with the symbol Ca for calcium versus Co for cobalt. The letters must be distinct, or else the symbol for the element may be misinterpreted.

Knowledge of symbols is essential for writing chemical formulas and equations, and will be used extensively in the remainder of this book and in any future chemistry courses you may take. One way to learn the symbols is to practice a few minutes a day by making side-by-side lists of names and symbols and then covering each list alternately and writing the corresponding name or symbol. Initially it is a good plan to learn the symbols of the most common elements shown in Table 3.4.

3.8   Metals, Nonmetals, and Metalloids

Three primary classifications of the elements are metals, nonmetals, and metalloids. Most of the elements are metals. We are familiar with them because of their widespread use in tools, materials of construction, automobiles, and so on. But nonmetals are equally useful in our everyday life as major components of clothing, food, fuel, glass, plastics, and wood. Metalloids are often used in the electronics industry.

metal

The **metals** are solids at room temperature (mercury is an exception). They have high luster, are good conductors of heat and electricity, are *malleable* (can be rolled or hammered into sheets), and are *ductile* (can be drawn into wires). Most metals have a high melting point and high density. Familiar metals are aluminum, chromium, copper, gold, iron, lead, magnesium, mercury, nickel, platinum, silver, tin, and zinc. Less familiar but still important metals are calcium, cobalt, potassium, sodium, uranium, and titanium.

Metals have little tendency to combine with each other to form compounds. But many metals readily combine with nonmetals such as chlorine, oxygen, and sulfur to form ionic compounds such as metallic chlorides, oxides, and sulfides. In nature the more reactive metals are found combined with other elements as minerals.

Hydrogen: Fuel of the Future?

Hydrogen, the lightest element on the periodic table, could provide the basis for a society powered by a fuel that is nearly inexhaustible, is environmentally benign, and is available domestically. The space program has used hydrogen for more than 20 years to power rockets as well as to provide electrical power and even drinking water. How can we harness the energy found in hydrogen and use it to form a hydrogen-based fuel economy?

A major hurdle to using hydrogen as a primary fuel source is the difficulty in storing it. As a gas, it has a low density and requires a bulky container; liquid hydrogen (b.p. $-253°C$) requires intensive refrigeration and well-insulated containers. The best solution to the storage problem may be to generate hydrogen as it is needed for fuel.

The most promising, simple, and inexpensive scheme for generating hydrogen is based on sponge iron. The most remarkable quality of sponge iron is its ability to rust easily. In the rusting process water is split into hydrogen and oxygen while iron is converted to iron oxide (rust). Usually rusting takes place very slowly so that the hydrogen is formed in amounts too small to be useful. However, the H Power Corp of Belleville, N. J. has developed a process that increases the speed of rusting

and cycles the iron oxide back to the sponge-iron state so it can produce more hydrogen.

This sponge-iron cycle could then be linked to fuel cells to eliminate the need for hydrogen storage units in transportation vehicles. The sponge-iron cycle could also be used in residential heating or in industrial processes. Since sponge iron is inexpensive and recyclable, it could be transported by ship and rail to many locations. The overall cost should be competitive to fossil fuels.

Changing over to hydrogen as a primary fuel source could be done gradually by using this type of technology. To use hydrogen effectively requires an infrastructure, which could take from 20 to 25 years to develop. Studies indicate that the transportation sector could be the first place to begin the use of hydrogen. Hydrogen-powered vehicles could have significant impact in places such as California where new regulations call for lower emission vehicles after 1998. Conventional cars could be converted to run on hydrogen, but the major difficulty is in getting enough hydrogen on board to provide a reasonable driving range. Another area of development for hydrogen fuels is for residential use. Studies at the University of Miami show that hydrogen can be used for cooking and heating.

The use of hydrogen as a fuel could have a fundamental impact on the way our society is organized. Instead of massive power plants, our energy needs would be met by on-site fuel cells that provide all of our heating, cooling, lighting, and other power requirements. There would be no losses over transmission lines and little capital investment. Best of all, the system would reduce the environmental impact of energy generation and take the cost of energy from a government-regulated commercial venture and put it into the hands of the individual consumer.

The LaserCel 1, unveiled in 1991, is more economical to operate on hydrogen than on gasoline. This is accomplished by replacing the internal combustion engine with a hydrogen fuel cell.

A few of the less reactive ones such as copper, gold, and silver are sometimes found in a native, or free, state.

Nonmetals, unlike metals, are not lustrous, have relatively low melting points and densities, and are generally poor conductors of heat and electricity. Carbon, phosphorus, sulfur, selenium, and iodine are solids; bromine is a liquid; the rest of the nonmetals are gases. Common nonmetals found uncombined in nature are carbon (graphite and diamond), nitrogen, oxygen, sulfur, and the noble gases (helium, neon, argon, krypton, xenon, and radon).

Nonmetals combine with one another to form molecular compounds such as carbon dioxide (CO_2), methane (CH_4), butane (C_4H_{10}), and sulfur dioxide (SO_2). Fluorine, the most reactive nonmetal, combines readily with almost all other elements.

nonmetal

Samples of various metals, ▶
including aluminum, copper,
mercury, titanium, beryllium,
cadmium, calcium, and nickel.

metalloid

Several elements (boron, silicon, germanium, arsenic, antimony, tellurium, and polonium) are classified as **metalloids** and have properties that are intermediate between those of metals and those of nonmetals. The intermediate position of these elements is shown in Table 3.6, which lists and classifies all the elements as metals, nonmetals, or metalloids. Certain metalloids, such as boron, silicon, and germanium, are the raw materials for the semiconductor devices that make our modern electronics industry possible.

3.9   Compounds

compound

A **compound** is a distinct substance containing two or more elements chemically combined in definite proportions by mass. Compounds, unlike elements, can be decomposed chemically into simpler substances—that is, into simpler compounds and/or elements. Atoms of the elements in a compound are combined in whole-number ratios, never as fractional parts. Compounds fall into two general types, *molecular* and *ionic*.

molecule

A **molecule** is the smallest uncharged individual unit of a compound formed by the union of two or more atoms. Water is a typical molecular compound. If we divide a drop of water into smaller and smaller particles, we finally obtain a single molecule of water consisting of two hydrogen atoms bonded to one oxygen atom. This molecule is the ultimate particle of water; it cannot be further subdivided without destroying the water molecule and forming hydrogen and oxygen.

TABLE 3.6 Classification of the Elements into Metals, Metalloids, and Nonmetals

1 H																	2 He
3 Li	4 Be											5 B	6 C	7 N	8 O	9 F	10 Ne
11 Na	12 Mg											13 Al	14 Si	15 P	16 S	17 Cl	18 Ar
19 K	20 Ca	21 Sc	22 Ti	23 V	24 Cr	25 Mn	26 Fe	27 Co	28 Ni	29 Cu	30 Zn	31 Ga	32 Ge	33 As	34 Se	35 Br	36 Kr
37 Rb	38 Sr	39 Y	40 Zr	41 Nb	42 Mo	43 Tc	44 Ru	45 Rh	46 Pd	47 Ag	48 Cd	49 In	50 Sn	51 Sb	52 Te	53 I	54 Xe
55 Cs	56 Ba	57 La*	72 Hf	73 Ta	74 W	75 Re	76 Os	77 Ir	78 Pt	79 Au	80 Hg	81 Tl	82 Pb	83 Bi	84 Po	85 At	86 Rn
87 Fr	88 Ra	89 Ac†	104 Unq	105 Unp	106 Unh	107 Uns	108 Uno	109 Une									

Metals
Metalloids
Nonmetals

	58 Ce	59 Pr	60 Nd	61 Pm	62 Sm	63 Eu	64 Gd	65 Tb	66 Dy	67 Ho	68 Er	69 Tm	70 Yb	71 Lu
*														
†	90 Th	91 Pa	92 U	93 Np	94 Pu	95 Am	96 Cm	97 Bk	98 Cf	99 Es	100 Fm	101 Md	102 No	103 Lr

◀ **Samples of various nonmetals, including iodine, bromine, oxygen, neon, and sulfur.**

ion

cation

anion

An **ion** is a positively or negatively charged atom or group of atoms. An ionic compound is held together by attractive forces that exist between positively and negatively charged ions. A positively charged ion is called a **cation** (pronounced *cat-eye-on*); a negatively charged ion is called an **anion** (pronounced *an-eye-on*).

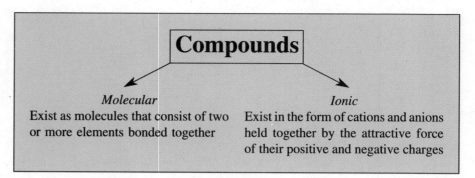

Compounds

Molecular	*Ionic*
Exist as molecules that consist of two or more elements bonded together	Exist in the form of cations and anions held together by the attractive force of their positive and negative charges

(a) H$_2$O

(b) NaCl

▲
FIGURE 3.5
Representation of molecular and ionic (nonmolecular) compounds. (a) Two hydrogen atoms combined with an oxygen atom to form a molecule of water. (b) A positively charged sodium ion and a negatively charged chloride ion form the compound sodium chloride.

Sodium chloride is a typical ionic compound. The ultimate particles of sodium chloride are positively charged sodium ions and negatively charged chloride ions. Sodium chloride is held together in a crystalline structure by the attractive forces existing between these oppositely charged ions. Although ionic compounds consist of large aggregates of cations and anions, their formulas are normally represented by the simplest possible ratio of the atoms in the compound. For example, in sodium chloride the ratio is one sodium ion to one chlorine ion, and the formula is NaCl. The two types of compounds, molecular and ionic, are illustrated in Figure 3.5.

There are more than 9 million known registered compounds, with no end in sight as to the number that will be prepared in the future. Each compound is unique and has characteristic properties. Let us consider two compounds, water and sodium chloride, in some detail. Water is a colorless, odorless, tasteless liquid that can be changed to a solid (ice) at 0°C and to a gas (steam) at 100°C. Composed of two atoms of hydrogen and one atom of oxygen per molecule, water is 11.2% hydrogen and 88.8% oxygen by mass. Water reacts chemically with sodium to produce hydrogen gas and sodium hydroxide, with lime to produce calcium hydroxide, and with sulfur trioxide to produce sulfuric acid. No other compound has all these exact physical and chemical properties; they are characteristic of water alone.

Sodium chloride is a colorless crystalline substance with a ratio of one atom of sodium to one atom of chlorine. Its composition by mass is 39.3% sodium and 60.7% chlorine. It does not conduct electricity in its solid state; it dissolves in water to produce a solution that conducts electricity. When a current is passed through molten sodium chloride, solid sodium and gaseous chlorine are produced. These specific properties belong to sodium chloride and to no other substance. Thus, a compound may be identified and distinguished from all other compounds by its characteristic properties.

3.10 Elements That Exist as Diatomic Molecules

diatomic molecules

Seven of the elements (all nonmetals) occur as **diatomic molecules**. These elements and their symbols, formulas, and brief descriptions are listed in Table 3.7. Whether found free in nature or prepared in the laboratory, the molecules of these elements

TABLE 3.7 Elements That Exist as Diatomic Molecules

Element	Symbol	Molecular formula	Normal state
Hydrogen	H	H_2	Colorless gas
Nitrogen	N	N_2	Colorless gas
Oxygen	O	O_2	Colorless gas
Fluorine	F	F_2	Pale yellow gas
Chlorine	Cl	Cl_2	Yellow-green gas
Bromine	Br	Br_2	Reddish-brown liquid
Iodine	I	I_2	Bluish-black solid

always contain two atoms. The formulas of the free elements are therefore always written to show this molecular composition: H_2, N_2, O_2, F_2, Cl_2, Br_2, and I_2.

It is important to see how symbols are used to designate either an atom or a molecule of an element. Consider hydrogen and oxygen. Hydrogen gas is present in volcanic gases and can be prepared by many chemical reactions. Regardless of their source, all samples of free hydrogen gas consist of diatomic molecules. Free hydrogen is designated by the formula H_2, which also expresses its composition. Oxygen makes up about 21% by volume of the air that we breathe. This free oxygen is constantly being replenished by photosynthesis; it can also be prepared in the laboratory by several reactions. The majority of free oxygen is diatomic and is designated by the formula O_2. Now consider water, a compound designated by the formula H_2O (sometimes HOH). Water contains neither free hydrogen (H_2) nor free oxygen (O_2). The H_2 part of the formula H_2O simply indicates that two atoms of hydrogen are combined with one atom of oxygen to form water.

Symbols are used to designate elements, show the composition of molecules of elements, and give the elemental composition of compounds.

3.11   Chemical Formulas

Chemical formulas are used as abbreviations for compounds. A **chemical formula** shows the symbols and the ratio of the atoms of the elements in a compound. Sodium chloride contains one atom of sodium per atom of chlorine; its formula is NaCl. The formula for water is H_2O; it shows that a molecule of water contains two atoms of hydrogen and one atom of oxygen.

chemical formula

The formula of a compound tells us which elements it is composed of and how many atoms of each element are present in a formula unit. For example, a molecule of sulfuric acid is composed of two atoms of hydrogen, one atom of sulfur, and four atoms of oxygen. We could express this compound as HHSOOOO, but the usual formula for writing sulfuric acid is H_2SO_4. The formula may be expressed verbally as "H-two-S-O-four." Numbers that appear partially below the line and to

FIGURE 3.6 ▶
Explanation of the formulas NaCl, H₂SO₄, and Ca(NO₃)₂.

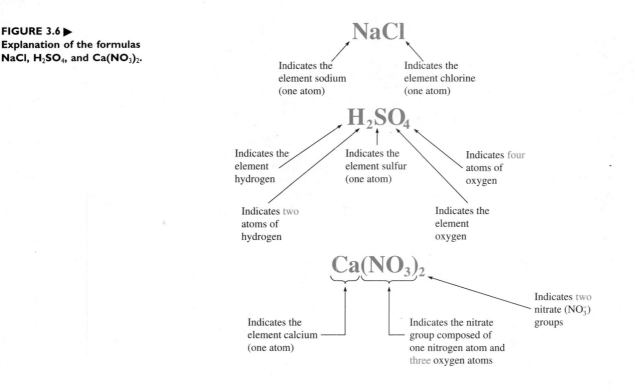

subscript   the right of a symbol of an element are called **subscripts.** Thus the 2 and the 4 in H₂SO₄ are subscripts (see Figure 3.6). Characteristics of chemical formulas are

1. The formula of a compound contains the symbols of all the elements in the compound.
2. When the formula contains one atom of an element, the symbol of that element represents that one atom. The number one (1) is not used as a subscript to indicate one atom of an element.
3. When the formula contains more than one atom of an element, the number of atoms is indicated by a subscript written to the right of the symbol of that atom. For example, the two (2) in H₂O indicates two atoms of H in the formula.
4. When the formula contains more than one of a group of atoms that occurs as a unit, parentheses are placed around the group, and the number of units of the group are indicated by a subscript placed to the right of the parentheses. Consider the nitrate group, NO₃. The formula for sodium nitrate, NaNO₃, has only one nitrate group; therefore no parentheses are needed. Calcium nitrate, Ca(NO₃)₂, has two nitrate groups, as indicated by the use of parentheses and the subscript 2. Ca(NO₃)₂ has a total of nine atoms: one Ca, two N, and six O atoms. The formula Ca(NO₃)₂ is read as "C-A [pause] N-O-three taken twice."
5. Formulas written as H₂O, H₂SO₄, Ca(NO₃)₂, and C₁₂H₂₂O₁₁ show only the number and kind of each atom contained in the compound; they do not show the arrangement of the atoms in the compound or how they are chemically bonded to one another.

Carbon—The Chameleon

One of the most diverse elements in the periodic table is carbon. Although it is much less abundant than many other elements, it is readily available. Carbon is found free in three forms called **allotropes**—as the mineral graphite, as diamonds, and as buckminsterfullerene, a form only recently discovered. The physical properties of the allotropes are quite distinct. Diamond consists of transparent, octahedral crystals that are colorless when pure—but may range from pale blue to jet black due to impurities. It is the hardest known substance and an excellent heat conductor. When a diamond has certain impurities added to the crystal intentionally it becomes an electrical semiconductor. Diamonds for cutting tools are both mined and produced synthetically. The majority of gem diamonds are mined in South Africa but some also come from South America. Graphite, on the other hand, consists of layers or sheets of carbon atoms, which are very soft and which easily slip over one another. It is an excellent conductor of electricity. Graphite is mined as massive crystals, or is obtained from heating coal and pitch in very high temperature furnaces. Buckminsterfullerene is composed of clusters of carbon atoms arranged in the shape of a soccer ball. The cage-like structure permits capture of other atoms leading to some interesting applications. More information on buckminsterfullerene can be found in the Chemistry in Action on page 232.

Carbon is an essential constituent of plant and animal life. For example, coal is formed by the gradual decay of plant life enriched in carbon through the loss of carbon dioxide and methane gas during the decaying process. If coal is heated without air present, *destructive distillation* occurs. This process decomposes the carbon compounds and produces coke, which is 90–95% graphite. Charcoal, consisting of tiny crystals of graphite, is prepared by the destructive distillation of wood. Bone black, also consisting of tiny graphite crystals, is made from the destructive distillation of bones and wastes from packing houses. Carbon black is formed when natural gas is burned with an insufficient quantity of air—a residue of graphite carbon is formed on a cold surface and then scraped off.

Free carbon finds a variety of uses in our society. Diamonds are collected and displayed for their gem quality, used in jewelry, and held as investments. They are also used in cutting and drilling tools. Graphite has many uses as a lubricant. Mixtures of clay and graphite are molded into the "lead" used in pencils. The higher the clay content, the harder the "lead." Graphite is also used as electrodes in dry cells and is found in some paints and stove polish.

Charcoal can absorb large quantities of substances onto its surface. For this reason it is very useful in water purification systems, in the manufacture of gas masks, and in removing color from solutions (as in the refining of sugar). Coke is an important fuel. It is also used to reduce iron from its ore. Alloys of iron and carbon form the major industrial product we call *steel.* Carbon black is used in making carbon paper, printer's ink, and some shoe polish. It is the additive in rubber that makes tires black.

In addition, carbon combines chemically with other elements to form a myriad of useful *compounds.* **Hydrocarbons** containing carbon and hydrogen are commonly found in petroleum and natural gas. There are so many hydrocarbons and their derivatives that an entire branch of chemistry, *organic chemistry,* is dedicated to studying and understanding them. Carbon is also found as carbon dioxide, CO_2, in our atmosphere. As such, it is one of the greenhouse gases that contributes to global warming and is of great concern to scientists (see Section 8.7). Carbon monoxide, CO, is an important fuel gas. It is colorless, odorless, tasteless, and extremely poisonous. Breathing even small quantities of carbon monoxide can be fatal (see Chemistry in Action, p. 422). Carbon can combine with chlorine and fluorine to make a group of compounds known as **chlorofluorocarbons**. These compounds are widely used as refrigerants. They are definite contributors to the destruction of the ozone layer in our atmosphere and are the topic of debate and regulation worldwide (see Section 12.17).

Diamonds and graphite.

Example 3.1   Write formulas for the following compounds, the atom composition of which is given. (a) Hydrogen chloride: 1 atom hydrogen + 1 atom chlorine; (b) Methane: 1 atom carbon + 4 atoms hydrogen; (c) Glucose: 6 atoms carbon + 12 atoms hydrogen + 6 atoms oxygen.

Solution   (a) First write the symbols of the atoms in the formula: H Cl. Since the ratio of atoms is one to one, we merely bring the symbols together to give the formula for hydrogen chloride as HCl.

(b) Write the symbols of the atoms: C H. Now bring the symbols together and place a subscript 4 after the hydrogen atom. The formula is CH_4.

(c) Write the symbols of the atoms: C H O. Now write the formula, bringing together the symbols followed by the correct subscripts according to the data given (six C, twelve H, six O). The formula is $C_6H_{12}O_6$.

3.12   Mixtures

Single substances—elements or compounds—seldom occur naturally in a pure state. Air is a mixture of gases; seawater is a mixture of a variety of dissolved minerals; ordinary soil is a complex mixture of minerals and various organic materials.

How is a mixture distinguished from a pure substance? A mixture always contains two or more substances that can be present in varying concentrations. Let us consider an example of a homogeneous mixture and an example of a heterogeneous mixture. Homogeneous mixtures (solutions) containing either 5% or 10% salt in water can be prepared simply by mixing the correct amounts of salt and water. These mixtures can be separated by boiling away the water, leaving the salt as a residue. The composition of a heterogeneous mixture of sulfur crystals and iron filings can be varied by merely blending in either more sulfur or more iron filings. This mixture can be separated physically by using a magnet to attract the iron or by adding carbon disulfide to dissolve the sulfur.

(a) When iron and sulfur exist as pure substances, only the iron is attracted to a magnet. (b) A mixture of iron and sulfur can be separated by using the difference in magnetic attraction. (c) The compound iron(II) sulfide cannot be separated into its elements with a magnet.

▼

(a)   (b)   (c)

TABLE 3.8 Comparison of Mixtures and Compounds		
	Mixture	**Compound**
Composition	May be composed of elements, compounds, or both in variable composition.	Composed of two or more elements in a definite, fixed proportion by mass.
Separation of components	Separation may be made by physical or mechanical means.	Elements can be separated by chemical changes only.
Identification of components	Components do not lose their identity.	A compound does not resemble the elements from which it is formed.

Iron(II) sulfide (FeS) contains 63.5% Fe and 36.5% S by mass. If we mix iron and sulfur in this proportion, do we have iron(II) sulfide? No, it is still a mixture; the iron is still attracted by a magnet. But if this mixture is heated strongly, a chemical change (reaction) occurs in which the reacting substances, iron and sulfur, form a new substance, iron(II) sulfide. Iron(II) sulfide, FeS, is a compound of iron and sulfur and has properties that are different from those of either iron or sulfur: It is neither attracted by a magnet nor dissolved by carbon disulfide. The differences between the iron and sulfur *mixture* and the iron(II) sulfide *compound* are as follows:

Iron(II) sulfide is the correct name for the compound formed from iron and sulfur. We will discuss the reason for the (II) in Chapter 6 when we learn to name compounds.

	Mixture of iron and sulfur	Compound of iron and sulfur
Formula	Has no definite formula; consists of Fe and S.	FeS
Composition	Contains Fe and S in any proportion by mass.	63.5% Fe and 36.5% S by mass.
Separation	Fe and S can be separated by physical means.	Fe and S can be separated only by chemical change.

The general characteristics of mixtures and compounds are compared in Table 3.8.

Concepts in Review

1. Identify the three physical states of matter.
2. Distinguish between substances and mixtures.
3. Classify common materials as elements, compounds, or mixtures.
4. Write the symbols when given the names, or write the names when given the symbols, of the common elements listed in Table 3.4.

5. Understand how symbols, including subscripts and parentheses, are used to write chemical formulas.

6. Differentiate among atoms, molecules, and ions.

7. List the characteristics of metals, nonmetals, and metalloids.

8. List the elements that occur as diatomic molecules.

Key Terms

The terms listed here have been defined within this chapter. Section numbers are referenced in parentheses for each term. More detailed definitions are given in the Glossary.

CIA is used to indicate a term found in a "Chemistry in Action" essay.

allotropes (CIA)
amorphous (3.2)
anion (3.9)
atom (3.4)
cation (3.9)
chemical formula (3.11)
chlorofluorocarbons (CIA)
compound (3.9)
diatomic molecules (3.10)
element (3.4)

gas (3.2)
heterogeneous (3.3)
homogeneous (3.3)
hydrocarbons (CIA)
ion (3.9)
liquid (3.2)
matter (3.1)
metal (3.8)
metalloid (3.8)
mixture (3.3)

molecule (3.9)
nonmetal (3.8)
phase (3.3)
solid (3.2)
subscripts (3.11)
substance (3.3)
symbol (3.7)
system (3.3)

Questions

Questions refer to tables, figures, and key words and concepts defined within the chapter. A particularly challenging question or exercise is indicated with an asterisk.

1. List four different substances in each of the three states of matter.

2. In terms of the properties of the ultimate particles of a substance, explain
   (a) why a solid has a definite shape but a liquid does not.
   (b) why a liquid has a definite volume but a gas does not.
   (c) why a gas can be compressed rather easily but a solid cannot be compressed appreciably.

3. What evidence can you find in Figure 3.1 that gases occupy space?

4. Which liquids listed in Table 3.1 are not mixtures?

5. Which of the gases listed in Table 3.1 are not pure substances?

6. When the stopper is removed from a partly filled bottle containing solid and liquid acetic acid at 16.7°C, a strong vinegarlike odor is noticeable immediately. How many acetic acid phases must be present in the bottle? Explain.

7. Is the system enclosed in the bottle of Question 6 homogeneous or heterogeneous? Explain.

8. Is a system that contains only one substance necessarily homogeneous? Explain.

9. Is a system that contains two or more substances necessarily heterogeneous? Explain.

10. Are there more atoms of silicon or hydrogen in the earth's crust, seawater, and atmosphere? Use Table 3.3 and the fact that the mass of a silicon atom is about 28 times that of a hydrogen atom.

11. What does the symbol of an element stand for?

12. Write down what you believe to be the symbols for the elements phosphorus, aluminum, hydrogen, potassium,

magnesium, sodium, nitrogen, nickel, silver, and plutonium. Now look up the correct symbols and rewrite them, comparing the two sets.

13. Interpret the difference in meanings for each of these pairs: (a) Si and SI (b) Pb and PB (c) 4 P and P_4

14. List six elements and their symbols in which the first letter of the symbol is different from that of the name. (Table 3.5)

15. Write the names and symbols for the 14 elements that have only one letter as their symbol. (See table on inside front cover.)

16. Distinguish between an element and a compound.

17. How many metals are there? Nonmetals? Metalloids? (Table 3.6)

18. Of the ten most abundant elements in the earth's crust, seawater, and atmosphere, how many are metals? Nonmetals? Metalloids? (Table 3.3)

19. Of the six most abundant elements in the human body, how many are metals? Nonmetals? Metalloids? (Figure 3.4)

20. Why is the symbol for gold Au rather than G or Go?

21. Give the names of (a) the solid diatomic nonmetal and (b) the liquid diatomic nonmetal. (Table 3.7)

22. Distinguish between a compound and a mixture.

23. What are the two general types of compounds? How do they differ from each other?

24. What is the basis for distinguishing one compound from another?

25. How many atoms are contained in (a) one molecule of hydrogen, (b) one molecule of water, and (c) one molecule of sulfuric acid?

26. What is the major difference between a cation and an anion?

27. Write the names and formulas of the elements that exist as diatomic molecules. (Table 3.7)

28. Distinguish between homogeneous and heterogeneous mixtures.

29. Tabulate the properties that characterize metals and nonmetals.

30. Which of the following are diatomic molecules?
(a) H_2 (c) HCl (e) NO (g) $MgCl_2$
(b) SO_2 (d) H_2O (f) NO_2

31. What is the major difficulty in using hydrogen as a primary source of fuel?

32. How can sponge iron be used to create a recyclable source of hydrogen for fuel?

33. Name the three allotropes of carbon.

34. List a minimum of three forms of graphite crystals and indicate a source for each.

35. List four uses of graphite in daily life.

36. Which of the following statements are correct? Rewrite the incorrect statements to make them correct. (Try to answer this question without referring to the text.)
(a) Liquids are the least compact state of matter.
(b) Liquids have a definite volume and a definite shape.
(c) Matter in the solid state is discontinuous; that is, it is made up of discrete particles.
(d) Wood is homogeneous.
(e) Wood is a substance.
(f) Dirt is a mixture.
(g) Seawater, although homogeneous, is a mixture.
(h) Any system made up of only one substance is homogeneous.
(i) Any system containing two or more substances is heterogeneous.
(j) A solution, although it contains dissolved material, is homogeneous.
(k) The smallest unit of an element that can exist and enter into a chemical reaction is called a molecule.
(l) The basic building blocks of all substances that cannot be decomposed into simpler substances by ordinary chemical change are compounds.
(m) The most abundant element in the earth's crust, seawater, and atmosphere by mass is oxygen.
(n) The most abundant element in the human body, by mass, is carbon.
(o) Most of the elements are represented by symbols consisting of one or two letters.
(p) The symbol for copper is Co.
(q) The symbol for sodium is Na.
(r) The symbol for potassium is P.
(s) The symbol for lead is Le.
(t) Early names for some elements led to unlikely symbols, such as Fe for iron.
(u) A compound is a distinct substance that contains two or more elements combined in a definite proportion by mass.
(v) The smallest uncharged individual unit of a compound formed by the union of two or more atoms is called a substance.
(w) An ion is a positive or negative electrically charged atom or group of atoms.
(x) Bromine is an element that occurs as a diatomic molecule, Br_2.
(y) The formula Na_2CO_3 indicates a total of six atoms, including three oxygen atoms.
(z) A general property of nonmetals is that they are good conductors of heat and electricity.

Paired Exercises

These exercises are paired. Each odd-numbered exercise is followed by a similar even-numbered exercise. Answers to the even-numbered exercises are given in Appendix V.

37. Given the following list of compounds and their formulas, what elements are present in each compound?
  (a) Potassium iodide KI
  (b) Sodium carbonate Na_2CO_3
  (c) Aluminum oxide Al_2O_3
  (d) Calcium bromide $CaBr_2$
  (e) Acetic acid $HC_2H_3O_2$

38. Given the following list of compounds and their formulas, what elements are present in each compound?
  (a) Magnesium bromide $MgBr_2$
  (b) Carbon tetrachloride CCl_4
  (c) Nitric acid HNO_3
  (d) Barium sulfate $BaSO_4$
  (e) Aluminum phosphate $AlPO_4$

39. Write the formula for each of the following compounds (the composition of the compound is given after each name):
  (a) Zinc oxide   1 atom Zn, 1 atom O
  (b) Potassium chlorate   1 atom K, 1 atom Cl, 3 atoms O
  (c) Sodium hydroxide   1 atom Na, 1 atom O, 1 atom H
  (d) Ethyl alcohol   2 atoms C, 6 atoms H, 1 atom O

40. Write the formula for each of the following compounds (the composition of the compound is given after each name):
  (a) Aluminum bromide 1 atom Al, 3 atoms Br
  (b) Calcium fluoride 1 atom Ca, 2 atoms F
  (c) Lead(II) chromate 1 atom Pb, 1 atom Cr, 4 atoms O
  (d) Benzene 6 atoms C, 6 atoms H

41. Explain the meaning of each symbol and number in the following formulas:
  (a) H_2O
  (b) Na_2SO_4
  (c) $HC_2H_3O_2$

42. Explain the meaning of each symbol and number in the following formulas:
  (a) $AlBr_3$
  (b) $Ni(NO_3)_2$
  (c) $C_{12}H_{22}O_{11}$ (sucrose)

43. How many atoms are represented in each of these formulas?
  (a) KF
  (b) $CaCO_3$
  (c) $K_2Cr_2O_7$
  (d) $NaC_2H_3O_2$
  (e) $(NH_4)_2C_2O_4$

44. How many atoms are represented in each of these formulas?
  (a) NaCl
  (b) N_2
  (c) $Ba(ClO_3)_2$
  (d) CCl_2F_2 (Freon)
  (e) $Al_2(SO_4)_3$

45. How many atoms of oxygen are represented in each formula?
  (a) H_2O
  (b) $CuSO_4$
  (c) H_2O_2
  (d) $Fe(OH)_3$
  (e) $Al(ClO_3)_3$

46. How many atoms of hydrogen are represented in each formula?
  (a) H_2
  (b) $Ba(C_2H_3O_2)_2$
  (c) $C_6H_{12}O_6$
  (d) $HC_2H_3O_2$
  (e) $(NH_4)_2Cr_2O_7$

47. Classify each of the following materials as an element, compound, or mixture:
  (a) Air
  (b) Oxygen
  (c) Sodium chloride
  (d) Wine

48. Classify each of the following materials as an element, compound, or mixture:
  (a) Platinum
  (b) Sulfuric acid
  (c) Iodine
  (d) Crude oil

49. Classify each of the following materials as an element, compound, or mixture:
  (a) Paint
  (b) Salt
  (c) Copper
  (d) Beer

50. Classify each of the following materials as an element, compound, or mixture:
  (a) Hydrochloric acid
  (b) Silver
  (c) Milk
  (d) Sodium hydroxide

51. Reduce each of the following chemical formulas to the smallest whole-number relationship among atoms. (In chemistry this is called the empirical formula. You will learn more about it later.)
    (a) $C_6H_{12}O_6$ glucose
    (b) C_8H_{18} octane
    (c) $C_{25}H_{52}$ paraffin wax

53. Is there a pattern to the location of the gaseous elements on the periodic table? If so describe it.

55. Of the first 36 elements on the periodic table what percent are metals?

52. Reduce each of the following chemical formulas to the smallest whole-number relationship among atoms. (In chemistry this is called the empirical formula. You will learn more about it later.)
    (a) H_2O_2 hydrogen peroxide
    (b) C_2H_6O ethyl alcohol
    (c) $Na_2Cr_2O_7$ sodium dichromate

54. Is there a pattern to the location of the liquid elements on the periodic table? If so describe it.

56. Of the first 36 elements on the periodic table what percent are solids?

Additional Exercises

These exercises are not paired or labeled by topic and provide additional practice on the concepts covered in this chapter.

57. Vitamin B_{12} has a formula that is $C_{63}H_{88}CoN_{14}O_{14}P$.
    (a) How many atoms make up one molecule of vitamin B_{12}?
    (b) What percentage of the total atoms are carbon?
    (c) What fraction of the total atoms are metallic?

58. How many total atoms are there in seven dozen molecules of nitric acid, HNO_3?

59. The following formulas look similar but represent different things. Compare and contrast them. How are they alike? How are they different?
    $$8\ S \qquad S_8$$

60. Calcium dihydrogen phosphate is an important fertilizer. How many atoms of hydrogen are there in ten molecules of $Ca(H_2PO_4)_2$?

61. How many total atoms are there in one molecule of $C_{145}H_{293}O_{168}$?

62. Name:
    (a) three elements, all metals, beginning with the letter M.
    (b) four elements, all solid nonmetals.
    (c) five elements, all solids in the first five rows of the periodic table, whose symbols start with letters different than the element name.

63. It has been estimated that there is 4×10^{-4} mg of gold/L of sea water. At a price of $19.40/g, what would be the value of the gold in 1 km³ (1×10^{15} cm³) of the ocean?

64. Make a graph using the data below. Plot the density of air in grams per liter along the x-axis and temperature along the y-axis.

Temperature (°C)	Density (g/L)
0	1.29
10	1.25
20	1.20
40	1.14
80	1.07

    (a) What is the relationship between density and temperature according to your graph?
    (b) From your plot find the density of air at these temperatures:
        5°C   25°C   70°C

The world we live in is a myriad of sights, sounds, smells, and tastes. Our senses help us to describe these objects in our lives. For example, the smell of freshly baked cinnamon rolls creates a mouth-watering desire to gobble down a fresh-baked sample. And so it is with each substance—its own unique properties allow us to identify it and predict its interactions.

These interactions produce both physical and chemical changes. When you eat an apple, the ultimate metabolic result is carbon dioxide and water. These same products are achieved by burning logs. Not only does a chemical change occur in these cases, but an energy change as well. Some reactions release energy (as does the apple or the log) whereas others require energy, such as the production of steel or the melting of ice. Over 90% of our current energy comes from chemical reactions.

4.1	Properties of Substances

How do we recognize substances? Each substance has a set of **properties** that is characteristic of that substance and gives it a unique identity. Properties are the personality traits of substances and are classified as either physical or chemical. **Physical properties** are the inherent characteristics of a substance that can be determined without altering its composition; they are associated with its physical existence. Common physical properties are color, taste, odor, state of matter (solid, liquid, or gas), density, melting point, and boiling point. **Chemical properties** describe the ability of a substance to form new substances, either by reaction with other substances or by decomposition.

properties

physical properties

chemical properties

We can select a few of the physical and chemical properties of chlorine as an example. Physically, chlorine is a gas about 2.4 times heavier than air. It is yellowish-green in color and has a disagreeable odor. Chemically, chlorine will not burn but will support the combustion of certain other substances. It can be used as a bleaching agent, as a disinfectant for water, and in many chlorinated substances such as refrigerants and insecticides. When chlorine combines with the metal sodium, it forms a salt called sodium chloride. These properties, among others, help to characterize and identify chlorine.

Substances, then, are recognized and differentiated by their properties. Table 4.1 lists four substances and several of their common physical properties. Information about common physical properties, such as that given in Table 4.1, is available in handbooks of chemistry and physics. Scientists do not pretend to know all the answers or to remember voluminous amounts of data, but it is important for them to know where to look for data in the literature. Handbooks are one of the most widely used resources for scientific data.*

No two substances have identical physical and chemical properties.

---

*Two such handbooks are David R. Lide, ed., *Handbook of Chemistry and Physics,* 75th Ed. (Cleveland: Chemical Rubber Company, 1995) and Norbert A. Lange, comp., *Handbook of Chemistry,* 13th ed. (New York: McGraw-Hill, 1985).

◄ **Chapter Opening Photo: Iron can be melted at very high temperatures and then cast into a variety of shapes.**

TABLE 4.1 Physical Properties of Chlorine, Water, Sugar, and Acetic Acid

Substance	Color	Odor	Taste	Physical state	Melting point (°C)	Boiling point (°C)
Chlorine	Yellowish-green	Sharp, suffocating	Sharp, sour	Gas	−101.6	−34.6
Water	Colorless	Odorless	Tasteless	Liquid	0.0	100.0
Sugar	White	Odorless	Sweet	Solid	—	Decomposes 170–186
Acetic acid	Colorless	Like vinegar	Sour	Liquid	16.7	118.0

4.2   Physical Changes

physical change

Matter can undergo two types of changes, physical and chemical. **Physical changes** are changes in physical properties (such as size, shape, and density) or changes in the state of matter without an accompanying change in composition. The changing of ice into water and water into steam are physical changes from one state of matter into another (Figure 4.1). No new substances are formed in these physical changes.

When a clean platinum wire is heated in a burner flame, the appearance of the platinum changes from silvery metallic to glowing red. This change is physical because the platinum can be restored to its original metallic appearance by cooling and, more importantly, because the composition of the platinum is not changed by heating and cooling.

4.3   Chemical Changes

chemical change

In a **chemical change**, new substances are formed that have different properties and composition from the original material. The new substances need not in any way resemble the initial material.

When a clean copper wire is heated in a burner flame, the appearance of the copper changes from coppery metallic to glowing red. Unlike the platinum previously mentioned, the copper is not restored to its original appearance by cooling but becomes instead a black material. This black material is a new substance called copper(II) oxide. It was formed by chemical change when copper combined with oxygen in the air during the heating process. The unheated wire was essentially 100% copper, but the copper(II) oxide is 79.9% copper and 20.1% oxygen. One gram of copper will yield 1.251 g of copper(II) oxide (see Figure 4.2). The platinum is changed only physically when heated, but the copper is changed both physically and chemically when heated.

When 1.00 g of copper reacts with oxygen to yield 1.251 g of copper(II) oxide, the copper must have combined with 0.251 g of oxygen. The percentage of copper

FIGURE 4.1
Ice melting into water or water turning into steam are physical changes from one state of matter to another.

and oxygen can be calculated from this data—the copper and oxygen each being a percent of the total mass of copper(II) oxide.

$$1.00 \text{ g copper} + 0.251 \text{ g oxygen} \longrightarrow 1.251 \text{ g copper(II) oxide}$$

$$\frac{1.00 \text{ g copper}}{1.251 \text{ g copper(II) oxide}} \times 100 = 79.9\% \text{ copper}$$

$$\frac{0.251 \text{ g oxygen}}{1.251 \text{ g copper(II) oxide}} \times 100 = 20.1\% \text{ oxygen}$$

Water can be decomposed chemically into hydrogen and oxygen. This is usually accomplished by passing electricity through the water in a process called *electrolysis*. Hydrogen collects at one electrode while oxygen collects at the other (see Figure 4.3). The composition and the physical appearance of the hydrogen and the oxygen are quite different from that of water. They are both colorless gases, but each behaves differently when a burning splint is placed into the sample: The hydrogen explodes with a pop while the flame brightens considerably in the oxygen (oxygen supports and intensifies the combustion of the wood). From these observations we conclude that a chemical change has taken place.

Before heating,
wire is copper-colored

Copper and oxygen
from the air combine
chemically on heating

After heating,
wire is black

Copper wire: 1.00 g
(100% copper)

Copper (II) oxide: 1.251 g
79.9% copper: 1.00 g
20.1% oxygen: 0.251 g

◀ **FIGURE 4.2**
Chemical change: formation of copper(II) oxide from copper and oxygen.

FIGURE 4.3 ▶
Electrolysis of water is forming
hydrogen in the left tube and
oxygen in the right tube.

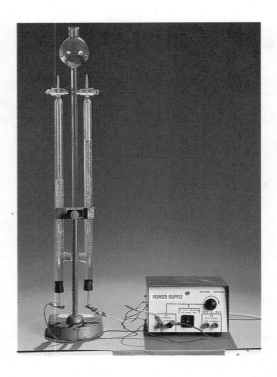

FIGURE 4.3 ▶
Electrolysis of water is forming hydrogen in the left tube and oxygen in the right tube.

Chemists have devised a shorthand method for expressing chemical changes in the form of **chemical equations**. The two previous examples of chemical changes can be represented by the following word equations:

chemical equations

$$\text{water} \xrightarrow{\text{electrical energy}} \text{hydrogen} + \text{oxygen}$$

$$\text{copper} + \text{oxygen} \xrightarrow{\Delta} \text{copper(II) oxide}$$

Equation (1) states that water decomposes into hydrogen and oxygen when electrolyzed. Equation (2) states that copper plus oxygen when heated produce copper(II) oxide. The arrow means "produces," and it points to the products. The greek letter delta (Δ) represents heat. The starting substances (water, copper, and oxygen) are called the **reactants** and the substances produced (hydrogen, oxygen, and copper(II) oxide) are called the **products**. These chemical equations can be presented in still more abbreviated form by using symbols to represent the substances:

reactants
products

$$2\,H_2O \xrightarrow{\text{electrical energy}} 2\,H_2 + O_2$$

$$2\,Cu + O_2 \xrightarrow{\Delta} 2\,CuO$$

We will learn more about writing chemical equations in later chapters.

Physical change usually accompanies a chemical change. Table 4.2 lists some common physical and chemical changes. In the examples given in the table, you will note that wherever a chemical change occurs, a physical change occurs also. However, wherever a physical change is listed, only a physical change occurs.

TABLE 4.2 Examples of Processes Involving Physical or Chemical Changes

Process taking place	Type of change	Accompanying observations
Rusting of iron	Chemical	Shiny, bright metal changes to reddish-brown rust.
Boiling of water	Physical	Liquid changes to vapor.
Burning of sulfur in air	Chemical	Yellow, solid sulfur changes to gaseous, choking sulfur dioxide.
Boiling an egg	Chemical	Liquid white and yolk change to solids.
Combustion of gasoline	Chemical	Liquid gasoline burns to gaseous carbon monoxide, carbon dioxide, and water.
Digesting food	Chemical	Food changes to liquid nutrients and partially solid wastes.
Sawing of wood	Physical	Smaller pieces of wood and sawdust are made from a larger piece of wood.
Burning of wood	Chemical	Wood burns to ashes, gaseous carbon dioxide, and water.
Heating of glass	Physical	Solid becomes pliable during heating, and the glass may change its shape.

4.4   Conservation of Mass

The **Law of Conservation of Mass** states that no change is observed in the total mass of the substances involved in a chemical change. This law, tested by extensive laboratory experimentation, is the basis for the quantitative mass relationships among reactants and products.

Law of Conservation of Mass

The decomposition of water into hydrogen and oxygen illustrates this law. One hundred grams of water decomposes into 11.2 g of hydrogen and 88.8 g of oxygen. For example,

$$\text{water} \longrightarrow \text{hydrogen} + \text{oxygen}$$

100.0 g   11.2 g   88.8 g

| 100.0 g Reactant | | 100.0 g Products |

mass of reactants = mass of products

4.5 | Energy

From the prehistoric discovery that fire could be used to warm shelters and cook food to the modern-day discovery that nuclear reactors can be used to produce vast amounts of controlled energy, our technical progress has been directed by our ability to produce, harness, and utilize energy. **Energy** is the capacity of matter to do work. Energy exists in many forms; some of the more familiar forms are mechanical, chemical, electrical, heat, nuclear, and radiant or light energy. Matter can have both potential and kinetic energy.

energy

potential energy

Potential energy is stored energy, or energy an object possesses due to its relative position. For example, a ball located 20 ft above the ground has more potential energy than when located 10 ft above the ground and will bounce higher when allowed to fall. Water backed up behind a dam represents potential energy that can be converted into useful work in the form of electrical or mechanical energy. Gasoline is a source of chemical potential energy. When gasoline burns (combines with oxygen), the heat released is associated with a decrease in potential energy. The new substances formed by burning have less chemical potential energy than the gasoline and oxygen did.

kinetic energy

Kinetic energy is energy that matter possesses due to its motion. When the water behind the dam is released and allowed to flow, its potential energy is changed into kinetic energy, which can be used to drive generators and produce electricity. All moving bodies possess kinetic energy. The pressure exerted by a confined gas is due to the kinetic energy of rapidly moving gas particles. We all know the results when two moving vehicles collide: Their kinetic energy is expended in the crash that occurs.

Energy can be converted from one form to another form. Some kinds of energy can be converted to other forms easily and efficiently. For example, mechanical energy can be converted to electrical energy with an electric generator at better than 90% efficiency. On the other hand, solar energy has thus far been directly converted to electrical energy at an efficiency of only about 15%. In chemistry, energy is most frequently expressed as heat.

4.6 | Heat: Quantitative Measurement

The SI-derived unit for energy is the joule (pronounced *jool* and abbreviated J). Another unit for heat energy, which has been used for many years, is the calorie (abbreviated cal). The relationship between joules and calories is

4.184 J = 1 cal (exactly)

joule

calorie

To give you some idea of the magnitude of these heat units, 4.184 **joule** or one **calorie** is the quantity of heat energy required to change the temperature of one g of water by 1°C, usually measured from 14.5°C to 15.5°C.

Since joule and calorie are rather small units, kilojoules (kJ) and kilocalories (kcal) are used to express heat energy in many chemical processes:

TABLE 4.3 Specific Heat of Selected Substances

Substance	Specific heat J/g°C	Specific heat cal/g°C
Water	4.184	1.00
Ethyl alcohol	2.138	0.511
Ice	2.059	0.492
Aluminum	0.900	0.215
Iron	0.473	0.113
Copper	0.385	0.0921
Gold	0.131	0.0312
Lead	0.128	0.0305

The mechanical energy of falling water is converted to electrical energy at this hydroelectric plant in the North Alps of Japan.

1 kJ = 1000 J
1 kcal = 1000 cal

The kilocalorie is also known as the nutritional or large Calorie (spelled with a capital C and abbreviated Cal). In this book heat energy will be expressed in joules with parenthetical values in calories.

The difference in the meanings of the terms *heat* and *temperature* can be seen by this example: Visualize two beakers, A and B. Beaker A contains 100 g of water at 20°C, and beaker B contains 200 g of water also at 20°C. The beakers are heated until the temperature of the water in each reaches 30°C. The temperature of the water in the beakers was raised by exactly the same amount, 10°C. But twice as much heat (8368 J or 2000 cal) was required to raise the temperature of the water in beaker B as was required in beaker A (4184 J or 1000 cal).

In the middle of the 18th century Joseph Black, a Scottish chemist, was experimenting with the heating of elements. He heated and cooled equal masses of iron and lead through the same temperature range. Black noted that much more heat was needed for the iron than for the lead. He had discovered a fundamental property of matter, namely, that every substance has a characteristic heat capacity. Heat capacities may be compared in terms of specific heats. The **specific heat** of a substance is the quantity of heat (lost or gained) required to change the temperature of one g of that substance by 1°C. It follows then that the specific heat of liquid water is 4.184 J/g°C (1.000 cal/g°C). The specific heat of water is high compared with that of most substances. Aluminum and copper, for example, have specific heats of 0.900 and 0.385 J/g°C, respectively (see Table 4.3). The relation of mass, specific heat, temperature change (Δt), and quantity of heat lost or gained by a system is expressed by this general equation:

specific heat

$$\left(\begin{array}{c} \text{mass of} \\ \text{substance} \end{array} \right) \times \left(\begin{array}{c} \text{specific heat} \\ \text{of substance} \end{array} \right) \times \Delta t = \text{energy (heat)} \qquad (1)$$

Thus, the amount of heat needed to raise the temperature of 200. g of water by 10.0°C can be calculated as follows:

$$200. \text{ g} \times \frac{4.184 \text{ J}}{\text{g°C}} \times 10.0 \text{°C} = 8.37 \times 10^3 \text{ J}$$

Examples of specific heat problems follow.

Fast Energy or Fast Fat?

Fast food restaurants have a special aroma. That aroma is the smell of fat. One whiff can stir the appetite for a juicy burger and fries.

Enjoying fat in our diets today has become a major problem since we now live longer than our ancestors. Diets that are high in fats have been connected with heart disease and to a variety of cancers. In the average American diet, fat supplies about 40% of the Calories (kcal). Nutritionists suggest that fat should be not more than 30% of our total daily Calories. For the average person eating about 2000 Calories per day, this is about 67 grams, 600 Calories, or 15 teaspoons of fat (e.g., a double hamburger with cheese). A small amount of fat is necessary in the diet to provide a natural source of vitamins A, D, E, and K, as well as polyunsaturated fats necessary for maintaining growth and good health.

Fats also supply us with energy. One gram of fat supplies 9 Calories of energy; the same amount of protein or carbohydrate supplies only 4 Calories. Human beings utilize energy in a variety of ways but one of the most important is to maintain body temperature, which for healthy individuals is around 37°C. The body tends to lose heat to the surroundings since heat flows from an area of higher temperature to an area of lower temperature. Additional heat energy is used to evaporate moisture and cool our bodies as we perspire. Still more energy is demanded by our daily physical activities. The source for all this energy is the chemical oxidation of the food we eat.

The energy content of food is determined by burning it in a calorimeter and measuring the heat released. Since the initial substances and the final products of the combustion in the calorimeter are the same as those accomplished in the human body, a calorie content can be assigned to each food. These calorie values are now found on food packages along with nutritional information regarding the contents of the food we eat.

Fats remain in the stomach longer giving us the "full" feeling we like after a

meal. But too much of a good thing can lead to weight and health problems.

How does fast food stack up in the nutrition department? Some sample meals are shown here from fast food restaurants that get less than 30% of their calories from fat. In general, chicken and turkey sandwiches that are not fried have less fat than hamburgers or roast beef. Salads with the lowest fat have little or no cheese. The best way to reduce fat in salads is to eliminate the dressing or to opt for a low-fat dressing.

- Broiled Chicken Salad
  (200 calories; 7 grams of fat)
- Newman's Own Light Italian dressing
  (30 calories; 1 gram of fat)
- Vanilla shake
  (310 calories; 7 grams of fat)

Total—540 calories; 15 grams of fat 25% calories from fat

- McLean Deluxe
  (340 calories; 12 grams of fat)
- Diet Coke, 12 ounces
  (1 calorie; 0 grams of fat)
- Hot Fudge Lowfat Frozen Yogurt Sundae
  (290 calories; 5 grams of fat)

Total—651 calories; 17 grams of fat 24% calories from fat

- Grilled Chicken Sandwich
  (290 calories; 7 grams of fat)
- Black coffee
  (0 calories; 0 grams fat)
- Chocolate Frosty Dairy Dessert (medium)
  (460 calories; 13 grams of fat)

Total—750 calories; 20 grams of fat 24% calories from fat

Believe it or not, most of the milkshakes and frozen desserts served in fast food restaurants have less than 30% Calories from fat per serving. This is primarily because they are made with skim milk. Fast food does not necessarily mean fast fat if the consumer is careful in selecting the particular food he or she consumes. Balancing a full day's diet is more important than worrying about each and every food eaten, although choosing foods that have less than 30% Calories from fat makes maintaining the balance easier.

Calculate the specific heat of a solid in J/g°C and cal/g°C if 1638 J raise the temperature of 125 g of the solid from 25.0°C to 52.6°C.

Example 4.1

First solve equation (1) to obtain an equation for specific heat:

Solution

$$\text{specific heat} = \frac{\text{energy}}{\text{g} \times \Delta t}$$

Now substitute in the data:

$$\text{energy} = 1638 \text{ J} \qquad \text{mass} = 125 \text{ g} \qquad \Delta t = 52.6°C - 25.0°C = 27.6°C$$

$$\text{specific heat} = \frac{1638 \text{ J}}{125 \text{ g} \times 27.6°C} = 0.475 \text{ J/g°C}$$

Now convert joules to calories using 1.000 cal/4.184 J:

$$\text{specific heat} = \frac{0.475 \text{ J}}{\text{g°C}} \times \frac{1.000 \text{ cal}}{4.184 \text{ J}} = 0.114 \text{ cal/g°C}$$

A sample of a metal with a mass of 212 g is heated to 125.0°C and then dropped into 375 g water at 24.0°C. If the final temperature of the water is 34.2°C, what is the specific heat of the metal? (Assume no heat losses to the surroundings.)

Example 4.2

When the metal enters the water it begins to cool, losing heat to the water. At the same time the temperature of the water rises. This process continues until the temperature of the metal and the temperature of the water are equal, at which point (34.2°C) no net flow of heat occurs.

Solution

The heat lost or gained by a system is given by equation (1). We use this equation first to calculate the heat gained by the water and then to calculate the specific heat of the metal:

$$\text{temperature rise of the water } (\Delta t) = 34.2°C - 24.0°C = 10.2°C$$

$$\text{heat gained by the water} = 375 \text{ g} \times \frac{4.184 \text{ J}}{\text{g°C}} \times 10.2°C = 1.60 \times 10^4 \text{ J}$$

The metal dropped into the water must have a final temperature the same as the water (34.2°C):

$$\text{temperature drop by the metal } (\Delta t) = 125.0°C - 34.2°C = 90.8°C$$

$$\text{heat lost by the metal} = \text{heat gained by the water} = 1.60 \times 10^4 \text{ J}$$

Rearranging equation (1) we get

$$\text{specific heat} = \frac{\text{energy}}{\text{g} \times \Delta t}$$

$$\text{specific heat of the metal} = \frac{1.60 \times 10^4 \text{ J}}{212 \text{ g} \times 90.8°C} = 0.831 \text{ J/g°C}$$

Practice 4.1

Calculate the quantity of energy needed to heat 8.0 g of water from 42.0°C to 45.0°C.

Energy from the sun is used to produce the chemical changes that occur during photosynthesis in the rain forests.

> **Practice 4.2**
> A 110.0-g sample of iron at 55.5°C raises the temperature of 150.0 mL of water from 23.0°C to 25.5°C. Determine the specific heat of the iron in cal/g°C.

4.7   Energy in Chemical Changes

In all chemical changes matter either absorbs or releases energy. Chemical changes can produce different forms of energy. For example, electrical energy to start automobiles is produced by chemical changes in the lead storage battery. Light energy is produced by the chemical change that occurs in a light stick. Heat and light energies are released from the combustion of fuels. All the energy needed for our life processes—breathing, muscle contraction, blood circulation, and so on—is produced by chemical changes occurring within the cells of our bodies.

Conversely, energy is used to cause chemical changes. For example, a chemical change occurs in the electroplating of metals when electrical energy is passed through a salt solution in which the metal is submerged. A chemical change also occurs when radiant energy from the sun is used by green plants in the process of photosynthesis. And, as we saw, a chemical change occurs when electricity is used to decompose water into hydrogen and oxygen. Chemical changes are often used primarily to produce energy rather than to produce new substances. The heat or thrust generated by the combustion of fuels is more important than the new substances formed.

4.8   Conservation of Energy

An energy transformation occurs whenever a chemical change occurs (see Figure 4.4). If energy is absorbed during the change, the products will have more chemical potential energy than the reactants. Conversely, if energy is given off in a chemical change, the products will have less chemical potential energy than the reactants. Water can be decomposed in an electrolytic cell by absorbing electrical energy. The products, hydrogen and oxygen, have a greater chemical potential energy level than that of water (see Figure 4.4a). This potential energy is released in the form of heat and light when the hydrogen and oxygen are burned to form water again (see Figure 4.4b). Thus, energy can be changed from one form to another or from one substance to another, and therefore is not lost.

The energy changes occurring in many systems have been thoroughly studied by many investigators. No system has been found to acquire energy except at the expense of energy possessed by another system. This principle is stated in other words as the **Law of Conservation of Energy**: Energy can be neither created nor destroyed, though it can be transformed from one form to another.

Law of Conservation of Energy

Hot to Go!

Instant hot and cold packs utilize the properties of matter to release or absorb energy. When a chemical dissolves in water, energy can be released or absorbed. In a cold pack a small sealed package of ammonium nitrate is placed in a separate sealed pouch containing water. As long as the substances remain separated, nothing happens. When the small package is broken (when the pack is activated), the substances mix and, as the temperature of the solution falls, the solution absorbs heat from the surroundings.

Instant hot packs work in one of two ways. The first type depends on a spontaneous chemical reaction that releases heat energy. In one product an inner paper bag perforated with tiny holes is contained in a plastic envelope. The inner bag contains a mixture of powdered iron, salt, activated charcoal, and dampened sawdust. The heat pack is activated by removing the inner bag, shaking to mix the chemicals, and replacing it in the outer envelope. The heat is the result of a chemical change—that of iron rusting (oxidizing) very rapidly. In the second type of heat pack a physical property of matter is responsible for the production of the heat. The hot pack consists of a tough sealed plastic envelope containing a solution of sodium acetate or sodium thiosulfate. A small crystal is added by squeezing a corner of the pack, or by bending a small metal activator. Crystals then form throughout the solution and release heat to the surroundings. This type of hot pack has the advantages of not being able to overheat and of being reusable. To reuse, it is simply heated in boiling water until the crystals dissolve, and then cooled and stored away until needed again.

Immediate application of hot and cold packs have helped reduce injuries to athletes.

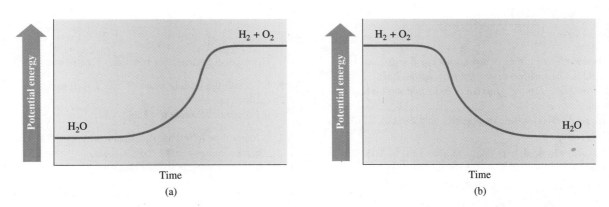

FIGURE 4.4
(a) In electrolysis of water, energy is absorbed by the system so the products H$_2$ and O$_2$ have a higher potential energy. (b) When hydrogen is burned (in O$_2$) energy is released and the product (H$_2$O) has lower potential energy.

Concepts in Review

1. List the physical properties used to characterize a substance.
2. Distinguish between the physical and chemical properties of matter.

3. Classify changes undergone by matter as either physical or chemical.

4. Distinguish between kinetic and potential energy.

5. State the Law of Conservation of Mass.

6. State the Law of Conservation of Energy.

7. Differentiate clearly between heat and temperature.

8. Make calculations using the equation:

$$\text{energy} = (\text{mass}) \times (\text{specific heat}) \times (\Delta t)$$

Key Terms

The terms listed here have been defined within this chapter. Section numbers are referenced in parentheses for each term. More detailed definitions are given in the Glossary.

calorie (4.6)
chemical change (4.3)
chemical equations (4.3)
chemical properties (4.1)
energy (4.5)
joule (4.6)

kinetic energy (4.5)
Law of Conservation of Energy (4.8)
Law of Conservation of Mass (4.4)
physical change (4.2)

physical properties (4.1)
potential energy (4.5)
products (4.3)
properties (4.1)
reactants (4.3)
specific heat (4.6)

Questions

Questions refer to tables, figures, and key words and concepts defined within the chapter. A particularly challenging question or exercise is indicated with an asterisk.

1. In what physical state does acetic acid exist at 10°C? (Table 4.1)

2. In what physical state does chlorine exist at 102 K? (Table 4.1)

3. What evidence of chemical change is visible when electricity is run through water? (Figure 4.2)

4. What physical changes occur during the electrolysis of water?

5. Distinguish between chemical and physical properties.

6. What is the fundamental difference between a chemical change and a physical change?

7. Explain how foods are assigned a specific energy (caloric) value.

8. Describe how a "reusable hot pack" works.

9. In a chemical change why can we consider that mass is neither gained nor lost (for practical purposes)?

10. Distinguish between potential and kinetic energy.

11. Calculate the boiling point of acetic acid in
    (a) Kelvins
    (b) degrees Fahrenheit. (Table 4.1)

12. Which of the following statements are correct? Rewrite the incorrect ones to make them correct.
    (a) An automobile rolling down a hill possesses both kinetic and potential energy.
    (b) When heated in the air, a platinum wire gains mass.
    (c) When heated in the air, a copper wire loses mass.
    (d) 4.184 cal is the equivalent of 1.000 J of energy.
    (e) Boiling water represents a chemical change, because a change of state occurs.
    (f) All the following represent chemical changes: baking a cake, frying an egg, leaves changing color, rusting iron.
    (g) All of the following represent physical changes: breaking a stick, melting wax, folding a napkin, burning hydrogen to form water.
    (h) Chemical changes can produce electrical energy.
    (i) Electrical energy can produce chemical changes.
    (j) A stretched rubber band possesses kinetic energy.

Evens

Paired Exercises

These exercises are paired. Each odd-numbered exercise is followed by a similar even-numbered exercise. Answers to the even-numbered exercises are given in Appendix V.

13. Classify each of the following as being primarily a physical or primarily a chemical change:
    (a) Formation of a snowflake
    (b) Freezing ice cream
    (c) Boiling water
    (d) Churning cream to make butter
    (e) Boiling an egg
    (f) Souring milk

14. Classify each of the following as being primarily a physical or primarily a chemical change:
    (a) Lighting a candle
    (b) Stirring cake batter
    (c) Dissolving sugar in water
    (d) Decomposition of limestone by heat
    (e) A leaf turning yellow
    (f) Formation of bubbles in a pot of water long before the water boils.

15. Cite the evidence that indicates that only physical changes occur when a platinum wire is heated in a Bunsen burner flame.

16. Cite the evidence that indicates that both physical and chemical changes occur when a copper wire is heated in a Bunsen burner flame.

17. Identify the reactants and products for heating a copper wire in a Bunsen burner flame.

18. Identify the reactants and products for the electrolysis of water.

19. What happens to the kinetic energy of a speeding car when the car is braked to a stop?

20. What energy transformation is responsible for the fiery reentry of the space shuttle?

21. Indicate with a plus sign ($+$) any of these processes that require energy, and a negative ($-$) sign for any that release energy:
    (a) melting ice
    (b) relaxing a taut rubber band
    (c) a rocket launching
    (d) striking a match
    (e) a Slinky toy (spring) "walking" down stairs

22. Indicate with a plus sign ($+$) any of these processes that require energy and a negative ($-$) sign for any that release energy:
    (a) boiling water
    (b) releasing a balloon full of air with the neck open
    (c) a race car crashing into the wall
    (d) cooking a potato in a microwave oven
    (e) ice cream freezing in an ice cream maker

23. How many joules of energy are required to raise the temperature of 75 g of water from 20.0°C to 70.0°C?

24. How many joules of energy are required to raise the temperature of 65 g of iron from 25°C to 95°C?

25. A 250.0-g metal bar requires 5.866 kJ to change its temperature from 22°C to 100.0°C. What is the specific heat of the metal?

26. A 1.00-kg sample of antimony absorbed 30.7 kJ, thus raising the temperature of the antimony from 20.0°C to its melting point (630.0°C). Calculate the specific heat of antimony.

*27. A 325-g piece of gold at 427°C is dropped into 200.0 mL of water at 22.0°C. The specific heat of gold is 0.131 J/g°C. Calculate the final temperature of the mixture. (Assume no heat loss to the surroundings.)

*28. A 500.0-g iron bar at 212°C is placed in 2.0 L of water at 24.0°C. What will be the change in temperature of the water? (Assume no heat is lost to the surroundings.)

Additional Exercises

These exercises are not paired or labeled by topic and provide additional practice on the concepts covered in this chapter.

*29. The specific heat of zinc is 0.096 cal/g°C. Determine the energy required to raise the temperature of 250.0 g of zinc from room temperature (24°C) to 150.0°C.

*30. If 40.0 kJ of energy is absorbed by 500.0 g of water at 10.0°C what would be the final temperature of the water?

*31. The heat of combustion of a sample of coal is 5500 cal/g. What quantity of this coal must be burned to heat 500.0 g of water from 20.0°C to 90.0°C?

32. One gram of anthracite coal gives off 7000. cal when burned. How many joules is this? If 4.0 L of water is heated from 20.0°C to 100.0°C, how many grams of anthracite are needed?

33. A 100.0-g sample of copper is heated from 10.0°C to 100.0°C.
  (a) Determine the number of calories needed. (Specific heat of copper is 0.0921 cal/g°C.)
  (b) The same amount of heat is added to 100.0 g of Al at 10.0°C. (Specific heat of Al is 0.215 cal/g°C.) Which metal gets hotter, the copper or the aluminum?

34. A 500.0-g piece of iron is heated in a flame and dropped into 400.0 g of water at 10.0°C. The temperature of the water rises to 90.0°C. How hot was the iron when it was first removed from the flame? (Specific heat of iron is 0.113 cal/g°C.)

*35. A 20.0-g piece of metal at 203°C is dropped into 100.0 g of water at 25.0°C. The water temperature rises to 29.0°C. Calculate the specific heat of the metal (J/g°C). Assume that all of the heat lost by the metal is transferred to the water and no heat is lost to the surroundings.

*36. Assuming no heat loss by the system, what will be the final temperature when 50.0 g of water at 10.0°C are mixed with 10.0 g of water at 50.0°C?

37. Three 500.0-g pans of iron, aluminum, and copper are each used to fry an egg. Which pan fries the egg (105°C) the quickest? Explain.

38. At 6:00 P.M. you put a 300.0-g copper pan containing 800.0 mL of water (all at room temperature, which is 25°C) on the stove. The stove supplies 150 cal/s. When will the water reach the boiling point? (Assume no heat loss.)

39. Why does blowing gently across the surface of a hot cup of coffee help to cool it? Why does inserting a spoon into the coffee do the same thing?

40. If you are boiling some potatoes in a pot of water, will they cook faster if the water is boiling vigorously than if the water is only gently boiling? Explain your reasoning.

41. Homogenized whole milk contains 4% butterfat by volume. How many milliliters of fat are there in a glass (250 mL) of milk? How many grams of butterfat ($d = 0.8$ g/mL) are in this glass of milk?

42. A 100.0-mL volume of mercury (density = 13.6 g/mL) is put into a container along with 100.0 g of sulfur. The two substances react when heated and result in 1460 g of a dark, solid matter. Is that material an element or a compound? Explain. How many grams of mercury were in the container? How does this support the Law of Conservation of Matter?

Answers to Practice Exercises

4.1 1.0×10^2 J = 24 cal
4.2 0.114 cal/g°C

Pure substances are classified into elements and compounds. But just what makes a substance possess its unique properties? Salt tastes salty, but how small a piece of salt will retain this property? Carbon dioxide puts out fires, is used by plants to produce oxygen, and forms "dry" ice when solidified. But how small a mass of this material still behaves like carbon dioxide? When substances finally reach the atomic, ionic, or molecular level they are in their simplest identifiable form. Further division produces a loss of characteristic properties.

What particles lie within an atom or ion? How are these tiny particles alike? How do they differ? How far can we continue to divide them? Alchemists began the quest, early chemists laid the foundation, and the modern chemist continues to build and expand on models of the atom.

5.1   Early Thoughts

The structure of matter has long intrigued and engaged the minds of people. The seed of modern atomic theory was sown during the time of the ancient Greek philosophers. About 440 B.C. Empedocles stated that all matter was composed of four "elements"—earth, air, water, and fire. Democritus (about 470–370 B.C.), one of the early atomic philosophers, thought that all forms of matter were divisible into invisible particles, which he called atoms, derived from the Greek word *atomos,* meaning indivisible. He held that atoms were in constant motion and that they combined with one another in various ways. This purely speculative hypothesis was not based on scientific observations. Shortly thereafter, Aristotle (384–322 B.C.) opposed the theory of Democritus and instead endorsed and advanced the Empedoclean theory. So strong was the influence of Aristotle that his theory dominated the thinking of scientists and philosophers until the beginning of the 17th century.

5.2   Dalton's Atomic Theory

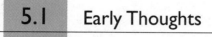

More than 2000 years after Democritus, the English schoolmaster John Dalton (1766–1844) revived the concept of atoms and proposed an atomic theory based on facts and experimental evidence. This theory, described in a series of papers published from 1803 to 1810, rested on the idea of a different kind of atom for each element. The essence of **Dalton's atomic theory** may be summed up as follows:

Dalton's atomic theory

1. Elements are composed of minute, indivisible particles called atoms.
2. Atoms of the same element are alike in mass and size.
3. Atoms of different elements have different masses and sizes.
4. Chemical compounds are formed by the union of two or more atoms of different elements.
5. Atoms combine to form compounds in simple numerical ratios, such as one to one, two to two, two to three, and so on.

◄ **Lightning occurs when electrons move to neutralize a charge difference between the clouds and the earth.**

6. Atoms of two elements may combine in different ratios to form more than one compound.

Dalton's atomic theory stands as a landmark in the development of chemistry. The major premises of his theory are still valid, but some of the statements must be modified or qualified because later investigations have shown that (1) atoms are composed of subatomic particles; (2) not all the atoms of a specific element have the same mass; and (3) atoms, under special circumstances, can be decomposed.

5.3 Composition of Compounds

John Dalton (1766–1844)

A large number of experiments extending over a long period of time have established the fact that a particular compound always contains the same elements in the same proportions by mass. For example, water will always contain 11.2% hydrogen and 88.8% oxygen by mass. The fact that water contains hydrogen and oxygen in this particular ratio does not mean that hydrogen and oxygen cannot combine in some other ratio, but that a compound with a different ratio would not be water. In fact, hydrogen peroxide is made up of two atoms of hydrogen and two atoms of oxygen per molecule and contains 5.9% hydrogen and 94.1% oxygen by mass; its properties are markedly different from those of water.

	Water	Hydrogen peroxide
Percent H	11.2	5.9
Percent O	88.8	94.1
Atomic composition	2 H + 1 O	2 H + 2 O

The **Law of Definite Composition** states:

Law of Definite Composition

A compound always contains two or more elements combined in a definite proportion by mass.

Let us consider two elements, oxygen and hydrogen, that form more than one compound. In water there are 8.0 g of oxygen for each gram of hydrogen. In hydrogen peroxide there are 16.0 g of oxygen for each gram of hydrogen. The masses of oxygen are in the ratio of small whole numbers, 16:8 or 2:1. Hydrogen peroxide has twice as much oxygen (by mass) as does water. Using Dalton's atomic theory, we deduce that hydrogen peroxide has twice as many oxygens per hydrogen as water. In fact, we now write the formulas for water as H_2O and for hydrogen peroxide as H_2O_2.

The **Law of Multiple Proportions** states:

Law of Multiple Proportions

Atoms of two or more elements may combine in different ratios to produce more than one compound.

The reliability of this law and the Law of Definite Composition is the cornerstone of the science of chemistry. In essence these laws state that (1) the composition of a particular substance will always be the same no matter what its origin or how it is formed, and (2) the composition of different compounds formed from the same elements will always be unique.

5.4   The Nature of Electric Charge

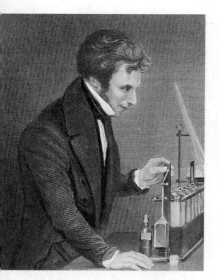

Michael Faraday (1791–1867)

Many of us have received a shock after walking across a carpeted area on a dry day. We have also experienced the static associated with combing our hair, and have had our clothing cling to us. All of these phenomena result from an accumulation of *electric charge*. This charge may be transferred from one object to another. The properties of electric charge follow:

1. Charge may be of two types, positive and negative.
2. Unlike charges attract (positive attracts negative) and like charges repel (negative repels negative and positive repels positive).
3. Charge may be transferred from one object to another, by contact or induction.
4. The less the distance between two charges, the greater the force of attraction between unlike charges (or repulsion between identical charges).

5.5   Discovery of Ions

The great English scientist Michael Faraday (1791–1867) made the discovery that certain substances when dissolved in water could conduct an electric current. He also noticed that certain compounds could be decomposed into their elements by passing an electric current through the compound. Atoms of some elements were attracted to the positive electrode, while atoms of other elements were attracted to the negative electrode. Faraday concluded that these atoms were electrically charged. He called them *ions* after the Greek word meaning "wanderer."

Any moving charge is an electric current. The electrical charge must travel through a substance known as a conducting medium. The most familiar conducting media are metals that are used in the form of wires.

The Swedish scientist Svante Arrhenius (1859–1927) extended Faraday's work. Arrhenius reasoned that an ion was an atom carrying a positive or negative charge. When a compound such as sodium chloride (NaCl) was melted, it conducted electricity. Water was unnecessary. Arrhenius's explanation of this conductivity was that upon melting, the sodium chloride dissociated, or broke up, into charged ions, Na^+ and Cl^-. The Na^+ ions moved toward the negative electrode (cathode) whereas the Cl^- migrated toward the positive electrode (anode). Thus, positive ions are *cations,* and negative ions are *anions.*

From Faraday's and Arrhenius's work with ions, Irish physicist G. J. Stoney (1826–1911) realized there must be some fundamental unit of electricity associated with atoms. He named this unit the *electron* in 1891. Unfortunately he had no means of supporting his idea with experimental proof. Evidence remained elusive until 1897 when English physicist Joseph Thomson (1856–1940) was able to show experimentally the existence of the electron.

Triboluminescence

Some substances give off light when they are rubbed, crushed, or broken in a phenomenon called triboluminescence. Examples of substances exhibiting this property include crystals of quartz, sugar cubes, adhesive tape torn off certain surfaces, and Wintergreen Lifesavers.

Linda M. Sweeting, a chemist at Towson State University in Maryland, investigated the sparks created by a Wintergreen Lifesaver when crushed inside one's mouth in a dark room. When a sugar crystal is fractured, separate areas of positive and negative charge form on opposite sides of the crack. Electrons tend to leap the gap and neutralize the charge. When the electrons collide with nitrogen molecules in the air, small amounts of light are emitted. (Lightning is a somewhat similar phenomenon but on a much grander scale.) The addition of wintergreen molecules changes the outcome though because they absorb some of the light energy from the electron leap and reemit it as bright blue-green flashes. These are the sparks we see as we crush a Lifesaver in the dark.

Triboluminescence occurs when Wintergreen Lifesavers are crushed in the mouth.

•••

5.6   Subatomic Parts of the Atom

The concept of the atom—a particle so small that until recently it could not be seen even with the most powerful microscope—and the subsequent determination of its structure stand among the very greatest creative intellectual human achievements.

Any visible quantity of an element contains a vast number of identical atoms. But when we refer to an atom of an element, we isolate a single atom from the multitude in order to present the element in its simplest form. Figure 5.1 shows individual atoms as we can see them today.

Let us examine this tiny particle we call the atom. The diameter of a single atom ranges from 0.1 to 0.5 nanometers (1 nm = 1×10^{-9} m). Hydrogen, the smallest atom, has a diameter of about 0.1 nm. To arrive at some idea of how small an atom is, consider this dot (•), which has a diameter of about 1 mm, or 1×10^{6} nm. It would take 10 million hydrogen atoms to form a line of atoms across this dot. As inconceivably small as atoms are, they contain smaller particles, the **subatomic particles**, such as electrons, protons, neutrons.

The development of atomic theory was helped in large part by the invention of new instruments. For example, the Crookes tube, developed by Sir William Crookes in 1875, opened the door to the subatomic structure of the atom (Figure 5.2). The emissions generated in a Crookes tube are called *cathode rays.* Joseph Thomson demonstrated in 1897 that cathode rays (1) travel in straight lines, (2) are negative in charge, (3) are deflected by electric and magnetic fields, (4) produce sharp shadows, and (5) are capable of moving a small paddle wheel. This was the experimental discovery of the fundamental unit of charge—the electron.

The **electron** (e^{-}) is a particle with a negative electrical charge and a mass of 9.110×10^{-28} g. This mass is 1/1837 the mass of a hydrogen atom and corresponds to 0.0005486 atomic mass unit (amu) (defined in Section 5.11). One atomic mass unit has a mass of 1.6606×10^{-24} g. Although the actual charge of an electron is known, its value is too cumbersome for practical use and therefore has been assigned a relative electrical charge of -1. The size of an electron has not been determined exactly, but its diameter is believed to be less than 10^{-12} cm.

subatomic particles

electron

87

FIGURE 5.1
A scanning tunneling microscope
shows an array of copper atoms.

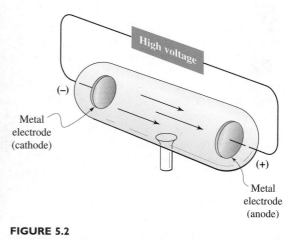

High voltage

(−)

Metal
electrode
(cathode)

(+)

Metal
electrode
(anode)

FIGURE 5.2
**Crookes tube. In a Crookes tube a gas is contained at
very low pressure. When a current is applied across
electrodes within the tube, a stream of electrons
travels from the cathode toward the anode.**

Protons were first observed by German physicist E. Goldstein (1850–1930) in
1886. However, it was Thomson who discovered the nature of the proton. He showed
that the proton was a particle, and he calculated its mass to be about 1837 times

proton that of an electron. The **proton** (p) is a particle with a relative mass of 1 amu and
an actual mass of 1.673×10^{-24} g. Its relative charge ($+1$) is equal in magnitude,
but of opposite sign, to the charge on the electron. The mass of a proton is only
very slightly less than that of a hydrogen atom.

Thomson had shown that atoms contained both negatively and positively
charged particles. Clearly, the Dalton model of the atom was no longer acceptable.
Atoms were not indivisible. They were composed of smaller parts. Thomson proposed a new model of the atom.

Thomson model of the atom In the **Thomson model of the atom** the electrons were negatively charged
particles embedded in the atomic sphere. Since atoms were electrically neutral the
sphere also contained an equal number of protons, or positive charges. A neutral
atom could become an ion by gaining or losing electrons.

Positive ions were explained by assuming that the neutral atom had lost electrons.
An atom with a net charge of $+1$ (for example, Na^+ or Li^+) would have lost one
electron. An atom with a net charge of $+3$ (for example, Al^{3+}) would have lost
three electrons.

Negative ions were explained by assuming that extra electrons had been added
to the atoms. A net charge of -1 (for example, Cl^- or F^-) would be produced
by the addition of one electron. A net charge of -2 (for example, O^{2-} or S^{2-})
would be explained as the addition of two electrons.

The third major subatomic particle was discovered in 1932 by James Chadwick
neutron (1891–1974). This particle, the **neutron** (n), bears neither a positive nor a negative
charge and has a relative mass of about 1 amu. Its actual mass (1.675×10^{-24} g)
is only very slightly greater than that of a proton. The properties of these three
subatomic particles are summarized in Table 5.1.

Nearly all the ordinary chemical properties of matter can be explained in terms
of atoms consisting of electrons, protons, and neutrons. The discussion of atomic

◀ Joseph Thomson (1856–1940)

structure that follows is based on the assumption that atoms contain only these principal subatomic particles. Many other subatomic particles, such as mesons, positrons, neutrinos, and antiprotons, have been discovered, but it is not yet clear whether all these particles are actually present in the atom or whether they are produced by reactions occurring within the nucleus. The fields of atomic and particle or high-energy physics have produced a long list of subatomic particles. Descriptions of the properties of many of these particles are to be found in recent physics textbooks, in various articles appearing in *Scientific American,* and in several of Isaac Asimov's books.

TABLE 5.1 Electrical Charge and Relative Mass of Electrons, Protons, and Neutrons

Particle	Symbol	Relative electrical charge	Relative mass (amu)	Actual mass (g)
Electron	e^-	-1	$\dfrac{1}{1837}$	9.110×10^{-28}
Proton	p	$+1$	1	1.673×10^{-24}
Neutron	n	0	1	1.675×10^{-24}

The mass of a helium atom is 6.65×10^{-24} g. How many atoms are in a 4.0-g sample of helium?

Example 5.1

$$4.0 \text{ g} \times \frac{1 \text{ atom He}}{6.65 \times 10^{-24} \text{ g}} = 6.0 \times 10^{23} \text{ atoms He}$$

Solution

Practice 5.1

The mass of an atom of hydrogen is 1.673×10^{-24} g. How many atoms are in a 10.0-g sample of hydrogen?

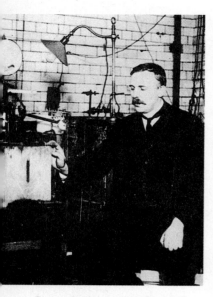

Ernest Rutherford (1871–1937)

5.7   The Nuclear Atom

The discovery that positively charged particles were present in atoms came soon after the discovery of radioactivity by Henri Becquerel in 1896. Radioactive elements spontaneously emit alpha particles, beta particles, and gamma rays from their nuclei (see Chapter 18).

Ernest Rutherford had, by 1907, established that the positively charged alpha particles emitted by certain radioactive elements were ions of the element helium. Rutherford used these alpha particles to establish the nuclear nature of atoms. In experiments performed in 1911, he directed a stream of positively charged helium ions (alpha particles) at a very thin sheet of gold foil (about 1000 atoms thick). See Figure 5.3(a). He observed that most of the alpha particles passed through the foil with little or no deflection; but a few of the particles were deflected at large angles, and occasionally one even bounced back from the foil (Figure 5.3b). It was known that like charges repel each other and that an electron with a mass of 1/1837 amu could not possibly have an appreciable effect on the path of a 4-amu alpha particle, which is about 7350 times more massive than an electron. Rutherford therefore reasoned that each gold atom must contain a positively charged mass occupying a relatively tiny volume and that, when an alpha particle approached close enough to this positive mass, it was deflected. Rutherford spoke of this positively charged mass as the *nucleus* of the atom. Because alpha particles have relatively high masses, the extent of the deflections (some actually bounced back) indicated to Rutherford that the nucleus is very heavy and dense. (The density of the nucleus of a hydrogen atom is about 10^{12} g/cm^3—about one trillion times the density of water.) Because most of the alpha particles passed through the thousand or so gold atoms without any apparent deflection, he further concluded that most of an atom consists of empty space.

When we speak of the mass of an atom, we are, for practical purposes, referring primarily to the mass of the nucleus. The nucleus contains all the protons and neutrons, which represent more than 99.9% of the total mass of any atom (see Table 5.1). By way of illustration, the largest number of electrons known to exist in an

FIGURE 5.3
(a) Diagram representing Rutherford's experiment on alpha-particle scattering. (b) Positive alpha particles (α), emanating from a radioactive source, were directed at a thin gold foil. Diagram illustrates the deflection and repulsion of the positive alpha particles by the positive nuclei of the gold atoms.
▼

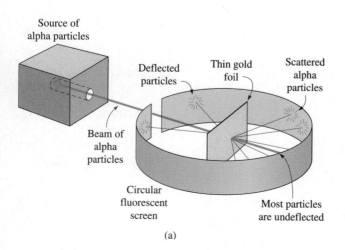

(a)

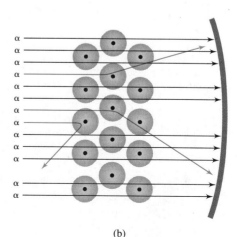

(b)

atom is 111. The mass of even 111 electrons is only about 1/17 of the mass of a single proton or neutron. The mass of an atom, therefore, is primarily determined by the combined masses of its protons and neutrons.

5.8 General Arrangement of Subatomic Particles

The alpha-particle scattering experiments of Rutherford established that the atom contains a dense, positively charged nucleus. The later work of Chadwick demonstrated that the atom contains neutrons, which are particles with mass but no charge. He also noted that light, negatively charged electrons are present and offset the positive charges in the nucleus. Based on this experimental evidence, a general description of the atom and the location of its subatomic particles was devised. Each atom consists of a **nucleus** surrounded by electrons. The nucleus contains protons and neutrons but not electrons. In a neutral atom the positive charge of the nucleus (due to protons) is exactly offset by the negative electrons. Because the charge of an electron is equal but of opposite sign to the charge of a proton, a neutral atom must contain exactly the same number of electrons as protons. However, this generalized picture of atomic structure provides no information on the arrangement of electrons within the atom.

nucleus

> **A neutral atom contains the same number of protons and electrons.**

5.9 Atomic Numbers of the Elements

The **atomic number** of an element is the number of protons in the nucleus of an atom of that element. The atomic number determines the identity of an atom. For example, every atom with an atomic number of 1 is a hydrogen atom; it contains one proton in its nucleus. Every atom with an atomic number of 8 is an oxygen atom; it contains 8 protons in its nucleus. Every atom with an atomic number of 92 is a uranium atom; it contains 92 protons in its nucleus. The atomic number tells us not only the number of positive charges in the nucleus, but also the number of electrons in the neutral atom.

atomic number

The number of protons *defines* the element.

> **atomic number = number of protons in the nucleus**

There is no need to memorize the atomic numbers of the elements because a periodic table is commonly provided in texts, laboratories, and on examinations. The atomic numbers of all elements are shown in the periodic table on the inside front cover of this book and are also listed in the table of atomic masses on the inside front cover.

5.10   Isotopes of the Elements

Shortly after Rutherford's conception of the nuclear atom, experiments were performed to determine the masses of individual atoms. These experiments showed that the masses of nearly all atoms were greater than could be accounted for by simply adding up the masses of all the protons and electrons that were known to be present in an atom. This fact led to the concept of the neutron, a particle with no charge but with a mass about the same as that of a proton. Because this particle has no charge, it was very difficult to detect, and the existence of the neutron was not proven experimentally until 1932. All atomic nuclei, except that of the simplest hydrogen atom, are now believed to contain neutrons.

All atoms of a given element have the same number of protons. Experimental evidence has shown that, in most cases, all atoms of a given element do not have identical masses. This is because atoms of the same element may have different numbers of neutrons in their nuclei.

isotopes   Atoms of an element having the same atomic number but different atomic masses are called **isotopes** of that element. Atoms of the various isotopes of an element, therefore, have the same number of protons and electrons but different numbers of neutrons.

Three isotopes of hydrogen (atomic number 1) are known. Each has one proton in the nucleus and one electron. The first isotope (protium), without a neutron, has a mass number of 1; the second isotope (deuterium), with one neutron in the nucleus, has a mass number of 2; the third isotope (tritium), with two neutrons, has a mass number of 3 (see Figure 5.4).

The three isotopes of hydrogen may be represented by the symbols ^1_1H, ^2_1H, ^3_1H, indicating an atomic number of 1 and mass numbers of 1, 2, and 3, respectively. This method of representing atoms is called *isotopic notation*. The subscript (Z) is

mass number   the atomic number; the superscript (A) is the **mass number**, which is the sum of the number of protons and the number of neutrons in the nucleus. The hydrogen isotopes may also be referred to as hydrogen-1, hydrogen-2, and hydrogen-3.

The *mass number* of an element is the sum of the protons and neutrons in the nucleus.

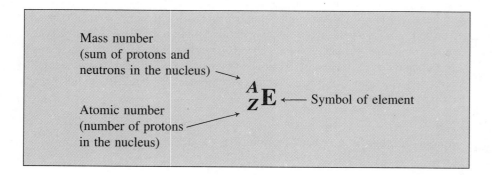

Mass number
(sum of protons and
neutrons in the nucleus)

Atomic number
(number of protons
in the nucleus)

$^A_Z E$ ← Symbol of element

Most of the elements occur in nature as mixtures of isotopes. However, not all isotopes are stable; some are radioactive and are continuously decomposing to form other elements. For example, of the seven known isotopes of carbon, only two, carbon-12 and carbon-13, are stable. Of the seven known isotopes of oxygen, only three, $^{16}_8O$, $^{17}_8O$, and $^{18}_8O$, are stable. Of the fifteen known isotopes of arsenic, $^{75}_{33}As$ is the only one that is stable.

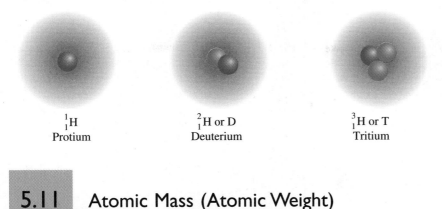

^1_1H
Protium

^2_1H or D
Deuterium

^3_1H or T
Tritium

◀ **FIGURE 5.4**
Schematic of the isotopes of hydrogen. The number of protons (purple) and neutrons (blue) are shown within the nucleus. The electron (e⁻) exists outside the nucleus (gray).

5.11	Atomic Mass (Atomic Weight)

The mass of a single atom is far too small to measure individually on a balance. But fairly precise determinations of the masses of individual atoms can be made with an instrument called a *mass spectrometer* (see Figure 5.5). The mass of a single hydrogen atom is 1.673×10^{-24} g. However, it is neither convenient nor practical to compare the actual masses of atoms expressed in grams; therefore, a table of relative atomic masses using *atomic mass units* was devised. (The term *atomic weight* is often used instead of atomic mass.) The carbon isotope having six protons and six neutrons and designated carbon-12, or $^{12}_6\text{C}$, was chosen as the standard for atomic masses. This reference isotope was assigned a value of exactly 12 atomic mass units (amu). Thus one **atomic mass unit** is defined as equal to exactly 1/12 of the mass of a carbon-12 atom. The actual mass of a carbon-12 atom is 1.9927×10^{-23} g, and that of one atomic mass unit is 1.6606×10^{-24} g. In the table of atomic masses all elements then have values that are relative to the mass assigned to the reference isotope carbon-12.

atomic mass unit

A table of atomic masses is given on the inside front cover of this book. Hydrogen atoms, with a mass of about 1/12 that of a carbon atom, have an average atomic mass of 1.00797 amu on this relative scale. Magnesium atoms, which are about twice as heavy as carbon, have an average mass of 24.305 amu. The average atomic mass of oxygen is 15.9994 amu.

Since most elements occur as mixtures of isotopes with different masses, the atomic mass determined for an element represents the average relative mass of all the naturally occurring isotopes of that element. The atomic masses of the individual isotopes are approximately whole numbers, because the relative masses of the protons and neutrons are approximately 1.0 amu each. Yet we find that the atomic masses given for many of the elements deviate considerably from whole numbers.

FIGURE 5.5
Schematic of a modern mass spectrometer. A beam of positive ions is produced from the sample as it enters the chamber. The positive ions are then accelerated as they pass through slits in an electric field. When the ions enter the magnetic field, they are deflected differently, depending on mass and charge. The ions are then detected at the end of the tube. From intensity and position of the lines on the mass spectrogram, the different isotopes of the elements and their relative amounts can be
◀ **determined.**

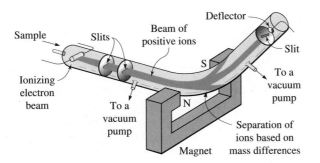

Chemical Fraud

The bottle of vanilla flavoring found in most kitchens may be labeled either as "vanilla extract" or as "imitation vanilla extract." Both substances taste the same because they contain the same molecules, and taste is related to molecular structure. What is the difference then between real vanilla extract and imitation vanilla extract?

Vanilla extract is an alcohol and water solution of materials extracted from vanilla beans. Imitation vanilla extract is made from lignin, a waste product in the wood pulp industry, which is converted to vanillin—the same molecule found in natural vanilla extract. The advantage of this process is monetary. The cost of producing natural vanilla is approximately $125 per gallon, whereas production of synthetic vanillin is approximately $60 per gallon.

One of the roles of a government chemist is to prevent fraudulent mislabeling of products. Since the compounds in both products are chemically the same, another method must be used to detect the source of the molecule. This is done by inspecting the carbon atoms in the vanillin, using a technique called stable isotope ratio analysis (SIRA). Two isotopes in organic compounds are carbon-12 and carbon-13. The ratio of these isotopes is slightly different for natural vanillin and imitation vanillin. It is from these slight differences that government scientists distinguish whether the vanillin is from the vanilla bean or from the lignin.

For example, the atomic mass of rubidium is 85.4678 amu, that of copper is 63.546 amu, and that of magnesium is 24.305 amu. The deviation of an atomic mass from a whole number is due mainly to the unequal occurrence of the various isotopes of an element. For example, the two principal isotopes of copper are $^{63}_{29}Cu$ and $^{65}_{29}Cu$. It is apparent that copper-63 atoms are the more abundant isotope, since the atomic mass of copper, 63.546 amu, is closer to 63 than to 65 amu (see Figure 5.6). The actual values of the copper isotopes observed by mass spectra determination are shown here:

Isotope	Isotopic mass (amu)	Abundance (%)	Average atomic mass (amu)
$^{63}_{29}Cu$	62.9298	69.09	63.55
$^{65}_{29}Cu$	64.9278	30.91	

The average atomic mass can be calculated by multiplying the atomic mass of each isotope by the fraction of each isotope present and adding the results. The calculation for copper is

$$62.9298 \text{ amu} \times 0.6909 = 43.48 \text{ amu}$$
$$64.9278 \text{ amu} \times 0.3091 = \underline{20.07} \text{ amu}$$
$$63.55 \text{ amu}$$

atomic mass

The **atomic mass** of an element is the average relative mass of the isotopes of that element referred to the atomic mass of carbon-12 (exactly 12.0000 . . . amu).

The relationship between mass number and atomic number is such that, if we subtract the atomic number from the mass number of a given isotope, we obtain the number of neutrons in the nucleus of an atom of that isotope. Table 5.2 shows

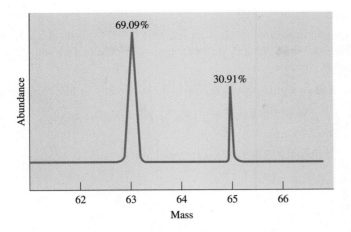

◀ FIGURE 5.6
A typical reading from a mass
spectrometer. The two principal
isotopes of copper are shown
with the abundance (%) given.

**TABLE 5.2 Determination of the Number of Neutrons in an Atom by
Subtracting Atomic Number from Mass Number**

	Hydrogen ($^1_1$H)	Oxygen ($^{16}_8$O)	Sulfur ($^{32}_{16}$S)	Fluorine ($^{19}_9$F)	Iron ($^{56}_{26}$Fe)
Mass number	1	16	32	19	56
Atomic number	(−)1	(−)8	(−)16	(−)9	(−)26
Number of neutrons	0	8	16	10	30

the application of this method of determining the number of neutrons. For example,
the fluorine atom ($^{19}_9$F), atomic number 9, having a mass of 19 amu, contains
10 neutrons:

Mass number − Atomic number = Number of neutrons

19 − 9 = 10

The atomic masses given in the table on the inside front cover of this book are
values accepted by international agreement. You need not memorize atomic masses.
In most of the calculations needed in this book, the use of atomic masses to four
significant figures will give results of sufficient accuracy.

How many protons, neutrons, and electrons are found in an atom of $^{14}_6$C?

Example 5.2

The element is carbon, atomic number 6. The number of protons or electrons equals
the atomic number and is 6. The number of neutrons is determined by subtracting
the atomic number from the mass number: 14 − 6 = 8.

Solution

Practice 5.2

How many protons, neutrons, and electrons are in each of the following?
(a) $^{16}_8$O, (b) $^{80}_{35}$Br, (c) $^{235}_{92}$U, (d) $^{64}_{29}$Cu

Example 5.3   Chlorine is found in nature as two isotopes, $^{37}_{17}Cl$ (24.47%) and $^{35}_{17}Cl$ (75.53%). The atomic masses are 36.96590 and 34.96885 amu, respectively. Determine the average atomic mass of chlorine.

Solution   Multiply each mass by its percentage and add the results to find the average.

$$(0.2447) \times (36.96590 \text{ amu}) + (0.7553) \times (34.96885 \text{ amu})$$
$$= 35.4575 \text{ amu}$$
$$= 35.46 \text{ amu (4 significant figures)}$$

Practice 5.3

Silver is found in two isotopes with atomic masses of 106.9041 and 108.9047 amu, respectively. The first isotope represents 51.82% and the second 48.18%. Determine the average atomic mass of silver.

Concepts in Review

1. State the major provisions of Dalton's atomic theory.
2. State the Law of Definite Composition and indicate its significance.
3. State the Law of Multiple Proportions and indicate its significance.
4. Give the names, symbols, and relative masses of the three principal subatomic particles.
5. Describe the Thomson model of the atom.
6. Describe the atom as conceived by Ernest Rutherford after his alpha-scattering experiment.
7. Determine the atomic number, mass number, or number of neutrons of an isotope when given the values of any two of these three items.
8. Name and distinguish among the three isotopes of hydrogen.
9. Calculate the average atomic mass of an element, given the isotopic masses and the abundance of its isotopes.
10. Determine the number of protons, neutrons, and electrons from the atomic number and atomic mass of an atom.

Key Terms

atomic mass (5.11)
atomic mass unit (5.11)
atomic number (5.9)
Dalton's atomic theory (5.2)
electron (5.6)

isotopes (5.10)
Law of Definite Composition (5.3)
Law of Multiple Proportions (5.3)
mass number (5.10)
neutron (5.6)

nucleus (5.8)
proton (5.6)
subatomic particles (5.6)
Thomson model of the atom (5.6)

Questions

Questions refer to tables, figures, and key words and concepts defined within the chapter. A particularly challenging question is indicated with an asterisk.

1. What are the atomic numbers of (a) copper, (b) nitrogen, (c) phosphorus, (d) radium, and (e) zinc?

2. A neutron is approximately how many times heavier than an electron?

3. From the point of view of a chemist, what are the essential differences among a proton, a neutron, and an electron?

4. Distinguish between an atom and an ion.

5. What letters are used to designate atomic number and mass number in isotopic notation of atoms?

6. Indicate the relationship between triboluminescence and Wintergreen Lifesavers.

7. Differentiate between synthetic and natural vanilla extract. Explain why this is important to our society.

8. Describe and indicate the value of the SIRA technique.

9. In what ways are isotopes alike? In what ways different?

10. Which of the following statements are correct? Rewrite the incorrect statements to make them correct.
    (a) John Dalton developed an important atomic theory in the early 1800s.
    (b) Dalton said that elements are composed of minute indivisible particles called atoms.
    (c) Dalton said that when atoms combine to form compounds, they do so in simple numerical ratios.
    (d) Dalton said that atoms are composed of protons, neutrons, and electrons.
    (e) All of Dalton's theory is still considered valid today.
    (f) Hydrogen is the smallest atom.

    (g) A proton is about 1837 times as heavy as an electron.
    (h) The nucleus of an atom contains protons, neutrons, and electrons.

11. Which of the following statements are correct? Rewrite the incorrect statements to make them correct.
    (a) An element with an atomic number of 29 has 29 protons, 29 neutrons, and 29 electrons.
    (b) An atom of the isotope $^{60}_{26}$Fe has 34 neutrons in its nucleus.
    (c) $^{2}_{1}$H is a symbol for the isotope deuterium.
    (d) An atom of $^{31}_{15}$P contains 15 protons, 16 neutrons, and 31 electrons.
    (e) In the isotope $^{6}_{3}$Li, $Z = 3$ and $A = 3$.
    (f) Isotopes of a given element have the same number of protons but differ in the number of neutrons.
    (g) The three isotopes of hydrogen are called protium, deuterium, and tritium.
    (h) $^{23}_{11}$Na and $^{24}_{11}$Na are isotopes.
    (i) $^{24}_{11}$Na has one more electron than $^{23}_{11}$Na.
    (j) $^{24}_{11}$Na has one more proton than $^{23}_{11}$Na.
    (k) $^{24}_{11}$Na has one more neutron than $^{23}_{11}$Na.
    (l) Only a few of the elements exist in nature as mixtures of isotopes.

12. Which of the following statements are correct? Rewrite the incorrect statements to make them correct.
    (a) One atomic mass unit has a mass 12 times as much as one carbon-12 atom.
    (b) The atomic masses of protium and deuterium differ by about 100%.
    (c) $^{23}_{11}$Na and $^{24}_{11}$Na have the same atomic masses.
    (d) The atomic mass of an element represents the average relative atomic mass of all the naturally occurring isotopes of that element.

Paired Exercises

These exercises are paired. Each odd-numbered exercise is followed by a similar even-numbered exercise. Answers to even-numbered exercises are given in Appendix V.

13. Explain why, in Rutherford's experiments, some alpha particles were scattered at large angles by the gold foil or even bounced back?

14. What experimental evidence led Rutherford to conclude each of the following?
    (a) The nucleus of the atom contains most of the atomic mass.
    (b) The nucleus of the atom is positively charged.
    (c) The atom consists of mostly empty space.

15. Describe the general arrangement of subatomic particles in the atom.

16. What part of the atom contains practically all its mass?

17. What contribution did each of the following scientists make to atomic theory?
    (a) Dalton
    (b) Thomson
    (c) Rutherford

19. Explain why the atomic masses of elements are not whole numbers.

21. What special names are given to the isotopes of hydrogen?

23. What is the symbol and name of the element that has an atomic number of 24 and a mass number of 52?

25. What is the nuclear composition of the six naturally occurring isotopes of calcium having mass numbers of 40, 42, 43, 44, 46, and 48?

27. Write isotopic notation symbols for the following:
    (a) $Z = 26$, $A = 55$
    (b) $Z = 12$, $A = 26$
    (c) $Z = 3$, $A = 6$
    (d) $Z = 79$, $A = 188$

29. Naturally occurring lead exists as four stable isotopes: ^{204}Pb with a mass of 203.973 amu (1.480%); ^{206}Pb, 205.974 amu (23.60%); ^{207}Pb, 206.9759 amu (22.60%); and ^{208}Pb, 207.9766 amu (52.30%). Calculate the average atomic mass of lead.

31. 68.9257 amu is the mass of 60.4% of the atoms of an element with only two naturally occurring isotopes. The atomic mass of the other isotope is 70.9249 amu. Determine the average atomic mass of the element. Identify the element.

33. An aluminum atom has an average diameter of about 3.0×10^{-8} cm. The nucleus has a diameter of about 2.0×10^{-13} cm. Calculate the ratio of these diameters.

18. Consider each of the following models of the atom: (a) Dalton, (b) Thomson, (c) Rutherford. How does the location of the electrons in an atom vary? How does the location of the atom's positive matter compare?

20. Is the isotopic mass of a given isotope ever an exact whole number? Is it always? In answering, consider the masses of $^{12}_{6}C$ and $^{63}_{29}Cu$.

22. List the similarities and differences in the three isotopes of hydrogen.

24. An atom of an element has a mass number of 201 and has 121 neutrons in its nucleus.
    (a) What is the electrical charge of the nucleus?
    (b) What is the symbol and name of the element?

26. Which of the isotopes of calcium in the previous exercise is the most abundant isotope? Can you be sure? Explain your choice.

28. Give the nuclear composition and isotopic notation for:
    (a) An atom containing 27 protons, 32 neutrons, and 27 electrons
    (b) An atom containing 15 protons, 16 neutrons, and 15 electrons
    (c) An atom containing 110 neutrons, 74 electrons, and 74 protons
    (d) An atom containing 92 electrons, 143 neutrons, and 92 protons

30. Naturally occurring magnesium consists of three stable isotopes: ^{24}Mg, 23.985 amu (78.99%); ^{25}Mg, 24.986 amu (10.00%); and ^{26}Mg, 25.983 amu (11.01%). Calculate the average atomic mass of magnesium.

32. A sample of enriched lithium contains 30.00% ^{6}Li (6.015 amu) and 70.00% ^{7}Li (7.016 amu). What is the average atomic mass of the sample?

*34. An average dimension for the radius of an atom is 1.0×10^{-8} cm, and the average radius of the nucleus is 1.0×10^{-13} cm. Determine the ratio of atomic volume to nuclear volume. Assume that the atom is spherical ($V = (4/3)\pi r^3$ for a sphere).

Additional Exercises

These exercises are not paired or labeled by topic and provide additional practice on the concepts covered in this chapter.

35. What experimental evidence supports each of the following statements?

    (a) The nucleus of an atom is small.
    (b) The atom consists of both positive and negative charges.
    (c) The nucleus of the atom is positive.

36. What is the relationship between the following two atoms:
    (a) one atom with 10 protons, 10 neutrons, and 10 electrons; and another atom with 10 protons, 11 neutrons, and 10 electrons
    (b) one atom with 10 protons, 11 neutrons, and 10 electrons; and another atom with 11 protons, 10 neutrons, and 11 electrons.

37. How will an atom's nucleus differ if it loses an alpha particle?

38. Suppose somewhere in the universe another kind of matter is discovered and your job is to determine the characteristics of this new matter. You first place a gaseous sample of this matter in a cathode-ray tube and find that, as with ordinary matter, cathode rays are seen. When they are deflected with a magnet, the rays emitted from the anode ($+$) are deflected thousands of times more than those emitted from the cathode ($-$).
    (a) What conclusions can you draw from this experiment? When a metal foil of the new matter is bombarded by alpha particles, most of them pass right on through, but one every few billion or so goes in and never comes out!
    (b) What further conclusions can you draw from this new information?

39. The radius of a carbon atom in many compounds is 0.77×10^{-8} cm. If the radius of a Styrofoam ball used to represent the carbon atom in a molecular model is 1.5 cm, how much of an enlargement is this?

40. How is it possible for there to be more than one kind of atom of the same element?

41. Which of the following contains the largest number of neutrons per atom: ^{210}Bi, ^{210}Po, ^{210}At, ^{211}At

*42. An unknown element Q has two known isotopes: ^{60}Q and ^{63}Q. If the average atomic mass is 61.5 amu, what are the relative percentages of the isotopes?

*43. The actual mass of one atom of an unknown isotope is 2.18×10^{-22} g. Calculate the atomic mass of this isotope.

44. The mass of an atom of argon is 6.63×10^{-24} g. How many atoms are in a 40.0-g sample of argon?

45. Using the periodic table inside the front cover of the book, determine which of the first twenty elements have isotopes that you would expect to have the same number of protons, neutrons, and electrons.

46. Complete the following table with the appropriate data for each isotope given:

Atomic number	Mass number	Symbol of element	Number of protons	Number of neutrons
(a) 8	16			
(b)		Ni		30
(c)	199		80	

47. Complete the following table (all are neutral atoms):

Element	Symbol	Atomic number	Number of protons	Number of neutrons	Number of electrons
(a) Platinum					
(b)	^{30}P				
(c)		53			
(d)			36		
(e)				45	34
(f)	^{40}Ca				

Answers to Practice Exercises

5.1 5.98×10^{24} atoms

5.2

	protons	neutrons	electrons
(a)	8	8	8
(b)	35	45	35
(c)	92	143	92
(d)	29	35	29

5.3 107.87 amu

6

As children, we begin to communicate with other people in our lives by learning the names of objects around us. As we continue to develop, we learn to speak and use language to complete a wide variety of tasks. As we enter school, we begin to learn of other languages—the languages of mathematics, of other cultures, of computers. In each case, we begin by learning the names of the building blocks, and then proceed to more abstract concepts. In chemistry, a new language beckons also—a whole new way of describing the objects so familiar to us in our daily lives—the nomenclature of compounds. Only after learning the language are we able to understand the complexities of the modern model of the atom and its applications to the various fields we have chosen as our careers.

6.1 Common and Systematic Names

Chemical nomenclature is the system of names that chemists use to identify compounds. When a new substance is formulated, it must be named in order to distinguish it from all other substances (see Figure 6.1). In this chapter, we will restrict our discussion to the nomenclature of inorganic compounds—compounds that do not generally contain carbon. The naming of organic compounds (carbon-containing compounds) will be covered separately in Chapter 19.

Common names are arbitrary names that are not based on the chemical composition of compounds. Before chemistry was systematized, a substance was given a name that generally associated it with one of its outstanding physical or chemical properties. For example, *quicksilver* was a common name for mercury, and nitrous oxide (N_2O), used as an anesthetic in dentistry, has been called *laughing gas* because it induces laughter when inhaled. Water and ammonia also are common names because neither provides any information about the chemical composition of the compounds. If every substance were assigned a common name, the amount of memorization required to learn over nine million names would be astronomical.

Common names have distinct limitations, but they remain in frequent use. Often common names continue to be used in industry because the systematic name is too long or too technical for everyday use. For example, calcium oxide (CaO) is called *lime* by plasterers; photographers refer to *hypo* rather than sodium thiosulfate

Water (H_2O) and ammonia (NH_3) are almost always referred to by their common

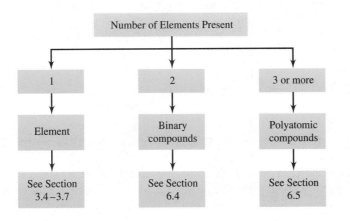

◀ **FIGURE 6.1**
Flow chart summarizing the text locations of rules for naming inorganic substances.

◀ **Chapter Opening Photo: This exquisite chambered nautilus shell is formed from calcium carbonate, commonly called limestone.**

TABLE 6.1 Common Names, Formulas, and Chemical Names of Familiar Substances

Common name	Formula	Chemical name
Acetylene	C_2H_2	Ethyne
Lime	CaO	Calcium oxide
Slaked lime	$Ca(OH)_2$	Calcium hydroxide
Water	H_2O	Water
Galena	PbS	Lead(II) sulfide
Alumina	Al_2O_3	Aluminum oxide
Baking soda	$NaHCO_3$	Sodium hydrogen carbonate
Cane or beet sugar	$C_{12}H_{22}O_{11}$	Sucrose
Borax	$Na_2B_4O_7 \cdot 10\ H_2O$	Sodium tetraborate decahydrate
Brimstone	S	Sulfur
Calcite, marble, limestone	$CaCO_3$	Calcium carbonate
Cream of tartar	$KHC_4H_4O_6$	Potassium hydrogen tartrate
Epsom salts	$MgSO_4 \cdot 7\ H_2O$	Magnesium sulfate heptahydrate
Gypsum	$CaSO_4 \cdot 2\ H_2O$	Calcium sulfate dihydrate
Grain alcohol	C_2H_5OH	Ethanol, ethyl alcohol
Hypo	$Na_2S_2O_3$	Sodium thiosulfate
Laughing gas	N_2O	Dinitrogen monoxide
Lye, caustic soda	NaOH	Sodium hydroxide
Milk of magnesia	$Mg(OH)_2$	Magnesium hydroxide
Muriatic acid	HCl	Hydrochloric acid
Plaster of Paris	$CaSO_4 \cdot \frac{1}{2} H_2O$	Calcium sulfate hemihydrate
Potash	K_2CO_3	Potassium carbonate
Pyrite (fool's gold)	FeS_2	Iron disulfide
Quicksilver	Hg	Mercury
Saltpeter (chile)	$NaNO_3$	Sodium nitrate
Table salt	NaCl	Sodium chloride
Vinegar	$HC_2H_3O_2$	Acetic acid
Washing soda	$Na_2CO_3 \cdot 10\ H_2O$	Sodium carbonate decahydrate
Wood alcohol	CH_3OH	Methanol, methyl alcohol

($Na_2S_2O_3$); and nutritionists use the name *vitamin D$_3$*, instead of 9,10-secocholesta-5,7,10(19)-trien-3-β-ol ($C_{27}H_{44}O$). Table 6.1 lists the common names, formulas, and systematic names of some familiar substances.

Chemists prefer systematic names that precisely identify the chemical composition of chemical compounds. The system for inorganic nomenclature was devised by the International Union of Pure and Applied Chemistry (IUPAC), which was founded in 1921. They continue to meet regularly and constantly review and update the system.

6.2   Elements and Ions

In Chapter 3 you learned the names and symbols for the elements as well as their general location on the periodic table. In Chapter 5 we investigated the composition of the atom and learned that all atoms are composed of protons, electrons, and neutrons; that a particular element is defined by the number of protons it contains; and that atoms are uncharged because they contain equal numbers of protons and electrons.

What's in a Name?

Sometimes it takes an extremely long time to name a new substance. Such was the case with element 106. It was discovered in 1974 by a team of scientists in California at Lawrence Berkeley Laboratories at the same time it was discovered by a team of Russian scientists at the Joint Institute for Nuclear Research in Dubna. Names for new elements are generally provided by its discoverers, but in this case neither team could reproduce their discovery since element 106 is a substance that disappears within a few seconds. Only in 1993 after almost 20 years did the group at Berkeley succeed in once more producing element 106, therefore giving Berkeley the claim for its discovery and the freedom to name it.

The eight-member team of scientists discussed naming the element after Sir Isaac Newton or Luis Alvarez (a Berkeley scientist who played a major role in developing the theory that the dinosaurs were killed by the impact of a comet). Seven members of the team finally met and suggested the name seaborgium after the eighth member, Dr. Glenn T. Seaborg. This is the first time that an element has ever been named for a living person.

Seaborg, now 81, is a former chancellor of UC Berkeley and a former chairman of the Atomic Energy Commission (now called the Department of Energy). He was first recognized as the prime mover in the discovery of plutonium, the radioactive element that is best known as the key component of atomic bombs. Since then, he has worked with a team of scientists to discover nine of the transuranium elements (those coming after uranium on the periodic table). They are created by firing streams of particles and nuclei into heavy elements in a nuclear accelerator. Elements credited to his team from the 1940s to the 1960s include neptunium (named after neptune), americium (America), fermium (Enrico Fermi, a famous physicist), mendelevium (Dimitri Mendeleev, creator of the modern peri-

Glenn T. Seaborg, discoverer of element 106.

odic table), nobelium (Alfred Nobel, inventor of dynamite and founder of the Nobel prize), berkelium (Berkeley, California, home of the group's lab), californium (California), curium (Marie Curie, discoverer of several elements), and einsteinium (Albert Einstein). In 1951, Seaborg received the Nobel prize for his role in the discoveries of these elements.

Since Berkeley's last discovery (element 106) in 1974, the creation of new

elements has been led by a team in Germany that has discovered elements 107, 108, and 109. They suggested neilsbohrium (Neils Bohr, a Finnish physicist), hassium (Hesse, a German state), and meitnerium (Lise Meitner, a German atomic fission pioneer).

All of these names must still be approved by the IUPAC before becoming official and placed into common use on the periodic table. The IUPAC group met in August 1994 and assigned different names to several of the elements above 100. The names for elements 101, 102, and 103 remained the same as they had been, but the name for 106 was changed because Glenn Seaborg is still alive. The names for elements 99 and 100 were also proposed (but not approved) when the scientists were still living. The proposed IUPAC names for 104 and 105 would change names that have been in wide use for over 20 years. The latest decisions by IUPAC have rekindled controversy over the naming of elements. If you look at the periodic table today you will find each of the elements (104–109) named only by the Latin version of its number. This was a provisional recommendation until names could be agreed upon. It appears that the newest elements will have to wait a little longer to be officially named.

Some of the Names Proposed for Elements Discovered in Recent Years

Number	American	German	Russian	IUPAC
101				mendelevium
102			joliotium	nobelium
103				lawrencium
104	rutherfordium			dubnium
105	hahnium		neilsbohrium	joliotium
106	seaborgium			rutherfordium
107		neilsbohrium		bohrium
108		hassium		hahnium
109		meitnerium		meitnerium

The formula for most elements is simply the symbol of the element. In chemical reactions or mixtures the element behaves as though it were a collection of individual particles. A small number of elements have formulas that are not single atoms at normal temperatures. Seven of the elements are *diatomic* molecules—that is, two atoms bonded together to form a molecule. These diatomic elements are hydrogen, H_2; oxygen, O_2; nitrogen, N_2; fluorine, F_2; chlorine, Cl_2; bromine, Br_2; and iodine, I_2. Two other elements that are commonly polyatomic are S_8, sulfur; and P_4, phosphorus.

Elements Occurring as Molecules					
Hydrogen	H_2	Chlorine	Cl_2	Sulfur	S_8
Oxygen	O_2	Fluorine	F_2	Phosphorus	P_4
Nitrogen	N_2	Bromine	Br_2		
		Iodine	I_2		

A charged atom, known as an *ion,* can be produced by adding or removing one or more electrons from a neutral atom. For example, potassium atoms contain 19 protons and 19 electrons. To make a potassium ion we remove one electron leaving 19 protons and only 18 electrons. This gives an ion with a positive one ($+1$) charge (see diagram).

Written in the form of an equation, $K \rightarrow K^+ + e^-$. A positive ion is called

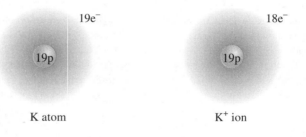

K atom   K$^+$ ion

a *cation*. Any neutral atom that *loses* an electron will form a cation. Sometimes an atom may lose one electron as we have seen in our potassium example. Other atoms may lose more than one electron as shown below:

$$Mg \rightarrow Mg^{2+} + 2e^-$$

or

$$Al \rightarrow Al^{3+} + 3e^-$$

Cations are named the same as their parent atoms, as shown below:

	Atom		Ion
K	potassium	K^+	potassium ion
Mg	magnesium	Mg^{2+}	magnesium ion
Al	aluminum	Al^{3+}	aluminum ion

Ions can also be formed by adding electrons to a neutral atom. For example, the chlorine atom contains 17 protons and 17 electrons. The equal number of positive charges and negative charges results in a net charge of zero for the atom. If one electron is added to the chlorine atom it now contains 17 protons and 18 electrons resulting in a net charge of negative one (-1) on the ion.

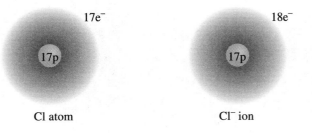

Summarizing the process in a chemical equation produces $Cl + e^- \rightarrow Cl^-$. A negative ion is called an *anion*. Any neutral atom that *gains* an electron will form an anion. Atoms may gain more than one electron to form anions with different charges as shown in the following examples:

$$O + 2e^- \rightarrow O^{2-}$$

$$N + 3e^- \rightarrow N^{3-}$$

Anions are named differently than cations. To name an anion consisting of only one element, use the stem part of the parent element name and change the ending to *-ide*. For example, the Cl^- ion is named in the following way: use the *chlor-* from chlorine and add *-ide* to form chloride ion. Other examples are shown below:

Symbol	Name of atom	Ion	Name of ion
F	fluorine	F^-	fluoride ion
Br	bromine	Br^-	bromide ion
Cl	chlorine	Cl^-	chloride ion
I	iodine	I^-	iodide ion
O	oxygen	O^{2-}	oxide ion
N	nitrogen	N^{3-}	nitride ion

Ions are always formed by adding or removing electrons from an atom. Atoms do not form ions on their own. Most often ions are formed when metals combine with nonmetals.

The charge on an ion can often be predicted from the position of the element on the periodic table. Figure 6.2 shows the charges of selected ions from several groups on the periodic table. Notice that all the metals in the far left column (Group IA) are $(1+)$, all those in the next column (Group IIA) are $(2+)$, and the metals in the next tall column (Group IIIA) form $(3+)$ ions. The elements in the lower center part of the table are called *transition metals*. These elements tend to form

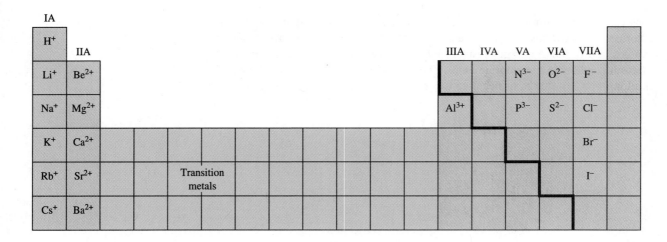

IA													IIIA	IVA	VA	VIA	VIIA	
H^+	IIA																	
Li^+	Be^{2+}														N^{3-}	O^{2-}	F^-	
Na^+	Mg^{2+}												Al^{3+}		P^{3-}	S^{2-}	Cl^-	
K^+	Ca^{2+}																Br^-	
Rb^+	Sr^{2+}			Transition metals													I^-	
Cs^+	Ba^{2+}																	

▲
FIGURE 6.2
Charges of selected ions in the periodic table.

FIGURE 6.3 ▶
Salt (NaCl) is a strong electrolyte as can be seen from the brightly glowing light bulb.

cations with various positive charges. There is no easy way to predict the charges on these cations. All metals lose electrons to form positive ions.

In contrast, the nonmetals form anions by gaining electrons. On the right side of the periodic table in Figure 6.2 you can see that the atoms in Group VIIA form $(1-)$ ions. The nonmetals in Group VIA form $(2-)$ ions. It is important to learn the charges on the ions shown in Figure 6.2 and their relationship to the group number at the top of the column. We will learn more about why these ions carry their particular charges later in the course.

6.3   Writing Formulas from Names of Compounds

In Chapters 3 and 5 we learned that compounds can be composed of ions. These substances are called *ionic compounds* and will conduct electricity when dissolved in water. An excellent example of an ionic compound is ordinary table salt. It is composed of crystals of sodium chloride. When dissolved in water, sodium chloride conducts electricity very well as shown in Figure 6.3.

A chemical compound must have a net charge of zero. If it contains ions, the charges on the ions must add up to zero in the formula for the compound. This is relatively easy in the case of sodium chloride. The sodium ion $(1+)$ and the chloride ion $(1-)$ add to zero, resulting in the formula NaCl. Now consider an ionic compound containing calcium (Ca^{2+}) and fluoride (F^-) ions. How can we write a formula with a net charge of zero? To do this we need one Ca^{2+} and two F^- ions. The correct formula is CaF_2. The subscript 2 indicates two fluoride ions are needed for each calcium ion. Aluminum oxide is a bit more complicated since it consists of Al^{3+} and O^{2-} ions. Since 6 is the least common multiple of 3 and 2, we have $2(3+) = 3(2-) = 0$ or a formula containing $2\ Al^{3+}$ ions and $3\ O^{2-}$ ions for Al_2O_3. The following table gives more examples of formula writing for ionic compounds.

Name of compound	Ions	Lowest common multiple	Sum of charges on ions	Formula
Sodium bromide	Na^+, Br^-	1	$(+1) + (-1) = 0$	NaBr
Potassium sulfide	K^+, S^{2-}	2	$2(+1) + (-2) = 0$	K_2S
Zinc sulfate	Zn^{2+}, SO_4^{2-}	2	$(+2) + (-2) = 0$	$ZnSO_4$
Ammonium phosphate	NH_4^+, PO_4^{3-}	3	$3(+1) + (-3) = 0$	$(NH_4)_3PO_4$
Aluminum chromate	Al^{3+}, CrO_4^{2-}	6	$2(+3) + 3(-2) = 0$	$Al_2(CrO_4)_3$

Write formulas for (a) calcium chloride, (b) magnesium oxide, (c) barium phosphide.

Example 6.1

Solution

(a) Use the following steps for calcium chloride.

Step 1 From the name we know that calcium chloride is composed of calcium and chloride ions. First write down the formulas of these ions:

$$Ca^{2+} \qquad Cl^-$$

Step 2 To write the formula of the compound, combine the smallest numbers of Ca^{2+} and Cl^- ions to give the charge sum equal to zero. In this case the lowest common multiple of the charges is 2:

$$(Ca^{2+}) + 2(Cl^-) = 0$$

$$(2+) \ + 2(1-) = 0$$

Therefore, the formula is $CaCl_2$.

(b) Use the same procedure for magnesium oxide.

Step 1 From the name we know that magnesium oxide is composed of magnesium and oxide ions. First write down the formulas of these ions:

$$Mg^{2+} \qquad O^{2-}$$

Step 2 To write the formula of the compound, combine the smallest numbers of Mg^{2+} and O^{2-} ions to give the charge sum equal to zero:

1. **Cobalt(II) chloride, CoCl₂**
2. **Sodium chloride, NaCl**
3. **Lead(II) sulfide, PbS**
4. **Sulfur, S**
5. **Zinc, Zn**
6. **Marble chips, CaCO₃**
7. **Logwood chips**
8. **Charcoal, C**
9. **Mercury(II) iodide, HgI₂**
10. **Pyrite, FeS**
11. **Chromium(III) oxide, Cr₂O₃**
12. **Iron(II) sulfate, FeSO₄**
13. **Sodium sulfite, Na₂SO₃**
14. **Rosin**
15. **Sodium thiosulfate, Na₂S₂O₃**
16. **Iron, Fe**
17. **Aluminum, Al**
18. **Potassium hexacyanoferrate, K₃Fe(CN)₆**
19. **Potassium chromium(III) sulfate, KCr(SO₄)₂**
20. **Menthol, C₁₀H₁₉OH**
21. **Potassium permanganate, KMnO₄**
22. **Ammonium nickel(II) sulfate, (NH₄)₂Ni(SO₄)₂**
23. **Copper(II) sulfate pentahydrate, CuSO₄·5H₂O**
24. **Sodium chromate, Na₂CrO₄**
25. **Trilead tetraoxide, Pb₃O₄**
26. **Hydroquinone, C₆H₄(OH)₂**
27. **Copper, Cu**

▲
Compounds of transition elements are typically very colorful and are useful as paint pigments. Compounds of elements on the left of the periodic table reflect all wavelengths of light, which gives them a white color. Carbon compounds take on a variety of colors.

$$(Mg^{2+}) + (O^{2-}) = 0$$

$$(2+) \quad + (2-) \ = 0$$

Therefore, the formula is MgO.

(c) Use the same procedure for barium phosphide.

Step 1 From the name we know that barium phosphide is composed of barium and phosphide ions. First write down the formulas of these ions:

$$Ba^{2+} \quad P^{3-}$$

Step 2 To write the formula of the compound, combine the smallest numbers of Ba^{2+} and P^{3-} ions to give the charge sum equal to zero. In this case the lowest common multiple of the charges is 6:

$$3(Ba^{2+}) + 2(P^{3-}) = 0$$

$$3(2+) \quad + 2(3-) \ = 0$$

Therefore, the formula is Ba_3P_2.

Practice 6.1

Write formulas for compounds containing the following ions:
(a) K^+ and F^-
(d) Na^+ and S^{2-}
(b) Ca^{2+} and Br^-
(e) Ba^{2+} and O^{2-}
(c) Mg^{2+} and N^{3-}

6.4   Binary Compounds

Binary compounds contain only two different elements. Many binary compounds are formed when a metal combines with a nonmetal to form a *binary ionic compound.* The metal loses one or more electrons to become a cation while the nonmetal gains one or more electrons to become an anion. The cation is written first in the formula and is followed by the anion.

A. Binary Ionic Compounds Containing a Metal Forming Only One Type of Cation

The chemical name is composed of the name of the metal followed by the name of the nonmetal, which has been modified to an identifying root plus the suffix *-ide.*

For example, sodium chloride, NaCl, is composed of one atom each of sodium and chlorine. The name of the metal, sodium, is written first and is not modified. The second part of the name is derived from the nonmetal, chlorine, by using the root *chlor-* and adding the ending *-ide;* it is named *chloride.* The compound name is *sodium chloride.*

NaCl

Elements: Sodium (metal)
                    Chlorine (nonmetal)
                       name modified to the stem *chlor-* + *-ide*
Name of compound: Sodium chloride

To name these compounds:
1. Write the name of the cation.
2. Write the root for the anion with the suffix *-ide.*

Stems of the more common negative-ion-forming elements are shown in Table 6.2. Table 6.3 shows some compounds with names ending in *-ide.*

Compounds may contain more than one atom of the same element, but as long as they contain only two different elements and only one compound of these two elements exists, the name follows the rule for binary compounds:

Examples:

$CaBr_2$ Mg_3N_2 Li_2O

Calcium bromide   Magnesium nitride   Lithium oxide

TABLE 6.2 Examples of Elements Forming Anions			
Symbol	**Element**	**Root**	**Anion name**
Br	Bromine	Brom	Bromide
Cl	Chlorine	Chlor	Chloride
F	Fluorine	Fluor	Fluoride
H	Hydrogen	Hydr	Hydride
I	Iodine	Iod	Iodide
N	Nitrogen	Nitr	Nitride
O	Oxygen	Ox	Oxide
P	Phosphorus	Phosph	Phosphide
S	Sulfur	Sulf	Sulfide

Example 6.6   Name the compound CaS.

Solution

Step 1 From the formula it is a two-element compound and follows the rules for binary compounds.

Step 2 The compound is composed of Ca, a metal, and S, a nonmetal. Elements in the IIA column form only one type of cation. Thus, we name the positive part of the compound *calcium.*

Step 3 Modify the name of the second element to the identifying root *sulf-* and add the binary ending *-ide* to form the name of the negative part, *sulfide.*

Step 4 The name of the compound, therefore, is *calcium sulfide.*

Practice 6.2

Write formulas for the following compounds:
(a) strontium chloride, (b) potassium iodide, (c) aluminum nitride, (d) calcium sulfide, (e) sodium oxide

---

B. Binary Ionic Compounds Containing a Metal That Can Form Two or More Types of Cations

The metals in the center of the periodic table (including the transition metals) often form more than one type of cation. For example, iron can form Fe^{2+} and Fe^{3+} ions, and copper can form Cu^+ and Cu^{2+} ions. This can lead to confusion when naming compounds. For example, copper chloride could be $CuCl_2$ or $CuCl$. To resolve this difficulty the IUPAC devised a system to name these compounds, which is known as the **Stock System.** This system is currently recognized as the official system to name these compounds although another older system is sometimes used. In the Stock System, when a compound contains a metal that can form more than one type of cation, the charge on the cation of the metal is designated by a Roman numeral placed in parentheses immediately following the name of the metal. The negative element is treated in the usual manner for binary compounds.

Stock System

TABLE 6.3 Examples of Compounds with Names Ending in *-ide*

Formula	Name	Formula	Name
$AlCl_3$	Aluminum chloride	BaS	Barium sulfide
Al_2O_3	Aluminum oxide	LiI	Lithium iodide
CaC_2	Calcium carbide	$MgBr_2$	Magnesium bromide
HCl	Hydrogen chloride	NaH	Sodium hydride
HI	Hydrogen iodide	Na_2O	Sodium oxide

Cation charge		+1	+2	+3	
Roman numeral		(I)	(II)	(III)	
Examples:	$FeCl_2$	Iron(II) chloride		Fe^{2+}	
	$FeCl_3$	Iron(III) chloride		Fe^{3+}	
	$CuCl$	Copper(I) chloride		Cu^+	
	$CuCl_2$	Copper(II) chloride		Cu^{2+}	

To name these compounds:
1. Write the name of the cation.
2. Write the charge on the cation as a Roman numeral in parentheses.
3. Write the root of the anion with the suffix *-ide*.

The fact that $FeCl_2$ has two chloride ions, each with a -1 charge, establishes that the charge of Fe is $+2$. To distinguish between the two iron chlorides, $FeCl_2$ is named iron(II) chloride and $FeCl_3$ is named iron(III) chloride.

$$\text{Iron(II) chloride} \qquad \text{Iron(III) chloride}$$
$$FeCl_2 \qquad\qquad FeCl_3$$

Charge:	+2	−1	+3	−1
Name:	Iron(II)	chloride	Iron(III)	chloride

When a metal forms only one possible cation, we need not distinguish one cation from another, so Roman numerals are not needed. Thus we do not say calcium(II) chloride for $CaCl_2$, but rather calcium chloride, since the charge of calcium is understood to be $+2$.

In classical nomenclature, when the metallic ion has only two cation types, the name of the metal (usually the Latin name) is modified with the suffixes *-ous* and *-ic* to distinguish between the two. The lower charge cation is given the *-ous* ending, and the higher one, the *-ic* ending.

Examples:	$FeCl_2$	Ferrous chloride	Fe^{2+}	(lower charge cation)
	$FeCl_3$	Ferric chloride	Fe^{3+}	(higher charge cation)
	$CuCl$	Cuprous chloride	Cu^+	(lower charge cation)
	$CuCl_2$	Cupric chloride	Cu^{2+}	(higher charge cation)

Table 6.4 lists some common metals that have more than one type of cation.

Notice that the *ous–ic* naming system does not give the charge of the cation of an element but merely indicates that at least two types of cations exist. The Stock System avoids any possible uncertainty by clearly stating the charge on the cation.

In this book we will use only the Stock System.

Name the compound FeS.

Step 1 This compound follows the rules for a binary compound and, like CaS, must be a sulfide.

Example 6.7

Solution

TABLE 6.4 Names and Charges of Some Common Metal Ions That Have More Than One Type of Cation

Formula	Stock System name	Classical name
Cu^{1+}	Copper(I)	Cuprous
Cu^{2+}	Copper(II)	Cupric
Hg^{1+} $(Hg_2)^{2+}$	Mercury(I)	Mercurous
Hg^{2+}	Mercury(II)	Mercuric
Fe^{2+}	Iron(II)	Ferrous
Fe^{3+}	Iron(III)	Ferric
Sn^{2+}	Tin(II)	Stannous
Sn^{4+}	Tin(IV)	Stannic
Pb^{2+}	Lead(II)	Plumbous
Pb^{4+}	Lead(IV)	Plumbic
As^{3+}	Arsenic(III)	Arsenous
As^{5+}	Arsenic(V)	Arsenic
Ti^{3+}	Titanium(III)	Titanous
Ti^{4+}	Titanium(IV)	Titanic

Step 2 It is a compound of Fe, a metal, and S, a nonmetal, and Fe is a transition metal that has more than one type of cation. In sulfides, the charge on the S is -2. Therefore, the charge on Fe must be $+2$, and the name of the positive part of the compound is *iron(II)*.

Step 3 We have already determined that the name of the negative part of the compound will be *sulfide*.

Step 4 The name of FeS is *iron(II) sulfide*.

Practice 6.3

Write the name for each of the following compounds using the official Stock System:
(a) PbI_2, (b) SnF_4, (c) Fe_2O_3, (d) CuO

C. Binary Compounds Containing Two Nonmetals

In a compound formed between two nonmetals, the element that occurs first in the series is written and named first.

Si, B, P, H, C, S, I, Br, N, Cl, O, F

The name of the second element retains the *-ide* ending as though it were an anion. A Latin or Greek prefix (*mono-, di-, tri-,* and so on) is attached to the name of each element to indicate the number of atoms of that element in the molecule. The prefix *mono-* is never used for naming the first element. Some common prefixes and their numerical equivalences follow.

mono = 1	*tetra* = 4	*hepta* = 7	*nona* = 9
di = 2	*penta* = 5	*octa* = 8	*deca* = 10
tri = 3	*hexa* = 6		

Here are some examples of compounds that illustrate this system:

$$N_2O_3$$

(Di)nitrogen   (Tri)oxide

Indicates two   Indicates three
nitrogen atoms   oxygen atoms

CO	Carbon monoxide		N_2O	Dinitrogen monoxide
CO_2	Carbon dioxide		N_2O_4	Dinitrogen tetroxide
PCl_3	Phosphorus trichloride		NO	Nitrogen monoxide
SO_2	Sulfur dioxide		N_2O_3	Dinitrogen trioxide
P_2O_5	Diphosphorus pentoxide		S_2Cl_2	Disulfur dichloride
CCl_4	Carbon tetrachloride		S_2F_{10}	Disulfur decafluoride

The example above illustrates that we sometimes drop the final *o*(mono) or *a*(penta) of the prefix when the second element is oxygen. This avoids creating a name that is awkward to pronounce. For example, CO is carbon monoxide instead of carbon mon*oo*xide.

To name these compounds:
1. Write the name for the first element using a prefix if there is more than one atom of this element.
2. Write the root for the second element with the suffix *-ide*. Use a prefix to indicate the number of atoms for the second element.

Name the compound PCl_5.

Example 6.8

Solution

Step 1 Phosphorus and chlorine are nonmetals, so the rules for naming binary compounds containing two nonmetals apply. Phosphorus is named first. Therefore, the compound is a chloride.

Step 2 No prefix is needed for phosphorus because each molecule has only one atom of phosphorus. The prefix *penta-* is used with chloride to indicate the five chlorine atoms. (PCl_3 is also a known compound.)

Step 3 The name for PCl_5 is *phosphorus pentachloride*.

Practice 6.4

Name the following compounds:
(a) Cl_2O, (b) SO_2, (c) CBr_4, (d) N_2O_5, (e) NH_3

D. Acids Derived from Binary Compounds

Certain binary hydrogen compounds, when dissolved in water, form solutions that have *acid* properties. Because of this property, these compounds are given acid names in addition to their regular *-ide* names. For example, HCl is a gas and is called *hydrogen chloride,* but its water solution is known as *hydrochloric acid.* Binary acids are composed of hydrogen and one other nonmetallic element. However, not all binary hydrogen compounds are acids. To express the formula of a binary acid it is customary to write the symbol of hydrogen first, followed by the symbol of the second element (for example, HCl, HBr, H_2S). When we see formulas such

TABLE 6.5 Names and Formulas of Selected Binary Acids

Formula	Acid name	Formula	Acid name
HF	Hydrofluoric acid	HI	Hydroiodic acid
HCl	Hydrochloric acid	H_2S	Hydrosulfuric acid
HBr	Hydrobromic acid	H_2Se	Hydroselenic acid

as CH_4 or NH_3, we understand that these compounds are not normally considered to be acids.

To name a binary acid, place the prefix *hydro-* in front of, and the suffix *-ic* after, the stem of the nonmetal name. Then add the word *acid*.

HCl H_2S

Examples: *Hydro-*chlor-*ic acid*   *Hydro-*sulfur-*ic acid*
            (hydrochloric acid)   (hydrosulfuric acid)

> To name these compounds:
> 1. Write the prefix *hydro-* with the root of the second element and add the suffix *-ic*.
> 2. Write the word *acid*.

Acids are hydrogen-containing substances that liberate hydrogen ions when dissolved in water. The same formula is often used to express binary hydrogen compounds, such as HCl, regardless of whether or not they are dissolved in water. Table 6.5 shows several examples of binary acids.

A summary of the approach to naming binary compounds is shown in Figure 6.4.

Practice 6.5

Name each of the following binary compounds:
(a) KBr, (b) Ca_3N_2, (c) SO_3, (d) SnF_2, (e) $CuCl_2$, (f) N_2O_4

6.5 Naming Compounds Containing Polyatomic Ions

polyatomic ion

A **polyatomic ion** is an ion that contains two or more elements. Compounds containing polyatomic ions are composed of three or more elements and usually consist of one or more cations combined with a negative polyatomic ion. In general, naming compounds containing polyatomic ions is similar to naming binary compounds. The cation is named first, followed by the name for the negative polyatomic ion.

In order to name these compounds you must learn to recognize the common polyatomic ions (Table 6.6) and know their charges. Consider the formula $KMnO_4$. You must be able to recognize that it consists of two parts K | MnO_4. These parts are composed of a K^+ ion and a MnO_4^- ions. The correct name for this compound is potassium permanganate. Many polyatomic ions that contain oxygen are called *oxy-anions,* and generally have the suffix *-ate* or *-ite*. Unfortunately, the suffix does not indicate the number of oxygen atoms present. The *-ate* form contains more oxygen atoms than the *-ite* form. Examples include sulfate (SO_4^{2-}), sulfite (SO_3^{2-}), nitrate (NO_3^-), and nitrite (NO_2^-).

> To name these compounds:
> 1. Write the name of the cation.
> 2. Write the name of the anion.

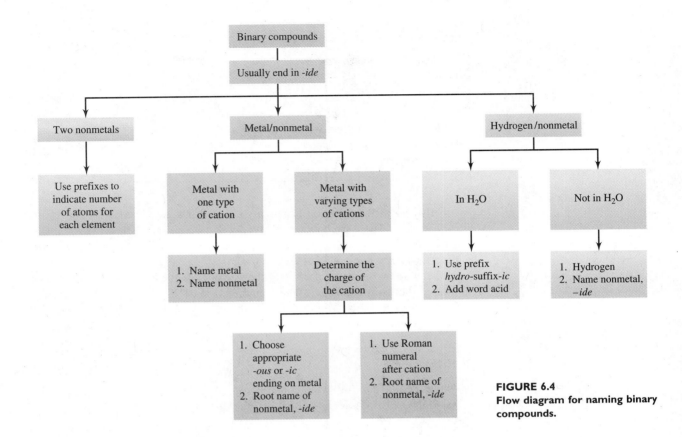

FIGURE 6.4
Flow diagram for naming binary compounds.

TABLE 6.6 Names, Formulas, and Charges of Some Common Polyatomic Ions

Name	Formula	Charge	Name	Formula	Charge
Acetate	$C_2H_3O_2^-$	-1	Cyanide	CN^-	-1
Ammonium	NH_4^+	$+1$	Dichromate	$Cr_2O_7^{2-}$	-2
Arsenate	AsO_4^{3-}	-3	Hydroxide	OH^-	-1
Hydrogen carbonate	HCO_3^-	-1	Nitrate	NO_3^-	-1
Hydrogen sulfate	HSO_4^-	-1	Nitrite	NO_2^-	-1
Bromate	BrO_3^-	-1	Permanganate	MnO_4^-	-1
Carbonate	CO_3^{2-}	-2	Phosphate	PO_4^{3-}	-3
Chlorate	ClO_3^-	-1	Sulfate	SO_4^{2-}	-2
Chromate	CrO_4^{2-}	-2	Sulfite	SO_3^{2-}	-2

Some elements form more than two different polyatomic ions containing oxygen. To name these ions, prefixes are used in addition to the suffix. To indicate more oxygen than in the *-ate* form we add the prefix *per-,* which is a short form of *hyper-,* meaning more. The prefix *hypo-,* meaning less (oxygen in this case), is used for the ion containing less oxygen than the *-ite* form. An example of this system is shown for the polyatomic ions containing chlorine and oxygen in Table 6.7. The prefixes are also used with other similar ions, such as iodate (IO_3^-), bromate (BrO_3^-), and phosphate (PO_4^{3-}).

TABLE 6.7 Oxy-Anions and Oxy-Acids of Chlorine

Anion formula	Anion name	Acid formula	Acid name
ClO^-	*Hypochlorite*	HClO	*Hypochlorous* acid
ClO_2^-	Chlor*ite*	$HClO_2$	Chlor*ous* acid
ClO_3^-	Chlor*ate*	$HClO_3$	Chlor*ic* acid
ClO_4^-	*Perchlorate*	$HClO_4$	*Perchloric acid*

Only two of the common negatively charged polyatomic ions do not use the *ate–ite* system. These exceptions are hydroxide (OH^-) and cyanide (CN^-). Care must be taken with these, as their endings can easily be confused with the *-ide* ending for binary compounds (Section 6.4).

There are two common positively charged polyatomic ions as well—the ammonium and the hydronium ions. The ammonium ion (NH_4^+) is frequently found in polyatomic compounds (Section 6.5), whereas the hydronium ion (H_3O^+) is usually seen in aqueous solutions of acids (Chapter 15).

Practice 6.6

Name each of the following compounds:
(a) $NaNO_3$, (b) $Ca_3(PO_4)_2$, (c) KOH, (d) Li_2CO_3, (e) $NaClO_3$

Inorganic compounds are also formed from more than three elements (see Table 6.8). In these cases one or more of the ions is often a polyatomic ion. Once you have learned to recognize the polyatomic ions, naming these compounds follows the patterns we have already learned. First identify the ions. Name the cations in the order given, and follow them with the names of the anions. Study the following examples:

Compound	Ions	Name
$NaHCO_3$	Na^+; HCO_3^-	Sodium hydrogen carbonate
NaHS	Na^+; HS^-	Sodium hydrogen sulfide
$MgNH_4PO_4$	Mg^{2+}; NH_4^+; PO_4^{3-}	Magnesium ammonium phosphate
$NaKSO_4$	Na^+; K^+; SO_4^{2-}	Sodium potassium sulfate

6.6   Acids

While we will learn much more about acids later (see Chapter 16), it is helpful to be able to recognize and name common acids both in the laboratory and in class. The simplest way to recognize many acids is to know that acid formulas often begin with hydrogen. Naming binary acids was covered in Section 6.4D. Inorganic

TABLE 6.8 Names of Selected Compounds That Contain More Than One Kind of Positive Ion

Formula	Name of Compound
$KHSO_4$	Potassium hydrogen sulfate
$Ca(HSO_3)_2$	Calcium hydrogen sulfite
NH_4HS	Ammonium hydrogen sulfide
$MgNH_4PO_4$	Magnesium ammonium phosphate
NaH_2PO_4	Sodium dihydrogen phosphate
Na_2HPO_4	Disodium hydrogen phosphate
KHC_2O_4	Potassium hydrogen oxalate
$KAl(SO_4)_2$	Potassium aluminum sulfate
$Al(HCO_3)_3$	Aluminum hydrogen carbonate

TABLE 6.9 Comparison of Acid and Anion Names for Selected Oxy-Acids

Acid	Anion	Acid	Anion
H_2SO_4 Sulfuric acid	SO_4^{2-} Sulfate ion	H_3PO_4 Phosphoric acid	PO_4^{3-} Phosphate ion
H_2SO_3 Sulfurous acid	SO_3^{2-} Sulfite ion	H_3PO_3 Phosphorous acid	PO_3^{3-} Phosphite ion
HNO_3 Nitric acid	NO_3^- Nitrate ion	HIO_3 Iodic acid	IO_3^- Iodate ion
HNO_2 Nitrous acid	NO_2^- Nitrite ion	$HC_2H_3O_2$ Acetic acid	$C_2H_3O_2^-$ Acetate ion
H_2CO_3 Carbonic acid	CO_3^{2-} Carbonate ion	$H_2C_2O_4$ Oxalic acid	$C_2O_4^{2-}$ Oxalate ion

compounds containing hydrogen, oxygen, and one other element are called *oxy-acids*. The element other than hydrogen or oxygen in these acids is often a nonmetal, but it can also be a metal.

The first step in naming these acids is to determine that the compound in question is really an oxy-acid. The keys to identification are that (1) hydrogen is the first element in the compound's formula; and (2) the second portion of the formula consists of a polyatomic ion containing oxygen.

Hydrogen in an oxy-acid is not specifically designated in the acid name. The presence of hydrogen in the compound is indicated by the use of the word *acid* in the name of the substance. To determine the particular type of acid, the polyatomic ion following hydrogen must be examined. The name of the polyatomic ion is modified in the following manner: (1) *-ate* changes to an *-ic* ending; (2) *-ite* changes to an *-ous* ending. (See Table 6.9.) The compound with the *-ic* ending contains more oxygen than the one with the *-ous* ending. Consider the following examples:

H_2SO_4 sulf*ate* $\longrightarrow$ sulfur*ic* acid (contains 4 oxygens)

H_2SO_3 sulf*ite* $\longrightarrow$ sulfur*ous* acid (3 oxygens)

HNO_3 nitr*ate* $\longrightarrow$ nitr*ic* acid (3 oxygens)

HNO_2 nitr*ite* $\longrightarrow$ nitr*ous* acid (2 oxygens)

FIGURE 6.5 ▶
Flow diagram for naming polyatomic compounds.

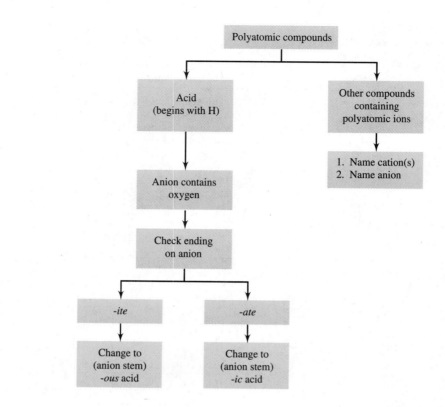

The complete system for naming oxy-acids is shown in Table 6.7 for the various acids containing chlorine. Examples of other oxy-acids and their names are shown in Table 6.10.

A summary of the approach to naming polyatomic compounds is shown in Figure 6.5. We have now looked at ways of naming a variety of inorganic compounds—binary compounds consisting of a metal and a nonmetal and of two nonmetals, and binary acids. These compounds are just a small part of the classified chemical compounds. Most of the remaining classes are in the broad field of organic chemistry under such categories as hydrocarbons, alcohols, ethers, aldehydes, ketones, phenols, and carboxylic acids.

TABLE 6.10 Formulas and Names of Selected Oxy-Acids

Formula	Acid name	Formula	Acid name
H_2SO_3	Sulfurous acid	$HC_2H_3O_2$	Acetic acid
H_2SO_4	Sulfuric acid	$H_2C_2O_4$	Oxalic acid
HNO_2	Nitrous acid	H_2CO_3	Carbonic acid
HNO_3	Nitric acid	$HBrO_3$	Bromic acid
H_3PO_3	Phosphorous acid	HIO_3	Iodic acid
H_3PO_4	Phosphoric acid	H_3BO_3	Boric acid

Charges in Your Life

Ions are used in living organisms to perform many important functions. For example, electrical neutrality must be maintained both inside and outside the body cells. Within the cell, neutrality is maintained by potassium ions (K^+) and hydrogen phosphate ions (HPO_4^{2-}). Outside the cell in the intercellular fluid the ions responsible for neutrality are sodium (Na^+) and chloride (Cl^-).

Another important ion within organisms is magnesium (Mg^{2+}), found in chlorophyll and used during nerve and muscle activity, as well as in conjunction with certain enzymes. Iron ions (Fe^{2+}) are incorporated within the hemoglobin molecule and are an integral part of the oxygen transport system within the body. Calcium ions (Ca^{2+}) are part of the matrix of both bones and teeth and play a significant role in the clotting of blood.

Ions also have a significant role in the detergent industry. Water is said to be "hard" when it contains relatively high concentrations of Ca^{2+} and Mg^{2+} ions.

In solution, soaps combine with these ions to form an insoluble scum. This material forms the common bathtub ring and, in laundry, settles on the clothes leaving them gray and dingy. Although soaps and detergents are similar in their cleansing actions, detergents are less likely to form this scum. For this reason, many people who live in areas with "hard" water use detergents instead of soaps.

Ions form an insoluble scum when combined with some products.

Concepts in Review

1. Write the formulas of compounds formed by combining the ions from Figure 6.2 (or from the inside back cover of this book) in the correct ratios.

2. Write the names or formulas for inorganic binary compounds in which the metal has only one type of cation.

3. Write the names or formulas for inorganic binary compounds that contain metals with multiple types of cations, using the Stock System.

4. Write the names or formulas for inorganic binary compounds that contain two nonmetals.

5. Write the names or formulas for binary acids.

6. Write the names or formulas for oxy-acids.

7. Write the names or formulas for compounds that contain polyatomic ions.

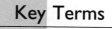

Key Terms

The terms listed here have been defined within this chapter. Section numbers are referenced in parenthesis for each term. More detailed definitions are given in the Glossary.

polyatomic ion (6.5)   Stock System (6.4)

Questions

Questions refer to tables, figures, and key words and concepts defined within the chapter. A particularly challenging question or exercise is indicated with an asterisk.

1. Use the common ion table on the inside back cover of your text to determine the formulas for compounds composed of the following ions:
   (a) Sodium and chlorate
   (b) Hydrogen and sulfate
   (c) Tin(II) and acetate
   (d) Copper(I) and oxide
   (e) Zinc and hydrogen carbonate
   (f) Iron(III) and carbonate

2. Does the fact that two elements combine in a one-to-one atomic ratio mean that the charges on their ions are both 1? Explain.

3. Explain how "hard water" may cause white clothing to become dingy and gray.

4. Why did the IUPAC vote not to use the name seaborgium for element 106?

5. What elements did Glenn Seaborg help to discover?

6. Which of these statements are correct? Rewrite each incorrect statement to make it correct.
   (a) The formula for calcium hydride is CaH_2.
   (b) The ions of all the following metals have a charge of $+2$: Ca, Ba, Sr, Cd, Zn.
   (c) The formulas for nitrous and sulfurous acids are HNO_2 and H_2SO_3.
   (d) The formula for the compound between Fe^{3+} and O^{2-} is Fe_3O_2.
   (e) The name for $NaNO_2$ is sodium nitrite.
   (f) The name for $Ca(ClO_3)_2$ is calcium chlorate.
   (g) The name for CuO is copper(I) oxide.
   (h) The name for SO_4^{2-} is sulfate ion.
   (i) The name for N_2O_4 is dinitrogen tetroxide.
   (j) The name for Na_2O is disodium oxide.
   (k) If the name of an anion ends with *-ide*, the name of the corresponding acid will start with *hydro-*.
   (l) If the name of an anion ends with *-ite*, the corresponding acid name will end with *-ic*.
   (m) If the name of an acid ends with *-ous*, the corresponding polyatomic ion will end with *-ate*.
   (n) In FeI_2, the iron is iron(II) because it is combined with two I^- ions.
   (o) In Cu_2SO_4, the copper is copper(II) because there are two copper ions.
   (p) When two nonmetals combine, prefixes of *di-, tri-, tetra-,* and so on, are used to specify how many atoms of each element are in a molecule.
   (q) N_2O_3 is called dinitrogen trioxide.
   (r) $Sn(CrO_4)_2$ is called tin dichromate.

Paired Exercises

These exercises are paired. Each odd-numbered exercise is followed by a similar even-numbered exercise. Answers to the even-numbered exercises are given in Appendix V.

7. Write the formula of the compound that would be formed between the given elements:
(a) Na and I
(b) Ba and F
(c) Al and O
(d) K and S
(e) Cs and Cl
(f) Sr and Br

8. Write the formula of the compound that would be formed between the given elements:
(a) Ba and O
(b) H and S
(c) Al and Cl
(d) Be and Br
(e) Li and Se
(f) Mg and P

9. Write formulas for the following cations (do not forget to include the charges): sodium, magnesium, aluminum, copper(II), iron(II), iron (III), lead(II), silver, cobalt(II), barium, hydrogen, mercury(II), tin(II), chromium(III), tin(IV), manganese(II), bismuth(III).

10. Write formulas for the following anions (do not forget to include the charges): chloride, bromide, fluoride, iodide, cyanide, oxide, hydroxide, sulfide, sulfate, hydrogen sulfate, hydrogen sulfite, chromate, carbonate, hydrogen carbonate, acetate, chlorate, permanganate, oxalate.

11. Complete the table, filling in each box with the proper formula.

Cations	Anions				
	Br^-	O^{2-}	NO_3^-	PO_4^{3-}	CO_3^{2-}
K^+	KBr				
Mg^{2+}					
Al^{3+}					
Zn^{2+}				$Zn_3(PO_4)_2$	
H^+					

12. Complete the table, filling in each box with the proper formula.

Cations	Anions				
	SO_4^{2-}	Cl^-	AsO_4^{3-}	$C_2H_3O_2^-$	CrO_4^{2-}
NH_4^+			$(NH_4)_3AsO_4$		
Ca^{2+}					
Fe^{3+}	$Fe_2(SO_4)_3$				
Ag^+					
Cu^{2+}					

13. Write formulas for the following binary compounds, all of which are composed of nonmetals:
(a) Carbon monoxide
(b) Sulfur trioxide
(c) Carbon tetrabromide
(d) Phosphorus trichloride
(e) Nitrogen dioxide
(f) Dinitrogen pentoxide
(g) Iodine monobromide
(h) Silicon tetrachloride
(i) Phosphorus pentaiodide
(j) Diboron trioxide

14. Name the following binary compounds, all of which are composed of nonmetals:
(a) CO_2
(b) N_2O
(c) PCl_5
(d) CCl_4
(e) SO_2
(f) N_2O_4
(g) P_2O_5
(h) OF_2
(i) NF_3
(j) CS_2

15. Write formulas for the following compounds:
(a) Sodium nitrate
(b) Magnesium fluoride
(c) Barium hydroxide
(d) Ammonium sulfate
(e) Silver carbonate
(f) Calcium phosphate
(g) Potassium nitrite
(h) Strontium oxide

16. Name the following compounds:
(a) K_2O
(b) NH_4Br
(c) CaI_2
(d) $BaCO_3$
(e) Na_3PO_4
(f) Al_2O_3
(g) $Zn(NO_3)_2$
(h) Ag_2SO_4

17. Name each of the following compounds by the Stock (IUPAC) System:
(a) $CuCl_2$
(b) $CuBr$
(c) $Fe(NO_3)_2$
(d) $FeCl_3$
(e) SnF_2
(f) $HgCO_3$

18. Write formulas for the following compounds:
(a) Tin(IV) bromide
(b) Copper(I) sulfate
(c) Iron(III) carbonate
(d) Mercury(II) nitrite
(e) Titanium(IV) sulfide
(f) Iron(II) acetate

19. Write formulas for the following acids:
    (a) Hydrochloric acid   (d) Carbonic acid
    (b) Chloric acid   (e) Sulfurous acid
    (c) Nitric acid   (f) Phosphoric acid

20. Write formulas for the following acids:
    (a) Acetic acid   (d) Boric acid
    (b) Hydrofluoric acid   (e) Nitrous acid
    (c) Hypochlorous acid   (f) Hydrosulfuric acid

21. Name the following acids:
    (a) HNO_2 (d) HBr (g) HF
    (b) H_2SO_4 (e) H_3PO_3 (h) $HBrO_3$
    (c) $H_2C_2O_4$ (f) $HC_2H_3O_2$

22. Name the following acids:
    (a) H_3PO_4 (d) HCl (g) HI
    (b) H_2CO_3 (e) $HClO$ (h) $HClO_4$
    (c) HIO_3 (f) HNO_3

23. Write formulas for the following compounds:
    (a) Silver sulfite
    (b) Cobalt(II) bromide
    (c) Tin(II) hydroxide
    (d) Aluminum sulfate
    (e) Manganese(II) fluoride
    (f) Ammonium carbonate
    (g) Chromium(III) oxide
    (h) Copper(II) chloride
    (i) Potassium permanganate
    (j) Barium nitrite
    (k) Sodium peroxide
    (l) Iron(II) sulfate
    (m) Potassium dichromate
    (n) Bismuth(III) chromate

24. Write formulas for the following compounds:
    (a) Sodium chromate
    (b) Magnesium hydride
    (c) Nickel(II) acetate
    (d) Calcium chlorate
    (e) Lead(II) nitrate
    (f) Potassium dihydrogen phosphate
    (g) Manganese(II) hydroxide
    (h) Cobalt(II) hydrogen carbonate
    (i) Sodium hypochlorite
    (j) Arsenic(V) carbonate
    (k) Chromium(III) sulfite
    (l) Antimony(III) sulfate
    (m) Sodium oxalate
    (n) Potassium thiocyanate

25. Write the name of each compound.
    (a) $ZnSO_4$
    (b) $HgCl_2$
    (c) $CuCO_3$
    (d) $Cd(NO_3)_2$
    (e) $Al(C_2H_3O_2)_3$
    (f) CoF_2
    (g) $Cr(ClO_3)_3$
    (h) Ag_3PO_4
    (i) NiS
    (j) $BaCrO_4$

26. Write the name of each compound.
    (a) $Ca(HSO_4)_2$
    (b) $As_2(SO_3)_3$
    (c) $Sn(NO_2)_2$
    (d) $FeBr_3$
    (e) $KHCO_3$
    (f) $BiAsO_4$
    (g) $Fe(BrO_3)_2$
    (h) $(NH_4)_2HPO_4$
    (i) $NaClO$
    (j) $KMnO_4$

27. Write the chemical formula for each of the following substances:
    (a) Baking soda
    (b) Lime
    (c) Epsom salts
    (d) Muriatic acid
    (e) Vinegar
    (f) Potash
    (g) Lye

28. Write the chemical formula for each of the following substances:
    (a) Fool's gold
    (b) Saltpeter
    (c) Limestone
    (d) Cane sugar
    (e) Milk of magnesia
    (f) Washing soda
    (g) Grain alcohol

Additional Exercises

These exercises are not paired or labeled by topic and provide additional practice on the concepts covered in this chapter.

29. Name the following compounds:
- **(a)** $Ba(NO_3)_2$
- **(d)** $MgSO_4$
- **(g)** NiS
- **(b)** $NaC_2H_3O_2$
- **(e)** $CdCrO_4$
- **(h)** $Sn(NO_3)_4$
- **(c)** PbI_2
- **(f)** $BiCl_3$
- **(i)** $Ca(OH)_2$

30. State how each of the following is used in naming inorganic compounds: *ide, ous, ic, hypo, per, ite, ate,* Roman numerals.

31. Translate each of the following formula sentences into unbalanced chemical equations:
- **(a)** Silver nitrate and sodium chloride react to form silver chloride and sodium nitrate.
- **(b)** Iron(III) sulfate and calcium hydroxide react to form iron(III) hydroxide and calcium sulfate.
- **(c)** Potassium hydroxide and sulfuric acid react to form potassium sulfate and water.

32. How many of each type of subatomic particle (protons and electrons) is in:
- **(a)** an atom of tin
- **(b)** an Sn^{2+} ion
- **(c)** an Sn^{4+} ion

33. The compound X_2Y_3 is a stable solid. What ionic charge do you expect for X and Y? Explain.

*34. The ferricyanide ion has the formula $Fe(CN)_6^{3-}$. Write the formula for the compounds that ferricyanide would form with the cations of elements 3, 13, and 30.

35. Compare and contrast the formulas of:
- **(a)** nitride with nitrite
- **(b)** nitrite with nitrate
- **(c)** nitrous acid with nitric acid

36. In the beaker below there is a pool of ions. Write all possible formulas for ionic compounds that could form using these ions. Write the name of each compound next to its formula.

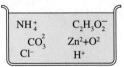

Answers to Practice Exercises

6.1 (a) KF, (b) $CaBr_2$, (c) Mg_3N_2, (d) Na_2S, (e) BaO

6.2 (a) $SrCl_2$, (b) KI, (c) AlN, (d) CaS, (e) Na_2O

6.3 (a) lead(II) iodide, (b) tin(IV) fluoride, (c) iron(III) oxide, (d) copper(II) oxide

6.4 (a) dichlorine monoxide, (b) sulfur dioxide, (c) carbon tetrabromide, (d) dinitrogen pentoxide, (e) ammonia

6.5 (a) potassium bromide, (b) calcium nitride, (c) sulfur trioxide, (d) tin(II) fluoride, (e) copper(II) chloride, (f) dinitrogen tetraoxide

6.6 (a) sodium nitrate, (b) calcium phosphate, (c) potassium hydroxide, (d) lithium carbonate, (e) sodium chlorate

7

Knowing the substances contained in a product is not sufficient information to produce the product. An artist can create an incredible array of colors from a limited number of pigments. A pharmacist can combine the same drugs in various amounts to produce different effects in patients. Cosmetics, cereals, cleaning products, and pain remedies all show a list of ingredients on the labels of the packages.

In each of these products the key to successful production lies in the quantity of each ingredient. The pharmaceutical industry maintains strict regulations on the amounts of the ingredients in the medicines we purchase. The formulas for soft drinks and most cosmetics are trade secrets. Small deviations in the composition of these products can result in large losses or lawsuits for these organizations.

The composition of compounds is an important concept in chemistry. The numerical relationships among elements within compounds and the measurement of exact quantities of particles are fundamental tasks for the chemist.

7.1 The Mole

In the laboratory we normally determine the mass of substances on a balance. When we run a chemical reaction, the reaction occurs between atoms and molecules. For example, in the reaction between magnesium and sulfur, one atom of sulfur reacts with one atom of magnesium:

$$Mg + S \longrightarrow MgS$$

However, when we measure the masses of these two elements, we find that 24.31 g of Mg are required to react with 32.06 g of S. Because magnesium and sulfur react in a 1:1 atom ratio, we can conclude from this experiment that 24.31 g of Mg contain the same number of atoms as 32.06 g of S. How many atoms are in 24.31 g of Mg or 32.06 g of S? These two amounts each contain 1 *mole* of atoms.

The mole (abbreviated mol) is one of the seven base units in the International System and is the unit for an amount of substance. The mole is a counting unit as in other things that we count, such as a dozen (12) eggs or a gross (144) of pencils. The mole is sometimes called the chemist's dozen. But a mole is a much larger number of things, namely 6.022×10^{23}. Thus 1 mole contains 6.022×10^{23} entities of anything. In reference to our reaction between magnesium and sulfur, 1 mol Mg (24.31 g) contains 6.022×10^{23} atoms of magnesium and 1 mol S (32.06 g) contains 6.022×10^{23} atoms of sulfur.

The number 6.022×10^{23} is known as **Avogadro's number** in honor of Amedeo Avogadro (1776–1856), an Italian physicist. Avogadro's number is an important constant in chemistry and physics and has been experimentally determined by several independent methods.

Avogadro's number

Avogadro's number = 6.022×10^{23}

◀ Chapter Opening Photo: For these chefs, preparing gourmet meals is an exercise in combining precise quantities of various ingredients to produce a delightful experience for their patrons.

It is difficult to imagine how large Avogadro's number really is, but the following analogy will help to express it: If 10,000 people started to count Avogadro's number and each counted at the rate of 100 numbers per minute each minute of the day, it would take them over 1 trillion (10^{12}) years to count the total number. So, even the most minute amount of matter contains extremely large numbers of atoms. For example, 1 mg (0.001 g) of sulfur contains 2×10^{19} atoms of sulfur.

Avogadro's number is the basis for the amount of substance used to express a particular number of chemical species, such as atoms, molecules, formula units, ions, or electrons. This amount of substance is the mole. We define a **mole** as an amount of a substance containing the same number of formula units as there are atoms in exactly 12 g of carbon-12. (Recall that carbon-12 is the reference isotope for atomic masses.) Other definitions of the mole are used, but they all relate to Avogadro's number of formula units of a substance. A **formula unit** is the atom or molecule indicated by the formula of the substance under consideration—for example, Mg, MgS, H_2O, O_2.

From the definition, we can say that the atomic mass in grams of any element contains 1 mol of atoms. The term *mole* is so commonplace in chemical jargon that chemists use it as freely as the words *atom* and *molecule*. The mole is used in conjunction with many different particles, such as atoms, molecules, ions, and electrons, to represent Avogadro's number of these particles. If we can speak of a mole of atoms, we can also speak of a mole of molecules, a mole of electrons, a mole of ions, understanding that in each case we mean 6.022×10^{23} formula units of these particles.

> **1 mol of atoms = 6.022×10^{23} atoms**
> **1 mol of molecules = 6.022×10^{23} molecules**
> **1 mol of ions = 6.022×10^{23} ions**

The atomic mass, in grams of an element, contains Avogadro's number of atoms and is defined as the **molar mass** of that element. To determine the molar mass of an element, change the units of the atomic mass found in the periodic table from atomic mass units to grams. For example, sulfur has an atomic mass of 32.06 amu. One molar mass of sulfur has a mass of 32.06 grams and contains 6.022×10^{23} atoms of sulfur. See the following table.

Element	Atomic mass	Molar mass	Number of atoms
H	1.008 amu	1.008 g	6.022×10^{23}
Mg	24.31 amu	24.31 g	6.022×10^{23}
Na	22.99 amu	22.99 g	6.022×10^{23}

> **1 molar mass (g) = 1 mol of atoms**
> **= Avogadro's number (6.022×10^{23}) of atoms**

Amedeo Avogadro (1776–1856).

In chemistry, we frequently encounter problems that require conversions involving quantities of mass (g), moles, and the number of atoms of an element. From the boxed equivalence statement above, many different conversion factors can be written. The key to writing a conversion factor is to remember that both sides of an equivalence statement represent equal amounts. This means that, when one of these amounts is placed in the numerator and another is placed in the denominator of a conversion factor, the resulting ratio has a value equal to one. To determine which amount should be placed in the numerator, consider the units that are desired in the result. The denominator should contain the units you are trying to eliminate. Study the following examples.

How many moles of iron does 25.0 g of iron, Fe, represent?   **Example 7.1**

The problem requires that we change grams of Fe to moles of Fe. We look up the atomic mass of Fe in the table of atomic masses on the periodic table or table of atomic masses and find it to be 55.85. Then we use the proper conversion factor to obtain moles: **Solution**

$$\text{grams Fe} \longrightarrow \text{moles Fe} \qquad \text{grams Fe} \times \frac{1 \text{ mol Fe}}{1 \text{ molar mass Fe}}$$

$$25.0 \ \cancel{\text{g Fe}} \times \frac{1 \text{ mol Fe}}{55.85 \ \cancel{\text{g Fe}}} = 0.448 \text{ mol Fe}$$

How many magnesium, Mg, atoms are contained in 5.00 g of Mg?   **Example 7.2**

The problem requires that we change grams of Mg to atoms of Mg.   **Solution**

$$\text{grams Mg} \longrightarrow \text{atoms Mg}$$

We find the atomic mass of magnesium to be 24.31 and set up the calculation using a conversion factor between atoms and molar mass:

$$\text{grams Mg} \times \frac{6.022 \times 10^{23} \text{ atoms Mg}}{1 \text{ molar mass Mg}}$$

$$5.00 \ \cancel{\text{g Mg}} \times \frac{6.022 \times 10^{23} \text{ atoms Mg}}{24.31 \ \cancel{\text{g Mg}}} = 1.24 \times 10^{23} \text{ atoms Mg}$$

An alternative solution is first to convert grams of Mg to moles of Mg, which are then changed to atoms of Mg.

$$\text{grams Mg} \longrightarrow \text{moles Mg} \longrightarrow \text{atoms Mg}$$

Use conversion factors for each step. The calculation setup is

$$5.00 \ \cancel{\text{g Mg}} \times \frac{1 \ \cancel{\text{mol Mg}}}{24.31 \ \cancel{\text{g Mg}}} \times \frac{6.022 \times 10^{23} \text{ atoms Mg}}{1 \ \cancel{\text{mol Mg}}} = 1.24 \times 10^{23} \text{ atoms Mg}$$

Thus 1.24×10^{23} atoms of Mg are contained in 5.00 g of Mg.

Example 7.3 What is the mass, in grams, of one atom of carbon, C?

Solution The molar mass of C is 12.01 g. Create a conversion factor between molar mass and atoms.

$$\text{atoms C} \longrightarrow \text{grams C}$$

$$\text{atoms C} \times \frac{1 \text{ molar mass}}{6.022 \times 10^{23} \text{ atoms C}}$$

$$1 \text{ atom C} \times \frac{12.01 \text{ g C}}{6.022 \times 10^{23} \text{ atoms C}} = 1.994 \times 10^{-23} \text{ g C}$$

Example 7.4 What is the mass of 3.01×10^{23} atoms of sodium (Na)?

Solution The information needed to solve this problem is the molar mass of Na (22.99 g) and a conversion factor between molar mass and atoms.

$$\text{atoms Na} \longrightarrow \text{grams Na}$$

$$\text{atoms Na} \times \frac{1 \text{ molar mass Na}}{6.022 \times 10^{23} \text{ atoms Na}}$$

$$3.01 \times 10^{23} \text{ atoms Na} \times \frac{22.99 \text{ g Na}}{6.022 \times 10^{23} \text{ atoms Na}} = 11.5 \text{ g Na}$$

Example 7.5 What is the mass of 0.252 mol of copper (Cu)?

Solution The information needed to solve this problem is the molar mass of Cu (63.55 g) and a conversion factor between molar mass and moles.

$$\text{moles Cu} \longrightarrow \text{grams Cu}$$

$$\text{moles Cu} \times \frac{1 \text{ molar mass Cu}}{1 \text{ mol Cu}}$$

$$0.252 \text{ mol Cu} \times \frac{63.55 \text{ g Cu}}{1 \text{ mol Cu}} = 16.0 \text{ g Cu}$$

Example 7.6 How many oxygen atoms are present in 1.00 mol of oxygen molecules?

Solution Oxygen is a diatomic molecule with the formula O_2. Therefore a molecule of oxygen contains two atoms of oxygen.

$$\frac{2 \text{ atoms O}}{1 \text{ molecule O}_2}$$

The sequence of conversions is

$$\text{moles O}_2 \longrightarrow \text{molecules O}_2 \longrightarrow \text{atoms O}$$

Two conversion factors are needed; they are

$$\frac{6.022 \times 10^{23} \text{ molecules O}_2}{1 \text{ mol O}_2} \quad \text{and} \quad \frac{2 \text{ atoms O}}{1 \text{ molecule O}_2}$$

The calculation is

$$1.00\ \cancel{\text{mol O}_2} \times \frac{6.022 \times 10^{23}\ \cancel{\text{molecules O}_2}}{1\ \cancel{\text{mol O}_2}} \times \frac{2\ \text{atoms O}}{1\ \cancel{\text{molecule O}_2}} = 1.20 \times 10^{24}\ \text{atoms O}$$

Practice 7.1

What is the mass of 2.50 mol of helium (He)?

Practice 7.2

How many atoms are present in 0.025 mol of iron?

7.2 Molar Mass of Compounds

One mole of a compound contains Avogadro's number of formula units of that compound. The terms *molecular weight, molecular mass, formula weight,* and *formula mass* have been used in the past to refer to the mass of one mole of a compound. However, the term *molar mass* is more inclusive, because it can be used for all types of compounds.

If the formula of a compound is known, its molar mass may be determined by adding the molar masses of all the atoms in the formula. If more than one atom of any element is present, its mass must be added as many times as it is used.

Example 7.7

The formula for water is H_2O. What is its molar mass?

Solution

Proceed by looking up the molar masses of H (1.008 g) and O (16.00 g) and adding the masses of all the atoms in the formula unit. Water contains two atoms of H and one atom of O. Thus,

$$\begin{aligned} 2\ \text{H} &= 2 \times 1.008\ \text{g} = 2.016\ \text{g} \\ 1\ \text{O} &= 1 \times 16.00\ \text{g} = \underline{16.00\ \text{g}} \\ & \qquad\qquad\qquad 18.02\ \text{g} = \text{molar mass} \end{aligned}$$

Example 7.8

Calculate the molar mass of calcium hydroxide, $Ca(OH)_2$.

Solution

The formula of this substance contains one atom of Ca and two atoms each of O and H. Proceed as in Example 7.7. Thus,

$$\begin{aligned} 1\ \text{Ca} &= 1 \times 40.08\ \text{g} = 40.08\ \text{g} \\ 2\ \text{O} &= 2 \times 16.00\ \text{g} = 32.00\ \text{g} \\ 2\ \text{H} &= 2 \times 1.008\ \text{g} = \underline{2.016\ \text{g}} \\ & \qquad\qquad\qquad 74.10\ \text{g} = \text{molar mass} \end{aligned}$$

Each sample contains one mole. *Clockwise from lower left:* magnesium, carbon, copper(II) sulfate, copper, mercury, potassium permanganate, cadmium, and sodium chloride *(center).*

Practice 7.3

Calculate the molar mass of KNO_3.

Remember in this text we round all molar masses to four significant figures, although you may need to use a different number of significant figures for other work (perhaps in the lab).

The mass of 1 mol of a compound contains Avogadro's number of formula units or molecules. Now consider the compound hydrogen chloride (HCl). One atom of H combines with one atom of Cl to form one molecule of HCl. When 1 mol of H (1.008 g of H representing 1 mol or 6.022×10^{23} H atoms) combines with 1 mol of Cl (35.45 g of Cl representing 1 mol or 6.022×10^{23} Cl atoms), 1 mol of HCl (36.46 g of HCl representing 1 mol or 6.022×10^{23} HCl molecules) is produced. These relationships are summarized in the following table:

H	Cl	HCl
6.022×10^{23} H *atoms*	6.022×10^{23} Cl *atoms*	6.022×10^{23} HCl *molecules*
1 mol H *atoms*	1 mol Cl *atoms*	1 mol HCl *molecules*
1.008 g H	35.45 g Cl	36.46 g HCl
1 molar mass H	1 molar mass Cl	1 molar mass HCl

In dealing with diatomic elements (H_2, O_2, N_2, F_2, Cl_2, Br_2, and I_2), we must take special care to distinguish between a mole of atoms and a mole of molecules.

For example consider 1 mol of oxygen molecules, which has a mass of 32.00 g. This quantity is equal to 2 mol of oxygen atoms. The key concept is that 1 mol represents Avogadro's number of the particular chemical entity—atoms, molecules, formula units, and so forth—that is under consideration.

$$1 \text{ mol } H_2O = 18.02 \text{ g } H_2O \quad = 6.022 \times 10^{23} \text{ molecules}$$

$$1 \text{ mol NaCl} = 58.44 \text{ g NaCl} \quad = 6.022 \times 10^{23} \text{ formula units}$$

$$1 \text{ mol } H_2 = 2.016 \text{ g } H_2 \quad = 6.022 \times 10^{23} \text{ molecules}$$

$$1 \text{ mol } HNO_3 = 63.02 \text{ g } HNO_3 \quad = 6.022 \times 10^{23} \text{ molecules}$$

$$1 \text{ mol } K_2SO_4 = 174.3 \text{ g } K_2SO_4 = 6.022 \times 10^{23} \text{ formula units}$$

> **1 mol = 6.022 × 10²³ formula units or molecules**
> **= 1 molar mass of a compound**

Remember to create the appropriate conversion factor by using the equivalence statement. Place the unit desired in the numerator and the unit to be eliminated in the denominator.

What is the mass of 1 mol of sulfuric acid (H_2SO_4)?

Example 7.9

One mole of H_2SO_4 is 1 molar mass of H_2SO_4. The problem, therefore, is solved in a similar manner to Examples 7.7 and 7.8. Look up the molar masses of hydrogen, sulfur, and oxygen, and solve.

Solution

$$2\,H = 2 \times 1.008 \text{ g} = 2.016 \text{ g}$$
$$1\,S = 1 \times 32.06 \text{ g} = 32.06 \text{ g}$$
$$4\,O = 4 \times 16.00 \text{ g} = \underline{64.00 \text{ g}}$$
$$98.08 \text{ g} = \text{mass of 1 mol of } H_2SO_4$$

How many moles of sodium hydroxide, NaOH, are there in 1.00 kg of sodium hydroxide?

Example 7.10

First we know that

Solution

$$1 \text{ molar mass} = (22.99 \text{ g} + 16.00 \text{ g} + 1.008 \text{ g}) \text{ or } 40.00 \text{ g NaOH}$$

To convert grams to moles we use the conversion factor

$$\frac{1 \text{ mol}}{1 \text{ molar mass}} \quad \text{or} \quad \frac{1 \text{ mol NaOH}}{40.00 \text{ g NaOH}}$$

Use this conversion sequence:

$$\text{kg NaOH} \longrightarrow \text{g NaOH} \longrightarrow \text{mol NaOH}$$

The calculation is

$$1.00 \text{ kg NaOH} \times \frac{1000 \text{ g NaOH}}{\text{kg NaOH}} \times \frac{1 \text{ mol NaOH}}{40.00 \text{ g NaOH}} = 25.0 \text{ mol NaOH}$$

$$1.00 \text{ kg NaOH} = 25.0 \text{ mol NaOH}$$

Example 7.11   What is the mass of 5.00 mol of water?

Solution   First we know that

$$1 \text{ mol } H_2O = 18.02 \text{ g (Example 7.7)}$$

The conversion is

$$\text{mol } H_2O \longrightarrow g \ H_2O$$

To convert moles to grams use the conversion factor

$$\frac{1 \text{ molar mass } H_2O}{1 \text{ mol } H_2O} \quad \text{or} \quad \frac{18.02 \text{ g } H_2O}{1 \text{ mol } H_2O}$$

The calculation is

$$5.00 \ \cancel{\text{mol } H_2O} \times \frac{18.02 \text{ g } H_2O}{1 \ \cancel{\text{mol } H_2O}} = 90.1 \text{ g } H_2O$$

Example 7.12   How many molecules of hydrogen chloride, HCl, are there in 25.0 g of hydrogen chloride?

Solution   From the formula, we find that the molar mass of HCl is 36.46 g (1.008 g + 35.45 g). The sequence of conversions is

$$\text{g HCl} \longrightarrow \text{mol HCl} \longrightarrow \text{molecules HCl}$$

Using the conversion factors

$$\frac{1 \text{ mol HCl}}{36.46 \text{ g HCl}} \quad \text{and} \quad \frac{6.022 \times 10^{23} \text{ molecules HCl}}{1 \text{ mol HCl}}$$

$$25.0 \ \cancel{\text{g HCl}} \times \frac{1 \ \cancel{\text{mol HCl}}}{36.46 \ \cancel{\text{g HCl}}} \times \frac{6.022 \times 10^{23} \text{ molecules HCl}}{1 \ \cancel{\text{mol HCl}}} =$$

$$4.13 \times 10^{23} \text{ molecules HCl}$$

Practice 7.4

What is the mass of 0.150 mol of Na_2SO_4?

Practice 7.5

How many moles are there in 500.0 g of $AlCl_3$?

7.3 Percent Composition of Compounds

percent composition of a compound

Percent means parts per 100 parts. Just as each piece of pie is a percent of the whole pie, each element in a compound is a percent of the whole compound. The **percent composition of a compound** is the *mass percent* of each element in the compound. The molar mass represents the total mass, or 100%, of the compound. Thus, the percent composition of water, H_2O, is 11.19% H and 88.79% O by mass. According to the Law of Definite Composition, the percent composition must be the same no matter what size sample is taken.

The percent composition of a compound can be determined (1) from knowing its formula or (2) from experimental data.

Percent Composition from Formula

If the formula is known, it is essentially a two-step process to determine the percent composition.

Step 1 Calculate the molar mass (Section 7.2).

Step 2 Divide the total mass of each element in the formula by the molar mass and multiply by 100. This gives the percent composition.

$$\frac{\text{total mass of the element}}{\text{molar mass}} \times 100 = \text{percent of the element}$$

Calculate the percent composition of sodium chloride, NaCl.

Example 7.13

Solution

Step 1 Calculate the molar mass of NaCl:

$1 \, Na = 1 \times 22.99 \, g = 22.99 \, g$

$1 \, Cl = 1 \times 35.45 \, g = \underline{35.45 \, g}$

$58.44 \, g$ (molar mass)

Step 2 Now calculate the percent composition. We know there are 22.99 g Na and 35.45 g Cl in 58.44 g NaCl.

Na: $\frac{22.99 \, g \, Na}{58.44 \, g \, NaCl} \times 100 = 39.34\% \, Na$

Cl: $\frac{35.45 \, g \, Cl}{58.44 \, g \, NaCl} \times 100 = \underline{60.66\% \, Cl}$

100.00% total

In any two-component system, if the percent of one component is known, the other is automatically defined by the difference; that is, if Na is 39.34%, then Cl is $100\% - 39.34\% = 60.66\%$. However, the calculation of the percent of each component should be carried out, since this provides a check against possible error. The percent composition data should add up to $100 \pm 0.2\%$.

Calculate the percent composition of potassium chloride, KCl.

Example 7.14

Solution

Step 1 Calculate the molar mass of KCl:

$1 \, K = 1 \times 39.10 \, g = 39.10 \, g$

$1 \, Cl = 1 \times 35.45 \, g = \underline{35.45 \, g}$

$74.55 \, g$ (molar mass)

Step 2 Now calculate the percent composition. We know there are 39.10 g K and 35.45 g Cl in 74.55 g KCl.

$$\text{K:} \quad \frac{39.10 \text{ g K}}{74.55 \text{ g KCl}} \times 100 = 52.45\% \text{ K}$$

$$\text{Cl:} \quad \frac{35.45 \text{ g Cl}}{74.55 \text{ g KCl}} \times 100 = \underline{47.55\% \text{ Cl}}$$
$$100.00\% \text{ total}$$

Comparing the data calculated for NaCl and for KCl, we see that NaCl contains a higher percentage of Cl by mass, although each compound has a one-to-one atom ratio of Cl to Na and Cl to K. The reason for this mass percent difference is that Na and K do not have the same atomic masses.

It is important to realize that, when we compare 1 mol of NaCl with 1 mol of KCl, each quantity contains the same number of Cl atoms—namely, 1 mol of Cl atoms. However, if we compare equal masses of NaCl and KCl, there will be more Cl atoms in the mass of NaCl since NaCl has a higher mass percent of Cl.

1 mol NaCl contains	100.00 g NaCl contains
1 mol Na	39.34 g Na
1 mol Cl	60.66 g Cl
60.66%Cl	

1 mol KCl contains	100.00 g KCl contains
1 mol K	52.45 g K
1 mol Cl	47.55 g Cl
47.55%Cl	

Example 7.15   Calculate the percent composition of potassium sulfate, K_2SO_4.

Solution   **Step 1** Calculate the molar mass of K_2SO_4:

$$2 \text{ K} = 2 \times 39.10 \text{ g} = 78.20 \text{ g}$$
$$1 \text{ S} = 1 \times 32.06 \text{ g} = 32.06 \text{ g}$$
$$4 \text{ O} = 4 \times 16.00 \text{ g} = \underline{64.00 \text{ g}}$$
$$174.3 \text{ g (molar mass)}$$

Step 2 Now calculate the percent composition. We know there are 78.20 g of K, 32.06 g of S, and 64.00 g of O in 174.3 g of K_2SO_4.

$$\text{K:} \quad \frac{78.20 \text{ g K}}{174.3 \text{ g K}_2\text{SO}_4} \times 100 = 44.87\% \text{ K}$$

$$\text{S:} \quad \frac{32.06 \text{ g S}}{174.3 \text{ g K}_2\text{SO}_4} \times 100 = 18.39\% \text{ S}$$

$$\text{O:} \quad \frac{64.00 \text{ g O}}{174.3 \text{ g K}_2\text{SO}_4} \times 100 = \underline{36.72\% \text{ O}}$$
$$99.98\% \text{ total}$$

The Taste of Chemistry

Flavorings, seasonings, and preservatives have been added to foods since ancient civilizations. Spices were originally added as preservatives when refrigeration was unavailable. These spices contained mild antiseptics and antioxidants and were effective in prolonging the time during which food could be eaten. Over the course of time a great variety of substances (food additives) came to be used in foods—preservatives, colorings, flavorings, antioxidants, sweeteners, and so on. Many of these substances are now regarded as virtual necessities for processing foods. However, there is genuine concern that some additives, particularly those that are not naturally present in foods, may be detrimental to one's health when consumed. This concern has led to the passage of federal and state laws that regulate the food industry.

The United States Food and Drug Administration (FDA) is the principal agency charged with enforcing federal laws concerning food and must approve the use of all food additives. The FDA divides food additives into several categories. These include a classification known as GRAS (Generally Regarded As Safe). The GRAS designation includes substances that were in use in 1958 and that met certain specifications for safety. All substances introduced after 1958 have been approved on an individual basis.

Before commercial use of a new food additive, a company must provide the FDA with satisfactory evidence that the chemical is safe for the proposed usage. Providing this evidence of safety is complex and expensive. Although mainly concerned with chemistry, this task requires the services of many people trained in a variety of disciplines—biochemistry, microbiology, medicine, physiology, and so on. Research and testing may have to be done in several laboratories before final FDA approval (or rejection) is obtained.

The amount of an additive that is considered to be safe in foods is determined by the maximum tolerable daily intake

These are a few foods that contain additives.

(MTDI). This is the amount of the food additive that can be eaten daily for a lifetime without adverse effects. It is calculated on the basis of body mass (mg/kg/day). If there is any doubt regarding the safety of an additive, a time limit is determined and a conditional MTDI is issued and reviewed with further testing at the end of the initial time period. To establish the MTDI, experiments are run on animals, increasing the quantities of the additive in successive experiments until acute and chronic toxicity occurs. A minimum of two species of animals must be tested, the most sensitive species forming the basis for the appropriate level for the additive. This quantity is then divided by 100 (for additional safety) to set the MTDI.

Once the MTDI is established, all foods in which the use of the additive is proposed must be considered. An estimate is made of the maximum amount of the additive that could be ingested. On the basis of this information the use may then be restricted to only certain foods (as in the exclusion of sulfites from most meats) or broadened to include additional foods.

Unfortunately, even with all the quantitative testing it is extremely difficult to determine safe levels of additives since the majority of toxicological data is from animal testing. The results are often not the same as in humans—chemicals toxic to one species may not be so for humans.

Children add a further complication. They cannot be considered simply as small adults. A child has a much greater energy demand per kilogram than an adult. Often children's chemical defense mechanisms are much different than those of an adult.

Flavorings are the largest class of food additives. There are 1100–1400 natural or synthetic flavorings used in foods. The task of checking for effects of these additives is herculean. Flavorings are complex mixtures of chemicals used in very small amounts making them difficult to analyze. Distinctions between artificial and natural flavorings must also be monitored (see Chemistry in Action, page 94). Chemists rely on quantitative analysis techniques (such as gas–liquid chromatography) to separate and identify components in these mixtures. It is possible to detect compounds in amounts as low as 10 μg/kg (parts per billion).

Practice 7.6

Calculate the percent composition of $Ca(NO_3)_2$.

Practice 7.7

Calculate the percent composition of K_2CrO_4.

Percent Composition from Experimental Data

The percent composition can be determined from experimental data without knowing the formula of a compound.

Step 1 Calculate the mass of the compound formed.

Step 2 Divide the mass of each element by the total mass of the compound and multiply by 100.

Example 7.16   When heated in the air, 1.63 g of zinc, Zn, combine with 0.40 g of oxygen, O_2, to form zinc oxide. Calculate the percent composition of the compound formed.

Solution   **Step 1** First, calculate the total mass of the compound formed.

$$\begin{array}{l} 1.63 \text{ g Zn} \\ \underline{0.40 \text{ g } O_2} \\ 2.03 \text{ g} \end{array} = \text{total mass of product}$$

Step 2 Divide the mass of each element by the total mass (2.03 g) and multiply by 100.

$$\frac{1.63 \cancel{g}}{2.03 \cancel{g}} \times 100 = 80.3\% \text{ Zn} \qquad \frac{0.40 \cancel{g}}{2.03 \cancel{g}} \times 100 = \frac{19.7\% \text{ O}}{100.0\% \text{ total}}$$

The compound formed contains 80.3% Zn and 19.7% O.

Practice 7.8

Aluminum chloride is formed by reacting 13.43 g aluminum with 53.18 g chlorine. What is the percent composition of the compound?

7.4   ## Empirical Formula Versus Molecular Formula

empirical formula   The **empirical formula**, or *simplest formula,* gives the smallest whole-number ratio of the atoms that are present in a compound. This formula gives the relative number of atoms of each element in the compound.

The **molecular formula** is the true formula, representing the total number of atoms of each element present in one molecule of a compound. It is entirely possible that two or more substances will have the same percent composition yet be distinctly different compounds. For example, acetylene, C_2H_2, is a common gas used in welding; benzene, C_6H_6, is an important solvent obtained from coal tar and is used in the synthesis of styrene and nylon. Both acetylene and benzene contain 92.3% C and 7.7% H. The smallest ratio of C and H corresponding to these percentages is CH (1:1). Therefore the empirical formula for both acetylene and benzene is CH, even though it is known that the molecular formulas are C_2H_2 and C_6H_6, respectively. Often the molecular formula is the same as the empirical formula. If the molecular formula is not the same, it will be an integral (whole number) multiple of the empirical formula. For example,

molecular formula

CH = empirical formula

$(CH)_2 = C_2H_2$ = acetylene   (molecular formula)

$(CH)_6 = C_6H_6$ = benzene   (molecular formula)

Table 7.1 summarizes the data concerning these CH formulas. Table 7.2 shows empirical and molecular formula relationships of other compounds.

TABLE 7.1 Molecular Formulas of Two Compounds Having an Empirical Formula with a 1:1 Ratio of Carbon and Hydrogen Atoms

Formula	Composition		Molar mass
	% C	% H	
CH (empirical)	92.3	7.7	13.02 (empirical)
C_2H_2 (acetylene)	92.3	7.7	26.04 (2 $\times$ 13.02)
C_6H_6 (benzene)	92.3	7.7	78.12 (6 $\times$ 13.02)

TABLE 7.2 Some Empirical and Molecular Formulas

Compound	Empirical formula	Molecular formula	Compound	Empirical formula	Molecular formula
Acetylene	CH	C_2H_2	Diborane	BH_3	B_2H_6
Benzene	CH	C_6H_6	Hydrazine	NH_2	N_2H_4
Ethylene	CH_2	C_2H_4	Hydrogen	H	H_2
Formaldehyde	CH_2O	CH_2O	Chlorine	Cl	Cl_2
Acetic acid	CH_2O	$C_2H_4O_2$	Bromine	Br	Br_2
Glucose	CH_2O	$C_6H_{12}O_6$	Oxygen	O	O_2
Hydrogen chloride	HCl	HCl	Nitrogen	N	N_2
Carbon dioxide	CO_2	CO_2			

	Calculation of
7.5	Empirical Formulas

It is possible to establish an empirical formula because (1) the individual atoms in a compound are combined in whole-number ratios, and (2) each element has a specific atomic mass.

In order to calculate the empirical formula we need to know (1) the elements that are combined, (2) their atomic masses, and (3) the ratio by mass or percentage in which they are combined. If elements A and B form a compound, we may represent the empirical formula as A_xB_y, where x and y are small whole numbers that represent the atoms of A and B. To write the empirical formula we must determine x and y. Three or four steps are required to do this.

Step 1 Assume a definite starting quantity (usually 100.0 g) of the compound, if not given, and express the mass of each element in grams.

Step 2 Convert the grams of each element into moles using the molar mass of each element. This conversion gives the number of moles of atoms of each element in the quantity assumed in Step 1. At this point these numbers will usually not be whole numbers.

Step 3 Divide each of the values obtained in Step 2 by the smallest of these values. If the numbers obtained by this procedure are whole numbers, use them as subscripts in writing the empirical formula. If the numbers obtained are not whole numbers, go on to Step 4.

Step 4 Multiply the values obtained in Step 3 by the smallest number that will convert them to whole numbers. Use these whole numbers as the subscripts in the empirical formula. For example, if the ratio of A to B is 1.0:1.5, multiply both numbers by 2 to obtain a ratio of 2:3. The empirical formula then is A_2B_3.

> **Some common fractions and their decimal equivalents are**
> $\frac{1}{4} = 0.25$
> $\frac{1}{3} = 0.333\ldots$
> $\frac{2}{3} = 0.666\ldots$
> $\frac{1}{2} = 0.5$
> $\frac{3}{4} = 0.75$
> **Multiply the decimal equivalent by the number in the denominator of the fraction to give a whole number:** $4(0.75) = 3$.

In many of these calculations results will vary somewhat from an exact whole number; this can be due to experimental errors in obtaining the data or from rounding off numbers. Calculations that vary by no more than ± 0.1 from a whole number usually can be rounded off to the nearest whole number. Deviations greater than about 0.1 unit usually mean that the calculated ratios need to be multiplied by a factor to make them all whole numbers. For example, an atom ratio of 1:1.33 should be multiplied by 3 to make the ratio 3:4.

Example 7.17 Calculate the empirical formula of a compound containing 11.19% hydrogen, H, and 88.79% oxygen, O.

Solution **Step 1** Express each element in grams. If we assume that there are 100.00 g of material, then the percent of each element is equal to the grams of each element in 100.00 g, and the percent sign can be omitted:

H = 11.19 g

O = 88.79 g

Step 2 Convert the grams of each element to moles:

$$\text{H:} \quad 11.19 \ \cancel{g \ H} \times \frac{1 \ \text{mol H atoms}}{1.008 \ \cancel{g \ H}} = 11.10 \ \text{mol H atoms}$$

$$\text{O:} \quad 88.79 \ \cancel{g \ O} \times \frac{1 \ \text{mol O atoms}}{16.00 \ \cancel{g \ O}} = 5.549 \ \text{mol O atoms}$$

The formula could be expressed as $H_{11.10}O_{5.549}$. However, it is customary to use the smallest whole-number ratio of atoms. This ratio is calculated in Step 3.

Step 3 Change these numbers to whole numbers by dividing each of them by the smaller number.

$$\text{H} = \frac{11.10 \ \text{mol}}{5.549 \ \text{mol}} = 2.000 \qquad \text{O} = \frac{5.549 \ \text{mol}}{5.549 \ \text{mol}} = 1.000$$

In this step, the ratio of atoms has not changed, because we divided the number of moles of each element by the same number.

The simplest ratio of H to O is 2:1.
Empirical formula $= H_2O$

The analysis of a salt shows that it contains 56.58% potassium, K, 8.68% carbon, C, and 34.73% oxygen, O. Calculate the empirical formula for this substance.

Example 7.18

Solution

Steps 1 and 2 After changing the percentage of each element to grams, find the relative number of moles of each element by multiplying by the proper mole/molar mass factor:

$$\text{K:} \quad 56.58 \ \cancel{g \ K} \times \frac{1 \ \text{mol K atoms}}{39.10 \ \cancel{g \ K}} = 1.447 \ \text{mol K atoms}$$

$$\text{C:} \quad 8.68 \ \cancel{g \ C} \times \frac{1 \ \text{mol C atoms}}{12.01 \ \cancel{g \ C}} = 0.723 \ \text{mol C atoms}$$

$$\text{O:} \quad 34.73 \ \cancel{g \ O} \times \frac{1 \ \text{mol O atoms}}{16.00 \ \cancel{g \ O}} = 2.171 \ \text{mol O atoms}$$

Step 3 Divide each number of moles by the smallest value:

$$\text{K} = \frac{1.447 \ \text{mol}}{0.723 \ \text{mol}} = 2.00$$

$$\text{C} = \frac{0.723 \ \text{mol}}{0.723 \ \text{mol}} = 1.00$$

$$\text{O} = \frac{2.171 \ \text{mol}}{0.723 \ \text{mol}} = 3.00$$

The simplest ratio of K:C:O is 2:1:3.
Empirical formula $= K_2CO_3$

Example 7.19 A sulfide of iron was formed by combining 2.233 g of iron, Fe, with 1.926 g of sulfur, S. What is the empirical formula of the compound?

Solution **Steps 1 and 2** The grams of each element are given, so we use them directly in our calculations. Calculate the relative number of moles of each element by multiplying the grams of each element by the proper mole/molar mass factor:

$$\text{Fe:} \quad 2.233 \text{ g Fe} \times \frac{1 \text{ mol Fe atoms}}{55.85 \text{ g Fe}} = 0.03998 \text{ mol Fe atoms}$$

$$\text{S:} \quad 1.926 \text{ g S} \times \frac{1 \text{ mol S atoms}}{32.06 \text{ g S}} = 0.06007 \text{ mol S atoms}$$

Step 3 Divide each number of moles by the smaller of the two numbers:

$$\text{Fe} = \frac{0.03998 \text{ mol}}{0.03998 \text{ mol}} = 1.000 \qquad \text{S} = \frac{0.06007 \text{ mol}}{0.03998 \text{ mol}} = 1.503$$

Step 4 We still have not reached a ratio that will give a formula containing whole numbers of atoms, so we must double each value to obtain a ratio of 2.000 atoms of Fe to 3.000 atoms of S. Doubling both values does not change the ratio of Fe and S atoms.

$$\text{Fe:} \quad 1.000 \times 2 = 2.000$$

$$\text{S:} \quad 1.503 \times 2 = 3.006$$

$$\text{Empirical formula} = \text{Fe}_2\text{S}_3$$

Practice 7.9

Calculate the empirical formula of a compound containing 53.33% C, 11.11% H, and 35.53% O.

Practice 7.10

Calculate the empirical formula of a compound that contains 43.7% phosphorus and 56.3% O by mass.

Calculation of the Molecular Formula
7.6 from the Empirical Formula

In addition to data for calculating the empirical formula, the molecular formula can be calculated from the empirical formula if the molar mass is known. The molecular formula, as stated in Section 7.4, will be either equal to or some multiple of the empirical formula. For example, if the empirical formula of a compound of hydrogen and fluorine is HF, the molecular formula can be expressed as $(\text{HF})_n$, where $n = 1, 2, 3, 4, \ldots$. This n means that the molecular formula could be HF, H_2F_2, H_3F_3, H_4F_4, and so on. To determine the molecular formula, we must evaluate n.

$$n = \frac{\text{molar mass}}{\text{mass of empirical formula}} = \text{number of empirical formula units}$$

What we actually calculate is the number of units of the empirical formula contained in the molecular formula.

A compound of nitrogen and oxygen with a molar mass of 92.00 g was found to have an empirical formula of NO_2. What is its molecular formula?

Example 7.20

Let n be the number of NO_2 units in a molecule; then the molecular formula is $(NO_2)_n$. Each NO_2 unit has a mass of $[14.01\text{ g} + (2 \times 16.00\text{ g})]$ or 46.01 g. The molar mass of $(NO_2)_n$ is 92.00 g and the number of 46.01 units in 92.00 is 2:

Solution

$$n = \frac{92.00\text{ g}}{46.01\text{ g}} = 2 \quad \text{(empirical formula units)}$$

The molecular formula is $(NO_2)_2$ or N_2O_4.

The hydrocarbon propylene has a molar mass of 42.00 g and contains 14.3% H and 85.7% C. What is its molecular formula?

Example 7.21

First find the empirical formula:

Solution

C: $85.7\text{ g C} \times \dfrac{1\text{ mol C atoms}}{12.01\text{ g C}} = 7.14\text{ mol C atoms}$

H: $14.3\text{ g H} \times \dfrac{1\text{ mol H atoms}}{1.008\text{ g H}} = 14.2\text{ mol H atoms}$

Divide each value by the smaller number of moles:

$$C = \frac{7.14\text{ mol}}{7.14\text{ mol}} = 1.00$$

$$H = \frac{14.2\text{ mol}}{7.14\text{ mol}} = 1.99$$

Empirical formula $= CH_2$

Then determine the molecular formula from the empirical formula and the molar mass:

Molecular formula $= (CH_2)_n$

Molar mass $= 42.00$ g

Each CH_2 unit has a mass of $(12.01\text{ g} + 2.016\text{ g})$ or 14.03 g. The number of CH_2 units in 42.00 g is 3:

$$n = \frac{42.00\text{ g}}{14.03\text{ g}} = 3 \quad \text{(empirical formula units)}$$

The molecular formula is $(CH_2)_3$ or C_3H_6.

Practice 7.11

Calculate the empirical and molecular formulas of a compound that contains 80.0% C, 20.0% H, and has a molar mass of 30.00 g.

Concepts in Review

1. Explain the meaning of the mole.
2. Discuss the relationship between a mole and Avogadro's number.
3. Convert grams, atoms, molecules, and molar masses to moles, and vice versa.
4. Determine the molar mass of a compound from the formula.
5. Calculate the percent composition of a compound from its formula.
6. Calculate the percent composition of a compound from experimental data.
7. Explain the relationship between an empirical formula and a molecular formula.
8. Determine the empirical formula for a compound from its percent composition.
9. Calculate the molecular formula of a compound from its percent composition and molar mass.

Key Terms

The terms listed here have been defined within this chapter. Section numbers are referenced in parenthesis for each term. More detailed definitions are given in the Glossary.

Avogadro's number (7.1)   mole (7.1)
empirical formula (7.4)   molecular formula (7.4)
formula unit (7.1)   percent composition of a compound (7.3)
molar mass (7.1)

Questions

Questions refer to tables, figures, and key words and concepts defined within the chapter. A particularly challenging question or exercise is indicated with an asterisk.

1. What is a mole?
2. Which would have a higher mass: a mole of K atoms or a mole of Au atoms?
3. Which would contain more atoms: a mole of K atoms, or a mole of Au atoms?

4. Which would contain more electrons: a mole of K atoms, or a mole of Au atoms?

* 5. If the atomic mass scale had been defined differently, with an atom of $^{12}_{6}C$ being defined as a mass of 50 amu, would this have any effect on the value of Avogadro's number? Explain.

6. What is the numerical value of Avogadro's number?

7. What is the relationship between Avogadro's number and the mole?

8. Complete the following statements, supplying the proper quantity.
   (a) A mole of O atoms contains _____ atoms.
   (b) A mole of O_2 molecules contains _____ molecules.
   (c) A mole of O_2 molecules contains _____ atoms.
   (d) A mole of O atoms has a mass of _____ grams.
   (e) A mole of O_2 molecules has a mass of _____ grams.

9. How many molecules are present in 1 molar mass of sulfuric acid (H_2SO_4)? How many atoms are present?

10. In calculating the empirical formula of a compound from its percent composition, why do we choose to start with 100.0 g of the compound?

11. List four characteristics of food additives.

12. Explain how the maximum tolerable daily intake of an additive is determined.

13. Why is it difficult to determine safe levels of food additives?

14. What is the difference between an empirical formula and a molecular formula?

15. Which of the following statements are correct? Rewrite the incorrect statements to make them correct.
   (a) One atomic mass of any element contains 6.022×10^{23} atoms.
   (b) The mass of 1 atom of Cl is $\dfrac{35.45 \text{ g}}{6.022 \times 10^{23} \text{ atoms}}$.
   (c) A mole of Mg atoms (24.31 g) contains the same number of atoms as a mole of Na atoms (22.99 g).
   (d) A mole of bromine atoms contains 6.022×10^{23} atoms of bromine.
   (e) A mole of Cl_2 molecules contains 6.022×10^{23} atoms of Cl.
   (f) A mole of Al atoms has the same mass as a mole of tin (Sn) atoms.
   (g) A mole of H_2O contains 6.022×10^{23} atoms.

   (h) A mole of H_2 molecules contains 1.204×10^{24} electrons.

16. Which of the following statements are correct? Rewrite the incorrect statements to make them correct.
   (a) A mole of Na and a mole of NaCl contain the same number of Na atoms.
   (b) One mole of nitrogen gas (N_2) has a mass of 14.01 g.
   (c) The percent of oxygen is higher in K_2CrO_4 than it is in Na_2CrO_4.
   (d) The number of Cr atoms is the same in a mole of K_2CrO_4 as it is in a mole of Na_2CrO_4.
   (e) Both K_2CrO_4 and Na_2CrO_4 contain the same percent by mass of Cr.
   (f) A molar mass of sucrose ($C_{12}H_{22}O_{11}$) contains 1 mol of sucrose molecules.
   (g) Two moles of nitric acid (HNO_3) contain 6 mol of O atoms.
   (h) The empirical formula of sucrose ($C_{12}H_{22}O_{11}$) is CH_2O.
   (i) A hydrocarbon that has a molar mass of 280 and an empirical formula of CH_2 has a molecular formula of $C_{22}H_{44}$.
   (j) The empirical formula is often called the simplest formula.
   (k) The empirical formula of a compound gives the smallest whole-number ratio of the atoms that are present in the compound.
   (l) If the molecular formula and the empirical formula of a compound are not the same, the empirical formula will be an integral multiple of the molecular formula.
   (m) The empirical formula of benzene (C_6H_6) is CH.
   (n) A compound having an empirical formula of CH_2O and a molar mass of 60 has a molecular formula of $C_3H_6O_3$.

Paired Exercises

These exercises are paired. Each odd-numbered exercise is followed by a similar even-numbered exercise. Answers to the even-numbered exercises are given in Appendix V.

Molar Masses

17. Determine the molar masses of the following compounds:
   (a) KBr
   (b) Na_2SO_4
   (c) $Pb(NO_3)_2$
   (d) C_2H_5OH
   (e) $HC_2H_3O_2$
   (f) Fe_3O_4
   (g) $C_{12}H_{22}O_{11}$
   (h) $Al_2(SO_4)_3$
   (i) $(NH_4)_2HPO_4$

18. Determine the molar masses of the following compounds:
   (a) NaOH
   (b) Ag_2CO_3
   (c) Cr_2O_3
   (d) $(NH_4)_2CO_3$
   (e) $Mg(HCO_3)_2$
   (f) C_6H_5COOH
   (g) $C_6H_{12}O_6$
   (h) $K_4Fe(CN)_6$
   (i) $BaCl_2 \cdot 2\,H_2O$

Moles and Avogadro's Number

19. How many moles of atoms are contained in the following?
(a) 22.5 g Zn
(b) 0.688 g Mg
(c) 4.5×10^{22} atoms Cu
(d) 382 g Co
(e) 0.055 g Sn
(f) 8.5×10^{24} molecules N_2

20. How many moles are contained in the following?
(a) 25.0 g NaOH
(b) 44.0 g Br_2
(c) 0.684 g $MgCl_2$
(d) 14.8 g CH_3OH
(e) 2.88 g Na_2SO_4
(f) 4.20 lb ZnI_2

21. Calculate the number of grams in each of the following:
(a) 0.550 mol Au
(b) 15.8 mol H_2O
(c) 12.5 mol Cl_2
(d) 3.15 mol NH_4NO_3

22. Calculate the number of grams in each of the following:
(a) 4.25×10^{-4} mol H_2SO_4
(b) 4.5×10^{22} molecules CCl_4
(c) 0.00255 mol Ti
(d) 1.5×10^{16} atoms S

23. How many molecules are contained in each of the following:
(a) 1.26 mol O_2
(b) 0.56 mol C_6H_6
(c) 16.0 g CH_4
(d) 1000. g HCl

24. How many molecules are contained in each of the following:
(a) 1.75 mol Cl_2
(b) 0.27 mol C_2H_6O
(c) 12.0 g CO_2
(d) 100. g CH_4

25. Calculate the mass in grams of each of the following:
(a) 1 atom Pb
(b) 1 atom Ag
(c) 1 molecule H_2O
(d) 1 molecule $C_3H_5(NO_3)_3$

26. Calculate the mass in grams of each of the following:
(a) 1 atom Au
(b) 1 atom U
(c) 1 molecule NH_3
(d) 1 molecule $C_6H_4(NH_2)_2$

27. Make the following conversions:
(a) 8.66 mol Cu to grams Cu
(b) 125 mol Au to kilograms Au
(c) 10 atoms C to moles C
(d) 5000 molecules CO_2 to moles CO_2

28. Make the following conversions:
(a) 28.4 g S to moles S
(b) 2.50 kg NaCl to moles NaCl
(c) 42.4 g Mg to atoms Mg
(d) 485 mL Br_2 ($d = 3.12$ g/mL) to moles Br_2

29. Exactly 1 mol of carbon disulfide contains
(a) how many carbon disulfide molecules?
(b) how many carbon atoms?
(c) how many sulfur atoms?
(d) how many total atoms of all kinds?

30. One mole of ammonia contains
(a) how many ammonia molecules?
(b) how many nitrogen atoms?
(c) how many hydrogen atoms?
(d) how many total atoms of all kinds?

31. How many atoms of oxygen are contained in each of the following?
(a) 16.0 g O_2
(b) 0.622 mol MgO
(c) 6.00×10^{22} molecules $C_6H_{12}O_6$

32. How many atoms of oxygen are contained in each of the following?
(a) 5.0 mol MnO_2
(b) 255 g $MgCO_3$
(c) 5.0×10^{18} molecules H_2O

33. Calculate the number of
(a) grams of silver in 25.0 g AgBr
(b) grams of nitrogen in 6.34 mol $(NH_4)_3PO_4$
(c) grams of oxygen in 8.45×10^{22} molecules SO_3

34. Calculate the number of
(a) grams of chlorine in 5.0 g $PbCl_2$
(b) grams of hydrogen in 4.50 mol H_2SO_4
(c) grams of iodine in 5.45×10^{22} molecules CaI_2

Percent Composition

35. Calculate the percent composition by mass of the following compounds:
(a) NaBr
(b) $KHCO_3$
(c) $FeCl_3$
(d) $SiCl_4$
(e) $Al_2(SO_4)_3$
(f) $AgNO_3$

36. Calculate the percent composition of the following compounds:
(a) $ZnCl_2$
(b) $NH_4C_2H_3O_2$
(c) MgP_2O_7
(d) $(NH_4)_2SO_4$
(e) $Fe(NO_3)_3$
(f) ICl_3

37. Calculate the percent of iron in the following compounds:
    (a) FeO
    (b) Fe_2O_3
    (c) Fe_3O_4
    (d) $K_4Fe(CN)_6$

38. Which of the following chlorides has the highest and which has the lowest percentage of chlorine, by mass, in its formula?
    (a) KCl
    (b) $BaCl_2$
    (c) $SiCl_4$
    (d) LiCl

39. A 6.20-g sample of phosphorus was reacted with oxygen to form an oxide with a mass of 14.20 g. Calculate the percent composition of the compound.

40. A sample of ethylene chloride was analyzed to contain 6.00 g of C, 1.00 g of H, and 17.75 g of Cl. Calculate the percent composition of ethylene chloride.

41. Answer the following by examining the formulas. Check your answers by calculation if you wish. Which compound has the
    (a) higher percent by mass of hydrogen, H_2O or H_2O_2?
    (b) lower percent by mass of nitrogen, NO or N_2O_3?
    (c) higher percent by mass of oxygen, NO_2 or N_2O_4?

42. Answer the following by examining the formulas. Check your answers by calculation if you wish. Which compound has the
    (a) lower percent by mass of chlorine, $NaClO_3$ or $KClO_3$?
    (b) higher percent by mass of sulfur, $KHSO_4$ or K_2SO_4?
    (c) lower percent by mass of chromium, Na_2CrO_4 or $Na_2Cr_2O_7$?

Empirical and Molecular Formulas

43. Calculate the empirical formula of each compound from the percent compositions given:
    (a) 63.6% N, 36.4% O
    (b) 46.7% N, 53.3% O
    (c) 25.9% N, 74.1% O
    (d) 43.4% Na, 11.3% C, 45.3% O
    (e) 18.8% Na, 29.0% Cl, 52.3% O
    (f) 72.02% Mn, 27.98% O

44. Calculate the empirical formula of each compound from the percent compositions given:
    (a) 64.1% Cu, 35.9% Cl
    (b) 47.2% Cu, 52.8% Cl
    (c) 51.9% Cr, 48.1% S
    (d) 55.3% K, 14.6% P, 30.1% O
    (e) 38.9% Ba, 29.4% Cr, 31.7% O
    (f) 3.99% P, 82.3% Br, 13.7% Cl

45. A sample of tin having a mass of 3.996 g was oxidized and found to have combined with 1.077 g of oxygen. Calculate the empirical formula of this oxide of tin.

46. A 3.054-g sample of vanadium (V) combined with oxygen to form 5.454 g of product. Calculate the empirical formula for this compound.

47. Hydroquinone is an organic compound commonly used as a photographic developer. It has a molar mass of 110.1 g/mol and a composition of 65.45% C, 5.45% H, and 29.09% O. Calculate the molecular formula of hydroquinone.

48. Fructose is a very sweet natural sugar that is present in honey, fruits, and fruit juices. It has a molar mass of 180.1 g/mol and a composition of 40.0% C, 6.7% H, and 53.3% O. Calculate the molecular formula of fructose.

Additional Exercises

These problems are not paired or labeled by topic and provide additional practice on the concepts covered in this chapter.

49. White phosphorus is one of several forms of phosphorus and exists as a waxy solid consisting of P_4 molecules. How many atoms are present in 0.350 mol of P_4?

50. How many grams of sodium contain the same number of atoms as 10.0 g of potassium?

51. One atom of an unknown element is found to have a mass of 1.79×10^{-23} g. What is the molar mass of this element?

*52. If a stack of 500 sheets of paper is 4.60 cm high, what will be the height, in meters, of a stack of Avogadro's number of sheets of paper?

53. There are about 5.0 billion (5.0×10^9) people on earth. If exactly 1 mol of dollars were distributed equally among these people, how many dollars would each person receive?

*54. If 20. drops of water equal 1.0 mL (1.0 cm³),
    (a) how many drops of water are there in a cubic mile of water?
    (b) what would be the volume in cubic miles of a mole of drops of water?

*55. Silver has a density of 10.5 g/cm^3. If 1.00 mol of silver were shaped into a cube,
   (a) what would be the volume of the cube?
   (b) what would be the length of one side of the cube?

*56. A sulfuric acid solution contains 65.0% H_2SO_4 by mass and has a density of 1.55 g/mL. How many moles of the acid are present in 1.00 L of the solution?

*57. A nitric acid solution containing 72.0% HNO_3 by mass has a density of 1.42 g/mL. How many moles of HNO_3 are present in 100. mL of the solution?

58. Given 1.00-g samples of each of the following compounds, CO_2, O_2, H_2O, and CH_3OH,
   (a) which sample will contain the largest number of molecules?
   (b) which sample will contain the largest number of atoms? Show proof for your answers.

59. How many grams of Fe_2S_3 will contain a total number of atoms equal to Avogadro's number?

60. How many grams of lithium will combine with 20.0 g of sulfur to form the compound Li_2S?

61. Calculate the percentage of
   (a) mercury in $HgCO_3$
   (b) oxygen in $Ca(ClO_3)_2$
   (c) nitrogen in $C_{10}H_{14}N_2$ (nicotine)
   (d) Mg in $C_{55}H_{72}MgN_4O_5$ (chlorophyll)

*62. Zinc and sulfur react to form zinc sulfide, ZnS. If we mix 19.5 g of zinc and 9.40 g of sulfur, have we added sufficient sulfur to fully react all the zinc? Show evidence for your answer.

63. Aspirin is well known as a pain reliever (analgesic) and as a fever reducer (antipyretic). It has a molar mass of 180.2 g/mol and a composition of 60.0% C, 4.48% H, and 35.5% O. Calculate the molecular formula of aspirin.

64. How many grams of oxygen are contained in 8.50 g $Al_2(SO_4)_3$?

65. Gallium arsenide is one of the newer materials used to make semiconductor chips for use in supercomputers. Its composition is 48.2% Ga and 51.8% As. What is the empirical formula?

66. Listed below are the compositions of four different compounds of carbon and chlorine. Determine both the empirical formula and the molecular formula for each.

	Percent C	Percent Cl	Molar mass (g)
(a)	7.79	92.21	153.8
(b)	10.13	89.87	236.7
(c)	25.26	74.74	284.8
(d)	11.25	88.75	319.6

67. How many years is a mole of seconds?

68. A normal penny has a mass of about 2.5 g. If we assume the penny to be pure copper (which means the penny is very old since newer pennies are a mixture of copper and zinc), how many atoms of copper does it contain?

69. What would be the mass (in grams) of one thousand trillion molecules of glycerin, $C_3H_8O_3$?

70. If we assume there are 5.0 billion people on the earth, how many moles of people is this?

71. An experimental catalyst used to make polymers has the following composition: Co, 23.3%; Mo, 25.3%; and Cl, 51.4%. What is the empirical formula for this compound?

72. If a student weighs 18 g of aluminum and needs twice as many atoms of magnesium as she has of aluminum, how many grams of Mg does she need?

*73. If 10.0 g of an unknown compound composed of carbon, hydrogen, and nitrogen contains 17.7% N and 3.8×10^{23} atoms of hydrogen, what is its empirical formula?

74. A substance whose formula is A_2O (A is a mystery element) is 60.0% A and 40.0% O. Identify the element A.

75. For the following compounds whose molecular formulas are given, indicate the empirical formula:
   (a) $C_6H_{12}O_6$ glucose
   (b) C_8H_{18} octane
   (c) $C_3H_6O_3$ lactic acid
   (d) $C_{25}H_{52}$ paraffin
   (e) $C_{12}H_4Cl_4O_2$ dioxin (a powerful poison)

Answers to Practice Exercises

7.1 10.0 g helium
7.2 1.5×10^{22} atoms
7.3 101.1 g KNO_3
7.4 21.3 g Na_2SO_4
7.5 3.751 mol $AlCl_3$
7.6 24.42% Ca; 17.07% N; 58.50% O

7.7 40.27% K; 26.78% Cr; 32.96% O
7.8 20.16% Al; 79.84% Cl
7.9 C_2H_5O
7.10 P_2O_5
7.11 The empirical formula is CH_3; the molecular formula is C_2H_6.

•••••••••••••••••••••••••••••

In the world today, much of our energy is directed toward expressing information in a concise, useful manner. From our earliest days in childhood, we are taught to translate ideas and desires into sentences. In mathematics, we learn to translate numerical relationships and situations into mathematical expressions and equations. A historian translates a thousand years of history into a 500-page textbook. A secretary translates an entire letter or document into a few lines of shorthand. A film maker translates an entire event, such as the Olympics, into several hours of entertainment.

And so it is with chemistry. A chemist uses a chemical equation to translate reactions that are observed in the laboratory or in nature. Chemical equations provide the necessary means (1) to summarize the reaction, (2) to display the substances that are reacting, (3) to show the products, and (4) to indicate the amounts of all component substances in the reaction.

8.1   The Chemical Equation

In a chemical reaction, the substances entering the reaction are called *reactants* and the substances formed are called the *products*. During a chemical reaction, atoms, molecules, or ions interact and rearrange themselves to form the products. In this process, chemical bonds are broken and new bonds are formed. The reactants and products may be in the solid, liquid, or gaseous state, or in solution.

chemical equation

A **chemical equation** is a shorthand expression for a chemical change or reaction. A chemical equation uses the chemical symbols and formulas of the reactants and products and other symbolic terms to represent a chemical reaction. The equations are written according to this general format:

1. The reactants are separated from the products by an arrow ($\rightarrow$) that indicates the direction of the reaction. The reactants are placed to the left and the products to the right of the arrow. A plus sign (+) is placed between reactants and between products when needed.
2. Coefficients (whole numbers) are placed in front of substances (e.g., 2 H_2O) to balance the equation and to indicate the number of units (atoms, molecules, moles, ions) of each substance reacting or being produced. When no number is shown, it is understood that one unit of the substance is indicated.
3. Conditions required to carry out the reaction may, if desired, be placed above or below the arrow or equality sign. For example, a delta sign placed over the arrow ($\overset{\Delta}{\rightarrow}$) indicates that heat is supplied to the reaction.
4. The physical state of a substance is indicated by the following symbols: (*s*) for solid state: (*l*) for liquid state; (*g*) for gaseous state; and (*aq*) for substances in aqueous solution. States are not always given in chemical equations.

The symbols commonly used in equations are given in Table 8.1.

◄ Chapter Opening Photo: The thermite reaction is a reaction between elemental aluminum and iron oxide. This reaction produces so much energy that the iron becomes molten. The thermite reaction is used to weld railroad rails.

TABLE 8.1 Symbols Commonly Used in Chemical Equations

Symbol	Meaning
→	Yields; produces (points to products)
(s)	Solid state (written after a substance)
(l)	Liquid state (written after a substance)
(g)	Gaseous state (written after a substance)
(aq)	Aqueous solution (substance dissolved in water)
Δ	Heat (written above arrow)
+	Plus or added to (placed between substances)

8.2 Writing and Balancing Equations

To represent the quantitative relationships of a reaction, the chemical equation must be balanced. A **balanced equation** contains the same number of each kind of atom on each side of the equation. The balanced equation, therefore, obeys the Law of Conservation of Mass.

balanced equation

Every chemistry student must learn to *balance* equations. Simple equations are easy to balance, but care and attention to detail are required. The way to balance an equation is to adjust the number of atoms of each element so that it is the same on each side of the equation, but a correct formula must not be changed in order to balance an equation. The following is a general procedure for balancing equations. Study this outline and refer to it as needed when working examples.

Step 1. Identify the Reaction for Which the Equation Is to Be Written. Formulate a description or word equation for the reaction if needed (e.g., mercury(II) oxide decomposes yielding mercury and oxygen).

Step 2. Write the Unbalanced, or Skeleton, Equation. Make sure that the formula for each substance is correct and that the reactants are written to the left and the products to the right of the arrow (e.g., $HgO \rightarrow Hg + O_2$). The correct formulas must be known or ascertained from the periodic table, lists of ions, or experimental data.

Step 3. Balance the Equation. Use the following steps as necessary:

(a) Count and compare the number of atoms of each element on each side of the equation and determine those that must be balanced.

(b) Balance each element, one at a time, by placing whole numbers (coefficients) in front of the formulas containing the unbalanced element. It is usually best to balance metals first, then nonmetals, then hydrogen and oxygen. Select the smallest coefficients that will give the same number of atoms of the element on each side. A coefficient placed before a formula multiplies every atom in the formula by that number (e.g., $2 H_2SO_4$ means two molecules of sulfuric acid and also means four H atoms, two S atoms, and eight O atoms.)

It is often helpful to leave elements that are in two or more formulas on the same side of the equation until just before balancing hydrogen and oxygen.

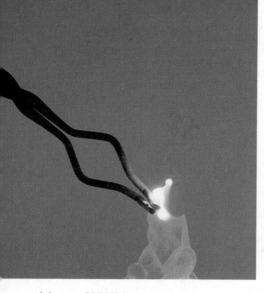

A burst of UV light is emitted as metallic magnesium combusts in air, producing magnesium oxide.

(c) Check all other elements after each individual element is balanced to see whether, in balancing one element, other elements have become unbalanced. Make adjustments as needed.

(d) Balance polyatomic ions such as SO_4^{2-}, which remain unchanged from one side of the equation to the other, in the same way as individual atoms.

(e) Do a final check, making sure that each element and/or polyatomic ion is balanced and that the smallest possible set of whole-number coefficients has been used.

$$4\ HgO \longrightarrow 4\ Hg + 2\ O_2 \quad \text{(incorrect form)}$$

$$2\ HgO \longrightarrow 2\ Hg + O_2 \quad \text{(correct form)}$$

Not all chemical equations can be balanced by the simple method of inspection just described. The following examples show *stepwise* sequences leading to balanced equations. Study each one carefully.

Example 8.1   Write the balanced equation for the reaction that takes place when magnesium metal is burned in air to produce magnesium oxide.

Solution   **Step 1** *Word equation:*

magnesium + oxygen ⟶ magnesium oxide

Step 2 *Skeleton equation:*

$$Mg + O_2 \longrightarrow MgO \quad \text{(unbalanced)}$$

Step 3 *Balance:*

(a) Mg is balanced.

(b) Oxygen is not balanced. Two O atoms appear on the left side and one on the right side.
Place the coefficient 2 before MgO:

$$Mg + O_2 \longrightarrow 2\ MgO \quad \text{(unbalanced)}$$

(c) Now Mg is not balanced. One Mg atom appears on the left side and two on the right side. Place a 2 before Mg:

$$2\ Mg + O_2 \longrightarrow 2\ MgO \quad \text{(balanced)}$$

(d) *Check:* Each side has two Mg and two O atoms.

Example 8.2   When methane, CH_4, undergoes complete combustion, it reacts with oxygen to produce carbon dioxide and water. Write the balanced equation for this reaction.

Solution   **Step 1** *Word equation:*

methane + oxygen ⟶ carbon dioxide + water

Step 2 *Skeleton equation:*

$$CH_4 + O_2 \longrightarrow CO_2 + H_2O \quad \text{(unbalanced)}$$

Step 3 *Balance:*
    (a) Carbon is balanced.
    (b) Hydrogen and oxygen are not balanced. Balance H atoms by placing a 2 before H_2O:

$$CH_4 + O_2 \longrightarrow CO_2 + 2\,H_2O \quad \text{(unbalanced)}$$

Each side of the equation has four H atoms; oxygen is still not balanced. Place a 2 before O_2 to balance the oxygen atoms:

$$CH_4 + 2\,O_2 \longrightarrow CO_2 + 2\,H_2O \quad \text{(balanced)}$$

    (c) *Check:* The equation is correctly balanced; it has one C, four O, and four H atoms on each side.

Oxygen and potassium chloride are formed by heating potassium chlorate. Write a balanced equation for this reaction.

Example 8.3

Solution

Step 1 *Word equation:*

potassium chlorate $\xrightarrow{\;\Delta\;}$ potassium chloride + oxygen

Step 2 *Skeleton equation:*

$$KClO_3 \xrightarrow{\;\Delta\;} KCl + O_2 \quad \text{(unbalanced)}$$

Step 3 *Balance:*
    (a) Potassium and chlorine are balanced.
    (b) Oxygen is unbalanced (three O atoms on the left and two on the right side).
    (c) How many oxygen atoms are needed? The subscripts of oxygen (3 and 2) in $KClO_3$ and O_2 have a least common multiple of 6. Therefore, coefficients for $KClO_3$ and O_2 are needed to give six O atoms on each side. Place a 2 before $KClO_3$ and a 3 before O_2 to give six O atoms on each side:

$$2\,KClO_3 \xrightarrow{\;\Delta\;} KCl + 3\,O_2 \quad \text{(unbalanced)}$$

Now K and Cl are not balanced. Place a 2 before KCl, which balances both K and Cl at the same time:

$$2\,KClO_3 \xrightarrow{\;\Delta\;} 2\,KCl + 3\,O_2 \quad \text{(balanced)}$$

    (d) *Check:* Each side now contains two K, two Cl, and six O atoms.

Silver nitrate reacts with hydrogen sulfide to produce silver sulfide and nitric acid. Write a balanced equation for this reaction.

Example 8.4

Solution

Step 1 *Word equation:*

silver nitrate + hydrogen sulfide $\longrightarrow$ silver sulfide + nitric acid

Step 2 *Skeleton equation:*

$$AgNO_3 + H_2S \longrightarrow Ag_2S + HNO_3 \quad \text{(unbalanced)}$$

Step 3 *Balance:*

(a) Ag and H are unbalanced.

(b) Place a 2 in front of $AgNO_3$ to balance Ag:

$$2\,AgNO_3 + H_2S \longrightarrow Ag_2S + HNO_3 \quad \text{(unbalanced)}$$

(c) H and NO_3^- are still unbalanced. Balance by placing a 2 in front of HNO_3:

$$2\,AgNO_3 + H_2S \longrightarrow Ag_2S + 2\,HNO_3 \quad \text{(balanced)}$$

(d) In this example N and O atoms are balanced by balancing the NO_3^- ion as a unit.

(e) *Check:* Each side has two Ag, two H, and one S atom. Also, each side has two NO_3^- ions.

Example 8.5   When aluminum hydroxide is mixed with sulfuric acid the products are aluminum sulfate and water. Write a balanced equation for this reaction.

Solution

Step 1 *Word equation:*

$$\text{aluminum hydroxide} + \text{sulfuric acid} \longrightarrow \text{aluminum sulfate} + \text{water}$$

Step 2 *Skeleton equation:*

$$Al(OH)_3 + H_2SO_4 \longrightarrow Al_2(SO_4)_3 + H_2O \quad \text{(unbalanced)}$$

Step 3 *Balance:*

(a) All elements are unbalanced.

(b) Balance Al by placing a 2 in front of $Al(OH)_3$. Treat the unbalanced SO_4^{2-} ion as a unit and balance by placing a 3 before H_2SO_4:

$$2\,Al(OH)_3 + 3\,H_2SO_4 \longrightarrow Al_2(SO_4)_3 + H_2O \quad \text{(unbalanced)}$$

(c) Balance the unbalanced H and O by placing a 6 in front of H_2O:

$$2\,Al(OH)_3 + 3\,H_2SO_4 \longrightarrow Al_2(SO_4)_3 + 6\,H_2O \quad \text{(balanced)}$$

(d) *Check:* Each side has two Al, twelve H, three S, and eighteen O atoms.

Example 8.6   When the fuel in a butane gas stove undergoes complete combustion, it reacts with oxygen to form carbon dioxide and water. Write the balanced equation for this reaction.

Solution

Step 1 *Word equation:*

$$\text{butane} + \text{oxygen} \longrightarrow \text{carbon dioxide} + \text{water}$$

Step 2 *Skeleton equation:*

$$C_4H_{10} + O_2 \longrightarrow CO_2 + H_2O \quad \text{(unbalanced)}$$

Step 3 *Balance:*

(a) All elements are unbalanced.

(b) Balance C by placing a 4 in front of CO_2:

$$C_4H_{10} + O_2 \longrightarrow 4\,CO_2 + H_2O \quad \text{(unbalanced)}$$

Balance H by placing a 5 in front of H_2O:

$$C_4H_{10} + O_2 \longrightarrow 4\,CO_2 + 5\,H_2O \quad \text{(unbalanced)}$$

Oxygen remains unbalanced. The oxygen atoms on the right side are fixed, because $4\,CO_2$ and $5\,H_2O$ are derived from the single C_4H_{10} molecule on the left. When we try to balance the O atoms, we find that there is no whole number that can be placed in front of O_2 to bring about a balance. The equation can be balanced if we use $6\frac{1}{2}\,O_2$ and then double the coefficients of each substance, including the $6\frac{1}{2}\,O_2$, to obtain the balanced equation:

$$C_4H_{10} + 6\tfrac{1}{2}\,O_2 \longrightarrow 4\,CO_2 + 5\,H_2O \quad \text{(balanced—incorrect form)}$$

$$2\,C_4H_{10} + 13\,O_2 \longrightarrow 8\,CO_2 + 10\,H_2O \quad \text{(balanced)}$$

(c) *Check:* Each side now has eight C, twenty H, and twenty-six O atoms.

Cooking with gas. The fuel in this butane stove undergoes complete combustion as it reacts with oxygen to form CO_2 and water.

Practice 8.1

Balance the following word equation.

   aluminum + oxygen $\longrightarrow$ aluminum oxide

Practice 8.2

Balance the following word equation.

   magnesium hydroxide + phosphoric acid $\longrightarrow$
   magnesium phosphate + water

8.3 What Information Does an Equation Tell Us?

Depending on the particular context in which it is used, a formula can have different meanings. The meanings refer either to an individual chemical entity (atom, ion, molecule, or formula unit) or to a mole of that chemical entity. For example, the formula H_2O can be used to indicate any of the following:

1. 2 H atoms and 1 O atom
2. 1 molecule of water
3. 1 mol of water
4. 6.022×10^{23} molecules of water
5. 18.02 g of water

Autumn Leaf Color

Chemical reactions can be the source of much natural beauty, even as they contribute to the metabolic cycle in plants. The wondrous array of color seen in the leaf display of autumn is the result of chemical reactions.

Chlorophylls are responsible for the common green color in plants and are necessary for the plant to produce food in the process called *photosynthesis*. Photosynthesis involves many chemical reactions but can be summarized by the following equation:

$$6 CO_2 + 6 H_2O \xrightarrow[\text{sunlight}]{\text{chlorophyll}}$$
$$\text{carbon} \quad \text{water}$$
$$\text{dioxide}$$

$$C_6H_{12}O_6 + 6 O_2$$
$$\text{glucose} \quad \text{oxygen}$$

There are several types of chlorophyll, including chlorophyll *a*, chlorophyll *b*, and chlorophyll *c*. All photosynthetic plants contain chlorophyll *a*, and some contain chlorophyll *b* and chlorophyll *c*, which are called *accessory pigments*. The additional pigments extend the range of colored light that plants can use for photosynthesis. In addition, plants also contain carotenoids (orange, yellow, and red pigments), which protect the plant from the

Chilly nights cause chlorophyll to decompose in leaves. As the green fades away, other pigments emerge, giving us our beautiful fall colors.

destructive potential of chlorophyll. The carotenoids absorb the high-energy oxygen that is released when the chlorophyll absorbs light energy, and they release the oxygen when it can be used construc-

tively. The least prevalent pigments are *anthocyanins* (red or blue). These pigments are responsible for the great diversity of combinations and concentrations of colors in the leaves. The carotenoids and anthocyanins are masked by the chlorophylls and can only be seen as the chlorophyll disintegrates in the autumn.

To produce brilliant autumn color, the proper set of conditions must exist. The trees need a long, vigorous growing season with plenty of water for photosynthesis. The autumn days should be bright to provide a longer photosynthetic period, with plenty of sugar production. When less intense sunlight and cold nights signal the leaves to stop photosynthesis, the protein bound to the chlorophyll begins to release from it. The protein breaks down into amino acids that go to the roots for storage. The chlorophyll decomposes and the green color fades away. The other pigments, especially the yellows and reds of the carotenoids, can now be seen. Sugar, trapped in the leaves during the very chilly nights, is changed through a set of complex reactions into anthocyanins to produce a variety of colors, mostly reds. The more productive a leaf has been, the greater the concentration of the other pigments and the more brilliant the color.

Formulas used in equations can be expressed in units of individual chemical entities or as moles, the latter being more commonly used. For example, in the reaction of hydrogen and oxygen to form water,

$$2 H_2 + O_2 \longrightarrow 2 H_2O$$

the 2 H_2 can represent 2 molecules or 2 mol of hydrogen; the O_2, 1 molecule or 1 mol of oxygen; and the 2 H_2O, 2 molecules or 2 mol of water. In terms of moles, this equation is stated: 2 mol of H_2 react with 1 mol of O_2 to give 2 mol of H_2O.

As indicated earlier, a chemical equation is a shorthand description of a chemical reaction. Interpretation of a balanced equation gives us the following information:

We usually use moles in equations because molecules are so small that we generally work with large collections at once.

1. What the reactants are and what the products are
2. The formulas of the reactants and products

3. The number of molecules or formula units of reactants and products in the reaction
4. The number of atoms of each element involved in the reaction
5. The number of moles of each substance
6. The number of grams of each substance used or produced

Consider the equation

$$H_2(g) + Cl_2(g) \longrightarrow 2\,HCl(g)$$

Here, hydrogen gas reacts with chlorine gas to produce hydrogen chloride, also a gas. Let's summarize all the information that can be stated about the relative amount of each substance, with respect to all other substances in the balanced equation:

Hydrogen		Chlorine		Hydrogen chloride
$H_2(g)$	+	$Cl_2(g)$	→	$2\,HCl(g)$
1 molecule		1 molecule		2 molecules
2 atoms H		2 atoms Cl		2 atoms H + 2 atoms Cl
1 molar mass		1 molar mass		2 molar masses
1 mol		1 mol		2 mol
2.016 g		70.90 g		2 × 36.46 g or 72.92 g

These data are very useful in calculating quantitative relationships that exist among substances in a chemical reaction. For example, if we react 2 mol of hydrogen (twice as much as is indicated by the equation) with 2 mol of chlorine, we can expect to obtain 4 mol, or 145.8 g, of hydrogen chloride as a product. We will study this phase of using equations in more detail in the next chapter.

Let us try another equation. When propane gas (C_3H_8) is burned in air, the products are carbon dioxide, CO_2, and water, H_2O. The balanced equation and its interpretation are as follows:

Propane		Oxygen		Carbon dioxide		Water
$C_3H_8(g)$	+	$5\,O_2(g)$	→	$3\,CO_2(g)$	+	$4\,H_2O(g)$
1 molecule		5 molecules		3 molecules		4 molecules
3 atoms C		10 atoms O		3 atoms C		8 atoms H
8 atoms H				6 atoms O		4 atoms O
1 molar mass		5 molar masses		3 molar masses		4 molar masses
1 mol		5 mol		3 mol		4 mol
44.09 g		5 × 32.00 g (160.0 g)		3 × 44.01 g (132.0 g)		4 × 18.02 g (72.08 g)

8.4   Types of Chemical Equations

Chemical equations represent chemical changes or reactions. Reactions are classified into types to assist in writing equations and in predicting other reactions. Many chemical reactions fit one or another of the four principal reaction types that we discuss in the following paragraphs. Reactions are also classified as oxidation–reduction. Special methods are used to balance complex oxidation–reduction equations.

Combination, or Synthesis, Reaction

combination reaction

In a **combination reaction**, two reactants combine to give one product. The general form of the equation is

$$A + B \longrightarrow AB$$

in which A and B are either elements or compounds and AB is a compound. The formula of the compound in many cases can be determined from a knowledge of the ionic charges of the reactants in their combined states. Some reactions that fall into this category are given here.

(a)   metal + oxygen $\longrightarrow$ metal oxide

$$2\ Mg(s) + O_2(g) \xrightarrow{\Delta} 2\ MgO(s)$$
$$4\ Al(s) + 3\ O_2(g) \xrightarrow{\Delta} 2\ Al_2O_3(s)$$

(b)   nonmetal + oxygen $\longrightarrow$ nonmetal oxide

$$S(s) + O_2(g) \xrightarrow{\Delta} SO_2(g)$$
$$N_2(g) + O_2(g) \xrightarrow{\Delta} 2\ NO(g)$$

(c)   metal + nonmetal $\longrightarrow$ salt:

$$2\ Na(s) + Cl_2(g) \longrightarrow 2\ NaCl(s)$$
$$2\ Al(s) + 3\ Br_2(l) \longrightarrow 2\ AlBr_3(s)$$

(d)   metal oxide + water $\longrightarrow$ metal hydroxide

$$Na_2O(s) + H_2O(l) \longrightarrow 2\ NaOH(aq)$$
$$CaO(s) + H_2O(l) \longrightarrow Ca(OH)_2(aq)$$

(e)   nonmetal oxide + water $\longrightarrow$ oxy-acid

$$SO_3(g) + H_2O(l) \longrightarrow H_2SO_4(aq)$$
$$N_2O_5(s) + H_2O(l) \longrightarrow 2\ HNO_3(aq)$$

Decomposition Reaction

decomposition reaction

In a **decomposition reaction**, a single substance is decomposed or broken down to give two or more different substances. The reaction may be considered the reverse of combination. The starting material must be a compound, and the products may be elements or compounds. The general form of the equation is

$$AB \longrightarrow A + B$$

Predicting the products of a decomposition reaction can be difficult and requires an understanding of each individual reaction. Heating oxygen-containing compounds often results in decomposition. Some reactions that fall into this category are:

(a) Metal oxides. Some metal oxides decompose to yield the free metal plus oxygen; others give another oxide, and some are very stable, resisting decomposition by heating:

$$2 \text{ HgO}(s) \xrightarrow{\Delta} 2 \text{ Hg}(l) + \text{O}_2(g)$$

$$2 \text{ PbO}_2(s) \xrightarrow{\Delta} 2 \text{ PbO}(s) + \text{O}_2(g)$$

(b) Carbonates and hydrogen carbonates decompose to yield CO_2 when heated:

$$\text{CaCO}_3(s) \xrightarrow{\Delta} \text{CaO}(s) + \text{CO}_2(g)$$

$$2 \text{ NaHCO}_3(s) \xrightarrow{\Delta} \text{Na}_2\text{CO}_3(s) + \text{H}_2\text{O}(g) + \text{CO}_2(g)$$

(c) Miscellaneous:

$$2 \text{ KClO}_3(s) \xrightarrow{\Delta} 2 \text{ KCl}(s) + 3 \text{ O}_2(g)$$

$$2 \text{ NaNO}_3(s) \xrightarrow{\Delta} 2 \text{ NaNO}_2(g) + \text{O}_2(g)$$

$$2 \text{ H}_2\text{O}_2(l) \xrightarrow{\Delta} 2 \text{ H}_2\text{O}(l) + \text{O}_2(g)$$

A single-displacement reaction occurs as elemental zinc reacts with hydrochloric acid to produce bubbles of hydrogen.

Single-Displacement Reaction

In a **single-displacement reaction** one element reacts with a compound to take the place of one of the elements of that compound, yielding a different element and a different compound. The general form of the equation is

$$A + BC \longrightarrow B + AC \quad \text{or} \quad A + BC \longrightarrow C + BA$$

If A is a metal, A will replace B to form AC, provided A is a more reactive metal than B. If A is a halogen, it will replace C to form BA, provided A is a more reactive halogen than C.

A brief activity series of selected metals (and hydrogen) and halogens are shown in Table 8.2. This series is listed in descending order of chemical reactivity, with the most active metals and halogens at the top. From such series it is possible to predict many chemical reactions. In an activity series, the atoms of any element in the series will replace the atoms of those elements below it. For example, zinc metal will replace hydrogen from a hydrochloric acid solution. But copper metal, which is underneath hydrogen on the list and thus less reactive than hydrogen, will not replace hydrogen from a hydrochloric acid solution. Some reactions that fall into this category follow:

(a) metal + acid $\longrightarrow$ hydrogen + salt

$$\text{Zn}(s) + 2 \text{ HCl}(aq) \longrightarrow \text{H}_2(g) + \text{ZnCl}_2(aq)$$

$$2 \text{ Al}(s) + 3 \text{ H}_2\text{SO}_4(aq) \longrightarrow 3 \text{ H}_2(g) + \text{Al}_2(\text{SO}_4)_3(aq)$$

(b) metal + water $\longrightarrow$ hydrogen + metal hydroxide or metal oxide

$$2 \text{ Na}(s) + 2 \text{ H}_2\text{O} \longrightarrow \text{H}_2(g) + 2 \text{ NaOH}(aq)$$

$$\text{Ca}(s) + 2 \text{ H}_2\text{O} \longrightarrow \text{H}_2(g) + \text{Ca(OH)}_2(aq)$$

$$\underset{\text{steam}}{3 \text{ Fe}(s) + 4 \text{ H}_2\text{O}(g)} \longrightarrow 4 \text{ H}_2(g) + \text{Fe}_3\text{O}_4(s)$$

single-displacement reaction

TABLE 8.2 Activity Series	
Metals	**Halogens**
K	F_2
Ca	Cl_2
Na	Br_2
Mg	I_2
Al	
Zn	
Fe	
Ni	
Sn	
Pb	
H	
Cu	
Ag	
Hg	
Au	

increasing activity →

(c) metal + salt $\longrightarrow$ metal + salt

$$Fe(s) + CuSO_4(aq) \longrightarrow Cu(s) + FeSO_4(aq)$$

$$Cu(s) + 2\,AgNO_3(aq) \longrightarrow 2\,Ag(s) + Cu(NO_3)_2(aq)$$

(d) halogen + halide salt $\longrightarrow$ halogen + halide salt

$$Cl_2(g) + 2\,NaBr(aq) \longrightarrow Br_2(l) + 2\,NaCl(aq)$$

$$Cl_2(g) + 2\,KI(aq) \longrightarrow I_2(s) + 2\,KCl(aq)$$

A common chemical reaction is the displacement of hydrogen from water or acids. This reaction is a good illustration of the relative reactivity of metals and the use of the activity series. For example,

- K, Ca, and Na displace hydrogen from cold water, steam, and acids.
- Mg, Al, Zn, and Fe displace hydrogen from steam and acids.
- Ni, Sn, and Pb displace hydrogen only from acids.
- Cu, Ag, Hg, and Au do not displace hydrogen.

Example 8.7 Will a reaction occur between (a) nickel metal and hydrochloric acid and (b) tin metal and a solution of aluminum chloride? Write balanced equations for the reactions.

Solution

(a) Nickel is more reactive than hydrogen, so it will displace hydrogen from hydrochloric acid. The products are hydrogen gas and a salt of Ni^{2+} and Cl^- ions:

$$Ni(s) + 2\,HCl(aq) \longrightarrow H_2(g) + NiCl_2(aq)$$

(b) According to the activity series, tin is less reactive than aluminum, so no reaction will occur:

$$Sn(s) + AlCl_3(aq) \longrightarrow \text{no reaction}$$

Practice 8.3

Write balanced equations for the reactions:
(a) iron metal and a solution of magnesium chloride
(b) zinc metal and a solution of lead(II) nitrate

Double-Displacement, or Metathesis, Reaction

double-displacement reaction

In a **double-displacement reaction**, two compounds exchange partners with each other to produce two different compounds. The general form of the equation is

$$AB + CD \longrightarrow AD + CB$$

This reaction can be thought of as an exchange of positive and negative groups, in which A combines with D and C combines with B. In writing formulas for the products, we must account for the charges of the combining groups.

It is also possible to write an equation in the form of a double-displacement reaction when a reaction has not occurred. For example, when solutions of sodium chloride and potassium nitrate are mixed, the following equation can be written:

$$NaCl(aq) + KNO_3(aq) \longrightarrow NaNO_3(aq) + KCl(aq)$$

When the procedure is carried out, no physical changes are observed, indicating that no chemical reaction has taken place.

A double-displacement reaction is accompanied by evidence of such reactions as the evolution of heat, the formation of an insoluble precipitate, or the production of gas bubbles. We will now look at some of these reactions.

Neutralization of an acid and a base. The production of a molecule of water from an H^+ and an OH^- ion is accompanied by a release of heat, which can be detected by touching the reaction container. For neutralization reactions, $H^+ + OH^- \longrightarrow H_2O$.

$$acid + base \longrightarrow salt + water$$

$$HCl(aq) + NaOH(aq) \longrightarrow NaCl(aq) + H_2O(l)$$

$$H_2SO_4(aq) + Ba(OH)_2(aq) \longrightarrow BaSO_4(s) + 2\ H_2O(l)$$

A double-displacement reaction results from pouring a clear, colorless solution of $Pb(NO_3)_2$ into a clear, colorless solution of KI forming a yellow precipitate of PbI_2.

Formation of an insoluble precipitate. The solubilities of the products can be determined by consulting the Solubility Table in Appendix IV. One or both of the products may be insoluble.

$$BaCl_2(aq) + 2\ AgNO_3(aq) \longrightarrow 2\ AgCl(s) + Ba(NO_3)_2(aq)$$

$$FeCl_3(aq) + 3\ NaOH(aq) \longrightarrow Fe(OH)_3(s) + 3\ NaCl(aq)$$

Metal oxide + acid. Heat is released by the production of a molecule of water.

$$metal\ oxide + acid \longrightarrow salt + water$$

$$CuO(s) + 2\ HNO_3(aq) \longrightarrow Cu(NO_3)_2(aq) + H_2O(l)$$

$$CaO(s) + 2\ HCl(aq) \longrightarrow CaCl_2(aq) + H_2O(l)$$

Formation of a gas. A gas such as HCl or H_2S may be produced directly, as in these two examples:

$$H_2SO_4(l) + NaCl(s) \longrightarrow NaHSO_4(s) + HCl(g)$$

$$2\ HCl(aq) + ZnS(s) \longrightarrow ZnCl_2(aq) + H_2S(g)$$

A gas can also be produced indirectly. Some unstable compounds formed in a double-displacement reaction, such as H_2CO_3, H_2SO_3, and NH_4OH, will decompose to form water and a gas:

$$2\ HCl(aq) + Na_2CO_3(aq) \longrightarrow 2\ NaCl(aq) + H_2CO_3(aq) \longrightarrow 2\ NaCl(aq) + H_2O(l) + CO_2(g)$$

$$2\ HNO_3(aq) + K_2SO_3(aq) \longrightarrow 2\ KNO_3(aq) + H_2SO_3(aq) \longrightarrow 2\ KNO_3(aq) + H_2O(l) + SO_2(g)$$

$$NH_4Cl(aq) + NaOH(aq) \longrightarrow NaCl(aq) + NH_4OH(aq) \longrightarrow NaCl(aq) + H_2O(l) + NH_3(g)$$

Example 8.8   Write the equation for the reaction between aqueous solutions of hydrobromic acid and potassium hydroxide.

Solution   First write the formulas for the reactants. (They are HBr and KOH.) Then classify the type of reaction that would occur between them. Because the reactants are compounds, one an acid and the other a base, the reaction will be of the neutralization type:

$$\text{acid} + \text{base} \longrightarrow \text{salt} + \text{water}$$

Now rewrite the equation using the formulas for the known substances:

$$HBr(aq) + KOH(aq) \longrightarrow \text{salt} + H_2O$$

In this reaction, which is a double-displacement type, the H^+ from the acid combines with the OH^- from the base to form water. The ionic compound must be composed of the other two ions, K^+ and Br^-. We determine the formula of the ionic compound to be KBr from the fact that K is a $+1$ cation and Br is a -1 anion. The final balanced equation is

$$HBr(aq) + KOH(aq) \longrightarrow KBr(aq) + H_2O(l)$$

Example 8.9   Complete and balance the equation for the reaction between aqueous solutions of barium chloride and sodium sulfate.

Solution   First determine the formula for the reactants. (They are $BaCl_2$ and Na_2SO_4.) Then classify these substances as acids, bases, or ionic compounds. (Both substances are salts.) Since both substances are compounds, the reaction will be of the double-displacement type. Start writing the equation with the reactants:

$$BaCl_2(aq) + Na_2SO_4(aq) \longrightarrow$$

If the reaction is double-displacement, Ba^{2+} will be written combined with SO_4^{2-}, and Na^+ with Cl^- as the products. The balanced equation is

$$BaCl_2(aq) + Na_2SO_4(aq) \longrightarrow BaSO_4 + 2\,NaCl$$

The final step is to determine the nature of the products, which controls whether or not the reaction will take place. If both products are soluble, we may merely have a mixture of all the ions in solution. But if an insoluble precipitate is formed, the reaction will definitely occur. We know from experience that NaCl is fairly soluble in water, but what about $BaSO_4$? The Solubility Table in Appendix IV can give us this information. From this table we see that $BaSO_4$ is insoluble in water, so it will be a precipitate in the reaction. Thus the reaction will occur, forming a precipitate. The equation is

$$BaCl_2(aq) + Na_2SO_4(aq) \longrightarrow BaSO_4(s) + 2\,NaCl(aq)$$

A double-displacement reaction occurs between solutions of barium chloride and sodium sulfate to form a sodium chloride solution (colorless) and a precipitate of barium sulfate (white).

Practice 8.4

Complete and balance the equations for the reactions:
(a) potassium phosphate + barium chloride
(b) hydrochloric acid + nickel carbonate
(c) ammonium chloride + sodium nitrate

Some of the reactions we attempt may fail because the substances are not reactive or because the proper conditions for reaction are not present. For example, mercury(II) oxide does not decompose until it is heated; magnesium does not burn in air or oxygen until the temperature reaches a certain point. When silver is placed in a solution of copper(II) sulfate, no reaction occurs. When a strip of copper is placed in a solution of silver nitrate, a single-displacement reaction takes place, because copper is a more reactive metal than silver.

The successful prediction of the products of a reaction is not always easy. The ability to predict products correctly comes with knowledge and experience. Although you may not be able to predict many reactions at this point, as you continue to experiment you will find that reactions can be categorized, and that prediction of the products thereby becomes easier, if not always certain.

We have a great deal yet to learn about which substances react with each other, how they react, and what conditions are necessary to bring about their reaction. It is possible to make accurate predictions concerning the occurrence of proposed reactions, but they require, in addition to appropriate data, a good knowledge of thermodynamics—a subject usually reserved for advanced courses in chemistry and physics. Even without the formal use of thermodynamics, your knowledge of the four general reaction types, the periodic table, atomic structure, and charges of ions can be put to good use in predicting reactions and in writing equations.

In this single-displacement reaction between a strip of copper and a solution of silver nitrate, crystals of silver metal are formed and the solution turns blue, indicating the presence of copper ions.

8.5 Heat in Chemical Reactions

Energy changes always accompany chemical reactions. One reason why reactions occur is that the products attain a lower, more stable energy state than the reactants. When the reaction leads to a more stable state, energy is released to the surroundings as heat (or as heat and work). When a solution of a base is neutralized by the addition of an acid, the liberation of heat energy is signaled by an immediate rise in the temperature of the solution. For example, when an automobile engine burns gasoline, heat is certainly liberated; at the same time, part of the liberated energy does the work of moving the automobile.

Reactions are either exothermic or endothermic. **Exothermic reactions** liberate heat; **endothermic reactions** absorb heat. In an exothermic reaction, heat is a product and may be written on the right side of the equation for the reaction. In an endothermic reaction, heat can be regarded as a reactant and is written on the left side of the equation. Here are two:

exothermic reaction
endothermic reaction

$$H_2(g) + Cl_2(g) \longrightarrow 2\,HCl(g) + 185\,kJ \quad (\textit{exothermic})$$
$$N_2(g) + O_2(g) + 181\,kJ \longrightarrow 2\,NO(g) \quad (\textit{endothermic})$$

The quantity of heat produced by a reaction is known as the **heat of reaction**. The units used can be kilojoules or kilocalories. Consider the reaction represented by this equation:

heat of reaction

$$C(s) + O_2(g) \longrightarrow CO_2(g) + 393\,kJ$$

When the heat liberated is expressed as part of the equation, the substances are expressed in units of moles. Thus, when 1 mol (12.01 g) of C combines with 1 mol (32.00 g) of O_2, 1 mol (44.01 g) of CO_2 is formed and 393 kJ of heat are liberated.

This cornfield is a good example of the endothermic reactions happening through photosynthesis in plants.

hydrocarbon

activation energy

In this reaction, as in many others, the heat energy is more useful than the chemical products.

Aside from relatively small amounts of energy from nuclear processes, the sun is the major provider of energy for life on earth. The sun maintains the temperature necessary for life and also supplies light energy for the endothermic photosynthetic reactions of green plants. In photosynthesis, carbon dioxide and water are converted to free oxygen and glucose:

$$6\ CO_2 + 6\ H_2O + 2519\ kJ \longrightarrow \underset{\text{glucose}}{C_6H_{12}O_6} + 6\ O_2$$

Nearly all of the chemical energy used by living organisms is obtained from glucose or compounds derived from glucose.

The major source of energy for modern technology is fossil fuel, such as coal, petroleum, and natural gas. The energy is obtained from the combustion (burning) of these fuels, which are converted to carbon dioxide and water. Fossil fuels are mixtures of **hydrocarbons**, compounds containing only hydrogen and carbon.

Natural gas is primarily methane, CH_4. Petroleum is a mixture of hydrocarbons (compounds of carbon and hydrogen). Liquefied petroleum gas (LPG) is a mixture of propane (C_3H_8) and butane (C_4H_{10}).

The combustion of these fuels releases a tremendous amount of energy, but reactions won't occur to a significant extent at ordinary temperatures. A spark or a flame must be present before methane will ignite. The amount of energy that must be supplied to start a chemical reaction is called the **activation energy**. Once this activation energy is provided, enough energy is then generated to keep the reaction going. Here are some examples:

$$CH_4(g) + 2\ O_2(g) \longrightarrow CO_2(g) + 2\ H_2O(g) + 890\ kJ$$

$$C_3H_8(g) + 5\ O_2(g) \longrightarrow 3\ CO_2(g) + 4\ H_2O(g) + 2200\ kJ$$

$$2\ C_8H_{18}(l) + 25\ O_2(g) \longrightarrow 16\ CO_2(g) + 18\ H_2O(g) + 10{,}900\ kJ$$

Be careful not to confuse an exothermic reaction that merely requires heat (activation energy) to get it started with a truly endothermic process. The combustion of magnesium, for example, is highly exothermic, yet magnesium must be heated to a fairly high temperature in air before combustion begins. Once started, however, the combustion reaction goes very vigorously until either the magnesium or the available supply of oxygen is exhausted. The electrolytic decomposition of water to hydrogen and oxygen is highly endothermic. If the electric current is shut off when this process is going on, the reaction stops instantly. The relative energy levels of reactants and products in exothermic and in endothermic processes are presented graphically in Figure 8.1.

In reaction (a) of Figure 8.1, the products are at a lower potential energy than the reactants. Energy (heat) is given off, producing an exothermic reaction. In reaction (b) the products are at a higher potential energy than the reactants. Energy has therefore been absorbed, and the reaction is endothermic.

Examples of endothermic and exothermic processes can be easily demonstrated in the laboratory. In Figure 8.2, solid $Ba(OH)_2$ and solid NH_4SCN are mixed in a beaker, which is standing in a puddle of water. The solids liquefy and absorb heat from the surroundings causing the beaker to freeze to the board. In another demonstration, potassium chlorate ($KClO_3$) and sugar are mixed and placed into a pile. A drop of concentrated sulfuric acid is added, creating a spectacular exothermic reaction (Figure 8.2).

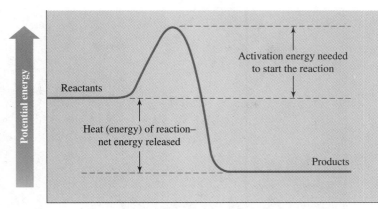

(a) Exothermic reaction

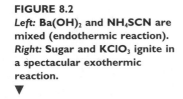

FIGURE 8.1
Energy changes in exothermic and endothermic reactions.

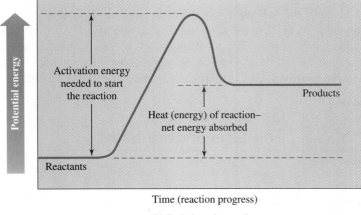

(b) Endothermic reaction

FIGURE 8.2
Left: Ba(OH)$_2$ and NH$_4$SCN are mixed (endothermic reaction). *Right:* Sugar and KClO$_3$ ignite in a spectacular exothermic reaction.

Carbon dioxide dissolves in ▶
the ocean, forming carbonates
and hydrogen carbonates.

Outside the laboratory you can experience an endothermic process when applying a cold pack to an injury. In this case ammonium chloride (NH_4Cl) dissolves in water. Temperature changes from 24.5°C to 18.1°C result when 10 g of NH_4Cl are added to 100 mL of water. Energy, in the form of heat, is taken from the immediate surroundings (water) causing the salt solution to become cooler.

8.6   Global Warming: The Greenhouse Effect

Fossil fuels, derived from coal and petroleum, provide the energy we use to power our industries, heat and light our homes and workplaces, and run our cars. As these fuels are burned they produce carbon dioxide and water, releasing over 50 billion tons of carbon dioxide into our atomosphere each year.

The concentration of carbon dioxide has been monitored by scientists since 1958. Analysis of the air trapped in a core sample of snow from Antartica provides data on carbon dioxide levels for the past 160,000 years. The results of this study show that as the carbon dioxide increased, the global temperature increased as well. The levels of carbon dioxide remained reasonably constant from the last ice age, 100,000 years ago, until the industrial revolution. Since then the concentration of carbon dioxide in our atmosphere has risen 15% to an all-time high.

Carbon dioxide is a minor component in our atmosphere and is not usually considered to be a pollutant. The concern expressed by scientists arises from the dramatic increase occurring in the earth's atmosphere. Without the influence of man in the environment, the exchange of carbon dioxide between plants and animals would be relatively balanced. Our continued use of fossil fuels has led to an increase of 7.4% in carbon dioxide between 1900 and 1970 and an additional 3.5% increase during the 1980s. See Figure 8.3.

In addition to the larger consumption of fossil fuels, there are still other factors that increase carbon dioxide levels in the atmosphere: Rain forests are being destroyed by cutting and burning to make room for increased population and agricul-

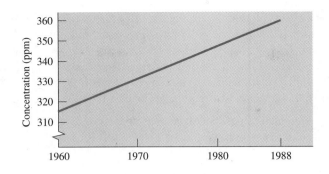

◄ FIGURE 8.3
Concentration of carbon dioxide in the atmosphere.

tural needs. Carbon dioxide is added to the atmosphere during the burning, and the loss of trees diminishes the uptake of carbon dioxide by plants.

About half of all the carbon dioxide released into our atmosphere each year remains there, thus increasing its concentration. The other half is absorbed by plants during photosynthesis or is dissolved in the ocean to form hydrogen carbonates and carbonates.

Carbon dioxide and other greenhouse gases, such as methane and water, act to warm our atmosphere by trapping heat near the surface of the earth. Solar radiation strikes the earth and warms the surface. The warmed surface then reradiates this energy as heat. The greenhouse gases absorb some of this heat energy from the surface, which warms our atmosphere. A similar principle is illustrated in a greenhouse where sunlight comes through the glass yet heat cannot escape. The air in the greenhouse warms, producing a climate considerably different than in nature. In the atmosphere these greenhouse gases are acting to warm our air and produce changes in our climate. See Figure 8.4.

FIGURE 8.4
Elements of a global temperature warming are caused by the greenhouse effect.
▼

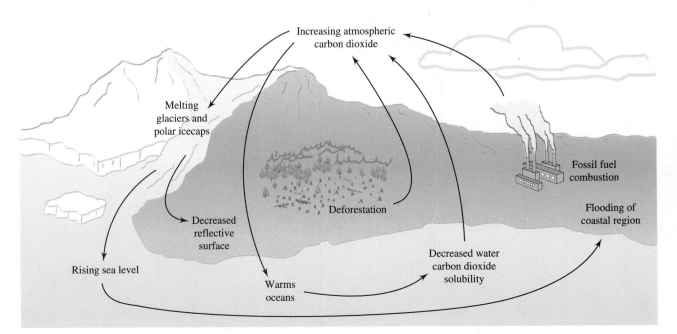

Long-term effects of global warming are still a matter of speculation and debate. One consideration is whether the polar ice caps would melt; this would cause a rise in sea level and lead to major flooding on the coasts of our continents. Further effects could include shifts in rainfall patterns, producing droughts and extreme seasonal changes in such major agricultural regions as California.

To reverse these trends will require major efforts in the following areas:

- The development of new energy sources to cut our dependence on fossil fuels
- An end to deforestation worldwide
- Intense efforts to improve conservation

On an individual basis each of us can play a significant role. For example, the simple conversion of a 100-watt incandescent light bulb to a compact fluorescent bulb can reduce the electrical consumption for that light 20%, and the bulb can last 10 times longer. Recycling, switching to more fuel-efficient cars, and energy-efficient applicances, heaters, and air conditioners all would result in decreased energy consumption and less carbon dioxide being released into our atmosphere.

Concepts in Review

1. Know the format used in setting up chemical equations.
2. Recognize the various symbols commonly used in writing chemical equations.
3. Be able to balance simple chemical equations.
4. Interpret a balanced equation in terms of the relative numbers or amounts of molecules, atoms, grams, or moles of each substance represented.
5. Classify equations as combination, decomposition, single-displacement, or double-displacement reactions.
6. Use the activity series to predict whether a single-displacement reaction will occur.
7. Complete and balance equations for simple combination, decomposition, single-displacement, and double-displacement reactions when given the reactants.
8. Distinguish between exothermic and endothermic reactions, and relate the quantity of heat to the amounts of substances involved in the reaction.
9. Identify the major sources of chemical energy and their uses.

Key Terms

The terms listed here have all been defined within the chapter. Section numbers are referenced in parenthesis for each term. More detailed definitions are given in the Glossary.

activation energy (8.5)	decomposition reaction (8.4)	heat of reaction (8.5)
balanced equation (8.2)	double-displacement	hydrocarbons (8.5)
chemical equation (8.1)	reaction (8.4)	single-displacement
combination reaction (8.4)	endothermic reaction (8.5)	reaction (8.4)
combustion (8.5)	exothermic reaction (8.5)	

Questions

Questions refer to tables, figures, and key words and concepts defined within the chapter. A particularly challenging question or exercise is indicated with an asterisk.

1. What is the purpose of balancing equations?

2. What is represented by the numbers (coefficients) that are placed in front of the formulas in a balanced equation?

3. In a balanced chemical equation:
   (a) are atoms conserved?
   (b) are molecules conserved?
   (c) are moles conserved?
   Explain your answers briefly.

4. Which of the following statements are correct? Rewrite the incorrect ones to make them correct.
   (a) The coefficients in front of the formulas in a balanced chemical equation give the relative number of moles of the reactants and products in the reaction.
   (b) A balanced chemical equation is one that has the same number of moles on each side of the equation.
   (c) In a chemical equation, the symbol $\xrightarrow{\Delta}$ indicates that the reaction is exothermic.
   (d) A chemical change that absorbs heat energy is said to be endothermic.
   (e) In the reaction $H_2 + Cl_2 \longrightarrow 2\,HCl$, 100 molecules of HCl are produced for every 50 molecules of H_2 reacted.
   (f) The symbol (aq) after a substance in an equation means that the substance is in a water solution.

   (g) The equation $H_2O \longrightarrow H_2 + O_2$ can be balanced by placing a 2 in front of H_2O.
   (h) In the equation $3\,H_2 + N_2 \longrightarrow 2\,NH_3$, there are fewer moles of product than there are moles of reactants.
   (i) The total number of moles of reactants and products represented by this equation is 5 mol:
       $$Mg + 2\,HCl \longrightarrow MgCl_2 + H_2$$
   (j) One mole of glucose, $C_6H_{12}O_6$, contains 6 mol of carbon atoms.
   (k) The reactants are the substances produced by the chemical reaction.
   (l) In a balanced equation, each side of the equation contains the same number of atoms of each element.
   (m) When a precipitate is formed in a chemical reaction, it can be indicated in the equation with an (s) immediately before the formula of the substance precipitated.
   (n) When a gas is involved in a chemical reaction, it can be indicated in the equation with a (g) immediately following the formula of the gas.
   (o) According to the equation $3\,H_2 + N_2 \longrightarrow 2\,NH_3$, 4 mol of NH_3 will be formed when 6 mol of H_2 and 2 mol of N_2 react.
   (p) The products of an exothermic reaction are at a lower potential energy than the reactants.
   (q) The combustion of hydrocarbons produces carbon dioxide and water as products.

Paired Exercises

These exercises are paired. Each odd-numbered exercise is followed by a similar even-numbered exercise. Answers to the even-numbered exercises are given in Appendix V.

5. Balance the following equations:
   (a) $H_2 + O_2 \longrightarrow H_2O$
   (b) $C + Fe_2O_3 \longrightarrow Fe + CO$
   (c) $H_2SO_4 + NaOH \longrightarrow H_2O + Na_2SO_4$
   (d) $Al_2(CO_3)_3 \xrightarrow{\Delta} Al_2O_3 + CO_2$
   (e) $NH_4I + Cl_2 \longrightarrow NH_4Cl + I_2$

6. Balance the following equations:
   (a) $H_2 + Br_2 \longrightarrow HBr$
   (b) $Al + C \xrightarrow{\Delta} Al_4C_3$
   (c) $Ba(ClO_3)_2 \xrightarrow{\Delta} BaCl_2 + O_2$
   (d) $CrCl_3 + AgNO_3 \longrightarrow Cr(NO_3)_3 + AgCl$
   (e) $H_2O_2 \longrightarrow H_2O + O_2$

7. Classify the reactions in Exercise 5 as combination, decomposition, single displacement, or double displacement.

8. Classify the reactions in Exercise 6 as combination, decomposition, single displacement, or double displacement.

9. Balance the following equations:
   (a) $SO_2 + O_2 \longrightarrow SO_3$
   (b) $Al + MnO_2 \xrightarrow{\Delta} Mn + Al_2O_3$
   (c) $Na + H_2O \longrightarrow NaOH + H_2$
   (d) $AgNO_3 + Ni \longrightarrow Ni(NO_3)_2 + Ag$
   (e) $Bi_2S_3 + HCl \longrightarrow BiCl_3 + H_2S$
   (f) $PbO_2 \xrightarrow{\Delta} PbO + O_2$
   (g) $LiAlH_4 \xrightarrow{\Delta} LiH + Al + H_2$
   (h) $KI + Br_2 \longrightarrow KBr + I_2$
   (i) $K_3PO_4 + BaCl_2 \longrightarrow KCl + Ba_3(PO_4)_2$

10. Balance the following equations:
    (a) $MnO_2 + CO \longrightarrow Mn_2O_3 + CO_2$
    (b) $Mg_3N_2 + H_2O \longrightarrow Mg(OH)_2 + NH_3$
    (c) $C_3H_5(NO_3)_3 \longrightarrow CO_2 + H_2O + N_2 + O_2$
    (d) $FeS + O_2 \longrightarrow Fe_2O_3 + SO_2$
    (e) $Cu(NO_3)_2 \longrightarrow CuO + NO_2 + O_2$
    (f) $NO_2 + H_2O \longrightarrow HNO_3 + NO$
    (g) $Al + H_2SO_4 \longrightarrow Al_2(SO_4)_3 + H_2$
    (h) $HCN + O_2 \longrightarrow N_2 + CO_2 + H_2O$
    (i) $B_5H_9 + O_2 \longrightarrow B_2O_3 + H_2O$

11. Change the following word equations into formula equations and balance them:
    (a) copper + sulfur $\xrightarrow{\Delta}$ copper(I) sulfide
    (b) phosphoric acid + calcium hydroxide $\longrightarrow$ calcium phosphate + water
    (c) silver oxide $\xrightarrow{\Delta}$ silver + oxygen
    (d) iron(III) chloride + sodium hydroxide $\longrightarrow$ iron(III) hydroxide + sodium chloride
    (e) nickel(II) phosphate + sulfuric acid $\longrightarrow$ nickel(II) sulfate + phosphoric acid
    (f) zinc carbonate + hydrochloric acid $\longrightarrow$ zinc chloride + water + carbon dioxide
    (g) silver nitrate + aluminum chloride $\longrightarrow$ silver chloride + aluminum nitrate

12. Change the following word equations into formula equations and balance them:
    (a) water $\longrightarrow$ hydrogen + oxygen
    (b) acetic acid + potassium hydroxide $\longrightarrow$ potassium acetate + water
    (c) phosphorus + iodine $\longrightarrow$ phosphorus triiodide
    (d) aluminum + copper(II) sulfate $\longrightarrow$ copper + aluminum sulfate
    (e) ammonium sulfate + barium chloride $\longrightarrow$ ammonium chloride + barium sulfate
    (f) sulfur tetrafluoride + water $\longrightarrow$ sulfur dioxide + hydrogen fluoride
    (g) chromium(III) carbonate $\xrightarrow{\Delta}$ chromium(III) oxide + carbon dioxide

13. Use the activity series to predict which of the following reactions will occur. Complete and balance the equations. Where no reaction will occur, write "no reaction" as the product.
    (a) $Ag(s) + H_2SO_4(aq) \longrightarrow$
    (b) $Cl_2(g) + NaBr(aq) \longrightarrow$
    (c) $Mg(s) + ZnCl_2(aq) \longrightarrow$
    (d) $Pb(s) + AgNO_3(aq) \longrightarrow$

14. Use the activity series to predict which of the following reactions will occur. Complete and balance the equations. Where no reaction will occur, write "no reaction" as the product.
    (a) $Cu(s) + FeCl_3(aq) \longrightarrow$
    (b) $H_2(g) + Al_2O_3(aq) \longrightarrow$
    (c) $Al(s) + HBr(aq) \longrightarrow$
    (d) $I_2(s) + HCl(aq) \longrightarrow$

15. Complete and balance the equations for these reactions. All reactions yield products.
    (a) $H_2 + I_2 \longrightarrow$
    (b) $CaCO_3 \xrightarrow{\Delta}$
    (c) $Mg + H_2SO_4 \longrightarrow$
    (d) $FeCl_2 + NaOH \longrightarrow$

16. Complete and balance the equations for these reactions. All reactions yield products.
    (a) $SO_2 + H_2O \longrightarrow$
    (b) $SO_3 + H_2O \longrightarrow$
    (c) $Ca + H_2O \longrightarrow$
    (d) $Bi(NO_3)_3 + H_2S \longrightarrow$

17. Complete and balance the equations for the following reactions. All reactions yield products.
    (a) $Ba + O_2 \longrightarrow$
    (b) $NaHCO_3 \xrightarrow{\Delta} Na_2CO_3 +$
    (c) $Ni + CuSO_4 \longrightarrow$
    (d) $MgO + HCl \longrightarrow$
    (e) $H_3PO_4 + KOH \longrightarrow$

18. Complete and balance the equations for the following reactions. All reactions yield products.
    (a) $C + O_2 \longrightarrow$
    (b) $Al(ClO_3)_3 \xrightarrow{\Delta} O_2 +$
    (c) $CuBr_2 + Cl_2 \longrightarrow$
    (d) $SbCl_3 + (NH_4)_2S \longrightarrow$
    (e) $NaNO_3 \xrightarrow{\Delta} NaNO_2 +$

19. Interpret the following chemical reactions in terms of the number of moles of each reactant and product:
    (a) $MgBr_2 + 2\,AgNO_3 \longrightarrow Mg(NO_3)_2 + 2\,AgBr$
    (b) $N_2 + 3\,H_2 \longrightarrow 2\,NH_3$
    (c) $2\,C_3H_7OH + 9\,O_2 \longrightarrow 6\,CO_2 + 8\,H_2O$

20. Interpret each of the following equations in terms of the relative number of moles of each substance involved and indicate whether the reaction is exothermic or endothermic:
    (a) $2\,Na + Cl_2 \longrightarrow 2\,NaCl + 822\ kJ$
    (b) $PCl_5 + 92.9\ kJ \longrightarrow PCl_3 + Cl_2$

21. Write balanced equations for each of these reactions, including the heat term:
    (a) Lime (CaO) is converted to slaked lime $Ca(OH)_2$ by reaction with water. The reaction liberates 65.3 kJ of heat for each mole of lime reacted.
    (b) The industrial production of aluminum metal from aluminum oxide is an endothermic electrolytic process requiring 1630 kJ per mole of Al_2O_3. Oxygen is also a product.

22. Write a balanced equation for each of the following descriptions. Include a heat term on the appropriate side of the equation.
    (a) Powdered aluminum will react with crystals of iodine when moistened with dishwashing detergent. The reaction produces violet sparks and flaming aluminum. The major product is aluminum iodide (AlI_3). The detergent is not a reactant.
    (b) Copper(II) oxide (CuO), a black powder, can be decomposed to produce pure copper by heating the powder in the presence of methane gas (CH_4). The products are copper, carbon dioxide and water vapor.
    (c) A form of rust, iron(III) oxide (Fe_2O_3), reacts with powdered aluminum to produce molten iron and aluminum oxide in the spectacular reaction shown in the opening photo for this chapter.

Additional Exercises

These problems are not paired or labeled by topic and provide additional practice on the concepts covered in this chapter.

23. Name one piece of evidence that a chemical reaction is actually taking place in each of these situations:
    (a) making a piece of toast
    (b) frying an egg
    (c) striking a match

24. Balance this equation, using the smallest possible whole numbers. Then determine how many atoms of oxygen appear on each side of the equation:

    $$P_4O_{10} + HClO_4 \longrightarrow Cl_2O_7 + H_3PO_4$$

25. Suppose that in a balanced equation the term $7\,Al_2(SO_4)_3$ appears.
    (a) How many atoms of aluminum are represented?
    (b) How many atoms of sulfur are represented?
    (c) How many atoms of oxygen are represented?
    (d) How many atoms of all kinds are represented?

26. Name two pieces of information that can be obtained from a balanced chemical equation. Name two pieces of information that the reaction does not provide.

27. Make a drawing to show six molecules of ammonia gas decomposing to form hydrogen and nitrogen gases.

28. Explain briefly why the following single-displacement reaction will not take place:

    $$Zn + Mg(NO_3)_2 \longrightarrow \text{no reaction}$$

29. A student does an experiment to determine where titanium metal should be placed on the activity series chart. He places newly cleaned pieces of titanium into solutions of nickel(II) nitrate, lead(II) nitrate, and magnesium nitrate. He finds that the titanium reacts with the nickel(II) nitrate and lead(II) nitrate solutions, but not with the magnesium nitrate solution. From this information place titanium in the activity series in a position relative to these ions.

30. Complete and balance the equations for these combination reactions:
    (a) $K + O_2 \longrightarrow$ (c) $CO_2 + H_2O \longrightarrow$
    (b) $Al + Cl_2 \longrightarrow$ (d) $CaO + H_2O \longrightarrow$

31. Complete and balance the equations for these decomposition reactions:
    (a) $HgO \xrightarrow{\Delta}$ (c) $MgCO_3 \xrightarrow{\Delta}$
    (b) $NaClO_3 \xrightarrow{\Delta}$ (d) $PbO_2 \xrightarrow{\Delta} PbO +$

32. Complete and balance the equations for these single-displacement reactions:
    (a) $Zn + H_2SO_4 \longrightarrow$ (c) $Mg + AgNO_3 \longrightarrow$
    (b) $AlI_3 + Cl_2 \longrightarrow$ (d) $Al + CoSO_4 \longrightarrow$

33. Complete and balance the equations for these double-displacement reactions:
    (a) $ZnCl_2 + KOH \rightarrow$ (d) $(NH_4)_3PO_4 + Ni(NO_3)_2 \rightarrow$
    (b) $CuSO_4 + H_2S \rightarrow$ (e) $Ba(OH)_2 + HNO_3 \rightarrow$
    (c) $Ca(OH)_2 + H_3PO_4 \rightarrow$ (f) $(NH_4)_2S + HCl \rightarrow$

34. Predict which of the following double-displacement reactions will occur. Complete and balance the equations. Where no reaction will occur, write "no reaction" as the product.
    (a) $AgNO_3(aq) + KCl(aq) \longrightarrow$
    (b) $Ba(NO_3)_2(aq) + MgSO_4(aq) \longrightarrow$
    (c) $H_2SO_4(aq) + Mg(OH)_2(aq) \longrightarrow$
    (d) $MgO(s) + H_2SO_4(aq) \longrightarrow$
    (e) $Na_2CO_3(aq) + NH_4Cl(aq) \longrightarrow$

35. Write balanced equations for the combustion of the following hydrocarbons:
  (a) ethane, C_2H_6 **(c)** heptane, C_7H_{16}
  (b) benzene, C_6H_6

36. List the various pigments found in plants and state the function of each.

37. State the four necessary requirements for successful photosynthesis.

38. Draw a flowchart illustrating how leaves "change color" in the autumn of the year.

39. List the factors that contribute to an increase in carbon dioxide in our atmosphere.

40. List three gases considered to be greenhouse gases. Explain why they are given this name.

41. How can the effects of global warming be reduced?

42. What happens to carbon dioxide released into our atmosphere?

Answers to Practice Exercises

8.1 $4\,Al + 3\,O_2 \longrightarrow 2\,Al_2O_3$

8.2 $3\,Mg(OH)_2 + 2\,H_3PO_4 \longrightarrow Mg_3(PO_4)_2 + 6\,H_2O$

8.3 (a) $Fe + MgCl_2 \longrightarrow$ no reaction
  (b) $Zn(s) + Pb(NO_3)_2(aq) \longrightarrow Pb(s) + Zn(NO_3)_2(aq)$

8.4 (a) $2\,K_3PO_4(aq) + 3\,BaCl_2(aq) \longrightarrow$
               $Ba_3(PO_4)_2(s) + 6\,KCl(aq)$
  (b) $2\,HCl(aq) + NiCO_3(aq) \longrightarrow$
               $NiCl_2(aq) + H_2O(l) + CO_2(g)$
  (c) $NH_4Cl(aq) + NaNO_3(aq) \longrightarrow$ no reaction

The old adage "waste not, want not" is appropriate in our daily life and in the laboratory. Determining correct amounts comes into play in most all professions. For example, a hostess determines the quantity of food and beverage necessary to serve her guests. These amounts are defined by specific recipes and a knowledge of the particular likes and dislikes of the guests. A seamstress determines the amount of material, lining, and trim necessary to produce a gown for her client by relying on a pattern or her own experience to guide the selection. A carpet layer determines the correct amount of carpet and padding necessary to recarpet a customer's house by calculating the floor area. The IRS determines the correct deduction for federal income taxes from your paycheck based on your expected annual income.

The chemist also finds it necessary to calculate amounts of products or reactants by using a balanced chemical equation. With these calculations, the chemist can control the amount of product by scaling the reaction up or down to fit the needs of the laboratory, and can thereby minimize waste or excess materials formed during the reaction.

9.1   A Short Review

Molar Mass. The molar mass is the sum of the atomic masses of all the atoms in a molecule. The molar mass also applies to the mass of a mole of any formula unit—atoms, molecules, or ions; it is the atomic mass of an atom, or the sum of the atomic masses in a molecule or an ion (in grams).

Relationship Between Molecule and Mole. A molecule is the smallest unit of a molecular substance (e.g., Cl_2), and a mole is Avogadro's number (6.022×10^{23}) of molecules of that substance. A mole of chlorine (Cl_2) has the same number of molecules as a mole of carbon dioxide, a mole of water, or a mole of any other molecular substance. When we relate molecules to molar mass, 1 molar mass is equivalent to 1 mol, or 6.022×10^{23} molecules.

In addition to referring to molecular substances, the term *mole* may refer to any chemical species. It represents a quantity (6.022×10^{23} particles) and may be applied to atoms, ions, electrons, and formula units of nonmolecular substances. In other words,

$$1 \text{ mole} = \begin{cases} 6.022 \times 10^{23} \text{ molecules} \\ 6.022 \times 10^{23} \text{ formula units} \\ 6.022 \times 10^{23} \text{ atoms} \\ 6.022 \times 10^{23} \text{ ions} \end{cases}$$

Other useful mole relationships are

$$\text{molar mass} = \frac{\text{grams of a substance}}{\text{number of moles of the substance}}$$

$$\text{molar mass} = \frac{\text{grams of a monatomic element}}{\text{number of moles of the element}}$$

$$\text{number of moles} = \frac{\text{number of molecules}}{6.022 \times 10^{23} \text{ molecules/mole}}$$

◀ **Chapter Opening Photo: A chemist must measure exact quantities of each rectant to produce new chemical compounds for use in our complex society.**

Balanced Equations. When using chemical equations for calculations of mole–mass–volume relationships between reactants and products, the equations must be balanced. Remember that the number in front of a formula in a balanced chemical equation can represent the number of moles of that substance in the chemical reaction.

9.2 Introduction to Stoichiometry: The Mole-Ratio Method

It is often necessary to calculate the amount of a substance that is produced from or needed to react with a given quantity of another substance. The area of chemistry that deals with the quantitative relationships among reactants and products is known as **stoichiometry** (*stoy-key-ah-meh-tree*). Although several methods are known, we firmly believe that the *mole* or *mole-ratio* method is generally best for solving problems in stoichiometry.

stoichiometry

A **mole ratio** is a ratio between the number of moles of any two species involved in a chemical reaction. For example, in the reaction

mole ratio

$$2\ H_2 + O_2 \longrightarrow 2\ H_2O$$
$$\text{2 mol}\quad\text{1 mol}\qquad\text{2 mol}$$

only six mole ratios apply to this reaction. They are

$$\frac{2\ \text{mol}\ H_2}{1\ \text{mol}\ O_2} \qquad \frac{2\ \text{mol}\ H_2}{2\ \text{mol}\ H_2O} \qquad \frac{1\ \text{mol}\ O_2}{2\ \text{mol}\ H_2}$$

$$\frac{1\ \text{mol}\ O_2}{2\ \text{mol}\ H_2O} \qquad \frac{2\ \text{mol}\ H_2O}{2\ \text{mol}\ H_2} \qquad \frac{2\ \text{mol}\ H_2O}{1\ \text{mol}\ O_2}$$

The mole ratio is a conversion factor used to convert the number of moles of one substance to the corresponding number of moles of another substance in a chemical reaction. For example, if we want to calculate the number of moles of H_2O that can be obtained from 4.0 mol of O_2, we use the mole ratio 2 mol H_2O/1 mol O_2:

$$4.0\ \cancel{\text{mol}\ O_2} \times \frac{2\ \text{mol}\ H_2O}{1\ \cancel{\text{mol}\ O_2}} = 8.0\ \text{mol}\ H_2O$$

Since stoichiometric problems are encountered throughout the entire field of chemistry, it is prudent to master this general method for their solution. The mole-ratio method makes use of three basic operations:

1. Convert the quantity of starting substance to moles (if it is not given in moles).
2. Convert the moles of starting substance to moles of desired substance.
3. Convert the moles of desired substance to the units specified in the problem.

Like learning to balance chemical equations, learning to make stoichiometric calculations requires practice. A detailed step-by-step description of the general method, together with a variety of worked examples, is given in the following paragraphs. Study this material and apply the method to the problems at the end of this chapter.

Use a balanced equation.

Step 1 Determine the number of moles of starting substance.
Identify the starting substance from the data given in the statement of the problem. If it is not in moles, convert the quantity of the starting substance to moles.

You may need to write the equation before beginning the problem.

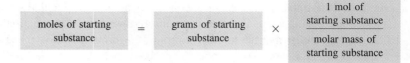

As in all problems with units, the desired quantity is in the numerator, and the quantity to be eliminated is in the denominator.

Step 2 Determine the mole ratio of the desired substance to the starting substance.

The number of moles of each substance in the balanced equation is indicated by the coefficient in front of each substance. Use these coefficients to set up the mole ratio:

$$\text{mole ratio} = \frac{\text{moles of desired substance in the equation}}{\text{moles of starting substance in the equation}}$$

Multiply the number of moles of starting substance (from Step 1) by the mole ratio to obtain the number of moles of desired substance:

Units of moles of starting substance cancel in the numerator and denominator.

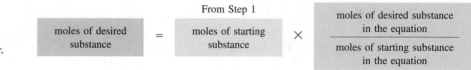

Step 3 Calculate the desired substance in the units specified in the problem.

If the answer is to be in moles, the problem is finished. If units other than moles are wanted, multiply the moles of the desired substance (from Step 2) by the appropriate factor to convert moles to the units required. For example, if grams of the desired substance are wanted,

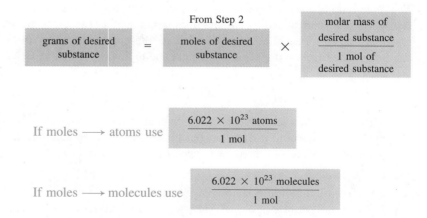

The steps for converting the mass of a starting substance *A* to either the mass, atoms, or molecules of desired substance *B* are summarized in Figure 9.1.

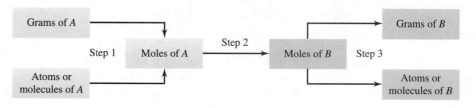

FIGURE 9.1 ▶
Steps for converting starting substance *A* to mass, atoms, or molecules of desired substance *B*.

9.3   Mole–Mole Calculations

The first application of the mole-ratio method of solving stoichiometric problems is that of mole–mole calculations.

The quantity of starting substance is given in moles and the quantity of desired substance is requested in moles. Understanding the use of mole ratios is very helpful in solving many problems. Some examples follow.

Example 9.1

How many moles of carbon dioxide will be produced by the complete reaction of 2.0 mol of glucose ($C_6H_{12}O_6$), according to the following reaction?

$$C_6H_{12}O_6 + 6\ O_2 \longrightarrow 6\ CO_2 + 6\ H_2O$$
$$\text{1 mol} \qquad \text{6 mol} \qquad \text{6 mol} \qquad \text{6 mol}$$

Solution

The balanced equation states that 6 mol of CO_2 will be produced from 1 mol of $C_6H_{12}O_6$. Even though we can readily see that 12 mol of CO_2 will be formed from 2.0 mol of $C_6H_{12}O_6$, the mole-ratio method of solving the problem is shown below.

Step 1 The number of moles of starting substance is 2.0 mol $C_6H_{12}O_6$.
Step 2 The conversion needed is

$$\text{moles } C_6H_{12}O_6 \longrightarrow \text{moles } CO_2$$

Multiply 2.0 mol of glucose (given in the problem) by this mole ratio:

$$2.0\ \text{mol } C_6H_{12}O_6 \times \frac{6\ \text{mol } CO_2}{1\ \text{mol } C_6H_{12}O_6} = 12\ \text{mol } CO_2$$

The numbers in the mole ratio are absolute and do not affect the number of significant figures in the answer.

Again note the use of units. The moles of $C_6H_{12}O_6$ cancel, leaving the answer in units of moles of CO_2.

Example 9.2

How many moles of ammonia can be produced from 8.00 mol of hydrogen reacting with nitrogen? The balanced equation is

$$3\ H_2 + N_2 \longrightarrow 2\ NH_3$$

Step 1 The starting substance is 8.00 mol of H_2.
Step 2 The conversion needed is

Solution

$$\text{moles } H_2 \longrightarrow \text{moles } NH_3$$

The balanced equation states that we get 2 mol of NH_3 for every 3 mol of H_2 that react. Set up the mole ratio of desired substance (NH_3) to starting substance (H_2):

$$\text{mole ratio} = \frac{2\ \text{mol } NH_3}{3\ \text{mol } H_2} \quad \text{(from equation)}$$

Multiply the 8.00 mol H_2 by the mole ratio:

$$8.00\ \text{mol } H_2 \times \frac{2\ \text{mol } NH_3}{3\ \text{mol } H_2} = 5.33\ \text{mol } NH_3$$

Example 9.3   Given the balanced equation

$$K_2Cr_2O_7 + 6 \text{ KI} + 7 \text{ H}_2SO_4 \longrightarrow Cr_2(SO_4)_3 + 4 \text{ K}_2SO_4 + 3 \text{ I}_2 + 7 \text{ H}_2O$$

$\qquad$ 1 mol $\qquad$ 6 mol $\qquad\qquad\qquad\qquad\qquad\qquad\qquad\qquad$ 3 mol

calculate (a) the number of moles of potassium dichromate ($K_2Cr_2O_7$) that will react with 2.0 mol of potassium iodide (KI); (b) the number of moles of iodine (I_2) that will be produced from 2.0 mol of potassium iodide.

Solution   Since the equation is balanced, we are concerned only with $K_2Cr_2O_7$, KI, and I_2, and we can ignore all the other substances. The equation states that 1 mol of $K_2Cr_2O_7$ will react with 6 mol of KI to produce 3 mol of I_2.

(a) Calculate the number of moles of $K_2Cr_2O_7$.

> **Step 1** The starting substance is 2.0 mol of KI.
>
> **Step 2** The conversion needed is
>
> $$\text{moles KI} \longrightarrow \text{moles } K_2Cr_2O_7$$
>
> Set up the mole ratio of desired substance to starting substance:
>
> $$\text{mole ratio} = \frac{1 \text{ mol } K_2Cr_2O_7}{6 \text{ mol KI}} \quad \text{(from equation)}$$
>
> Multiply the moles of starting material by this ratio:
>
> $$2.0 \text{ mol KI} \times \frac{1 \text{ mol } K_2Cr_2O_7}{6 \text{ mol KI}} = 0.33 \text{ mol } K_2Cr_2O_7$$

(b) Calculate the number of moles of I_2.

> **Step 1** The moles of starting substance are 2.0 mol KI as in part (a).
>
> **Step 2** The conversion needed is
>
> $$\text{moles KI} \longrightarrow \text{moles } I_2$$
>
> Set up the mole ratio of desired substance to starting substance:
>
> $$\text{mole ratio} = \frac{3 \text{ mol } I_2}{6 \text{ mol KI}} \quad \text{(from equation)}$$
>
> Multiply the moles of starting material by this ratio:
>
> $$2.0 \text{ mol KI} \times \frac{3 \text{ mol } I_2}{6 \text{ mol KI}} = 1.0 \text{ mol } I_2$$

Example 9.4   How many molecules of water can be produced by reacting 0.010 mol of oxygen with hydrogen?

Solution   The balanced equation is $2 \text{ H}_2 + \text{O}_2 \rightarrow 2 \text{ H}_2O$.
The sequence of conversions needed in the calculation is

$$\text{moles O}_2 \longrightarrow \text{moles H}_2O \longrightarrow \text{molecules H}_2O$$

◄ The space shuttle is powered by H_2 and O_2, which react to produce H_2O.

Step 1 The starting substance is 0.010 mol O_2.

Step 2 The conversion needed is moles $O_2 \longrightarrow$ moles H_2O. Set up the mole ratio of desired substance to starting substance:

$$\text{mole ratio} = \frac{2 \text{ mol } H_2O}{1 \text{ mol } O_2} \quad \text{(from equation)}$$

Multiply 0.010 mol O_2 by the mole ratio:

$$0.010 \text{ mol } O_2 \times \frac{2 \text{ mol } H_2O}{1 \text{ mol } O_2} = 0.020 \text{ mol } H_2O$$

Step 3 Since the problem asks for molecules instead of moles of H_2O, we must convert moles to molecules. Use the conversion factor $(6.022 \times 10^{23} \text{ molecules})/\text{mole}$:

$$0.020 \text{ mol } H_2O \times \frac{6.022 \times 10^{23} \text{ molecules}}{1 \text{ mol}} = 1.2 \times 10^{22} \text{ molecules } H_2O$$

Note that 0.020 mol is still quite a large number of water molecules.

Practice 9.1

How many moles of aluminum oxide will be produced from 0.50 mol of oxygen?

$$4\ Al\ +\ 3\ O_2\ \longrightarrow\ 2\ Al_2O_3$$

Practice 9.2

How many moles of aluminum hydroxide are required to produce 22.0 mol of water?

$$2\ Al(OH)_3\ +\ 3\ H_2SO_4\ \longrightarrow\ Al_2(SO_4)_3\ +\ 6\ H_2O$$

9.4   Mole–Mass Calculations

The object of this type of problem is to calculate the mass of one substance that reacts with or is produced from a given number of moles of another substance in a chemical reaction. If the mass of the starting substance is given, it is necessary to convert it to moles. The mole ratio is used to convert from moles of starting substance to moles of desired substance. Moles of desired substance can then be changed to mass if required. Each example is solved in two ways:

- **Method 1:** Step-by-Step.
- **Method 2:** Continuous Calculation where all the individual steps are combined in a single line.

Select the method that is easiest for you and focus your attention on solving problems that way.

Example 9.5   What mass of hydrogen can be produced by reacting 6.0 mol of aluminum with hydrochloric acid?

Solution   The balanced equation is $2\ Al(s)\ +\ 6\ HCl(aq)\ \longrightarrow\ 2\ AlCl_3(aq)\ +\ 3\ H_2(g)$.

Method 1: Step-by-Step
First calculate the moles of hydrogen produced, using the mole-ratio method; then calculate the mass of hydrogen by multiplying the moles of hydrogen by its grams per mole. The sequence of conversions in the calculation is

moles Al $\longrightarrow$ moles H_2 $\longrightarrow$ grams H_2

Step 1 The starting substance is 6.0 mol of aluminum.
Step 2 Calculate moles of H_2 by the mole-ratio method:

$$6.0\ \cancel{mol\ Al} \times \frac{3\ mol\ H_2}{2\ \cancel{mol\ Al}} = 9.0\ mol\ H_2$$

Step 3 Convert moles of H_2 to grams [g = mol × (g/mol)]:

$$9.0 \text{ mol } H_2 \times \frac{2.016 \text{ g } H_2}{1 \text{ mol } H_2} = 18 \text{ g } H_2 \quad \text{(Answer)}$$

We see that 18 g of H_2 can be produced by reacting 6.0 mol of Al with HCl.

Method 2: Continuous Calculation

$$6.0 \text{ mol } Al \times \frac{3 \text{ mol } H_2}{2 \text{ mol } Al} \times \frac{2.016 \text{ g } H_2}{1 \text{ mol } H_2} = 18 \text{ g } H_2$$

How many moles of water can be produced by burning 325 g of octane (C_8H_{18})? The balanced equation is $2 C_8H_{18}(g) + 25 O_2(g) \rightarrow 16 CO_2(g) + 18 H_2O(g)$.

Example 9.6

Method 1: Step-by-Step
The sequence of conversions in the calculation is

Solution

$$\text{grams } C_8H_{18} \longrightarrow \text{moles } C_8H_{18} \longrightarrow \text{moles } H_2O$$

Step 1 The starting substance is 325 g C_8H_{18}. Convert 325 g of C_8H_{18} to moles:

$$325 \text{ g } C_8H_{18} \times \frac{1 \text{ mol } C_8H_{18}}{114.2 \text{ g } C_8H_{18}} = 2.85 \text{ mol } C_8H_{18}$$

Step 2 Calculate the moles of water by the mole-ratio method:

$$2.85 \text{ mol } C_8H_{18} \times \frac{18 \text{ mol } H_2O}{2 \text{ mol } C_8H_{18}} = 25.7 \text{ mol } H_2O \quad \text{(Answer)}$$

Method 2: Continuous Calculation

$$325 \text{ g } C_8H_{18} \times \frac{1 \text{ mol } C_8H_{18}}{114.2 \text{ g } C_8H_{18}} \times \frac{18 \text{ mol } H_2O}{2 \text{ mol } C_8H_{18}} = 25.6 \text{ mol } H_2O$$

Notice the answers for the different methods vary in the last digit. This results from rounding off at different times in the calculation. Check with your instructor to find the appropriate rules for your course.

Practice 9.3

How many moles of potassium chloride can be produced from 100.0 g of potassium chlorate?

$$2 KClO_3 \longrightarrow 2 KCl + 3 O_2$$

Practice 9.4

How many grams of silver nitrate are required to produce 0.25 mol of silver sulfide?

$$2 AgNO_3 + H_2S \longrightarrow Ag_2S + 2 HNO_3$$

9.5   Mass–Mass Calculations

Remember to select either step-by-step method or continuous calculation for solving these problems.

Solving mass–mass stoichiometry problems requires all the steps of the mole-ratio method. The mass of starting substance is converted to moles. The mole ratio is then used to determine moles of desired substance, which in turn is converted to mass.

Example 9.7

What mass of carbon dioxide is produced by the complete combustion of 100. g of the hydrocarbon pentane, C_5H_{12}? The balanced equation is
$C_5H_{12} + 8 O_2 \rightarrow 5 CO_2 + 6 H_2O$.

Solution

Method 1: Step-by-Step
The sequence of conversions in the calculation is

$$\text{grams } C_5H_{12} \longrightarrow \text{moles } C_5H_{12} \longrightarrow \text{moles } CO_2 \longrightarrow \text{grams } CO_2$$

Step 1 The starting substance is 100. g of C_5H_{12}. Convert 100. g of C_5H_{12} to moles:

$$100. \text{ g } C_5H_{12} \times \frac{1 \text{ mol } C_5H_{12}}{72.15 \text{ g } C_5H_{12}} = 1.39 \text{ mol } C_5H_{12}$$

Step 2 Calculate the moles of CO_2 by the mole-ratio method:

$$1.39 \text{ mol } C_5H_{12} \times \frac{5 \text{ mol } CO_2}{1 \text{ mol } C_5H_{12}} = 6.95 \text{ mol } CO_2$$

Step 3 Convert moles of CO_2 to grams:

$$\text{mol } CO_2 \times \frac{\text{molar mass } CO_2}{1 \text{ mol } CO_2} = \text{grams } CO_2$$

$$6.95 \text{ mol } CO_2 \times \frac{44.01 \text{ g } CO_2}{1 \text{ mol } CO_2} = 306 \text{ g } CO_2$$

Method 2: Continuous Calculation

Remember to round off as appropriate for your particular course.

$$100. \text{ g } C_5H_{12} \times \frac{1 \text{ mol } C_5H_{12}}{72.15 \text{ g } C_5H_{12}} \times \frac{5 \text{ mol } CO_2}{1 \text{ mol } C_5H_{12}} \times \frac{44.01 \text{ g } CO_2}{1 \text{ mol } CO_2} = 305 \text{ g } CO_2$$

Example 9.8

How many grams of nitric acid, HNO_3, are required to produce 8.75 g of dinitrogen monoxide (N_2O) according to the following equation?

$$4 Zn(s) + 10 HNO_3(aq) \longrightarrow 4 Zn(NO_3)_2(aq) + N_2O(g) + 5 H_2O(l)$$
$$\qquad\qquad\quad 10 \text{ mol} \qquad\qquad\qquad\qquad\qquad 1 \text{ mol}$$

Solution

Method 1: Step-by-Step
The sequence of conversions in the calculation is

$$\text{grams } N_2O \longrightarrow \text{moles } N_2O \longrightarrow \text{moles } HNO_3 \longrightarrow \text{grams } HNO_3$$

A Shrinking Technology

In the high-tech world of computers, the microchip has miniaturized the field of electronics. In order to produce ever smaller computers, calculators, and even microbots (microsized robots) precise quantities of chemicals in exact proportions are required.

Engineers at Bell Laboratories, Massachusetts Institute of Technology, University of California, and Stanford University are busy trying to produce parts for tiny machines and robots. New techniques are allowing them to produce gears smaller than a grain of sand, and motors lighter than a speck of dust.

The secret behind these tiny circuits is to print the entire circuit or blueprint at one time. Computers are used to draw a design of how the chip will look. This image is then transferred into a pattern, or mask, with details finer than a human hair. Light is then shone through the mask onto a silicon-coated surface. The process is similar to photography. The areas created on the silicon exhibit high or low resistance to chemical etching. Chemicals are then applied to etch away the silicon.

Micromachinery is produced in the same way. First a thin layer of silicon dioxide is applied (sacrificial material). Then a layer of polysilicon is carefully applied (structural material). A mask is then applied and the whole structure is covered with plasma (excited gas). The plasma acts as a tiny sandblaster removing everything the mask doesn't protect. This process is repeated as the entire machine is constructed. When the entire assembly is complete, the whole machine is placed in hydrofluoric acid, which dissolves all the sacrificial material and permits the various parts of the machine to move.

More development is necessary to turn these micromachines into true microbots, but research is currently underway to design methods of locomotion and sensing imaging systems. Possible uses for these microbots include "smart" pills, which could contain sensors (see Chemistry in Action, page 34), or drug reservoirs (currently done for birth control). Tiny pumps, once inside the body, will be able to dispense the proper amount of medication at precisely the correct site.

These microbots are currently in production for the treatment of diabetes (to release insulin).

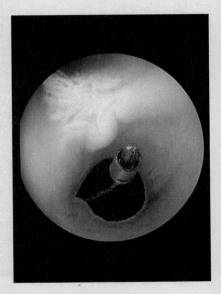

An endoscopic image of a micromotor in an artery.

Step 1 The starting substance is 8.75 g of N_2O. Convert 8.75 g of N_2O to moles:

$$8.75 \ \text{g } N_2O \times \frac{1 \ \text{mol } N_2O}{44.02 \ \text{g } N_2O} = 0.199 \ \text{mol } N_2O$$

Step 2 Calculate the moles of HNO_3 by the mole-ratio method:

$$0.199 \ \text{mol } N_2O \times \frac{10 \ \text{mol } HNO_3}{1 \ \text{mol } N_2O} = 1.99 \ \text{mol } HNO_3$$

Step 3 Convert moles of HNO_3 to grams:

$$1.99 \ \text{mol } HNO_3 \times \frac{63.02 \ \text{g } HNO_3}{1 \ \text{mol } HNO_3} = 125 \ \text{g } HNO_3$$

Method 2: Continuous Calculation

$$8.75 \ \text{g } N_2O \times \frac{1 \ \text{mol } N_2O}{44.02 \ \text{g } N_2O} \times \frac{10 \ \text{mol } HNO_3}{1 \ \text{mol } N_2O} \times \frac{63.02 \ \text{g } HNO_3}{1 \ \text{mol } HNO_3} = 125 \ \text{g } HNO_3$$

Practice 9.5

How many grams of chromium(III) chloride are required to produce 75.0 g of silver chloride?

$$CrCl_3 + 3\ AgNO_3 \longrightarrow Cr(NO_3)_3 + 3\ AgCl$$

Practice 9.6

What mass of water is produced by the complete combustion of 225.0 g of butane (C_4H_{10})?

$$2\ C_4H_{10} + 13\ O_2 \longrightarrow 8\ CO_2 + 10\ H_2O$$

9.6   Limiting-Reactant and Yield Calculations

limiting reactant

In many chemical processes the quantities of the reactants used are such that the amount of one reactant is in excess of the amount of a second reactant in the reaction. The amount of the product(s) formed in such a case will depend on the reactant that is not in excess. Thus the reactant that is not in excess is known as **the limiting reactant** (sometimes called the limiting reagent), because it limits the amount of product that can be formed.

Consider the case illustrated in Figure 9.2. How many bicycles can be assembled

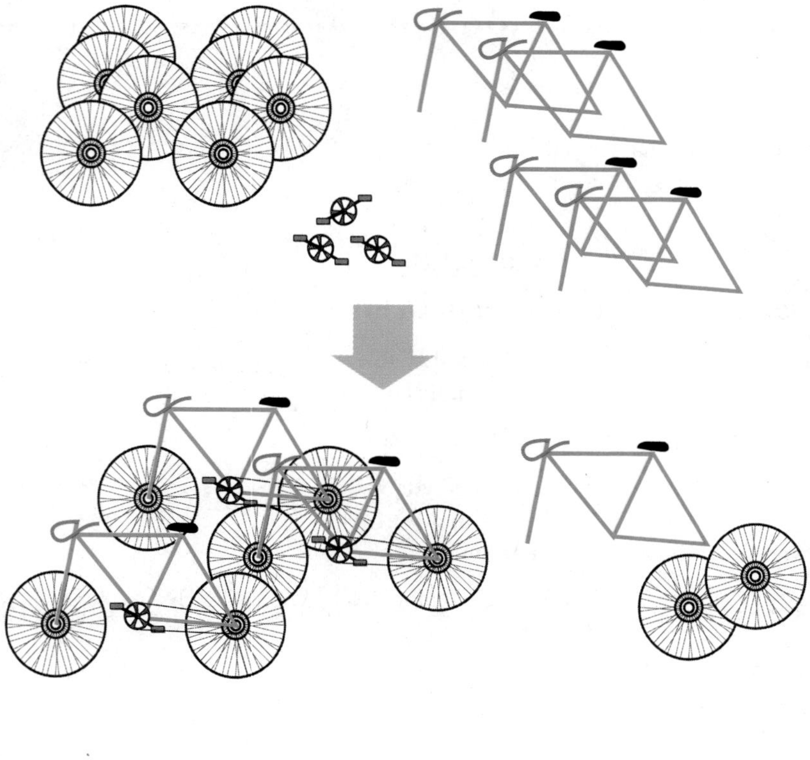

The number of bicycles that ▶ can be built from this pile of parts is determined by the limiting reactant (the pedal assemblies).

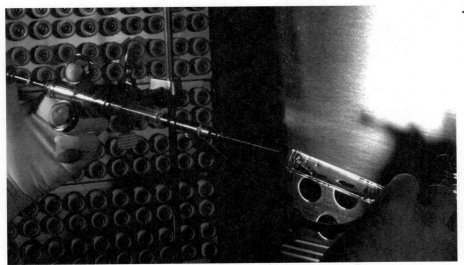

◀ **The amount of product formed into each medicine is determined by the limiting reactant.**

The limiting reactant in this case is the number of pedal assemblies. This part limits the number of bicycles that can be built. The wheels and frames are reactants in excess.

Now let's consider a chemical situation in which solutions containing 1.0 mol of sodium hydroxide and 1.5 mol of hydrochloric acid are mixed:

$$NaOH + HCl \longrightarrow NaCl + H_2O$$

　1 mol　　1 mol　　　1 mol　　1 mol

According to the equation it is possible to obtain 1.0 mol of NaCl from 1.0 mol of NaOH, and 1.5 mol of NaCl from 1.5 mol of HCl. However, we cannot have two different yields of NaCl from the reaction. When 1.0 mol of NaOH and 1.5 mol of HCl are mixed, there is insufficient NaOH to react with all of the HCl. Therefore, HCl is the reactant in excess and NaOH is the limiting reactant. Since the amount of NaCl formed is dependent on the limiting reactant, only 1.0 mol of NaCl will be formed. Because 1.0 mol of NaOH reacts with 1.0 mol of HCl, 0.5 mol of HCl remains unreacted:

$$\left.\begin{array}{l} 1.0 \text{ mol NaOH} \\ 1.5 \text{ mol HCl} \end{array}\right\} \longrightarrow \begin{array}{l} 1.0 \text{ mol NaCl} \\ 1.0 \text{ mol H}_2\text{O} \end{array} + 0.5 \text{ mol HCl unreacted}$$

Problems giving the amounts of two reactants are generally of the limiting-reactant type, and there are several methods used to identify them in a chemical reaction. In the most direct method two steps are needed to determine the limiting reactant and the amount of product formed.

1. Calculate the amount of product (moles or grams, as needed) that can be formed from each reactant.
2. Determine which reactant is limiting. (The reactant that gives the least amount of product is the limiting reactant; the other reactant is in excess. The limiting reactant will determine the amount of product formed in the reaction.)

Sometimes, however, it is necessary to calculate the amount of excess reactant.

3. This can be done by first calculating the amount of excess reactant required to react with the limiting reactant. Then subtract the amount that reacts from the starting quantity of the reactant in excess. This result is the amount of that substance that remains unreacted.

Example 9.9

How many moles of Fe_3O_4 can be obtained by reacting 16.8 g Fe with 10.0 g H_2O? Which substance is the limiting reactant? Which substance is in excess?

Solution

$$3\ Fe(s)\ +\ 4\ H_2O(g) \xrightarrow{\Delta} Fe_3O_4(s)\ +\ 4\ H_2(g)$$

Step 1 Calculate the moles of Fe_3O_4 that can be formed from each reactant:

$$g\ reactant \longrightarrow mol\ reactant \longrightarrow mol\ Fe_3O_4$$

The continuous calculation method is shown here. You can also use the step-by-step method to determine the requested substance.

$$16.8\ g\ Fe \times \frac{1\ mol\ Fe}{55.85\ g\ Fe} \times \frac{1\ mol\ Fe_3O_4}{3\ mol\ Fe} = 0.100\ mol\ Fe_3O_4$$

$$10.0\ g\ H_2O \times \frac{1\ mol\ H_2O}{18.02\ g\ H_2O} \times \frac{1\ mol\ Fe_3O_4}{4\ mol\ H_2O} = 0.139\ mol\ Fe_3O_4$$

Step 2 Determine the limiting reactant. The limiting reactant is Fe because it produces less Fe_3O_4; the H_2O is in excess. The yield of product is 0.100 mol of Fe_3O_4.

Example 9.10

How many grams of silver bromide (AgBr) can be formed when solutions containing 50.0 g of $MgBr_2$ and 100.0 g of $AgNO_3$ are mixed together? How many grams of the excess reactant remain unreacted?

Solution

$$MgBr_2(aq)\ +\ 2\ AgNO_3(aq) \longrightarrow 2\ AgBr(s)\ +\ Mg(NO_3)_2(aq)$$

Step 1 Calculate the grams of AgBr that can be formed from each reactant.

$$g\ reactant \longrightarrow mol\ reactant \longrightarrow mol\ AgBr \longrightarrow g\ AgBr$$

$$50.0\ g\ MgBr_2 \times \frac{1\ mol\ MgBr_2}{184.1\ g\ MgBr_2} \times \frac{2\ mol\ AgBr}{1\ mol\ MgBr_2} \times \frac{187.8\ g\ AgBr}{1\ mol\ AgBr} = 102\ g\ AgBr$$

$$100.0\ g\ AgNO_3 \times \frac{1\ mol\ AgNO_3}{169.9\ g\ AgNO_3} \times \frac{2\ mol\ AgBr}{2\ mol\ AgNO_3} \times \frac{187.8\ g\ AgBr}{1\ mol\ AgBr} = 110.5\ g\ AgBr$$

Step 2 Determine the limiting reactant. The limiting reactant is $MgBr_2$ because it gives less AgBr; $AgNO_3$ is in excess. The yield is 102 g AgBr.

Step 3 Calculate the grams of unreacted $AgNO_3$. Calculate the grams of $AgNO_3$ that will react with 50.0 g of $MgBr_2$:

$$g\ MgBr_2 \longrightarrow mol\ MgBr_2 \longrightarrow mol\ AgNO_3 \longrightarrow g\ AgNO_3$$

$$50.0\ g\ MgBr_2 \times \frac{1\ mol\ MgBr_2}{184.1\ g\ MgBr_2} \times \frac{2\ mol\ AgNO_3}{1\ mol\ MgBr_2} \times \frac{169.9\ g\ AgNO_3}{1\ mol\ AgNO_3} = 92.3\ g\ AgNO_3$$

Thus, 92.3 g of $AgNO_3$ react with 50.0 g of $MgBr_2$. The amount of $AgNO_3$ that remains unreacted is

$$100.0 \text{ g AgNO}_3 - 92.3 \text{ g AgNO}_3 = 7.7 \text{ g AgNO}_3 \text{ unreacted}$$

The final mixture will contain 102 g $AgBr(s)$, 7.7 g $AgNO_3$ and an undetermined amount of $Mg(NO_3)_2$ in solution.

Practice 9.7

How many grams of hydrogen chloride can be produced from 0.490 g of hydrogen and 50.0 g of chlorine?

$$H_2(g) + Cl_2(g) \longrightarrow 2 \text{ HCl}(g)$$

Practice 9.8

How many grams of barium sulfate will be formed from 200.0 g of barium nitrate and 100.0 grams of sodium sulfate?

$$Ba(NO_3)_2(aq) + Na_2SO_4(aq) \longrightarrow BaSO_4(s) + 2 \text{ NaNO}_3(aq)$$

The quantities of the products we have been calculating from equations represent the maximum yield (100%) of product according to the reaction represented by the equation. Many reactions, especially those involving organic substances, fail to give a 100% yield of product. The main reasons for this failure are the side reactions that give products other than the main product and the fact that many reactions are reversible. In addition, some product may be lost in handling and transferring from one vessel to another. The **theoretical yield** of a reaction is the calculated amount of product that can be obtained from a given amount or reactant, according to the chemical equation. The **actual yield** is the amount of product that we finally obtain.

The **percent yield** is the ratio of the actual yield to the theoretical yield multiplied by 100. Both the theoretical and the actual yields must have the same units to obtain a percent:

$$\frac{\text{actual yield}}{\text{theoretical yield}} \times 100 = \text{percent yield}$$

For example, if the theoretical yield calculated for a reaction is 14.8 g and the amount of product obtained is 9.25 g, the percent yield is

$$\text{percent yield} = \frac{9.25 \text{ g}}{14.8 \text{ g}} \times 100 = 62.5\%$$

theoretical yield

actual yield
percent yield

Remember to round off as appropriate for your particular course.

Carbon tetrachloride (CCl_4) was prepared by reacting 100. g of carbon disulfide and 100. g of chlorine. Calculate the percent yield if 65.0 g of CCl_4 was obtained from the reaction:

$$CS_2 + 3 \text{ Cl}_2 \longrightarrow CCl_4 + S_2Cl_2$$

Example 9.11

Solution   In this problem we need to determine the limiting reactant in order to calculate the quantity of CCl_4 (theoretical yield) that can be formed. Then we can compare that amount with the 65.0 g CCl_4 (actual yield) to calculate the percent yield.

Step 1 Determine the theoretical yield. Calculate the grams of CCl_4 that can be formed from each reactant:

$$g \text{ reactant} \longrightarrow mol \text{ reactant} \longrightarrow mol \ CCl_4 \longrightarrow g \ CCl_4$$

$$100. \ g \ CS_2 \times \frac{1 \ mol \ CS_2}{76.13 \ g \ CS_2} \times \frac{1 \ mol \ CCl_4}{1 \ mol \ CS_2} \times \frac{153.8 \ g \ CCl_4}{1 \ mol \ CCl_4} = 202 \ g \ CCl_4$$

$$100. \ g \ Cl_2 \times \frac{1 \ mol \ Cl_2}{70.90 \ g \ Cl_2} \times \frac{1 \ mol \ CCl_4}{3 \ mol \ Cl_2} \times \frac{153.8 \ g \ CCl_4}{1 \ mol \ CCl_4} = 72.3 \ g \ CCl_4$$

Step 2 Determine the limiting reactant. The limiting reactant is Cl_2 because it gives less CCl_4. The CS_2 is in excess. The theoretical yield is 72.3 g CCl_4.

Step 3 Calculate the percent yield. According to the equation, 72.3 g of CCl_4 is the maximum amount or theoretical yield of CCl_4 possible from 100. g of Cl_2. Actual yield is 65.0 g of CCl_4:

$$\text{percent yield} = \frac{65.0 \ g}{72.3 \ g} \times 100 = 89.9\%$$

Example 9.12   Silver bromide was prepared by reacting 200.0 g of magnesium bromide and an adequate amount of silver nitrate. Calculate the percent yield if 375.0 g of silver bromide was obtained from the reaction:

$$MgBr_2 + 2 \ AgNO_3 \longrightarrow Mg(NO_3)_3 + 2 \ AgBr$$

Solution   **Step 1** Determine the theoretical yield. Calculate the grams of AgBr that can be formed:

$$200.0 \ g \ MgBr_2 \times \frac{1 \ mol \ MgBr_2}{184.1 \ g \ MgBr_2} \times \frac{2 \ mol \ AgBr}{1 \ mol \ MgBr_2} \times \frac{187.8 \ g \ AgBr}{1 \ mol \ AgBr} = 408.0 \ g \ AgBr$$

The theoretical yield is 408.0 g AgBr.

Step 2 Calculate the percent yield. According to the equation, 408.0 g AgBr is the maximum amount of AgBr possible from 200.0 g $MgBr_2$. Actual yield is 375.0 g AgBr:

$$\text{Percent yield} = \frac{375.0 \ g \ AgBr}{408.0 \ g \ AgBr} \times 100 = 91.91\%$$

Practice 9.9

Aluminum oxide was prepared by heating 225 g of chromium(II) oxide with 125 g of aluminum. Calculate the percent yield if 100.0 g of aluminum oxide was obtained:

$$2 \ Al + 3 \ CrO \longrightarrow Al_2O_3 + 3 \ Cr$$

When solving problems, you will achieve better results if you work in an organized manner.

1. Write the data and numbers in a logical, orderly manner.
2. Make certain that the equations are balanced and that the computations are accurate and expressed to the correct number of significant figures.
3. Remember that units are very important; a number without units has little meaning.

Concepts in Review

1. Write mole ratios for any two substances involved in a chemical reaction.
2. Outline the mole or mole-ratio method for making stoichiometric calculations.
3. Calculate the number of moles of a desired substance obtainable from a given number of moles of a starting substance in a chemical reaction (mole–mole calculations).
4. Calculate the mass of a desired substance obtainable from a given number of moles of a starting substance in a chemical reaction, and vice versa (mole–mass and mass–mole calculations).
5. Calculate the mass of a desired substance involved in a chemical reaction from a given mass of a starting substance (mass–mass calculation).
6. Deduce the limiting reactant or reagent when given the amounts of starting substances, and then calculate the moles or mass of desired substance obtainable from a given chemical reaction (limiting reactant calculation).
7. Apply theoretical yield or actual yield to any of the foregoing types of problems, or calculate theoretical and actual yields of a chemical reaction.

Key Terms

The terms listed here have all been defined within the chapter. Section numbers are referenced in parenthesis for each term. More detailed definitions are given in the Glossary.

actual yield (9.6)   percent yield (9.6)
limiting reactant (9.6)   stoichiometry (9.2)
mole ratio (9.2)   theoretical yield (9.6)

Questions

Questions refer to tables, figures, and key words and concepts defined within the chapter. A particularly challenging question or exercise is indicated with an asterisk.

1. Phosphine (PH_3) can be prepared by the hydrolysis of calcium phosphide, Ca_3P_2:

$$Ca_3P_2 + 6\ H_2O \longrightarrow 3\ Ca(OH)_2 + 2\ PH_3$$

   Based on this equation, which of the following statements are correct? Show evidence to support your answer.
   (a) One mole of Ca_3P_2 produces 2 mol of PH_3.
   (b) One gram of Ca_3P_2 produces 2 g of PH_3.
   (c) Three moles of $Ca(OH)_2$ are produced for each 2 mol of PH_3 produced.
   (d) The mole ratio between phosphine and calcium phosphide is $\dfrac{2\ \text{mol}\ PH_3}{1\ \text{mol}\ Ca_3P_2}$.
   (e) When 2.0 mol of Ca_3P_2 and 3.0 mol of H_2O react, 4.0 mol of PH_3 can be formed.
   (f) When 2.0 mol of Ca_3P_2 and 15.0 mol of H_2O react, 6.0 mol of $Ca(OH)_2$ can be formed.
   (g) When 200. g of Ca_3P_2 and 100. g of H_2O react, Ca_3P_2 is the limiting reactant.
   (h) When 200. g of Ca_3P_2 and 100. g of H_2O react, the theoretical yield of PH_3 is 57.4 g.

2. The equation representing the reaction used for the commercial preparation of hydrogen cyanide is

$$2\ CH_4 + 3\ O_2 + 2\ NH_3 \longrightarrow 2\ HCN + 6\ H_2O$$

   Based on this equation, which of the following statements are correct? Rewrite incorrect statements to make them correct.
   (a) Three moles of O_2 are required for 2 mol of NH_3.
   (b) Twelve moles of HCN are produced for every 16 mol of O_2 that react.
   (c) The mole ratio between H_2O and CH_4 is $\dfrac{6\ \text{mol}\ H_2O}{2\ \text{mol}\ CH_4}$
   (d) When 12 mol of HCN are produced, 4 mol of H_2O will be formed.
   (e) When 10 mol of CH_4, 10 mol of O_2, and 10 mol of NH_3 are mixed and reacted, O_2 is the limiting reactant.
   (f) When 3 mol each of CH_4, O_2, and NH_3 are mixed and reacted, 3 mol of HCN will be produced.

Paired Exercises

These exercises are paired. Each odd-numbered exercise is followed by a similar even-numbered exercise. Answers to the even-numbered exercises are given in Appendix V.

Mole Review Exercises

3. Calculate the number of moles in each of the following quantities:
   (a) 25.0 g KNO_3
   (b) 56 millimol NaOH
   (c) 5.4×10^2 g $(NH_4)_2C_2O_4$
   (d) 16.8 mL H_2SO_4 solution ($d = 1.727$ g/mL, 80.0% H_2SO_4 by mass)

4. Calculate the number of moles in each of the following quantities:
   (a) 2.10 kg $NaHCO_3$
   (b) 525 mg $ZnCl_2$
   (c) 9.8×10^{24} molecules CO_2
   (d) 250 mL ethyl alcohol, C_2H_5OH ($d = 0.789$ g/mL)

5. Calculate the number of grams in each of the following quantities:
   (a) 2.55 mol $Fe(OH)_3$
   (b) 125 kg $CaCO_3$
   (c) 10.5 mol NH_3
   (d) 72 millimol HCl
   (e) 500.0 mL of liquid Br_2 ($d = 3.119$ g/mL)

6. Calculate the number of grams in each of the following quantities:
   (a) 0.00844 mol $NiSO_4$
   (b) 0.0600 mol $HC_2H_3O_2$
   (c) 0.725 mol Bi_2S_3
   (d) 4.50×10^{21} molecules glucose, $C_6H_{12}O_6$
   (e) 75 mL K_2CrO_4 solution ($d = 1.175$ g/mL, 20.0% K_2CrO_4 by mass)

7. Which contains the larger number of molecules, 10.0 g H_2O or 10.0 g H_2O_2?

8. Which contains the larger numbers of molecules, 25.0 g HCl or 85.0 g $C_6H_{12}O_6$?

Mole-Ratio Exercises

9. Given the equation for the combustion of isopropyl alcohol

$$2\ C_3H_7OH\ +\ 9\ O_2\ \longrightarrow\ 6\ CO_2\ +\ 8\ H_2O$$

what is the mole ratio of
   (a) CO_2 to C_3H_7OH
   (b) C_3H_7OH to O_2
   (c) O_2 to CO_2
   (d) H_2O to C_3H_7OH
   (e) CO_2 to H_2O
   (f) H_2O to O_2

10. For the reaction

$$3\ CaCl_2\ +\ 2\ H_3PO_4\ \longrightarrow\ Ca_3(PO_4)_2\ +\ 6\ HCl$$

set up the mole ratio of
   (a) $CaCl_2$ to $Ca_3(PO_4)_2$
   (b) HCl to H_3PO_4
   (c) $CaCl_2$ to H_3PO_4
   (d) $Ca_3(PO_4)_2$ to H_3PO_4
   (e) HCl to $Ca_3(PO_4)_2$
   (f) H_3PO_4 to HCl

11. How many moles of Cl_2 can be produced from 5.60 mol HCl?

$$4\ HCl\ +\ O_2\ \longrightarrow\ 2\ Cl_2\ +\ 2\ H_2O$$

12. How many moles of CO_2 can be produced from 7.75 mol C_2H_5OH? (Balance the equation first.)

$$C_2H_5OH\ +\ O_2\ \longrightarrow\ CO_2\ +\ H_2O$$

13. Given the equation

$$Al_4C_3\ +\ 12\ H_2O\ \longrightarrow\ 4\ Al(OH)_3\ +\ 3\ CH_4$$

   (a) How many moles of water are needed to react with 100. g of Al_4C_3?
   (b) How many moles of $Al(OH)_3$ will be produced when 0.600 mol of CH_4 is formed?

14. An early method of producing chlorine was by the reaction of pyrolusite (MnO_2) and hydrochloric acid. How many moles of HCl will react with 1.05 mol of MnO_2? (Balance the equation first.)

$$MnO_2(s)\ +\ HCl(aq)\ \longrightarrow\ Cl_2(g)\ +\ MnCl_2(aq)\ +\ H_2O(l)$$

15. How many grams of sodium hydroxide can be produced from 500 g of calcium hydroxide according to this equation?

$$Ca(OH)_2\ +\ Na_2CO_3\ \longrightarrow\ 2\ NaOH\ +\ CaCO_3$$

16. How many grams of zinc phosphate, $Zn_3(PO_4)_2$, are formed when 10.0 g of Zn are reacted with phosphoric acid?

$$3\ Zn\ +\ 2\ H_3PO_4\ \longrightarrow\ Zn_3(PO_4)_2\ +\ 3\ H_2$$

17. In a blast furnace, iron(III) oxide reacts with coke (carbon) to produce molten iron and carbon monoxide:

$$Fe_2O_3\ +\ 3\ C\ \longrightarrow\ 2\ Fe\ +\ 3\ CO$$

How many kilograms of iron would be formed from 125 kg of Fe_2O_3?

18. How many grams of steam and iron must react to produce 375 g of magnetic iron oxide, Fe_3O_4?

$$3\ Fe(s)\ +\ 4\ H_2O(g)\ \longrightarrow\ Fe_3O_4(s)\ +\ 4\ H_2(g)$$

19. Ethane gas, C_2H_6, burns in air (i.e., reacts with the oxygen in air) to form carbon dioxide and water:

$$2\ C_2H_6\ +\ 7\ O_2\ \longrightarrow\ 4\ CO_2\ +\ 6\ H_2O$$

   (a) How many moles of O_2 are needed for the complete combustion of 15.0 mol of ethane?
   (b) How many grams of CO_2 are produced for each 8.00 g of H_2O produced?
   (c) How many grams of CO_2 will be produced by the combustion of 75.0 g of C_2H_6?

20. Given the equation

$$4\ FeS_2\ +\ 11\ O_2\ \longrightarrow\ 2\ Fe_2O_3\ +\ 8\ SO_2$$

   (a) How many moles of Fe_2O_3 can be made from 1.00 mol of FeS_2?
   (b) How many moles of O_2 are required to react with 4.50 mol of FeS_2?
   (c) If the reaction produces 1.55 mol of Fe_2O_3, how many moles of SO_2 are produced?
   (d) How many grams of SO_2 can be formed from 0.512 mol of FeS_2?
   (e) If the reaction produces 40.6 g of SO_2, how many moles of O_2 were reacted?
   (f) How many grams of FeS_2 are needed to produce 221 g of Fe_2O_3?

Limiting-Reactant and Percent-Yield Exercises

21. In the following equations, determine which reactant is the limiting reactant and which reactant is in excess. The amounts mixed together are shown below each reactant. Show evidence for your answers.

(a) $KOH + HNO_3 \longrightarrow KNO_3 + H_2O$
      16.0 g   12.0 g

(b) $2\ NaOH + H_2SO_4 \longrightarrow Na_2SO_4 + 2\ H_2O$
      10.0 g   10.0 g

22. In the following equations, determine which reactant is the limiting reactant and which reactant is in excess. The amounts mixed together are shown below each reactant. Show evidence for your answers.

(a) $2\ Bi(NO_3)_3 + 3\ H_2S \longrightarrow Bi_2S_3 + 6\ HNO_3$
      50.0 g   6.00 g

(b) $3\ Fe + 4\ H_2O \longrightarrow Fe_3O_4 + 4\ H_2$
      40.0 g   16.0 g

23. The reaction for the combustion of propane is

$$C_3H_8 + 5\ O_2 \longrightarrow 3\ CO_2 + 4\ H_2O$$

(a) If 5.0 mol of C_3H_8 and 5.0 mol of O_2 are reacted, how many moles of CO_2 can be produced?

(b) If 3.0 mol of C_3H_8 and 20.0 mol of O_2 are reacted, how many moles of CO_2 are produced?

(c) If 20.0 mol of C_3H_8 and 3.0 mol of O_2 are reacted, how many moles of CO_2 can be produced?

24. The reaction for the combustion of propane is

$$C_3H_8 + 5\ O_2 \longrightarrow 3\ CO_2 + 4\ H_2O$$

(a) If 20.0 g of C_3H_8 and 20.0 g of O_2 are reacted, how many moles of CO_2 can be produced?

(b) If 20.0 g of C_3H_8 and 80.0 g of O_2 are reacted, how many moles of CO_2 are produced?

(c) If 2.0 mol of C_3H_8 and 14.0 mol of O_2 are placed in a closed container and they react to completion (until one reactant is completely used up), what compounds will be present in the container after the reaction, and how many moles of each compound are present?

*25. When a particular metal X reacts with HCl, the resulting products are XCl_2 and H_2. Write and balance the equation. When 78.5 g of the metal completely react, 2.42 g of hydrogen gas results. Identify the element X.

27. Aluminum reacts with bromine to form aluminum bromide:

$$2\ Al + 3\ Br_2 \longrightarrow 2\ AlBr_3$$

If 25.0 g of Al and 100. g of Br_2 are reacted, and 64.2 g of $AlBr_3$ product are recovered, what is the percent yield for the reaction?

*29. Carbon disulfide, CS_2, can be made from coke, C, and sulfur dioxide, SO_2:

$$3\ C + 2\ SO_2 \longrightarrow CS_2 + 2\ CO_2$$

If the actual yield of CS_2 is 86.0% of the theoretical yield, what mass of coke is needed to produce 950 g of CS_2?

*26. When a certain nonmetal whose formula is X_8 combusts in air, XO_3 forms. Write a balanced equation for this reaction. If 120.0 g of oxygen gas are consumed completely, along with 80.0 g of X_8, identify element X.

28. Iron was reacted with a solution containing 400. g of copper(II) sulfate. The reaction was stopped after 1 hour, and 151 g of copper were obtained. Calculate the percent yield of copper obtained:

$$Fe(s) + CuSO_4(aq) \longrightarrow Cu(s) + FeSO_4(aq)$$

*30. Acetylene (C_2H_2) can be manufactured by the reaction of water and calcium carbide, CaC_2:

$$CaC_2(s) + 2\ H_2O(l) \longrightarrow C_2H_2(g) + Ca(OH)_2(s)$$

When 44.5 g of commercial grade (impure) calcium carbide is reacted, 0.540 mol of C_2H_2 is produced. Assuming that all of the CaC_2 was reacted to C_2H_2, what is the percent of CaC_2 in the commercial grade material?

Additional Exercises

These problems are not paired or labeled by topic and provide additional practice on the concepts covered in this chapter.

31. A tool set contains 6 wrenches, 4 screwdrivers, and 2 pliers. The manufacturer has 1000 pliers, 2000 screwdrivers, and 3000 wrenches in stock. Can an order for 600 tool sets be filled? Explain briefly.

32. What is the difference between using a number as a subscript and using a number as a coefficient in a chemical equation?

33. Oxygen masks for producing O_2 in emergency situations contain potassium superoxide (KO_2). It reacts according to the following equation:

$$4\ KO_2 + 2\ H_2O + 4\ CO_2 \longrightarrow 4\ KHCO_3 + 3\ O_2$$

(a) If a person wearing such a mask exhales 0.85 g of CO_2 every minute, how many moles of KO_2 are consumed in 10.0 minutes?

(b) How many grams of oxygen are produced in 1.0 hour?

34. Ethyl alcohol is the result of fermentation of sugar, $C_6H_{12}O_6$:

$$C_6H_{12}O_6 \longrightarrow 2\ C_2H_5OH + 2\ CO_2$$

(a) How many grams of ethyl alcohol and how many grams of carbon dioxide can be produced from 750 g of sugar?

(b) How many milliliters of alcohol ($d = 0.79$ g/mL) can be produced from 750 g of sugar?

35. Phosphoric acid, H_3PO_4, can be synthesized from phosphorus, oxygen, and water according to these two reactions

$$4\ P + 5\ O_2 \longrightarrow P_4O_{10}$$

$$P_4O_{10} + 6\ H_2O \longrightarrow 4\ H_3PO_4$$

Starting with 20.0 g P, 30.0 g O_2, and 15.0 g H_2O, what is the mass of phosphoric acid that can be formed?

36. The methyl alcohol (CH_3OH) used in alcohol burners combines with oxygen gas to form carbon dioxide and water. How many grams of oxygen are required to burn 60.0 mL of methyl alcohol ($d = 0.72$ g/mL)?

37. Hydrazine (N_2H_4) and hydrogen peroxide (H_2O_2) have been used as rocket propellants. They react according to the following equation:

$$7\ H_2O_2 + N_2H_4 \longrightarrow 2\ HNO_3 + 8\ H_2O$$

(a) How many moles of HNO_3 are formed from 0.33 mol of hydrazine?

(b) How many moles of hydrogen peroxide are required if 2.75 mol of water are to be produced?

(c) How many moles of water are produced if 8.72 mol of HNO_3 are also produced?

(d) How many grams of hydrogen peroxide are needed to completely react with 120 g of hydrazine?

38. Silver tarnishes in the presence of hydrogen sulfide (which smells like rotten eggs) and oxygen because of the reaction:

$$4\ Ag + 2\ H_2S + O_2 \longrightarrow 2\ Ag_2S + 2\ H_2O$$

How many grams of silver sulfide can be formed from a mixture of 1.1 g Ag, 0.14 g H_2S, and 0.080 g O_2?

39. After 180.0 g of zinc were dropped into a beaker of hydrochloric acid and the reaction ceased, 35 g of unreacted zinc remained in the beaker:

$$Zn + HCl \longrightarrow ZnCl_2 + H_2$$

Balance the equation first.

(a) How many moles of hydrogen gas were produced?

(b) How many grams of HCl were reacted?

40. Given the following equation, answer (a) and (b) below:

$$Fe(s) + CuSO_4(aq) \longrightarrow Cu(s) + FeSO_4(aq)$$

(a) When 2.0 mol of Fe and 3.0 mol of $CuSO_4$ are reacted, what substances will be present when the reaction is over? How many moles of each substance are present?

(b) When 20.0 g of Fe and 40.0 g of $CuSO_4$ are reacted, what substances will be present when the reaction is over? How many grams of each substance are present?

41. Methyl alcohol (CH_3OH) is made by reacting carbon monoxide and hydrogen in the presence of certain metal oxide catalysts. How much alcohol can be obtained by reacting 40.0 g of CO and 10.0 g of H_2? How many grams of excess reactant remain unreacted?

$$CO(g) + 2\ H_2(g) \longrightarrow CH_3OH(l)$$

*42. Ethyl alcohol (C_2H_5OH), also called grain alcohol, can be made by the fermentation of sugar, which often comes from starch in grain:

$$\underset{\text{glucose}}{C_6H_{12}O_6} \longrightarrow 2\ \underset{\text{ethyl alcohol}}{C_2H_5OH} + 2\ CO_2$$

If an 84.6% yield of ethyl alcohol is obtained,

(a) what mass of ethyl alcohol will be produced from 750 g glucose?

(b) what mass of glucose should be used to produce 475 g of C_2H_5OH?

*43. Both $CaCl_2$ and $MgCl_2$ react with $AgNO_3$ to precipitate AgCl. When solutions containing equal masses of $CaCl_2$ and $MgCl_2$ are reacted with $AgNO_3$ solutions that will produce the larger amount of AgCl? Show proof.

*44. An astronaut excretes about 2500 g of water a day. If lithium oxide (Li_2O) is used in the spaceship to absorb this water, how many kilograms of Li_2O must be carried for a 30-day space trip for three astronauts?

$$Li_2O + H_2O \longrightarrow 2\ LiOH$$

*45. Much commercial hydrochloric acid is prepared by the reaction of concentrated sulfuric acid with sodium chloride:

$$H_2SO_4 + 2\ NaCl \longrightarrow Na_2SO_4 + 2\ HCl$$

How many kilograms of concentrated H_2SO_4, 96% H_2SO_4 by mass, are required to produce 20.0 L of concentrated hydrochloric acid ($d = 1.20$ g/mL, 42.0% HCl by mass)?

*46. Gastric juice contains about 3.0 g HCl per liter. If a person produces about 2.5 L of gastric juice per day, how many antacid tablets, each containing 400. mg of $Al(OH)_3$, are needed to neutralize all the HCl produced in 1 day?

$$Al(OH)_3(s) + 3\ HCl(aq) \longrightarrow AlCl_3(aq) + 3\ H_2O(l)$$

*47. When 12.82 g of a mixture of $KClO_3$ and NaCl is heated strongly, the $KClO_3$ reacts according to the following equation:

$$2\ KClO_3(s) \longrightarrow 2\ KCl(s) + 3\ O_2(g)$$

The NaCl does not undergo any reaction. After the heating, the mass of the residue (KCl and NaCl) is 9.45 g. Assuming that all the loss of mass represents loss of oxygen gas, calculate the percent of $KClO_3$ in the original mixture.

48. What is the purpose of a mask in the manufacture of tiny circuits and machines?

49. Why is silicon dioxide called a sacrificial material in the production of micromachinery?

50. What are some possible uses for micromachines in our society? Indicate some of the difficulties that must be overcome to implement these uses.

Answers to Practice Exercises

9.1 0.33 mol Al_2O_3

9.2 7.33 mol $Al(OH)_3$

9.3 0.8157 mol KCl

9.4 85 g $AgNO_3$

9.5 27.6 g $CrCl_3$

9.6 348.8 g H_2O

9.7 17.7 g HCl

9.8 164.4 g $BaSO_4$

9.9 88.5% yield

10

How do we go about studying an object that is too small to see? Think back to that birthday when you had a present you could look at but not yet open. Simply judging from the wrapping and size of the box was not very useful. Shaking, turning, and lifting the package all gave clues, indirectly, to the contents. After all the experiments were done on the package, a fairly good hypothesis of the contents could be made. But was it correct? The only means of absolute verification would be to open the package.

The same is true for chemists in their study of the atom. Atoms are so very small that it is not possible to use the normal senses and rules to describe them. Chemists are essentially working in the dark with this package we call the atom. Yet, as instruments (x-ray machines and scanning-tunneling microscopes) and measuring devices (spectrophotometers and magnetic resonance imaging, MRI) have progressed, along with our skill in mathematics and probability, the secrets of the atom are beginning to unravel.

10.1 A Brief History

In the last 200 years, vast amounts of data have been accumulated to support atomic theory. When atoms were originally suggested by the early Greeks, no evidence existed to physically support their ideas. Faraday, Arrhenius, and others did a variety of experiments, which culminated in Dalton's theory of the atom. Because of the limitations of Dalton's model, modifications were proposed by Thomson and then by Rutherford, which eventually led to our modern concept of the nuclear atom. These early models of the atom work reasonably well—in fact, we continue to use them today to visualize a variety of chemical concepts. There remain, however, questions that these models cannot answer, including an explanation of how atomic structure relates to the periodic table. In this chapter, we will look at our modern model of the atom to see how it varies from and improves upon the earlier atomic theories.

10.2 Electromagnetic Radiation

Scientists have studied energy and light for centuries, and several models have been proposed to explain how energy is transferred from place to place. One of the ways that energy travels through space is by *electromagnetic radiation.* Examples of electromagnetic radiation include light from the sun, X rays in your dentist's office, microwaves from your microwave oven, radio and television waves, and radiant heat from your fireplace. While these examples seem quite different, they are all similar in some important ways. Each shows wave-like behavior, and all travel at the same speed in a vacuum (3.00×10^8 m/s).

The study of wave behavior is a topic for another course, but we shall need some basic terminology in order to understand atoms. Waves have three basic characteristics: wavelength, frequency, and speed. **Wavelength** (lambda, λ) is the distance between consecutive peaks (or troughs) in a wave, as shown in Figure 10.1.

wavelength

◀ **Chapter Opening Photo:
The brilliance of these neon
signs is the result of electrons
being transferred between
energy levels.**

Atomic Clocks

Imagine a clock that keeps time to within one second over a million years. The National Institute of Standards and Technology in Boulder, Colorado, has an atomic clock that does just that—a little better than your average alarm, grandfather, or cuckoo clock! This atomic clock serves as the international standard for time and frequency. How does it work?

Within the glistening case are several layers of magnetic shielding. In the heart of the clock is a small oven that heats cesium metal to release cesium atoms, which are collected into a narrow beam (1 mm wide). The beam of atoms passes down a long evacuated tube while being excited by a laser until all the cesium atoms are in the same electron state.

The atomic clock at the National Institute of Standards and Technology in Boulder, CO, loses only 1 second in a million years.

The atoms then enter another chamber filled with reflecting microwaves. The frequency of the microwaves (9,192,631,770 cycles per second) is exactly the same frequency required to excite a cesium atom from its ground state to the next higher energy level. These excited cesium atoms then release electromagnetic radiation in a process known as fluorescence. Electronic circuits maintain the microwave frequency at precisely the right level to keep the cesium atoms moving from one level to the next. One second is equal to 9,192,631,770 of these vibrations. The clock is set to this frequency and can keep accurate time for over a million years.

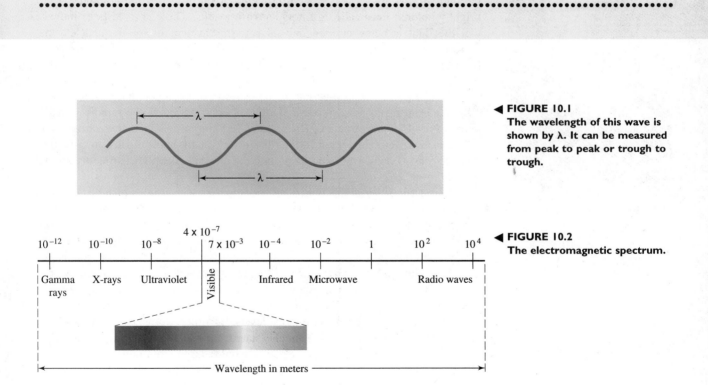

◀ **FIGURE 10.1**
The wavelength of this wave is shown by λ. It can be measured from peak to peak or trough to trough.

◀ **FIGURE 10.2**
The electromagnetic spectrum.

Frequency (nu, ν) tells how many waves pass a particular point per second. **Speed** (v) tells how fast a wave moves through space.

Light is one form of electromagnetic radiation and is usually classified by its wavelength, as shown in Figure 10.2. Notice that visible light is only a tiny part of the electromagnetic spectrum. Some examples of electromagnetic radiation involved

frequency
speed

in energy transfer outside the visible region are hot coals in your backyard grill, which transfer infrared radiation to cook your food; and microwaves, which transfer energy to water molecules in the food, causing them to move more quickly and thus raise the temperature of your food.

photons

Scientists have evidence for the wave-like nature of light. They also know that a beam of light behaves like a stream of tiny packets of energy called **photons**. So what is light exactly? Is it a particle? Is it a wave? Scientists have agreed to explain the properties of electromagnetic radiation by using both wave and particle properties. Neither explanation is ideal, but currently these are our best models.

10.3   The Bohr Atom

Niels Bohr (1885–1962).

line spectrum

quanta

ground state

As scientists struggled to understand the properties of electromagnetic radiation, evidence began to accumulate that atoms could radiate light. At high temperatures, or when subjected to high voltages, elements in the gaseous state give off colored light. Brightly colored neon signs illustrate this property of matter very well. When the light emitted by a gas is passed through a prism or diffraction grating, a set of brightly colored lines called a **line spectrum** results. These colored lines indicate that the light is being emitted only at certain wavelengths, or frequencies, that correspond to specific colors. Each element possesses a unique set of these spectral lines that is different from the sets of all the other elements. This is illustrated in Figure 10.3.

In 1912–1913, while studying the line spectra of hydrogen, Niels Bohr (1885–1962), a Danish physicist, made a significant contribution to the rapidly growing knowledge of atomic structure. His research led him to believe that electrons in an atom exist in specific regions at various distances from the nucleus. He also visualized the electrons as revolving in orbits around the nucleus, like planets rotating around the sun.

Bohr's first paper in this field dealt with the hydrogen atom, which he described as a single electron revolving in an orbit about a relatively heavy nucleus. He applied the concept of energy quanta, proposed in 1900 by the German physicist Max Planck (1858–1947), to the observed line spectra of hydrogen. Planck stated that energy is never emitted in a continuous stream but only in small discrete packets called **quanta** (Latin, *quantus,* how much). From this, Bohr theorized that electrons have several possible energies corresponding to several possible orbits at different distances from the nucleus. Therefore, an electron has to be in one specific energy level; it cannot exist between energy levels. In other words, the energy of the electron is said to be quantized. Bohr also stated that when a hydrogen atom absorbed one or more quanta of energy, its electron would "jump" to a higher energy level.

Bohr was able to account for spectral lines of hydrogen this way. A number of energy levels are available, the lowest of which is called the **ground state**. When an electron falls from a high-energy level to a lower one (say, from the fourth to the second), a quantum of energy is emitted as light at a specific frequency, or wavelength. This light corresponds to one of the lines visible in the hydrogen spectrum (see Figure 10.3). Several lines are visible in this spectrum, each one corresponding to a specific electron energy-level shift within the hydrogen atom.

The chemical properties of an element and its position in the periodic table depend on electron behavior within the atoms. In turn, much of our knowledge of the behavior of electrons within atoms is based on spectroscopy. Niels Bohr contributed a great deal to our knowledge of atomic structure by (1) suggesting quantized energy levels for electrons and (2) showing that spectral lines result from the radiation of small increments of energy (Planck's quanta) when electrons shift from one energy level to another. Bohr's calculations succeeded very well in correlating the experimentally observed spectral lines with electron energy levels for the hydrogen atom. However, Bohr's methods of calculation did not succeed for heavier atoms. More theoretical work on atomic structure was needed.

In 1924, the French physicist Louis de Broglie suggested a surprising hypothesis: All objects have wave properties. De Broglie used sophisticated mathematics to show that the wave properties for an object of ordinary size, such as a baseball, are too small to be observed. But for small objects, such as an electron, the wave properties become significant. Other scientists confirmed de Broglie's hypothesis, showing that electrons do exhibit wave properties. In 1926, Erwin Schrödinger, an Austrian physicist, created a mathematical model that described electrons as waves. Using Schrödinger's wave mechanics, it is possible to determine the probability of finding an electron in a certain region around the atom.

This treatment of the atom led to a new branch of physics called *wave mechanics* or *quantum mechanics,* which forms the basis for the modern understanding of atomic structure. Although the wave-mechanical description of the atom is mathematical it can be translated, at least in part, into a visual model. It is important to recognize that we cannot locate an electron precisely within an atom; however, it is clear that electrons are not revolving around the nucleus in *orbits* as Bohr postulated. The electrons are instead found in *orbitals.* An **orbital** can be pictured as a region in space around the nucleus where there is a high probability of finding a given electron. We will have more to say about the meaning of orbitals in the next section.

orbital

This stamp commemorates the work of Max Planck (1858–1947).

10.4 Energy Levels of Electrons

One of the ideas Bohr contributed to the modern concept of the atom was that the energy of the electron is quantized—that is, the electron is restricted to only certain allowed energies. The wave-mechanical model of the atom also predicts discrete **principal energy levels** within the atom. These energy levels are designated by the letter n, where n is a positive integer. The lowest principal energy level corresponds to $n = 1$, the next to $n = 2$, and so on. As n increases, the energy of the electron increases, and the electron is found on average farther from the nucleus.

principal energy levels

◀ FIGURE 10.3
Line spectrum of hydrogen. Each line corresponds to the wavelength of the energy emitted when the electron of a hydrogen atom, which has absorbed energy, falls back to a lower principal energy level.

Ripples on the Surface

Modern physicists have astounded us with the idea that particles have wave properties. Protons, electrons, and all other elementary particles sometimes behave like waves and sometimes like particles. Donald Eigler at the IBM Almaden Research Center in San Jose, California, used a scanning-tunneling microscope at 4 kelvins to produce the picture shown here. The surface of this copper crystal surprised even the physicists. The waves in the photo are produced by electrons moving around on the surface of the crystal and bouncing off impurities (the two pits in the photo) in the copper. Since each electron behaves like a wave, it interferes with itself after reflecting off an impurity in the copper. The interference pattern is called a standing wave, which is a wave that vibrates up and down without visible transverse movement, similar to a plucked violin string. The crests of the waves represent regions where the electron is most probably in its particle form.

Metal atoms easily lose one or more electrons, which move about freely within the metal crystal, forming what is often called an "electron sea." At the surface of the metal these loose electrons are confined to a single layer, which is free to move in only two dimensions. Within these constraints the particles also behave as waves. The electron layer responsible for this beautiful pattern is only 0.02 Å thick. Similar images of standing electron waves have also been produced with gold at room temperature.

This STM of the surface of a copper crystal beautifully illustrates the wave nature of matter. Magnification is 215,000,000×.

Number of sublevels

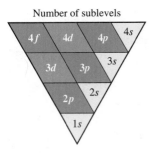

FIGURE 10.4
The types of orbitals on each of the first four principal energy levels.

orbital or sublevel

spin

Pauli exclusion principle

Corresponding to each energy level are orbitals (the regions of space where the electrons are found). The first principal energy level has one type of **orbital or sublevel**. The second principal energy level has two sublevels, and so on. A good way to visualize this is to use an inverted triangle as shown in Figure 10.4.

In each sublevel the electrons are found within orbitals. Let's consider each principal energy level in turn. The first principal energy level ($n = 1$) has one sublevel. This orbital is spherical in shape and is designated as $1s$. It is important to understand what the spherical shape of the $1s$ orbital means. The electron does *not* move around on the surface of the sphere, but rather the surface encloses a space where there is a 90% probability that the electron may be found. It might help to consider these orbital shapes in the same way we consider the atmosphere. There is no distinct dividing line between the atmosphere and "space." The boundary is quite fuzzy. The same is true for atomic orbitals. Each has a region of highest density roughly corresponding to its shape. The probability of finding the electron outside this region drops rapidly but never quite reaches zero. Scientists often speak of orbitals as electron "clouds" to emphasize the fuzzy nature of their boundaries.

How many electrons can fit into a $1s$ orbital? To answer this question we need to consider one more property of electrons. This property is called **spin**. Each electron appears to be spinning on an axis, like a globe. It can only spin in two directions. We represent this spin with an arrow, ↑ or ↓. In order to occupy the same orbital, electrons must have *opposite* spins. That is, two electrons with the same spin cannot occupy the same orbital. This gives us the answer to our question: An atomic orbital can hold a maximum of two electrons, which must have opposite spins. This rule is called the **Pauli exclusion principle**. To summarize, the first principal energy level contains one type of orbital ($1s$) that holds a maximum of two electrons.

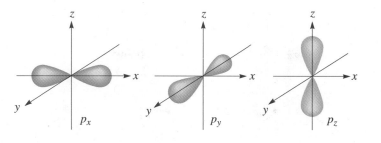

◀ **FIGURE 10.5**
Perspective representation of
the p_x, p_y, and p_z atomic orbitals.

What happens with the second principal energy level ($n = 2$)? Here we find two types of orbitals (sublevels), 2s and 2p. The 2s orbital is spherical in shape as it is for the first principal energy level but is larger in size and higher in energy than the 1s. It can also hold a maximum of two electrons. The second type of orbital is designated by 2p. The shape of these orbitals is quite different from the s orbitals. The shapes of the p orbitals are shown in Figure 10.5.

◀ The light in this neon sign is the
result of electrons falling from
one principal energy level to
another.

FIGURE 10.6 ▶
Orbitals on the second principal energy level are one 2s and three 2p orbitals.

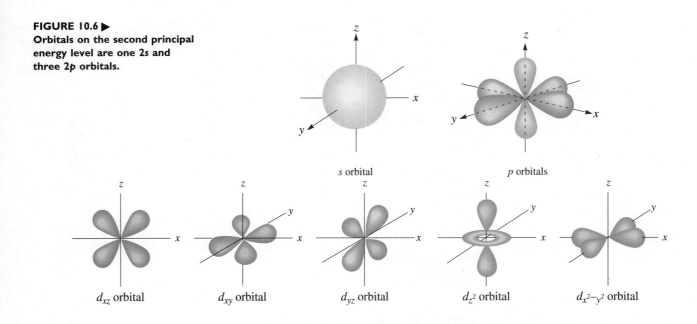

s orbital   p orbitals

d_{xz} orbital d_{xy} orbital d_{yz} orbital d_{z^2} orbital $d_{x^2-y^2}$ orbital

FIGURE 10.7
The five d orbitals are found in the third principal energy level along with one 3s orbital and three 3p orbitals.

Each p orbital has two "lobes." Remember, the space enclosed by these surfaces represents the regions of probability for finding the electrons 90% of the time. Notice there are three separate p orbitals, each oriented in a different direction. Each p orbital can hold a maximum of two electrons. Thus the total number of electrons that can reside in all three p orbitals is six. To summarize once again, the first principal energy level of an atom has a $1s$ orbital. The second principal energy level has a $2s$ and three $2p$ orbitals labeled $2p_x$, $2p_y$, and $2p_z$ as shown in Figure 10.6.

The third principal energy level has three types of orbitals (sublevels) labeled $3s$, $3p$, and $3d$. The $3s$ orbital is spherical and larger than the $1s$ and $2s$ orbitals. The $3p_x$, $3p_y$, $3p_z$ orbitals are shaped like those of the second level, only larger. There are five $3d$ orbitals with the shapes shown in Figure 10.7. You do not need to memorize these shapes, but notice that they look different from the s or p orbitals.

Each time a new principal energy level is added, we also add a new type of orbital. This makes sense when we realize that each energy level corresponds to a larger average distance from the nucleus, which provides more room on each level for new orbitals.

The pattern continues with the fourth principal energy level. It has $4s$, $4p$, $4d$, and $4f$ orbitals. There are one $4s$, three $4p$, five $4d$, and seven $4f$ orbitals. The shapes of the s, p, and d orbitals are the same as those for lower levels, only larger. We will not consider the shapes of the f orbitals. To summarize atomic structure, we can show the types of orbitals associated with each principal energy level:

$n = 1$	$1s$			
$n = 2$	$2s$	$2p\ 2p\ 2p$		
$n = 3$	$3s$	$3p\ 3p\ 3p$	$3d\ 3d\ 3d\ 3d\ 3d$	
$n = 4$	$4s$	$4p\ 4p\ 4p$	$4d\ 4d\ 4d\ 4d\ 4d$	$4f\ 4f\ 4f\ 4f\ 4f\ 4f\ 4f$

The hydrogen atom consists of a nucleus (containing one proton) and an electron moving around the outside of the nucleus. In its ground state the electron occupies a $1s$ orbital, but by absorbing energy the electron can become *excited* and move to a higher energy level.

1p

FIGURE 10.8
The modern concept of a hydrogen atom is shown here. It consists of an electron in an s orbital as a cloud of negative charge surrounding the proton in the nucleus.

Yes, We Can See Atoms!

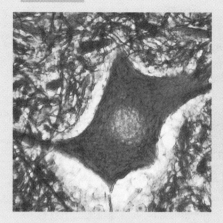

Neurons, optical microscope.

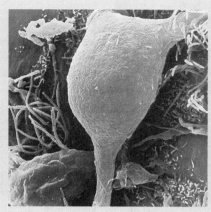

Neurons, electron microscope.

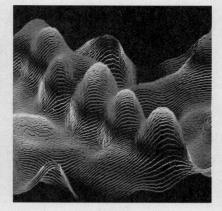

DNA molecule, scanning-tunneling microscope.

For centuries, scientists have argued and theorized over the nature and existence of atoms. Today, physicists and chemists can produce pictures of atoms and even move them individually from place to place. This new-found ability to see atoms, molecules, and even watch chemical reactions occur is the direct result of the evolution of the microscope.

An optical microscope is capable of viewing objects as small as the size of a cell. To see smaller objects an electron microscope is necessary. Since the eye cannot respond to a beam of electrons, the image is produced on a fluorescent screen or on film. These microscopes have been used for some time to photograph large molecules. To see tiny objects, however, the objects must be placed in a vacuum and the electrons must be in a high energy state. If the sample is fragile, as are most molecules, it can be destroyed before the image is formed.

In 1981, Gerd Binnig and Heinrich Rohrer, two scientists from IBM, invented the first scanning-probe microscope. These instruments are fundamentally different from previous microscopes. In a scanning-probe microscope a probe is placed near the surface of a sample and a parameter of some sort (voltage, magnetic field, etc.) is measured. As the probe is moved across the surface, an image is produced—in the same manner a child would determine the identity of an object sealed in an opaque bag. The first of the scanning-probe instruments was called a scanning-tunneling microscope. It produced the first clear pictures of silicon atoms in January 1983. The greatest limitation for the scanning-tunneling microscope is that, in order to be viewed, organic molecules must be given a thin coating of metal so that electrons are free to jump from the surface to the probe.

In 1985, a team of physicists from Stanford University and IBM found a solution to this problem. In a new instrument known as an atomic-force microscope, the probe measures tiny electric forces between electrons instead of the actual movement of electrons from surface to probe. The greatest advantage of this approach is that the probe is so gentle that even very fragile molecules remain intact. The probe is a tiny shard of a diamond attached to a tiny piece of silicon, and it works like a phonograph needle. At University of California, Santa Barbara, a group of scientists has even succeeded in making a movie of the formation of a blood clot on the molecular level.

In industry, another type of scanning-probe instrument has been developed to check the quality of microelectronic equipment. Researchers at IBM have developed a laser-force microscope in which a tiny wire probe measures small attractive forces (surface tension of water on the sample) to show imperfections as small as 25 atoms across.

The hydrogen atom can be represented as shown in Figure 10.8. The diameter of the nucleus is about 10^{-13} cm, and the diameter of the electron orbital is about 10^{-8} cm. The diameter of the electron cloud of a hydrogen atom is about 100,000 times greater than the diameter of the nucleus.

10.5 Atomic Structures of the First 18 Elements

We have seen that hydrogen has one electron that can occupy a variety of orbitals in different principal energy levels. Now we must consider the structure of atoms with more than one electron. We have learned that all atoms contain orbitals similar to those found in hydrogen. Therefore, we can describe the structures of atoms beyond hydrogen by systematically placing electrons in these hydrogen-like orbitals. The following guidelines apply to this process:

1. No more than two electrons can occupy one orbital.
2. Electrons occupy the lowest energy orbitals available. They enter a higher energy orbital only when the lower orbitals are filled. For the atoms beyond hydrogen, orbital energies vary as $s < p < d < f$ for a given value of n.
3. Each one of a given type of orbital is occupied by a single electron before a second electron enters. For example, all three p orbitals must contain one electron before a second electron enters a p orbital.

We can use several methods to represent the atomic structures of atoms, depending on what we are trying to illustrate. When we want to show both the nuclear make-up and the electron structure of each principal energy level (without orbital detail), we can use a diagram such as Figure 10.9.

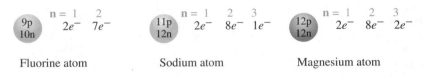

Fluorine atom   Sodium atom   Magnesium atom

FIGURE 10.9
Atomic structure diagrams of fluorine, sodium, and magnesium atoms. The number of protons and neutrons is shown in the nucleus. The number of electrons is shown in each principal energy level outside the nucleus. ▶

Often we are interested in showing the arrangement of the electrons in an atom in their orbitals. There are two ways to show this. The first method is called the **electron configuration**. In this method we list each type of orbital, showing the number of electrons in it as an exponent. The following diagram shows how to read these:

electron configuration

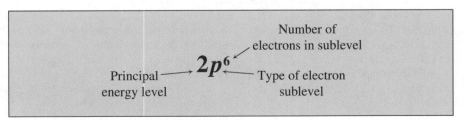

orbital diagram

We can also represent this configuration by using an **orbital diagram** in which the orbitals are represented by boxes grouped by sublevel with small arrows indicating the electrons. When the orbital contains one electron, an arrow, pointing upward ($\uparrow$), is placed in the box. A second arrow, pointing downward ($\downarrow$), indicates the second electron in that orbital.

Let's consider each of the first 20 elements on the periodic table in turn. The order of filling for the orbitals in these elements is $1s$, $2s$, $2p$, $3s$, $3p$, and $4s$.

Hydrogen, the first element, has only one electron. The electron will be in the $1s$ orbital since this is the most favorable position (where it will have the greatest attraction for the nucleus). Both representations for hydrogen are shown below:

H: ↑ $1s^1$
      Orbital   Electron
      diagram   configuration

Helium with two electrons can be shown as

He: ↑↓ $1s^2$
       Orbital   Electron
       diagram   configuration

The first energy level is now full. An atom with three electrons will have its third electron in the second energy level. Thus in lithium (atomic number 3), the third electron is in the $2s$ orbital of the second energy level. Lithium has the following structure:

Li   ↑↓   ↑ $1s^2 2s^1$

All four electrons of beryllium are s electrons:

Be   ↑↓   ↑↓ $1s^2 2s^2$

The next six elements illustrate the filling of the p orbitals. Boron has the first p electron. Because all of the p orbitals have the same energy, it does not matter which of these orbitals fills first:

B   ↑↓   ↑↓   ↑ $1s^2 2s^2 2p^1$

Carbon is the sixth element. It has two electrons in the $1s$ orbital, 2 electrons in the $2s$ orbital and 2 electrons to place in $2p$ orbitals. Because it is energetically more difficult for the p electrons to pair up than to occupy a second p orbital, the second p electron is located in a different p orbital. We could show this by writing $2p_x^1 2p_y^1$, but usually it is written as $2p^2$ and it is *understood* that the electrons are in different p orbitals. The spins on these electrons are alike for reasons we will not explain here.

C   ↑↓   ↑↓   ↑ ↑ $1s^2 2s^2 2p^2$

Nitrogen has 7 electrons. They are in $1s$, $2s$, and $2p$ orbitals. The third p electron in nitrogen is still unpaired and is found in the $2p_z$ orbital:

N   ↑↓   ↑↓   ↑ ↑ ↑ $1s^2 2s^2 2p^3$

Oxygen is the eighth element. It has 2 electrons in both the $1s$ and $2s$ orbitals, and 4 electrons in the $2p$ orbitals. One of the $2p$ orbitals is now occupied by a second electron, which has a spin opposite the electron already in that orbital:

O   ↑↓   ↑↓   ↑↓ ↑ $1s^2 2s^2 2p^4$

The next two elements are fluorine with 9 electrons and neon with 10 electrons:

F   ↑↓   ↑↓   ↑↓ ↑↓ ↑ $1s^2 2s^2 2p^5$

Ne   ↑↓   ↑↓   ↑↓ ↑↓ ↑↓ $1s^2 2s^2 2p^6$

With neon, the first and second energy levels are filled.

Gerd Bennig and Heinrich Rohrer won the Nobel Prize in 1986 for inventing the scanning-tunneling microscope, which measures electron patterns.

TABLE 10.1 Orbital Diagrams and Electron Configurations for Elements 11–18

Number	Element	Orbital					Electron configuration
		$1s$	$2s$	$2p$	$3s$	$3p$	
11	Na	⇅	⇅	⇅ ⇅ ⇅	↑		$1s^22s^22p^63s^1$
12	Mg	⇅	⇅	⇅ ⇅ ⇅	⇅		$1s^22s^22p^63s^2$
13	Al	⇅	⇅	⇅ ⇅ ⇅	⇅	↑	$1s^22s^22p^63s^23p^1$
14	Si	⇅	⇅	⇅ ⇅ ⇅	⇅	↑ ↑	$1s^22s^22p^63s^23p^2$
15	P	⇅	⇅	⇅ ⇅ ⇅	⇅	↑ ↑ ↑	$1s^22s^22p^63s^23p^3$
16	S	⇅	⇅	⇅ ⇅ ⇅	⇅	⇅ ↑ ↑	$1s^22s^22p^63s^23p^4$
17	Cl	⇅	⇅	⇅ ⇅ ⇅	⇅	⇅ ⇅ ↑	$1s^22s^22p^63s^23p^5$
18	Ar	⇅	⇅	⇅ ⇅ ⇅	⇅	⇅ ⇅ ⇅	$1s^22s^22p^63s^23p^6$

Sodium, element 11, has 2 electrons in the first energy level and 8 electrons in the second energy level, with the remaining electron occupying the 3s orbital in the third energy level:

Na   ⇅   ⇅   ⇅ ⇅ ⇅   ↑ $1s^22s^22p^63s^1$

     $1s$ $2s$ $2p$ $3s$

Magnesium (12), aluminum (13), silicon (14), phosphorus (15), sulfur (16), chlorine (17), and argon (18) follow in order. Table 10.1 summarizes the filling of the orbitals for elements 11–18.

The electrons in the outermost (highest) energy level of an atom are called the **valence electrons**. For example, oxygen, which has the electron configuration of $1s^22s^22p^4$, has electrons in the first and second energy levels. Therefore, the second level (2) is the valence level for oxygen. The 2s and 2p electrons are the valence electrons. In the case of magnesium ($1s^22s^22p^63s^2$), the valence electrons are in the 3s orbital since the outermost level containing electrons is the third energy level. The valence electrons are involved in bonding atoms together to form compounds and are of particular interest to chemists, as we will see in Chapter 11.

valence electrons

10.6   Electron Structures and the Periodic Table

We have seen how the electrons are assigned for the atoms of elements 1–18. How do the electron structures of these atoms relate to their position on the periodic table? To answer this question we need to look at the periodic table more closely.

The periodic table represents the efforts of chemists to organize the elements logically. Chemists of the early 19th century had sufficient knowledge of the properties of elements to recognize similarities among groups of elements. In 1869 Dimitri Mendeleev (1834–1907) of Russia and Lothar Meyer (1830–1895) of Germany independently published periodic arrangements of the elements that were based on increasing atomic masses. Mendeleev's arrangement is the precursor to the modern periodic table and his name is associated with it. The periodic table is shown in Figure 10.10.

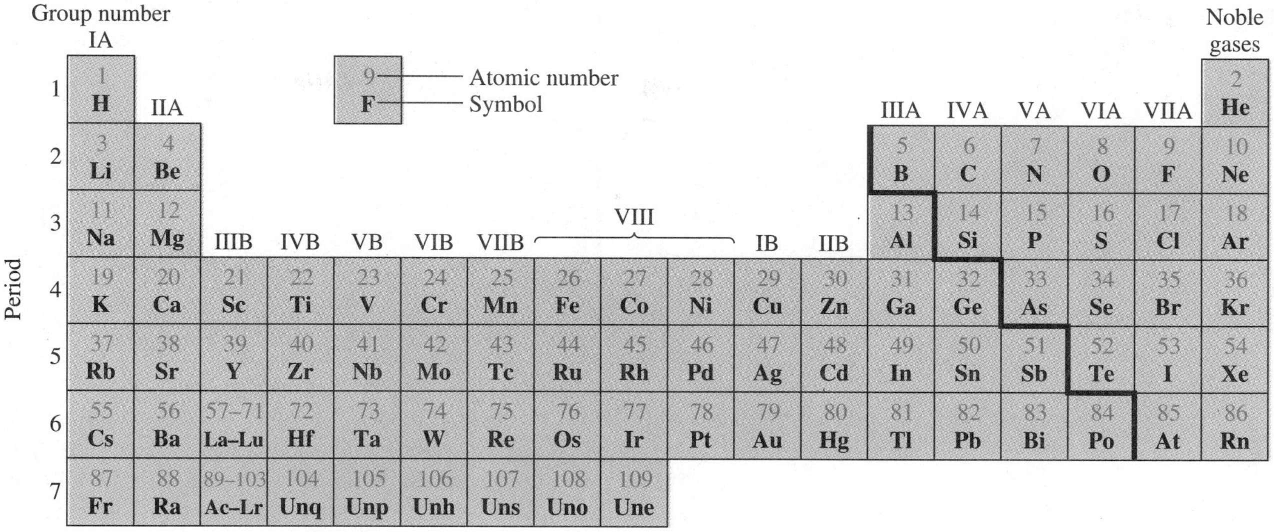

Group number																	Noble gases
IA																	
1 H	**IIA**											**IIIA**	**IVA**	**VA**	**VIA**	**VIIA**	2 He
3 Li	4 Be											5 B	6 C	7 N	8 O	9 F	10 Ne
11 Na	12 Mg	**IIIB**	**IVB**	**VB**	**VIB**	**VIIB**	**VIII**			**IB**	**IIB**	13 Al	14 Si	15 P	16 S	17 Cl	18 Ar
19 K	20 Ca	21 Sc	22 Ti	23 V	24 Cr	25 Mn	26 Fe	27 Co	28 Ni	29 Cu	30 Zn	31 Ga	32 Ge	33 As	34 Se	35 Br	36 Kr
37 Rb	38 Sr	39 Y	40 Zr	41 Nb	42 Mo	43 Tc	44 Ru	45 Rh	46 Pd	47 Ag	48 Cd	49 In	50 Sn	51 Sb	52 Te	53 I	54 Xe
55 Cs	56 Ba	57–71 La–Lu	72 Hf	73 Ta	74 W	75 Re	76 Os	77 Ir	78 Pt	79 Au	80 Hg	81 Tl	82 Pb	83 Bi	84 Po	85 At	86 Rn
87 Fr	88 Ra	89–103 Ac–Lr	104 Unq	105 Unp	106 Unh	107 Uns	108 Uno	109 Une									

Atomic number: 9
Symbol: F

Period (rows 1–7)

▲
FIGURE 10.10
The periodic table of the elements.

period

groups, families

representative elements
transition elements

Each horizontal row in the periodic table is called a **period** as shown in Figure 10.10. The number of each period corresponds with the number of the outermost energy level that contains electrons for elements in that period. Those in the first row contain electrons only in energy level 1, while those in the second row contain electrons in levels 1 and 2. In the third row, electrons are found in levels 1, 2, and 3, and so on.

Elements that behave in a similar manner are found in **groups** or **families**. These form the vertical columns on the periodic table. There are several systems used for numbering the groups. In one system the columns are numbered from left to right using the numbers 1–18. We use a system that numbers the columns with Roman numerals and the letters A and B, as shown in Figure 10.10. The A groups are known as the **representative elements**. The B groups and Group VIII are called the **transition elements**. In this book we will focus on the representative elements. In addition, the groups (columns) of the periodic table often have family names. For example, the group on the far right side of the periodic table (He, Ne, Ar, Kr, Xe, and Rn) is called the *noble gases*. Group IA is called the *alkali metals,* Group IIA the *alkaline earth metals,* and Group VIIA the *halogens.*

How does the structure of the periodic table connect to the atomic structures of the elements? We've just seen that the rows of the periodic table are associated with the energy level of the outermost electrons of the atoms in that row. Let's summarize the valence electron configurations for the elements we have examined so far. Figure 10.11 gives the valence electron configurations for the first eighteen elements. Notice that a pattern is emerging. The valence configuration for the elements in columns is the same. Only the number for the energy level is different. This is expected since each new row is associated with a different energy level for the valence electrons. The chemical behavior and properties of elements in a particular family must therefore be associated with the electronic configuration of the elements.

The electron configurations for elements beyond these eighteen become long and tedious to write. We often abbreviate the electron configuration by using the following notation:

Na $[Ne]3s^1$

FIGURE 10.11 ▶
Valence shell electron configurations for the first 18 elements.

IA							Noble gases
1 **H** $1s^1$	IIA	IIIA	IVA	VA	VIA	VIIA	2 **He** $1s^2$
3 **Li** $2s^1$	4 **Be** $2s^2$	5 **B** $2s^22p^1$	6 **C** $2s^22p^2$	7 **N** $2s^22p^3$	8 **O** $2s^22p^4$	9 **F** $2s^22p^5$	10 **Ne** $2s^22p^6$
11 **Na** $3s^1$	12 **Mg** $3s^2$	13 **Al** $3s^23p^1$	14 **Si** $3s^23p^2$	15 **P** $3s^23p^3$	16 **S** $3s^23p^4$	17 **Cl** $3s^23p^5$	18 **Ar** $3s^23p^6$

Look carefully at Figure 10.11. Notice that the p orbitals are full at the noble gases on the periodic table. By placing the symbol for the noble gas in square brackets we can abbreviate the complete electron configuration and focus our attention on the valence electrons (the electrons we will be interested in when we discuss bonding in Chapter 11). To write the abbreviated electron configuration for any element, go back to the previous noble gas and place it in square brackets. Then list the valence electrons. Several examples are given below:

B	$1s^2\ 2s^2\ 2p^1$	$[\text{He}]2s^22p^1$
Cl	$1s^22s^22p^63s^23p^5$	$[\text{Ne}]3s^23p^5$
Na	$1s^22s^22p^63s^2$	$[\text{Ne}]3s^2$

The sequence for filling the orbitals is exactly as expected up through the $3p$ orbitals. The third energy level might logically be expected to fill with $3d$ electrons before electrons enter the $4s$ orbital, but this is not the case. The behavior and properties of the next two elements, potassium and calcium, are very similar to the elements in Groups IA and IIA. They clearly belong in these groups. The other elements in Group IA and Group IIA have electron configurations that indicate valence electrons in the s orbitals. For example, since the electron configuration is connected to the element's properties, we should place the last electrons for potassium and calcium in the $4s$ orbital. Their electron configurations are

K	$1s^22s^22p^63s^23p^64s^1$	or	$[\text{Ar}]4s^1$
Ca	$1s^22s^22p^63s^23p^64s^2$	or	$[\text{Ar}]4s^2$

Elements 21–30 belong to the elements known as *transition elements* on the periodic table. Electrons are placed in the $3d$ orbitals for each of these elements. When the $3d$ orbitals are full, the electrons fill the $4p$ orbitals to complete the fourth row of the periodic table. Let's consider the overall relationship between filling of the orbitals and the periodic table. Figure 10.12 illustrates the type of orbital filling and its location on the periodic table. The tall columns on the table (labeled IA–VIIA, and noble gases) are often called the *representative elements*. Valence electrons in these elements occupy s and p orbitals. The row number of the periodic table corresponds to the energy level of the valence electrons. The elements in the center of the periodic table (shown in ■) are the transition elements where the d orbitals are being filled. Notice that the number for the d orbitals is one behind the row number on the periodic table. The two rows shown at the bottom of the table in Figure 10.12 are called the *inner transition elements*. The last electrons in these elements are placed in the f orbitals. The number for the f orbitals is always two behind that of the s and p orbitals. A periodic table is almost always available to

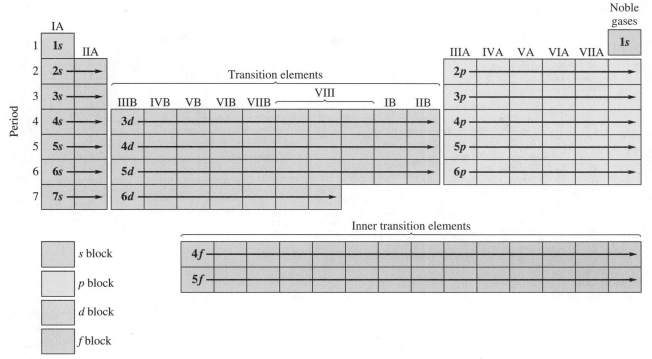

FIGURE 10.12
Arrangement of elements according to the sublevel of electrons being filled in their atomic structure.

you, so if you understand the relationship between the orbitals and the periodic table you can write the electron configuration for any element. There are several minor variations to these rules. We shall not concern ourselves with these variations in this course.

Use the periodic table to write the electron configuration for phosphorus and tin.

Example 10.1

Solution

Phosphorus is element 15 and is located in Period 3, Group VA. The electron configuration must have a full first and second energy level:

P $1s^2 2s^2 2p^6 3s^2 3p^3$

You can determine the electron configuration by looking across the row and counting the element blocks.

Tin is element 50 in Period 5, Group IVA, two places after the transition metals. It must have two electrons in the $5p$ series. Its electron configuration is

Sn $1s^2 2s^2 2p^6 3s^2 3p^6 4s^2 3d^{10} 4p^6 5s^2 4d^{10} 5p^2$

Notice that the d series of electrons is always one energy level behind the period number.

Practice 10.1

Use the periodic table to write the electron configuration for (a) O, (b) Ca, and (c) Ti.

IA	IIA	IIIB	IVB	VB	VIB	VIIB	VIII	VIII	VIII	IB	IIB	IIIA	IVA	VA	VIA	VIIA	Noble gases
1 H $1s^1$																	2 He $1s^2$
3 Li $2s^1$	4 Be $1s^2$											5 B $2s^22p^1$	6 C $2s^22p^2$	7 N $2s^22p^3$	8 O $2s^22p^4$	9 F $2s^22p^5$	10 Ne $2s^22p^6$
11 Na $3s^1$	12 Mg $2s^2$											13 Al $3s^23p^1$	14 Si $3s^23p^2$	15 P $3s^23p^3$	16 S $3s^23p^4$	17 Cl $3s^23p^5$	18 Ar $3s^23p^6$
19 K $4s^1$	20 Ca $4s^2$	21 Sc $4s^23d^1$	22 Ti $4s^23d^2$	23 V $4s^23d^3$	24 Cr $4s^13d^5$	25 Mn $4s^23d^5$	26 Fe $4s^23d^6$	27 Co $4s^23d^7$	28 Ni $4s^23d^8$	29 Cu $4s^13d^{10}$	30 Zn $4s^23d^{10}$	31 Ga $4s^24p^1$	32 Ge $4s^24p^2$	33 As $4s^24p^3$	34 Se $4s^24p^4$	35 Br $4s^24p^5$	36 Kr $4s^24p^6$
37 Rb $5s^1$	38 Sr $5s^2$	39 Y $5s^24d^1$	40 Zr $5s^24d^2$	41 Nb $5s^14d^4$	42 Mo $5s^14d^5$	43 Te $5s^14d^6$	44 Ru $5s^14d^7$	45 Rh $5s^14d^8$	46 Pd $5s^04d^{10}$	47 Ag $5s^14d^{10}$	48 Cd $5s^24d^{10}$	49 In $5s^25p^1$	50 Sn $5s^25p^2$	51 Sb $5s^25p^3$	52 Te $5s^25p^4$	53 I $5s^25p^5$	54 Xe $5s^25p^6$
55 Cs $6s^1$	56 Ba $6s^2$	57 La $6s^25d^1$	72 Hf $6s^25d^2$	73 Ta $6s^25d^3$	74 W $6s^25d^4$	75 Re $6s^25d^5$	76 Os $6s^25d^6$	77 Ir $6s^25d^7$	78 Pt $6s^15d^9$	79 Au $6s^15d^{10}$	80 Hg $6s^25d^{10}$	81 Tl $6s^26p^1$	82 Pb $6s^26p^2$	83 Bi $6s^26p^3$	84 Po $6s^26p^4$	85 At $6s^26p^5$	86 Rn $6s^26p^6$
87 Fr $7s^1$	88 Ra $7s^2$	89 Ac $7s^26d^1$	104 Unq $7s^26d^2$	105 Unp $7s^26d^3$	106 Unh $7s^26d^4$	107 Uns $7s^26d^5$	108 Uno $7s^26d^6$	109 Une $7s^26d^7$									

Group number — Noble gases (headers). Period (left axis), periods 1–7.

FIGURE 10.13
Outermost electron configurations of the elements.

The early chemists classified the elements based only on their observed properties, but modern atomic theory gives us an explanation for why the properties of elements vary periodically. For example, as we build up atoms by filling orbitals with electrons, the same type of orbitals occur on each energy level. This means that the same electron configuration reappears regularly for each level. Groups of elements show similar chemical properties because of these outermost electron similarities.

Consider the periodic table shown in Figure 10.13. For the representative elements in this table only the electron configuration of the outer shell electrons is given. This table illustrates the following important points:

1. In each row the number of the period corresponds with the highest energy level occupied by electrons.
2. The group numbers for the representative elements are equal to the total number of outer shell electrons in the atoms of the group. For example, the elements in Group VIIA always have the electron configuration ns^2np^5. The d and f electrons are always in a lower energy level than the highest energy level and so are not considered as outermost electrons.
3. The elements of a family have the same outer shell electron configuration except that the electrons are in different energy levels.
4. The elements within each of the s, p, d, f blocks are filling the s, p, d, f orbitals, as shown in Figure 10.12.
5. Within the transition elements some discrepancies in the order of filling occur. (Explanation of these discrepancies and the similar ones that occur in the inner transition elements are beyond the scope of this book.)

Example 10.2   Write the electron configuration of a zinc atom and a rubidium atom.

Solution   The atomic number of zinc is 30; therefore it has 30 protons and 30 electrons in a neutral atom. Using Figure 10.10, the electron configuration of a zinc atom is $1s^22s^22p^63s^23p^64s^23d^{10}$. Check by adding the superscripts, which should equal 30.

The atomic number of rubidium is 37; therefore it has 37 protons and 37 electrons in a neutral atom. With a little practice using a periodic table, the electron configuration may be written directly. The electron configuration of a rubidium atom is $1s^2 2s^2 2p^6 3s^2 3p^6 4s^2 3d^{10} 4p^6 5s^1$. Check by adding the superscripts, which should equal 37.

Concepts in Review

1. Describe the atom as conceived by Niels Bohr.

2. Discuss the contributions to atomic theory made by Dalton, Thomson, Rutherford, Bohr, and Schrödinger. (See Chapter 5 as well.)

3. Explain what is meant by an electron orbital.

4. Explain how an electron configuration can be determined from the periodic table.

5. Write the electron configuration for any of the first 56 elements.

6. Indicate the locations of the metals, nonmetals, metalloids, and noble gases in the periodic table. (See Chapter 3 for help.)

7. Indicate the areas in the periodic table where the *s, p, d,* and *f* orbitals are being filled.

8. Determine the number of valence electrons in any atom in the Group A elements.

9. Distinguish between representative elements and transition elements.

10. Identify groups of elements by their special family names.

11. Describe the changes in outer shell electron structure (a) when moving from left to right in a period, and (b) when going from top to bottom in a group.

12. Explain the relationship between group number and the number of outer shell electrons for the representative elements.

Key Terms

The terms listed here have been defined within this chapter. Section numbers are referenced in parenthesis for each term. More detailed definitions are given in the Glossary.

electron configuration (10.5)
frequency (10.2)
ground state (10.3)
groups, families (10.6)
line spectrum (10.3)
orbital (10.3), (10.4)
orbital diagram (10.5)

Pauli exclusion principle (10.4)
period (10.6)
photons (10.2)
principal energy levels (10.4)
quanta (10.3)
representative elements (10.6)

speed (10.2)
spin (10.4)
transition elements (10.6)
valence electrons (10.5)
wavelength (10.2)

Questions

Questions refer to tables, figures, and key words and concepts defined within the chapter. A particularly challenging question or exercise is indicated with an asterisk.

1. What is an electron orbital?

2. Under what conditions can a second electron enter an orbital already containing one electron?

3. What is meant when we say that the electron structure of an atom is in its ground state?

4. How do $1s$ and $2s$ orbitals differ? How are they alike?

5. What letters are used to designate the types of orbitals?

6. List the following sublevels in order of increasing energy: $2s$, $2p$, $4s$, $1s$, $3d$, $3p$, $4p$, $3s$.

7. How many s electrons, p electrons, and d electrons are possible in any energy level?

8. What is the major difference between an orbital and a Bohr orbit?

9. Explain how and why Bohr's model of the atom was modified to include the cloud model of the atom.

10. Sketch the s, p_x, p_y, and p_z orbitals.

11. In the designation $3d^7$, give the significance of 3, d, and 7.

12. Describe the difference between transition and representative elements.

13. From the standpoint of electron structure, what do the elements in the s block have in common?

14. Write symbols for elements with atomic numbers 8, 16, 34, 52, and 84. What do these elements have in common?

15. Write the symbols of the family of elements that have seven electrons in their outermost energy level.

16. What is the greatest number of elements to be found in any period? Which periods have this number?

17. From the standpoint of energy level, how does the placement of the last electron in the Group A elements differ from that of the Group B elements?

18. Find the places in the modern periodic table where elements are not in proper sequence according to atomic mass. (See inside of front cover for periodic table.)

19. Which of the following statements are correct? Rewrite each incorrect statement to make it correct.
   (a) In the ground state, electrons tend to occupy orbitals having the lowest possible energy.
   (b) The maximum number of p electrons in the first energy level is six.
   (c) A $2s$ electron is in a lower energy state than a $2p$ electron.
   (d) The electron structure for a carbon atom is $1s^2 2s^2 2p^2$.
   (e) The $2p_x$, $2p_y$, and $2p_z$ electron orbitals are all in the same energy state.
   (f) The energy level of a $3d$ electron is higher than that of a $4s$ electron.
   (g) The electron structure for a calcium atom is $1s^2 2s^2 2p^6 3s^2 3p^6 3d^2$.
   (h) The third energy level can have a maximum of 18 electrons.
   (i) The number of possible d electrons in the third energy level is ten.
   (j) The first f electron occurs in the fourth principal energy level.
   (k) Atoms of all the noble gases (except helium) have eight electrons in their outermost energy level.
   (l) A p orbital is spherically symmetrical around the nucleus.
   (m) An atom of nitrogen has two electrons in a $1s$ orbital, two electrons in a $2s$ orbital, and one electron in each of three different $2p$ orbitals.
   (n) When an orbital contains two electrons, the electrons have parallel spins.
   (o) The Bohr theory proposed that electrons move around the nucleus in circular orbits.
   (p) Bohr concluded from his experiment that the positive charge and almost all the mass were concentrated in a very small nucleus.

20. Which of the following statements are correct? Rewrite the incorrect statements to make them correct.
   (a) Properties of the elements are periodic functions of their atomic numbers.
   (b) There are more nonmetallic elements than metallic elements.
   (c) Calcium is a member of the alkaline earth metal family.
   (d) Iron belongs to the alkali metal family.
   (e) Bromine belongs to the halogen family.
   (f) Neon is a noble gas.
   (g) Group A elements do not contain partially filled d or f sublevels.
   (h) An atom of aluminum (Group IIIA) has five electrons in its outer shell.
   (i) The element $[Ar]4s^2 3d^{10} 4p^5$ is a halogen.
   (j) The element $[Kr]5s^2$ is a nonmetal.
   (k) The element with $Z = 12$ forms compounds similar to the element with $Z = 37$.
   (l) Nitrogen, fluorine, neon, gallium, and bromine are all nonmetals.
   (m) The atom having an outer shell electron structure of $5s^2 5p^2$ would be in period 6, Group IVA.
   (n) The yet-to-be-discovered element with an atomic number of 118 would be a noble gas.

Paired Exercises

These exercises are paired. Each odd-numbered exercise is followed by a similar even-numbered exercise. Answers to the even-numbered exercises are given in Appendix V.

21. How many protons are in the nucleus of an atom of each of these elements?
    (a) H
    (b) B
    (c) Sc
    (d) U

22. How many protons are in the nucleus of an atom of each of these elements?
    (a) F
    (b) Ag
    (c) Br
    (d) Sb

23. Give the electron configuration for the following:
    (a) B
    (b) Ti
    (c) Zn
    (d) Br
    (e) Sr

24. Give the electron configuration for the following:
    (a) chlorine
    (b) silver
    (c) lithium
    (d) iron
    (e) iodine

25. Explain how the spectral lines of hydrogen occur.

26. Explain how Bohr used the data from the hydrogen spectrum to support his model of the atom.

27. How many orbitals exist in the third energy level? What are they?

28. How many electrons can be present in the fourth energy level?

29. Write orbital diagrams for the following:
    (a) N
    (b) Cl
    (c) Zn
    (d) Zr
    (e) I

30. Write orbital diagrams for the following:
    (a) Si
    (b) S
    (c) Ar
    (d) V
    (e) P

31. Which atoms have the following electron configurations?
    (a) $1s^2 2s^2 2p^6 3s^2$
    (b) $1s^2 2s^2 2p^6 3s^2 3p^1$
    (c) $1s^2 2s^2 2p^6 3s^2 3p^6 4s^2 3d^8$
    (d) $1s^2 2s^2 2p^6 3s^2 3p^6 4s^2 3d^5$

32. Which atoms have the following electron configurations?
    (a) $[Ar]4s^2 3d^1$
    (b) $[Ar]4s^2 3d^{10}$
    (c) $[Kr]5s^2 4d^{10} 5p^2$
    (d) $[Xe]6s^1$

33. Show the electron configurations for elements with atomic numbers of:
    (a) 8
    (b) 11
    (c) 17
    (d) 23
    (e) 28
    (f) 34

34. Show the electron configurations for elements with atomic numbers of:
    (a) 9
    (b) 26
    (c) 31
    (d) 39
    (e) 52
    (f) 10

35. Identify these atoms from their atomic structure diagrams:

    (a) $\left(\begin{array}{c} 16p \\ 16n \end{array}\right)$ $2e^-$ $8e^-$ $6e^-$

    (b) $\left(\begin{array}{c} 28p \\ 32n \end{array}\right)$ $2e^-$ $8e^-$ $16e^-$ $2e^-$

36. Diagram the atomic structures (as you see in Exercise 35) for these atoms:
    (a) $^{27}_{13}Al$

    (b) $^{48}_{22}Ti$

37. Why is the eleventh electron of the sodium atom located in the third energy level rather than in the second energy level?

38. Why is the last electron in potassium located in the fourth energy level rather than in the third energy level?

39. What electron structure do the noble gases have in common?

40. What is unique about the noble gases, from an electron point of view?

41. How are elements in a period related to one another?

42. How are elements in a group related to one another?

43. What is common about the electron structures of the alkali metals?

44. Why would you expect the elements zinc, cadmium, and mercury to be in the same chemical family?

45. Pick the electron structures below that represent elements in the same chemical family:
  (a) $1s^2 2s^1$
  (b) $1s^2 2s^2 2p^4$
  (c) $1s^2 2s^2 2p^2$
  (d) $1s^2 2s^2 2p^6 3s^2 3p^4$
  (e) $1s^2 2s^2 2p^6 3s^2 3p^6$
  (f) $1s^2 2s^2 2p^6 3s^2 3p^6 4s^2$
  (g) $1s^2 2s^2 2p^6 3s^2 3p^6 4s^1$
  (h) $1s^2 2s^2 2p^6 3s^2 3p^6 4s^2 3d^1$

46. Pick the electron structures below that represent elements in the same chemical family:
  (a) $[He]2s^2 2p^6$
  (b) $[Ne]3s^1$
  (c) $[Ne]3s^2$
  (d) $[Ne]3s^2 3p^3$
  (e) $[Ar]4s^2 3d^{10}$
  (f) $[Ar]4s^2 3d^{10} 4p^6$
  (g) $[Ar]4s^2 3d^5$
  (h) $[Kr]5s^2 4d^{10}$

47. In the periodic table, calcium, element 20, is surrounded by elements 12, 19, 21, and 38. Which of these have physical and chemical properties most resembling calcium?

48. In the periodic table, phosphorus, element 15, is surrounded by elements 14, 7, 16, and 33. Which of these have physical and chemical properties most resembling phosphorus?

49. Classify each of the following elements as metals, nonmetals, or metalloids (review Chapter 3 if you need help):
  (a) potassium
  (b) plutonium
  (c) sulfur
  (d) antimony

50. Classify each of the following elements as metals, nonmetals, or metalloids (review Chapter 3 if you need help):
  (a) iodine
  (b) tungsten
  (c) molybdenum
  (d) germanium

51. In which period and group does an electron first appear in a d orbital?

52. In which period and group does an electron first appear in an f orbital?

53. How many electrons occur in the valence level of Group IIIA and IIIB elements? Why are they different?

54. How many electrons occur in the valence level of Group VIIA and VIIB elements? Why are they different?

Additional Exercises

These exercises are not paired or labeled by topic and provide additional practice on the concepts covered in this chapter.

55. If all the orbitals within an atom could hold three electrons rather than two, what would be the atomic numbers and identities of the first three noble gases?

56. Why does the emission spectrum for nitrogen reveal many more spectral lines than that for hydrogen?

57. Among the first 100 elements on the periodic table, how many have
  (a) at least one s electron?
  (b) at least one p electron?
  (c) at least one d electron?
  (d) at least one f electron?

58. For each of the following elements, what percent of their electrons are s electrons?
  (a) He   **(d)** Se
  (b) Be   **(e)** Cs
  (c) Xe

59. How many pairs of valence electrons do these elements have?
  (a) O   **(d)** Xe
  (b) P   **(e)** Rb
  (c) I

60. Suppose we use a foam ball to represent a typical atom. If the radius of the ball is 1.5 cm and the radius of a typical atom is 1.0×10^{-8} cm, how much of an enlargement is this? Use a ratio to express your answer.

61. List the first element on the periodic table that satisfies each of these conditions:
  (a) a completed set of p orbitals
  (b) two $4p$ electrons
  (c) seven valence electrons
  (d) three unpaired electrons

62. Explain how scanning-probe microscopes are different from earlier microscopes.

63. Describe the functioning of a scanning-tunneling microscope.

64. What is the difference between a scanning-tunneling microscope and an atomic-force microscope?

65. Oxygen is a gas. Sulfur is a solid. What is it about their electron structures that causes them to be grouped in the same chemical family?

66. In which groups are the transition elements located?

67. How do the electron structures of the transition elements differ from the representative elements?

Try to answer Exercises 68–71 without referring to the periodic table.

68. The atomic numbers of the noble gases are 2, 10, 18, 36, 54, and 86. What are the atomic numbers for the elements with six electrons in their outer electron shells?

69. Element number 87 is in Group IA, Period 7. Describe its outermost energy level. How many energy levels of electrons does it have?

70. If element 36 is a noble gas, in which groups would you expect elements 35 and 37 to occur?

71. Write a paragraph describing the general features of the periodic table.

72. Some scientists have proposed the existence of element 117. If it were to exist:
    (a) what would its electron configuration be?
    (b) how many valence electrons would it have?
    (c) what element would it likely resemble?
    (d) to what family and period would it belong?

73. What is the relationship between two elements if
    (a) one of them has 10 electrons, 10 protons, and 10 neutrons and the other has 10 electrons, 10 protons, and 12 neutrons?
    (b) one of them has 23 electrons, 23 protons, and 27 neutrons and the other has 24 electrons, 24 protons, and 26 neutrons?

74. Is there any pattern to the location of gases on the periodic table? Is there a pattern for the location of liquids? Is there a pattern for the location of solids?

Answers to Practice Exercises

10.1 (a) O $1s^2 2s^2 2p^4$
     (b) Ca $1s^2 2s^2 2p^6 3s^2 3p^6 4s^2$
     (c) Ti $1s^2 2s^2 2p^6 3s^2 3p^6 4s^2 3d^2$

Chemical Bonds: The Formation of Compounds from Atoms

11

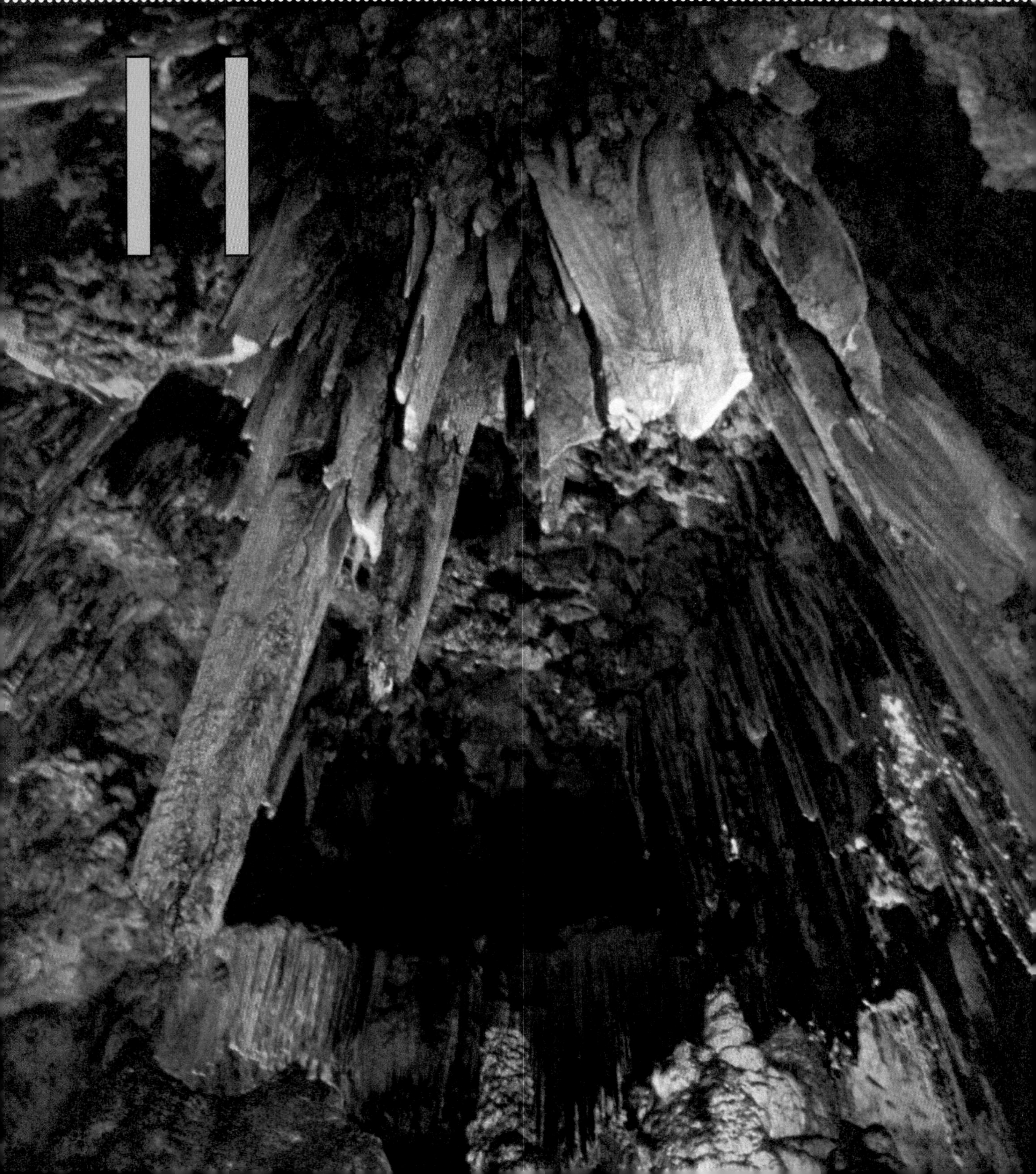

For centuries, we have been aware that certain metals cling to a magnet. Balloons may stick to a wall. Why? The activity of a superconductor floating in air has been the subject of television commercials. High-speed levitation trains are heralded to be the wave of the future. How do they function? In each case, forces of attraction and repulsion are at work.

Interestingly, human interactions also suggest that "opposites attract" and "likes repel." Attractions draw us into friendships and significant relationships, whereas repulsive forces may produce debate and antagonism. We form and break apart interpersonal bonds throughout our lives.

In chemistry, we also see this phenomenon. Substances form chemical bonds as a result of electrical attractions. These bonds provide the tremendous diversity of compounds found in nature.

11.1   Periodic Trends in Atomic Properties

Although atomic theory and electron configuration can help us to understand the arrangement and behavior of the elements, it is important to remember that the design of the periodic table is based on the observation of properties of the elements. Before we proceed to use the concept of atomic structure to explain how and why atoms combine to form compounds, we need to understand the characteristic properties of the elements and the trends that occur in these properties on the periodic table. These trends in observed properties allow us to use the periodic table to accurately predict properties and reactions of a diversity of substances without needing to possess the substance or complete the reaction.

Metals and Nonmetals

In Section 3.8 we began our study of the periodic table by classifying elements as metals, nonmetals, or metalloids. The heavy stair-step line beginning at boron and running diagonally down the periodic table separates the elements into metals and nonmetals. Metals are usually lustrous, malleable, and good conductors of heat and electricity. Nonmetals are just the opposite—nonlustrous, brittle, and poor conductors. Metalloids are found bordering the heavy diagonal line and may have properties of both metals and nonmetals.

Most elements are classified as metals (see Figure 11.1). Metals are found on the left side of the stair-step line, while the nonmetals are located toward the upper right of the table. Note that hydrogen does not fit into the division of metals and nonmetals. It displays nonmetallic properties under normal conditions, even though it has only one outermost electron like the alkali metals. Hydrogen is considered to be a unique element.

It is the chemical properties of metals and nonmetals that are most interesting. Metals tend to lose electrons and form positive ions, while nonmetals tend to gain electrons and form negative ions. When a metal reacts with a nonmetal a transfer of electrons from the metal to the nonmetal frequently occurs.

◀ **Chapter Opening Photo: This colorfully lighted limestone cave reveals dazzling stalactites and stalagmites, formed from calcium carbonate.**

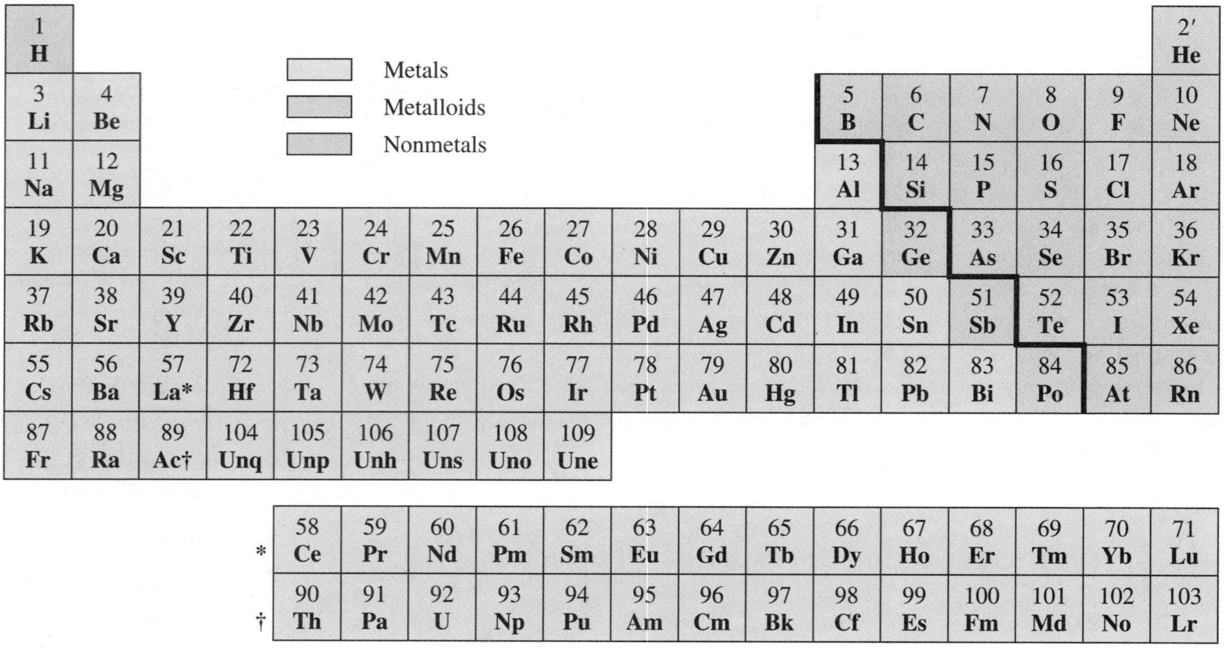

1 H																	2' He
3 Li	4 Be		Metals									5 B	6 C	7 N	8 O	9 F	10 Ne
11 Na	12 Mg		Metalloids									13 Al	14 Si	15 P	16 S	17 Cl	18 Ar
19 K	20 Ca	21 Sc	22 Ti	23 V	24 Cr	25 Mn	26 Fe	27 Co	28 Ni	29 Cu	30 Zn	31 Ga	32 Ge	33 As	34 Se	35 Br	36 Kr
37 Rb	38 Sr	39 Y	40 Zr	41 Nb	42 Mo	43 Tc	44 Ru	45 Rh	46 Pd	47 Ag	48 Cd	49 In	50 Sn	51 Sb	52 Te	53 I	54 Xe
55 Cs	56 Ba	57 La*	72 Hf	73 Ta	74 W	75 Re	76 Os	77 Ir	78 Pt	79 Au	80 Hg	81 Tl	82 Pb	83 Bi	84 Po	85 At	86 Rn
87 Fr	88 Ra	89 Ac†	104 Unq	105 Unp	106 Unh	107 Uns	108 Uno	109 Une									

	58 Ce	59 Pr	60 Nd	61 Pm	62 Sm	63 Eu	64 Gd	65 Tb	66 Dy	67 Ho	68 Er	69 Tm	70 Yb	71 Lu
*	58 Ce	59 Pr	60 Nd	61 Pm	62 Sm	63 Eu	64 Gd	65 Tb	66 Dy	67 Ho	68 Er	69 Tm	70 Yb	71 Lu
†	90 Th	91 Pa	92 U	93 Np	94 Pu	95 Am	96 Cm	97 Bk	98 Cf	99 Es	100 Fm	101 Md	102 No	103 Lr

▲
FIGURE 11.1
Classification of the elements into metals, nonmetals, and metalloids.

Atomic Radius

The relative radii of the representative elements are shown in Figure 11.2. Notice that the radii of the atoms tend to increase down each group and that they tend to decrease from left to right across a period.

The increase in radius down a column can be understood if we consider the electron structure of the atoms. For each step down a column an additional energy level is added to the atom. The average distance from the nucleus to the outside edge of the atom must increase as each new energy level is added. The atoms get bigger as electrons are placed in these new higher energy levels.

Understanding the decrease in atomic radius across a row requires more thought, however. As we move from left to right across a period electrons within the same block are being added to the same energy level. Within a given energy level we expect the orbitals to have about the same size. We would then expect the atoms to be about the same size across the period. But each time an electron is added a proton is added to the nucleus as well. The increase in positive charge (in the nucleus) pulls the electrons closer to the nucleus resulting in a gradual decrease in atomic radius across a row.

Ionization Energy

ionization energy

The **ionization energy** of an atom is the energy required to remove an electron from the atom. For example,

$$\text{Na} + \text{ionization energy} \longrightarrow \text{Na}^+ + e^-$$

The first ionization energy is the amount of energy required to remove the first electron from an atom, the second is the amount required to remove the second electron from that atom, and so on.

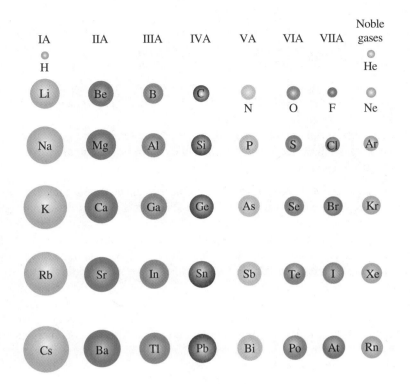

▲ FIGURE 11.2
Relative atomic radii for representative elements. Atomic radius decreases across a period and increases down a group.

TABLE 11.1 Ionization Energies for Selected Elements*

	Required amounts of energy (kJ/mol)				
Element	1st e⁻	2nd e⁻	3rd e⁻	4th e⁻	5th e⁻
H	1,314				
He	2,372	5,247			
Li	520	7,297	11,810		
Be	900	1,757	14,845	21,000	
B	800	2,430	3,659	25,020	32,810
C	1,088	2,352	4,619	6,222	37,800
Ne	2,080	3,962	6,276	9,376	12,190
Na	496	4,565	6,912	9,540	13,355

* Values are expressed in kilojoules per mole, showing energies required to remove 1 to 5 electrons per atom. Color indicates the energy needed to remove an electron from a noble-gas electron structure.

Table 11.1 gives the ionization energies for the removal of one to five electrons from several elements. The table shows that even higher amounts of energy are needed to remove the second, third, fourth, and fifth electrons. This makes sense because removing electrons does not decrease the positive charge in the nucleus; the remaining electrons are held even more tightly. The data in Table 11.1 also show that an extra large ionization energy (red) is needed when an electron is removed from a noble gas structure, clearly showing the stability of this configuration.

First ionization energies have been experimentally determined for most elements. Figure 11.3 is a graphic plot of these ionization energies for representative elements in the first four periods. Note these important points:

1. Ionization energy in Group A elements decreases from top to bottom in a group. For example, in Group IA the ionization energy changes from 520 kJ/mol for Li to 419 kJ/mol for K.
2. From left to right across a period the ionization energy gradually increases. Noble gases have a relatively high value, confirming the nonreactive nature of these elements.

All metals do not behave in exactly the same manner. Some metals give up electrons much more easily than others. In the alkali metal family, cesium gives up its 6s electron much more easily than the metal lithium gives up its 2s electron. This makes sense when we consider that the size of the atoms increases down the group. The distance between the nucleus and the outer electrons increases and the ionization energy decreases. The most chemically active metals are located on the lower left of the periodic table.

Nonmetals have relatively large ionization energies compared to metals. Nonmetals tend to gain electrons and form anions. Since the nonmetals are located at the right side of the periodic table it is not surprising that ionization energies tend to increase from left to right across a period. The most active nonmetals are found in the *upper* right corner of the periodic table.

11.2 Lewis Structures of Atoms

Metals tend to form cations (positively charged ions) and nonmetals form anions (negatively charged ions) in order to attain a stable valence electron structure. For

FIGURE 11.3 ▶
Periodic relationship of the first ionization energy for representative elements in the first four periods.

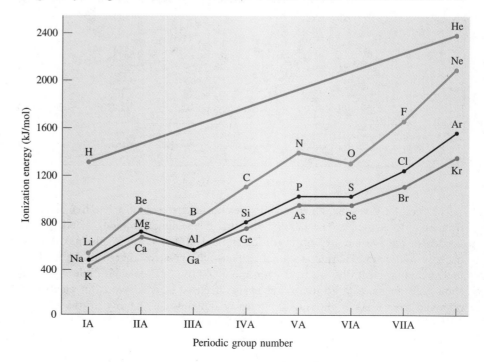

◀ **Gilbert N. Lewis (1875–1946) in his laboratory.**

many elements this stable valence level contains eight electrons (two *s* and six *p*), identical to the valence electron configuration of the noble gases. Atoms undergo rearrangements of electron structure to lower their chemical potential energy (or become more stable). These rearrangements are accomplished by losing, gaining, or sharing electrons with other atoms. For example, a hydrogen atom could accept a second electron and attain an electron structure the same as the noble gas helium. A fluorine atom could gain an electron and attain an electron structure like neon. A sodium atom could lose one electron to attain an electron structure like neon.

The valence electrons in the outermost energy level of an atom are responsible for the electron activity that occurs to form chemical bonds. The **Lewis structure** of an atom is a representation that shows the valence electrons for that atom. American chemist, Gilbert N. Lewis (1875–1946) proposed using the symbol for the element and dots for electrons. The number of dots placed around the symbol equals the number of *s* and *p* electrons in the outermost energy level of the atom. Paired dots represent paired electrons; unpaired dots represent unpaired electrons. For example, **H·** is the Lewis symbol for a hydrogen atom, $1s^1$; **:Ḃ** is the Lewis symbol for a boron atom, with valence electrons $2s^2 2p^1$. In the case of boron, the symbol B represents the boron nucleus and the $1s^2$ electrons; the dots represent only the $2s^2 2p^1$ electrons.

Lewis structure

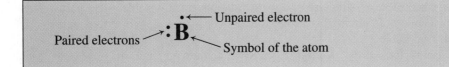

The Lewis method is often used not only because of its simplicity of expression but also because much of the chemistry of the atom is directly associated with the electrons in the outermost energy level. Figure 11.4 shows Lewis structures for the elements hydrogen through calcium.

FIGURE 11.4▶
FIGURE 11.4▶
Lewis structures of the first 20 elements. Dots represent electrons in the outermost energy level only.

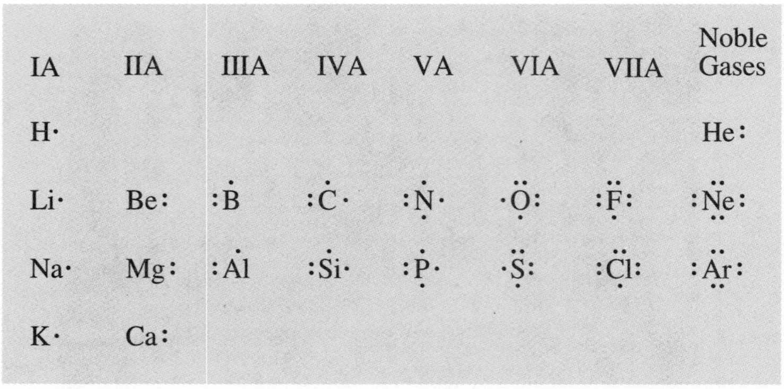

IA	IIA	IIIA	IVA	VA	VIA	VIIA	Noble Gases
H·							He:
Li·	Be:	:B·	:C·	:N·	·Ö:	:F̈:	:N̈e:
Na·	Mg:	:Al·	:Si·	:P·	·S̈:	:C̈l:	:Är:
K·	Ca:						

Example 11.1   Write the Lewis structure for a phosphorus atom.

Solution   First establish the electron structure for a phosphorus atom. It is $1s^2 2s^2 2p^6 3s^2 3p^3$. Note that there are five electrons in the outermost energy level; they are $3s^2 3p^3$. Write the symbol for phosphorus and place the five electrons as dots around it.

A quick way to determine the correct number of dots (electrons) for a Lewis structure is to use the group number. For the A groups on the periodic table, the roman numeral is the same as the number of electrons in the Lewis structure.

$$:\overset{\cdot}{\underset{\cdot}{P}}\cdot$$

The $3s^2$ electrons are paired and are represented by the paired dots. The $3p^3$ electrons, which are unpaired, are represented by the single dots.

Practice 11.1

Write the Lewis structure for the following elements:
(a) N (b) Al (c) Sr (d) Br

11.3   The Ionic Bond: Transfer of Electrons from One Atom to Another

The chemistry of many elements, especially the representative ones, is to attain an outer shell electron structure like that of the chemically stable noble gases. With the exception of helium, this stable structure consists of eight electrons in the outer shell (see Table 11.2).

Let us look at the electron structures of sodium and chlorine to see how each element can attain a structure of eight electrons in its outer shell. A sodium atom has eleven electrons: two in the first energy level, eight in the second energy level, and one in the third energy level. A chlorine atom has seventeen electrons: two in the first energy level, eight in the second energy level, and seven in the third energy level. If a sodium atom transfers or loses its $3s$ electron, its third energy level becomes vacant, and it becomes a sodium ion with an electron configuration identical to that of the noble gas neon. This process requires energy:

An atom that has lost or gained electrons will have a positive or negative charge, depending on which particles, protons or electrons are in excess. Remember that a charged atom or group of atoms is called an *ion.*

By losing a negatively charged electron, the sodium atom becomes a positively charged particle known as a sodium ion. The charge, $+1$, results because the nucleus still contains eleven positively charged protons, and the electron orbitals contain only ten negatively charged electrons. The charge is indicated by a plus sign ($+$) and is written with a superscript after the symbol of the element (Na^+).

A chlorine atom with seven electrons in the third energy level needs one electron to pair up with its one unpaired $3p$ electron to attain the stable outer shell electron structure of argon. By gaining one electron the chlorine atom becomes a chloride ion (Cl^-), a negatively charged particle containing seventeen protons and eighteen electrons. This process releases energy:

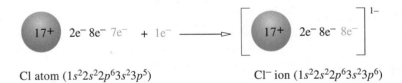

Consider sodium and chlorine atoms reacting with each other. The $3s$ electron from the sodium atom transfers to the half-filled $3p$ orbital in the chlorine atom to form a positive sodium ion and a negative chloride ion. The compound sodium chloride results because the Na^+ and Cl^- ions are strongly attracted to each other

TABLE 11.2 Arrangement of Electrons in the Noble Gases*

		Electron structure					
Noble gas	Symbol	$n = 1$	2	3	4	5	6
Helium	He	$1s^2$					
Neon	Ne	$1s^2$	$2s^22p^6$				
Argon	Ar	$1s^2$	$2s^22p^6$	$3s^23p^6$			
Krypton	Kr	$1s^2$	$2s^22p^6$	$3s^23p^63d^{10}$	$4s^24p^6$		
Xenon	Xe	$1s^2$	$2s^22p^6$	$3s^23p^63d^{10}$	$4s^24p^64d^{10}$	$5s^25p^6$	
Radon	Rn	$1s^2$	$2s^22p^6$	$3s^23p^63d^{10}$	$4s^24p^64d^{10}4f^{14}$	$5s^25p^65d^{10}$	$6s^26p^6$

* Each gas except helium has eight electrons in its outermost energy level.

by their opposite electrostatic charges. The force holding the oppositely charged ions together is an ionic bond:

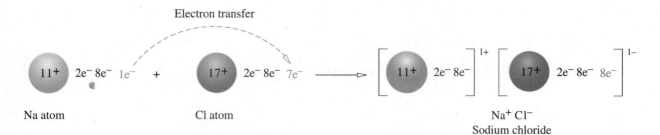

Electron transfer

Na atom   Cl atom

$Na^+ Cl^-$
Sodium chloride

The Lewis representation of sodium chloride formation is shown here:

$$Na + \overset{\cdot\cdot}{\underset{\cdot\cdot}{\cdot}Cl}: \longrightarrow [Na]^+ \left[:\overset{\cdot\cdot}{\underset{\cdot\cdot}{Cl}}:\right]^-$$

The chemical reaction between sodium and chlorine is a very vigorous one, producing considerable heat in addition to the salt formed. When energy is released in a chemical reaction, the products are more stable than the reactants. Note that in NaCl both atoms attain a noble-gas electron structure.

Sodium chloride is made up of cubic crystals in which each sodium ion is surrounded by six chloride ions and each chloride ion by six sodium ions, except at the crystal surface. A visible crystal is a regularly arranged aggregate of millions of these ions, but the ratio of sodium to chloride ions is 1:1, hence the formula NaCl. The cubic crystalline lattice arrangement of sodium chloride is shown in Figure 11.5.

Figure 11.6 contrasts the relative sizes of sodium and chlorine atoms with those of their ions. The sodium ion is smaller than the atom due primarily to two factors: (1) The sodium atom has lost its outer shell of one electron, thereby reducing its size; and (2) the ten remaining electrons are now attracted by eleven protons and are thus drawn closer to the nucleus. Conversely, the chloride ion is larger than the atom because (1) it has eighteen electrons but only seventeen protons; and (2) the nuclear attraction on each electron is thereby decreased, allowing the chlorine atom to expand as it forms an ion.

We have seen that when sodium reacts with chlorine, each atom becomes an ion. Sodium chloride, like all ionic substances, is held together by the attraction existing between positive and negative charges. An **ionic bond** is the attraction between oppositely charged ions.

Ionic bonds are formed whenever one or more electrons are transferred from one atom to another. Metals, which have relatively little attraction for their valence electrons, tend to form ionic bonds when they combine with nonmetals.

It is important to recognize that substances with ionic bonds do not exist as molecules. In sodium chloride, for example, the bond does not exist solely between a single sodium ion and a single chloride ion. Each sodium ion in the crystal attracts six near-neighbor negative chloride ions; in turn, each negative chloride ion attracts six near-neighbor positive sodium ions (see Figure 11.5).

A metal will usually have one, two, or three electrons in its outer energy level. In reacting, metal atoms characteristically lose these electrons, attain the electron

It is helpful to remember that a cation is always smaller than its parent atom whereas an anion is always larger than its parent atom.

ionic bond

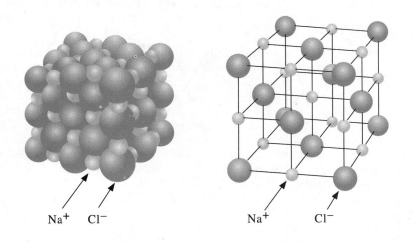

◀ **FIGURE 11.5**
Sodium chloride crystal. Diagram represents a small fragment of sodium chloride, which forms cubic crystals. Each sodium ion is surrounded by six chloride ions, and each chloride ion is surrounded by six sodium ions.

structure of a noble gas, and become positive ions. A nonmetal, on the other hand, is only a few electrons short of having a noble gas electron structure in its outer energy level and thus has a tendency to gain electrons. In reacting with metals, nonmetal atoms characteristically gain one, two or three electrons, attain the electron structure of a noble gas, and become negative ions. The ions formed by loss of electrons are much smaller than the corresponding metal atoms; the ions formed by gaining electrons are larger than the corresponding nonmetal atoms. The actual dimensions of the atomic and ionic radii of several metals and nonmetals are given in Table 11.3.

Study the following examples. Note the loss and gain of electrons between atoms; also note that the ions in each compound have a noble-gas electron structure.

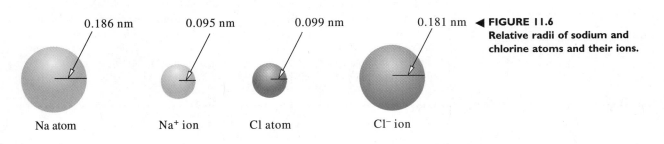

◀ **FIGURE 11.6**
Relative radii of sodium and chlorine atoms and their ions.

TABLE 11.3 Change in Atomic Radii of Selected Metals and Nonmetals*

Atomic radius (nm)		Ionic radius (nm)		Atomic radius (nm)		Ionic radius (nm)	
Li	0.152	Li$^+$	0.060	F	0.071	F$^-$	0.136
Na	0.186	Na$^+$	0.095	Cl	0.099	Cl$^-$	0.181
K	0.227	K$^+$	0.133	Br	0.114	Br$^-$	0.195
Mg	0.160	Mg^{2+}	0.065	O	0.074	O^{2-}	0.140
Al	0.143	Al^{3+}	0.050	S	0.103	S^{2-}	0.184

* The metals shown lose electrons to become positive ions. The nonmetals gain electrons to become negative ions.

Superconductors—A New Frontier

When electric current flows through a wire, resistance slows the current and heats the wire. In order to keep the current flowing this electrical "friction" must be overcome by adding energy to the system. In fact the existence of electrical resistance limits the efficiency of all electrical devices.

When chilled in liquid nitrogen, the superconductor acts as a perfect mirror to the magnet, causing it to levitate as it "sees" its reflection in the superconductor.

In 1911, a Dutch scientist, Heike Kamerlingh Onnes, discovered that at very cold temperatures (near 0 K), electrical resistance disappears. Onnes named this phenomenon *superconductivity*. Scientists have been fascinated by it ever since. Unfortunately, because such low temperatures were required, liquid helium was necessary to cool the wires. With helium costing $7 per liter, commercial applications were far too expensive to be considered.

For many years scientists were convinced that superconductivity was not possible at higher temperatures (not even as high as 77 K, the boiling point of liquid nitrogen, a bargain at $0.17 per liter). The first high-temperature superconductor, developed in 1986, was superconducting at 30 K. This material is a complex metal oxide that has a sandwich-like crystal structure with copper and oxygen atoms on the inside and barium and lanthanum atoms on the outside.

Scientists immediately tried to develop materials that would be superconducting at even higher temperatures. To do this they relied on their knowledge of the periodic table and the properties of chemical families. Paul Chu at the University of Houston, Texas, found the critical temperature could be raised by compressing the superconducting oxide. The pressure was too intense to be useful commercially, so Chu looked for another way to bring the layers closer together. He recognized that this could be accomplished by replacing the barium with strontium, an element in the same family with similar chemical properties and a smaller ionic radius. The idea was successful—the critical temperature changed from 30 K to 40 K. He then tried replacing the strontium with calcium (same family, smaller still) but to no avail. The new material had a lower critical temperature! But Chu persisted, and on January 12, 1987, by substituting yttrium for lanthanum (same family, smaller radius) he produced a new superconductor having a critical temperature of 95 K, well above the 77 K boiling point of liquid nitrogen. This material has the formula $YBa_2Cu_3O_7$ and is a good candidate for commercial applications.

There are several barriers to cross before superconductors are in wide use. The material is brittle and easily broken, nonmalleable, and does not carry as high a current per unit cross section as conventional conductors. Many researchers are currently working to surmount these problems and develop potential uses for superconductors, including high-speed levitation trains, tiny efficient electric motors, and smaller, faster computers.

Example 11.2   Explain how magnesium and chlorine combine to form magnesium chloride, $MgCl_2$.

Solution   A magnesium atom of electron structure $1s^2 2s^2 2p^6 3s^2$ must lose two electrons or gain six to reach a stable electron structure. If magnesium reacts with chlorine and each chlorine atom can accept only one electron, two chlorine atoms will be needed for the two electrons from each magnesium atom. The compound formed will contain one magnesium ion and two chloride ions. The magnesium atom, having lost two electrons, becomes a magnesium ion with a $+2$ charge. Each chloride ion will have a -1 charge. The transfer of electrons from a magnesium atom to two chlorine atoms is shown in the following illustration:

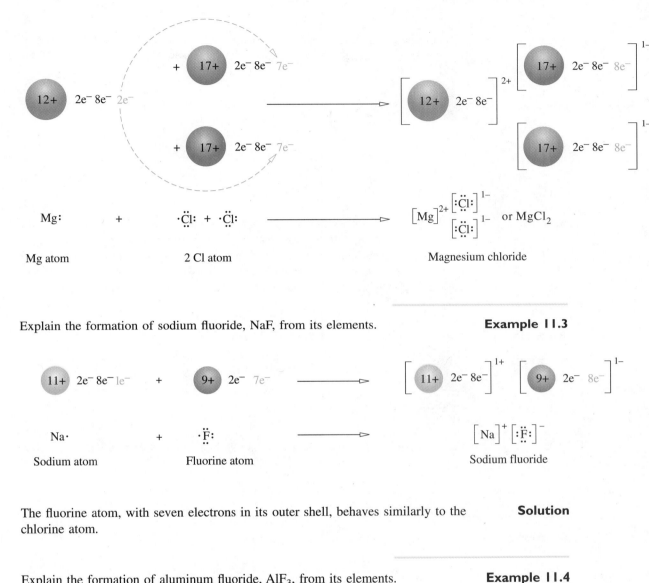

Explain the formation of sodium fluoride, NaF, from its elements.

Example 11.3

Solution

The fluorine atom, with seven electrons in its outer shell, behaves similarly to the chlorine atom.

Explain the formation of aluminum fluoride, AlF_3, from its elements.

Example 11.4

Solution

Each fluorine atom can accept only one electron. Therefore three fluorine atoms are needed to combine with the three outer shell electrons of one aluminum atom. The aluminum atom has lost three electrons to become an aluminum ion, Al^{3+}, with a +3 charge.

Example 11.5 Explain the formation of magnesium oxide, MgO, from its elements.

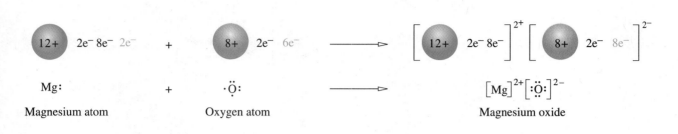

Mg:

Magnesium atom

+

·Ö:

Oxygen atom

$\longrightarrow$

$\left[\text{Mg}\right]^{2+}\left[:\ddot{\text{O}}:\right]^{2-}$

Magnesium oxide

Solution The magnesium atom, with two electrons in the outer energy level, exactly fills the need of one oxygen atom for two electrons. The resulting compound has a ratio of one atom of magnesium to one atom of oxygen. The oxygen (oxide) ion has a −2 charge, having gained two electrons. In combining with oxygen, magnesium behaves the same way as when combining with chlorine; it loses two electrons.

Example 11.6 Explain the formation of sodium sulfide, Na_2S, from its elements.

Solution

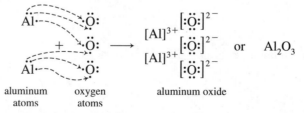

Two sodium atoms supply the electrons that one sulfur atom needs to make eight in its outer shell.

Example 11.7 Explain the formation of aluminum oxide, Al_2O_3, from its elements.

Solution

Al·------·Ö:
                          $[Al]^{3+}$ $\left[:\ddot{\text{O}}:\right]^{2-}$
          + ·Ö: $\longrightarrow$ $\left[:\ddot{\text{O}}:\right]^{2-}$ or Al_2O_3
Al·------·Ö: $[Al]^{3+}$ $\left[:\ddot{\text{O}}:\right]^{2-}$

aluminum   oxygen   aluminum oxide
atoms   atoms

One oxygen atom, needing two electrons, cannot accommodate the three electrons from one aluminum atom. One aluminum atom falls one electron short of the four electrons needed by two oxygen atoms. A ratio of two atoms of aluminum to three atoms of oxygen, involving the transfer of six electrons (two to each oxygen atom), gives each atom a stable electron configuration.

Note that in each of the examples above, outer shells containing eight electrons were formed in all the negative ions. This formation resulted from the pairing of all the *s* and *p* electrons in these outer shells.

11.4   Predicting Formulas of Ionic Compounds

In the previous examples we have seen that when a metal and a nonmetal react to form an ionic compound, the metal loses one or more electrons to the nonmetal. In Chapter 6, where we learned to name compounds and write formulas, we saw that Group IA metals always formed $+1$ cations, whereas Group IIA formed $+2$ cations. Group VIIA elements formed -1 anions and Group VIA formed -2 anions.

We now realize that this pattern is directly related to the stability of the noble gas configuration. Metals lose electrons to attain the electron configuration of a noble gas (the previous one on the periodic table). A nonmetal forms an ion by gaining enough electrons to achieve the electron configuration of the noble gas following it on the periodic table. These observations lead us to an important chemical principle: In almost all stable chemical compounds of representative elements, each atom attains a noble gas electron configuration. This concept forms the basis for our understanding of chemical bonding.

We can apply this principle in predicting the formulas of ionic compounds. To predict the formula of an ionic compound we must recognize that chemical compounds are always electrically neutral. In addition, the metal will lose electrons to achieve noble gas configuration and the nonmetal will gain electrons to achieve noble gas configuration. Consider the compound formed between barium and sulfur. Barium has two valence electrons whereas sulfur has six valence electrons:

$$Ba \quad [Xe]6s^2 \qquad S \quad [Ne]3s^23p^4$$

If barium loses two electrons it will achieve the configuration of Xenon. By gaining two electrons sulfur achieves the configuration of argon. Consequently a pair of electrons is transferred between atoms. Now we have Ba^{2+} and S^{2-}. Since compounds are electrically neutral there must be a ratio of one Ba to one S, giving the empirical formula BaS.

The same principle works for many other cases. Since the key to the principle lies in the electron configuration, the periodic table can be used to extend the prediction even further. Because of similar electron structures, the elements in a family generally form compounds with the same atomic ratios. In general, if we know the atomic ratio of a particular compound, say NaCl, we can predict the atomic ratios and formulas of the other alkali metal chlorides. These formulas are LiCl, KCl, RbCl, CsCl, and FrCl (see Table 11.4).

In a similar way, if we know that the formula of the oxide of hydrogen is H_2O, we can predict that the formula of the sulfide will be H_2S, because sulfur has the same valence electron structure as oxygen. It must be recognized, however, that these are only predictions; it does not necessarily follow that every element in a group will behave like the others or even that a predicted compound will actually exist. Knowing the formulas for potassium chlorate, bromate, and iodate to be $KClO_3$, $KBrO_3$, and KIO_3, we can correctly predict the corresponding sodium compounds to have the formulas $NaClO_3$, $NaBrO_3$, and $NaIO_3$. Fluorine belongs to the same family of elements (Group VIIA) as chlorine, bromine, and iodine. So we could predict that potassium and sodium fluorates would have the formulas

TABLE 11.4 Formulas of Compounds Formed by Alkali Metals

Lewis structure	Monoxides	Chlorides	Bromides	Sulfates
Li·	Li_2O	LiCl	LiBr	Li_2SO_4
Na·	Na_2O	NaCl	NaBr	Na_2SO_4
K·	K_2O	KCl	KBr	K_2SO_4
Rb·	Rb_2O	RbCl	RbBr	Rb_2SO_4
Cs·	Cs_2O	CsCl	CsBr	Cs_2SO_4

KFO_3 and $NaFO_3$. But this prediction would not be correct, because potassium and sodium fluorates are not known to exist. However, if they did exist, the formulas could very well be correct, for these predictions are based on comparisons with known formulas and similar electron structures.

In the discussion in this section we refer only to representative metals (Groups IA, IIA, IIIA). The transition metals (Group B) show more complicated behavior (they form multiple ions) and their formulas are not as easily predicted.

Example 11.8   The formula for calcium sulfide is CaS and that for lithium phosphide is Li_3P. Predict formulas for (a) magnesium sulfide, (b) potassium phosphide, and (c) magnesium selenide.

Solution

(a) Look in the periodic table for calcium and magnesium. They are both in Group IIA. Since the formula for calcium sulfide is CaS, it is reasonable to predict that the formula for magnesium sulfide is MgS.

(b) Find lithium and potassium in the periodic table. They are in Group IA. Since the formula for lithium phosphide is Li_3P, it is reasonable to predict that K_3P is the formula for potassium phosphide.

(c) Find selenium in the periodic table. It is in Group VIA just below sulfur. Therefore, it is reasonable to assume that selenium forms selenide in the same way that sulfur forms sulfide. Since MgS was the predicted formula for magnesium sulfide in part (a), it is reasonable to assume that the formula for magnesium selenide is MgSe.

Practice 11.2

The formula for sodium oxide is Na_2O. Predict the formula for
(a) sodium sulfide
(b) rubidium oxide

Practice 11.3

The formula for barium phosphide is Ba_3P_2. Predict the formula for
(a) magnesium nitride
(b) barium arsenide

11.5   The Covalent Bond: Sharing Electrons

Some atoms do not transfer electrons from one atom to another to form ions. Instead they form a chemical bond by sharing pairs of electrons between them.

A **covalent bond** consists of a pair of electrons shared between two atoms. This bonding concept was introduced in 1916 by G. N. Lewis. In the millions of compounds that are known, the covalent bond is the predominant chemical bond.

covalent bond

True molecules exist in substances in which the atoms are covalently bonded. It is proper to refer to molecules of such substances as hydrogen, chlorine, hydrogen chloride, carbon dioxide, water, or sugar. These substances contain only covalent bonds and exist as aggregates of molecules. We do not use the term *molecule* when talking about ionically bonded compounds such as sodium chloride, because such substances exist as large aggregates of positive and negative ions, not as molecules.

A study of the hydrogen molecule gives us an insight into the nature of the covalent bond and its formation. The formation of a hydrogen molecule, H_2, involves the overlapping and pairing of $1s$ electron orbitals from two hydrogen atoms. This overlapping and pairing is shown in Figure 11.7. Each atom contributes one electron of the pair that is shared jointly by two hydrogen nuclei. The orbital of the electrons now includes both hydrogen nuclei, but probability factors show that the most likely place to find the electrons (the point of highest electron density) is between the two nuclei. The two nuclei are shielded from each other by the pair of electrons, allowing the two nuclei to be drawn very close to each other.

The formula for chlorine gas is Cl_2. When the two atoms of chlorine combine to form this molecule, the electrons must interact in a manner that is similar to that shown in the preceding example. Each chlorine atom would be more stable with eight electrons in its outer shell. But chlorine atoms are identical, and neither is able to pull an electron away from the other. What happens is this: The unpaired $3p$ electron orbital of one chlorine atom overlaps the unpaired $3p$ electron orbital of the other atom, resulting in a pair of electrons that are mutually shared between the two atoms. Each atom furnishes one of the pair of shared electrons. Thus, each atom attains a stable structure of eight electrons by sharing an electron pair with the other atom. The pairing of the p electrons and the formation of a chlorine molecule are illustrated in Figure 11.8. Neither chlorine atom has a positive or negative charge, since both contain the same number of protons and have equal attraction for the pair of electrons being shared. Other examples of molecules in which electrons are equally shared between two atoms are hydrogen, H_2; oxygen,

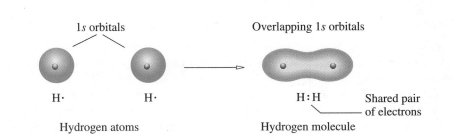

1s orbitals   Overlapping 1s orbitals

H·   H·   H : H   Shared pair
                                               of electrons

Hydrogen atoms   Hydrogen molecule

◀ **FIGURE 11.7**
The formation of a hydrogen molecule from two hydrogen atoms. The two 1s orbitals overlap, forming the H_2 molecule. In this molecule the two electrons are shared between the atoms, forming a covalent bond.

FIGURE 11.8 ▶
Pairing of p electrons in the formation of a chlorine molecule.

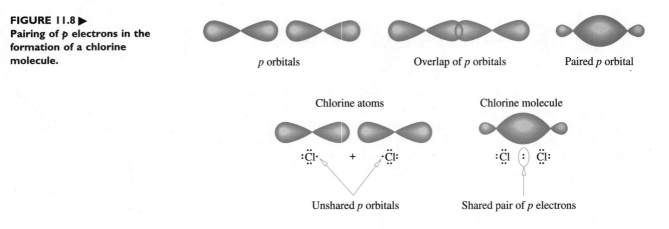

p orbitals   Overlap of p orbitals   Paired p orbital

Chlorine atoms   Chlorine molecule

Unshared p orbitals   Shared pair of p electrons

O_2; nitrogen, N_2; fluorine, F_2; bromine, Br_2; and iodine, I_2. Note that more than one pair of electrons may be shared between atoms:

H:H	:F:F:	:Br:Br:	:I:I:	:O::O:	:N:::N:
hydrogen	fluorine	bromine	iodine	oxygen	nitrogen

The Lewis structure given for oxygen does not adequately account for all the properties of the oxygen molecule. Other theories explaining the bonding in oxygen molecules have been advanced, but they are complex and beyond the scope of this book.

Remember that a dash represents a shared pair of electrons.

A common practice in writing structures is to replace the pair of dots used to represent a shared pair of electrons by a dash (—). One dash represents a single bond; two dashes, a double bond; and three dashes, a triple bond. The six structures just shown may be written thus:

H—H   :F—F: :Br—Br: :I—I: :O=O: :N≡N:

The ionic bond and the covalent bond represent two extremes. In ionic bonding the atoms are so different that electrons are transferred between them, forming a charged pair of ions. In covalent bonding two identical atoms share electrons equally. The bond is the mutual attraction of the two nuclei for the shared electrons. Between these extremes lie many cases in which the atoms are not different enough for a transfer of electrons, but are different enough that the electrons cannot be shared equally. This unequal sharing of electrons results in the formation of a **polar covalent bond**.

polar covalent bond

11.6   Electronegativity

When two *different* kinds of atoms share a pair of electrons a bond forms in which electrons are shared unequally. One atom assumes a partial positive charge and the other a partial negative charge with respect to each other. This difference in charge occurs because the two atoms exert unequal attraction for the pair of shared electrons. The attractive force that an atom of an element has for shared electrons in a molecule or polyatomic ion is known as its **electronegativity**. Elements differ in

electronegativity

◄ **Nobel laureate Linus Pauling (1901–1994).**

their electronegativities. For example, both hydrogen and chlorine need one electron to form stable electron configurations. They share a pair of electrons in hydrogen chloride, HCl. Chlorine is more electronegative and therefore has a greater attraction for the shared electrons than does hydrogen. As a result, the pair of electrons is displaced toward the chlorine atom, giving it a partial negative charge and leaving the hydrogen atom with a partial positive charge. It should be understood that the electron is not transferred entirely to the chlorine atom, as in the case of sodium chloride, and no ions are formed. The entire molecule, HCl, is electrically neutral. A partial charge is usually indicated by the Greek letter delta, δ. Thus, a partial positive charge is represented by $\delta +$ and a partial negative charge by $\delta -$.

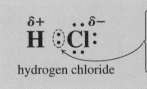

hydrogen chloride

The pair of shared electrons in HCl is closer to the more electronegative chlorine atom than to the hydrogen atom, giving chlorine a partial negative charge with respect to the hydrogen atom.

A scale of relative electronegativities, in which the most electronegative element, fluorine, is assigned a value of 4.0, was developed by the Nobel laureate (1954 and 1962) Linus Pauling (1901–1994). Table 11.5 shows that the relative electronegativity of the nonmetals is high and that of the metals is low. These electronegativities indicate that atoms of metals have a greater tendency to lose electrons than do atoms of nonmetals and that nonmetals have a greater tendency to gain electrons than do metals. The higher the electronegativity value, the greater the attraction for electrons. Notice that electronegativity generally increases from left to right across a period and decreases down a group for the representative elements. The highest electronegativity is 4.0 for fluorine, and the lowest is 0.7 for francium and cesium. It is important to remember that the higher the electronegativity the stronger an atom holds electrons.

The polarity of a bond is determined by the difference in electronegativity

Goal! A Spherical Molecule

One of the most diverse elements in the periodic table is carbon. Although it is much less abundant than many other elements, it is readily available. Two well-known forms of elemental carbon are graphite and diamond. Both contain extended arrays of carbon atoms. In graphite, the carbon atoms are arranged in sheets, the bonds forming hexagons that look like chicken wire (see figure). The bonding between the sheets is very weak and the atoms can slide past each other. This slippery property makes graphite useful as a lubricant. Diamond consists of transparent octahedral crystals in which each carbon atom is bonded to four other carbon atoms. This three-dimensional network of bonds gives diamond the property of hardness for which it is noted. In the 1980s, a new form of carbon was discovered in which the atoms are arranged in relatively small clusters.

Just how do scientists discover a new form of an element? Harold Kroto of the University of Sussex, England, and Richard Smalley of Rice University, Texas, were investigating the effect of light on the surface of graphite. As they analyzed the surface clusters with a mass spectrometer, they discovered a strange molecule whose formula is C_{60}. What could be the structure of such a molecule? They deduced that the most stable arrangement for the atoms would be in the shape of a soccer ball. In thinking about possible arrangements, the scientists considered the geodesic domes designed by R. Buckminster Fuller in the 1960s. The C_{60} form of carbon was named *buckminsterfullerene*. Its structure is shown in the diagram.

The amounts of buckminsterfullerene (now known as buckyballs or fullerenes) prepared by laser were very small; thus much effort went into finding a way to prepare larger amounts of the new allo-

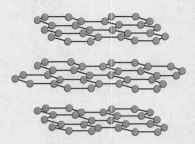

Graphite

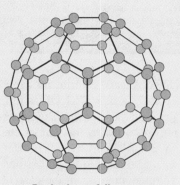

Buckminsterfullerene

trope. In 1990, a group led by Donald Huffman of the University of Arizona discovered that by vaporizing graphite electrodes, fullerenes could be manufactured in large quantities. Now that relatively large amounts of C_{60} are available, buckyballs have captured the imagination of a variety of chemists.

Research on fullerenes has led to a host of possible applications for the molecules. If metals are bound to the carbon atoms the fullerenes become superconducting; that is, they conduct electrical current without resistance, at very low temperature. Scientists are now able to make buckyball compounds that superconduct at temperatures of 45 K. Other fullerenes

are being used in lubricants and optical materials.

Chemists at Yale University have managed to trap helium and neon inside buckyballs. This is the first time chemists have ever observed helium or neon in a compound of any kind. They found that at temperatures from 1000 °F to 1500 °F one of the covalent bonds linking neighboring carbon atoms in the buckyball breaks. This opens a window in the fullerene molecule through which a helium or neon atom can enter the buckyball. When the fullerene is allowed to cool, the broken bond between carbon atoms re-forms, shutting the window and trapping the helium or neon atom inside the buckyball. Since the visiting helium or neon cannot react or share electrons with its host, the resulting compound has forced scientists to invent a new kind of chemical formula to describe the compound. The relationship between the "prisoner" helium or neon and the host buckyball is shown with an @ sign. Therefore, a helium fullerene containing 60 carbon atoms would be He@C_{60}.

The noble-gas fullerenes could be used to "label" crude oil and other pollutants in order to identify and track down the polluter. Two different isotopes of helium could be trapped with buckyballs in a specific ratio to create a coding system for each manufacturer. This ratio of fullerenes could then be detected even in a small quantity of oil recovered from an oil slick at sea and could be used to identify the source of the oil.

Buckyballs can also be tailored to fit a particular size requirement. Raymond Schinazi of the Emory University School of Medicine made a buckyball to fit the active site of a key HIV enzyme that paralyzes the virus, making it noninfectious

continued →

Goal! A Spherical Molecule continued

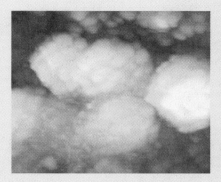

"Raspberries" of worn fullerene lubricant.

in human cells. The key to making this compound was preparing a water-soluble buckyball that would fit in the active site of the enzyme. Eventually, scientists created a fullerene molecule that is water soluble and has two charged arms to grasp

the binding site of the enzyme. It is also toxic to the virus but does not appear to harm the host cells.

Fullerenes also are being tested as lubricants to protect surfaces under conditions present in space. Bharat Bhushan of Ohio State University is depositing thin fullerene films on silicon surfaces, and then testing the film by sliding steel balls across the surface while measuring the friction. Engineers have evaluated the films under a nitrogen atmosphere and in vacuum. The film breaks down slightly as the fullerene molecules cluster together to form larger balls resembling raspberries (see photo). These larger clusters roll like ball bearings between the silicon surface and the steel ball. Fullerene lubricants seem to work best at 110°C, in low humidity, and inert environments.

Scandium atoms trapped in a buckyball.

TABLE 11.5 Relative Electronegativity of the Elements*

Key:
9 — Atomic number
F — Symbol
4.0 — Electronegativity

1 H 2.1																	2 He
3 Li 1.0	4 Be 1.5											5 B 2.0	6 C 2.5	7 N 3.0	8 O 3.5	9 F 4.0	10 Ne
11 Na 0.9	12 Mg 1.2											13 Al 1.5	14 Si 1.8	15 P 2.1	16 S 2.5	17 Cl 3.0	18 Ar
19 K 0.8	20 Ca 1.0	21 Sc 1.3	22 Ti 1.4	23 V 1.6	24 Cr 1.6	25 Mn 1.5	26 Fe 1.8	27 Co 1.8	28 Ni 1.8	29 Cu 1.9	30 Zn 1.6	31 Ga 1.6	32 Ge 1.8	33 As 2.0	34 Se 2.4	35 Br 2.8	36 Kr
37 Rb 0.8	38 Sr 1.0	39 Y 1.2	40 Zr 1.4	41 Nb 1.6	42 Mo 1.8	43 Tc 1.9	44 Ru 2.2	45 Rh 2.2	46 Pd 2.2	47 Ag 1.9	48 Cd 1.7	49 In 1.7	50 Sn 1.8	51 Sb 1.9	52 Te 2.1	53 I 2.5	54 Xe
55 Cs 0.7	56 Ba 0.9	57–71 La–Lu 1.1–1.2	72 Hf 1.3	73 Ta 1.5	74 W 1.7	75 Re 1.9	76 Os 2.2	77 Ir 2.2	78 Pt 2.2	79 Au 2.4	80 Hg 1.9	81 Tl 1.8	82 Pb 1.8	83 Bi 1.9	84 Po 2.0	85 At 2.2	86 Rn
87 Fr 0.7	88 Ra 0.9	89–103 Ac–Lr 1.1–1.7	104 Unq	105 Unp	106 Unh	107 Uns	108 Uno	109 Une									

* The electronegativity value is given below the symbol of each element.

FIGURE 11.9 ▶
Nonpolar, polar covalent, and
ionic compounds.

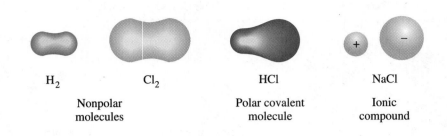

values of the atoms forming the bond (see Figure 11.9). If the electronegativities

nonpolar bond

are the same the bond is **nonpolar** and the electrons are shared equally. If the atoms have greatly different electronegativities the bond is very *polar*. At the extreme, one or more electrons are actually transferred and an ionic bond results.

dipole

A **dipole** is a molecule that is electrically asymmetrical, causing it to be oppositely charged at two points. A dipole is often written as ⊕⊖. A hydrogen chloride molecule is polar and behaves as a small dipole. The HCl dipole may be written as H ↔ Cl. The arrow points toward the negative end of the dipole. Molecules of H_2O, HBr, and ICl are polar:

$$H \longmapsto Cl \qquad H \longmapsto Br \qquad I \longmapsto Cl \qquad H \overset{O}{\underset{H}{\diagup\diagdown}}$$

How do we know whether a bond between two atoms is ionic or covalent? The difference in electronegativity between two atoms determines the character of the bond formed between them. As the difference in electronegativity increases, the polarity of the bond (or percent ionic character) increases. As a rule, if the electronegativity difference between two bonded atoms is greater than 1.7–1.9, the bond will be more ionic than covalent. If the electronegativity difference is greater than 2.0, the bond is strongly ionic. If the electronegativity difference is less than 1.5, the bond is strongly covalent.

Care must be taken to distinguish between polar bonds and polar molecules. A covalent bond between different kinds of atoms is always polar. But a molecule containing different kinds of atoms may or may not be polar, depending on its shape or geometry. Molecules of HF, HCl, HBr, HI, and ICl are all polar because each contains a single polar bond. However, CO_2, CH_4, and CCl_4 are nonpolar molecules despite the fact that all three contain polar bonds. The carbon dioxide molecule O=C=O is nonpolar because the carbon–oxygen dipoles cancel each other by acting in opposite directions.

$$\overset{\longleftarrow + \quad + \longrightarrow}{O=C=O}$$

dipoles in opposite directions

Methane (CH_4) and carbon tetrachloride (CCl_4) are nonpolar because the four C—H and C—Cl polar bonds are identical, and, because these bonds emanate from the center to the corners of a tetrahedron in the molecule, the effect of their polarities cancel one another. We shall discuss the shapes of molecules later in this chapter.

We have said that water is a polar molecule. If the atoms in water were linear like those in carbon dioxide, the two O—H dipoles would cancel each other, and the molecule would be nonpolar. However, water is definitely polar and has a nonlinear (bent) structure with an angle of 105° between the two O—H bonds.

Bond type

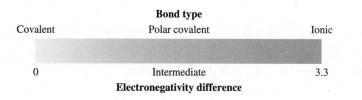

The relationships among types of bonds are summarized in Figure 11.10. It is important to realize that bonding is a continuum, the difference between ionic and covalent is a gradual change.

11.7   Lewis Structures of Compounds

As we have seen, Lewis structures are a convenient way of showing the covalent bonds in many molecules or ions of the representative elements. In writing Lewis structures the most important consideration for forming a stable compound is that the atoms attain a noble gas configuration.

The most difficult part of writing Lewis structures is determining the arrangement of the atoms in a molecule or an ion. In simple molecules with more than two atoms, one atom will be the central atom surrounded by the other atoms. Thus Cl_2O has two possible arrangements, Cl—Cl—O or Cl—O—Cl. Usually, but not always, the single atom in the formula (except H) will be the central atom.

Athough Lewis structures for many molecules and ions can be written by inspection, the following procedure is helpful for learning to write these structures:

Step 1. Obtain the total number of valence electrons to be used in the structure by adding the number of valence electrons in all of the atoms in the molecule or ion. If you are writing the structure of an ion, add one electron for each negative charge or subtract one electron for each positive charge on the ion.

Step 2. Write down the skeletal arrangement of the atoms and connect them with a single covalent bond (two dots or one dash). Hydrogen, which contains only one bonding electron, can form only one covalent bond. Oxygen atoms are not normally bonded to each other, except in compounds known to be peroxides. Oxygen atoms normally have a maximum of two covalent bonds, two single bonds or one double bond.

Step 3. Subtract two electrons for each single bond you used in Step 2 from the total number of electrons calculated in Step 1. This calculation gives you the net number of electrons available for completing the structure.

Step 4. Distribute pairs of electrons (pairs of dots) around each atom (except hydrogen) to give each atom a noble gas structure.

Step 5. If there are not enough electrons to give these atoms eight electrons, change single bonds between atoms to double or triple bonds by shifting unbonded pairs of electrons as needed. Check to see that each atom has a noble gas electron structure (two electrons for hydrogen and eight for the others). A double bond counts as four electrons for each atom to which it is bonded.

Remember, the number of valence electrons of Group A elements is the same as their group number in the periodic table.

Example 11.9 How many valence electrons are in each of these atoms: Cl, H, C, O, N, S, P, I?

Solution You can look at the periodic table to determine the electron structure, or, if the element is in Group A of the periodic table, the number of valence electrons is equal to the group number:

Atom	Periodic group	Valence electrons
Cl	VIIA	7
H	IA	1
C	IVA	4
O	VIA	6
N	VA	5
S	VIA	6
P	VA	5
I	VIIA	7

Example 11.10 Write the Lewis structure for water, H_2O.

Solution

Step 1 The total number of valence electrons is eight, two from the two hydrogen atoms and six from the oxygen atom.

Step 2 The two hydrogen atoms are connected to the oxygen atom. Write the skeletal structure:

H O   or   H O H
 H

Place two dots between the hydrogen and oxygen atoms to form the covalent bonds:

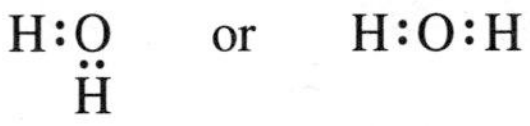

H:O   or   H:O:H
 ··
 H

Step 3 Subtract the four electrons used in Step 2 from eight to obtain four electrons yet to be used.

Step 4 Distribute the four electrons around the oxygen atom. Hydrogen atoms cannot accommodate any more electrons:

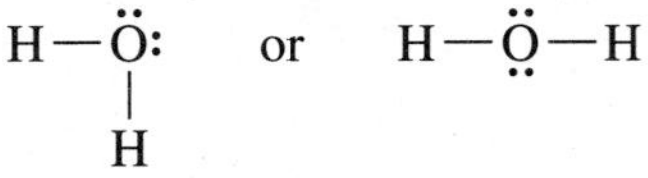

H—Ö: or   H—Ö—H
  | ··
  H

This arrangement is the Lewis structure. Each atom has a noble-gas electron structure. Notice the shape of the molecule is not shown by the Lewis structure.

Example 11.11 Write Lewis structures for a molecule of (a) methane, CH_4, and (b) carbon tetrachloride, CCl_4.

Solution **Part A**

Step 1 The total number of valence electrons is eight, one from each hydrogen atom and four from the carbon atom.

Step 2 The skeletal structure contains four H atoms around a central C atom. Place two electrons between the C and each H.

$$\begin{array}{ccc} & H & & & H \\ H & C & H & & H:\overset{\cdot\cdot}{\underset{\cdot\cdot}{C}}:H \\ & H & & & \ddot{H} \end{array}$$

Step 3 Subtract the eight electrons used in Step 2 from eight to obtain zero electrons yet to be placed. Therefore, the Lewis structure must be as written in Step 2:

$$\begin{array}{ccc} \overset{H}{\underset{}{}} & & \overset{H}{\underset{|}{}} \\ H:\overset{\cdot\cdot}{\underset{\cdot\cdot}{C}}:H & \text{or} & H-C-H \\ \ddot{H} & & \overset{|}{H} \end{array}$$

Part B

Step 1 The total number of valence electrons to be used is 32, four from the carbon atom and seven from each of the four chlorine atoms.

Step 2 The skeletal structure contains the four Cl atoms around a central C atom. Place two electrons between the C and each Cl:

$$\begin{array}{ccc} & Cl & & & Cl \\ Cl & C & Cl & & Cl:\overset{\cdot\cdot}{\underset{\cdot\cdot}{C}}:Cl \\ & Cl & & & \ddot{Cl} \end{array}$$

Step 3 Subtract the eight electrons used in Step 2 from 32, to obtain 24 electrons yet to be placed.

Step 4 Distribute the 24 electrons (12 pairs) around the Cl atoms so that each Cl atom has eight electrons around it:

$$\begin{array}{ccc} :\ddot{C}l: & & :\ddot{C}l: \\ :\ddot{C}l:\overset{\cdot\cdot}{\underset{\cdot\cdot}{C}}:\ddot{C}l: & \text{or} & :\ddot{C}l-\overset{|}{\underset{|}{C}}-\ddot{C}l: \\ :\ddot{C}l: & & :\ddot{C}l: \end{array}$$

This arrangement is the Lewis structure; CCl_4 contains four covalent bonds.

Write a Lewis structure for CO_2.   **Example 11.12**

Solution

Step 1 The total number of valence electrons is 16, four from the C atom and six from each O atom.

Step 2 The two O atoms are bonded to a central C atom. Write the skeletal structure and place two electrons between the C and each O atom.

O:C:O

Step 3 Subtract the four electrons used in Step 2 from 16 to obtain 12 electrons yet to be placed.

Step 4 Distribute the 12 electrons around the C and O atoms. Several possibilities exist:

$$\begin{array}{ccc} :\ddot{O}:C:\ddot{O}: & :\ddot{O}:C:\ddot{O}: & :\ddot{O}:\ddot{C}:O: \\ I & II & III \end{array}$$

Step 5 All the atoms do not have eight electrons around them (noble gas structure). Remove one pair of unbonded electrons from each O atom in structure I and place one pair between each O and the C atom forming two double bonds:

$$:\ddot{O}::C::\ddot{O}: \quad \text{or} \quad :\ddot{O}=C=\ddot{O}:$$

Each atom now has eight electrons around it. The carbon is sharing four pairs of electrons, and each oxygen is sharing two pairs. These bonds are known as double bonds since each involves sharing two pairs of electrons.

Practice 11.4

Write the Lewis structure for each of the following:
(a) PBr_3, (b) $CHCl_3$, (c) HF, (d) H_2CO.

Although many compounds attain a noble gas structure in covalent bonding, there are numerous exceptions. Sometimes it is impossible to write a structure in which each atom has eight electrons around it. For example, in BF_3 the boron atom has only six electrons around it, and in SF_6 the sulfur atom has 12 electrons around it.

Even though there are exceptions, many molecules can be described using Lewis structures where each atom has a noble gas electron configuration. This is a useful model for understanding chemistry.

11.8   Complex Lewis Structures

Most Lewis structures give bonding pictures that are consistent with experimental information on bond strength and length. There are some molecules and polyatomic ions for which no single Lewis structure consistent with all characteristics and bonding information can be written. For example, consider the nitrate ion, NO_3^-. To write a Lewis structure for this polyatomic ion we would use the following steps.

Step 1 The total number of valence electrons is 24, five from the nitrogen atom, and six from each oxygen atom, plus one electron from the − 1 charge.

Step 2 The three O atoms are bonded to a central N atom. Write the skeletal structure and place two electrons between each pair of atoms:

$$\left[\begin{array}{c} O \\ O:N:O \end{array} \right]^-$$

Step 3 Subtract the six electrons used in Step 2 from 24 to obtain eighteen electrons yet to be placed.

Step 4 Distribute the eighteen electrons around the N and O atoms:

:Ö: ⟵ electron deficient
:Ö:N:Ö:

Step 5 One pair of electrons is still needed to give all the N and O atoms a noble gas structure. Move the unbonded pair of electrons from the N atom and place it between the N and the electron-deficient O atom, making a double bond.

$$\left[\begin{array}{c} :\ddot{O} \\ \| \\ :\ddot{O}-N-\ddot{O}: \end{array}\right]^{-} \quad \text{or} \quad \left[\begin{array}{c} :\ddot{O}: \\ | \\ :\ddot{O}-N=\ddot{O}: \end{array}\right]^{-} \quad \text{or} \quad \left[\begin{array}{c} :\ddot{O}: \\ | \\ \ddot{O}=N-\ddot{O}: \end{array}\right]^{-}$$

Are these all valid Lewis structures? Yes, so there are really three possible Lewis structures for NO_3^-.

A molecule or ion that has multiple correct Lewis structures shows *resonance*. Each of these Lewis structures is called a **resonance structure**. In this book, however, we will not be concerned with how to choose the correct resonance structure for a molecule or ion. Therefore, any of the possible resonance structures may be used to represent the ion or molecule.

resonance structure

Write the Lewis structure for a carbonate ion, CO_3^{2-}.

Example 11.13

Solution

Step 1 These four atoms have 22 valence electrons plus two electrons from the -2 charge, which makes 24 electrons to be placed.

Step 2 In the carbonate ion the carbon is the central atom surrounded by the three oxygen atoms. Write the skeletal structure and place two electrons between each pair of atoms:

$$\begin{array}{c} O \\ | \\ C-O \\ | \\ O \end{array}$$

Step 3 Subtract the six electrons used in Step 2 from 24 to give 18 electrons yet to be placed.

Step 4 Distribute the 18 electrons around the three oxygen atoms and indicate that the carbonate ion has a -2 charge:

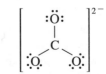

The difficulty with this structure is that the carbon atom has only six electrons around it instead of a noble gas octet.

Step 5 Move one of the nonbonding pairs of electrons from one of the oxygens and place them between the carbon and the oxygen. These Lewis structures are possible:

$$\left[\begin{array}{c} :\ddot{O}: \\ | \\ \ddot{O}^{C}\ddot{O}: \end{array}\right]^{2-} \quad \text{or} \quad \left[\begin{array}{c} :\ddot{O}: \\ | \\ :\ddot{O}^{C}\dot{O}: \end{array}\right]^{2-} \quad \text{or} \quad \left[\begin{array}{c} :O: \\ \| \\ :\ddot{O}^{C}\ddot{O}: \end{array}\right]^{2-}$$

Practice 11.5

Write the Lewis structure for each of the following: (a) NH_3, (b) H_3O^+, (c) NH_4^+, (d) HCO_3^-.

11.9 Compounds Containing Polyatomic Ions

A polyatomic ion is a stable group of atoms that has either a positive or a negative charge and behaves as a single unit in many chemical reactions. Sodium carbonate, Na_2CO_3, contains two sodium ions and a carbonate ion. The carbonate ion, CO_3^{2-}, is a polyatomic ion composed of one carbon atom and three oxygen atoms and has a charge of -2. One carbon and three oxygen atoms have a total of 22 electrons in their outer shells. The carbonate ion contains 24 outer shell electrons and therefore has a charge of -2. In this case, the two additional electrons come from the two sodium atoms, which are now sodium ions:

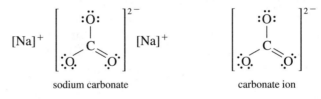

sodium carbonate   carbonate ion

Sodium carbonate has both ionic and covalent bonds. Ionic bonds exist between each of the sodium ions and the carbonate ion. Covalent bonds are present between the carbon and oxygen atoms within the carbonate ion. One important difference between the ionic and covalent bonds in this compound can be demonstrated by dissolving sodium carbonate in water. It dissolves in water forming three charged particles, two sodium ions and one carbonate ion, per formula unit of Na_2CO_3:

$$Na_2CO_3 \xrightarrow{\text{water}} 2\,Na^+ \;+\; CO_3^{2-}$$

sodium carbonate   sodium ions   carbonate ion

The CO_3^{2-} ion remains as a unit, held together by covalent bonds; but where the bonds are ionic, dissociation of the ions takes place. Do not think, however, that polyatomic ions are so stable that they cannot be altered. Chemical reactions by which polyatomic ions can be changed to other substances do exist.

11.10 Molecular Structure

So far in our discussion of bonding we have used Lewis structures to represent valence electrons in molecules and ions, but they do not indicate anything regarding the molecular or geometric structure of a molecule. The three-dimensional arrangement of the atoms within a molecule is a significant feature in understanding molecular interactions. Let's consider several examples. Water is known to have the geometric structure:

H⧵   ⧸H
  O

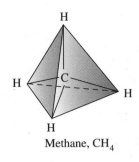
Methane, CH₄

known as "bent" or "V-shaped." Carbon dioxide exhibits a linear shape:

O=C=O

whereas BF_3 forms a third molecular structure:

F⧵   ⧸F
  B
  |
  F

This last structure is called *trigonal planar* since all the atoms lie in one plane in a triangular arrangement. One of the more common molecular structures is the tetrahedron illustrated by the molecule methane, CH_4, shown in the margin.

How can chemists predict the geometric structure of a molecule? We will now study a model developed to assist in making predictions from the Lewis structure.

11.11 The Valence Shell Electron Pair Repulsion (VSEPR) Model

The chemical properties of a substance are closely related to the structure of its molecules. A change in a single site on a large biomolecule can make a difference in whether or not a particular reaction occurs.

Instrumental analysis can be used to determine exact spatial arrangements of atoms. Quite often, though, we only need to be able to predict the approximate structure of a molecule. A relatively simple model has been developed by chemists to allow us to make predictions of shape from Lewis structures.

The VSEPR model is based on the idea that electron pairs will repel each other electrically and will seek to minimize this repulsion. In order to accomplish this minimization, the electron pairs will be arranged around a central atom as far apart as possible. Consider $BeCl_2$, a molecule with only two pairs of electrons surrounding the central atom. These electrons are arranged 180° apart for maximum separation:

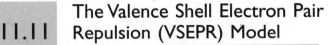

The molecular structure can now be labeled as a **linear structure**. When only two pairs of electrons surround an atom they should be placed 180° apart to give a linear structure.

linear structure

What occurs when there are three pairs of electrons on the central atom? Consider the BF_3 molecule. The greatest separation of electron pairs occurs when the angles between atoms are 120°:

F⧵120°⧸F
   B
120° | 120°
   F

This arrangement of atoms is flat (planar) and is usually called **trigonal planar**. When three pairs of electrons surround an atom they should be placed 120° apart to show the trigonal planar structure.

trigonal planar

Now consider the most common situation, CH_4, with four pairs of electrons on the central carbon atom. In this case, the central atom exhibits a noble gas electron structure. What arrangement will best minimize the electron pair repulsions? At first it seems that an obvious choice is a 90° angle with all the atoms in a single plane:

However, we must consider that molecules are three-dimensional. This concept results in a structure in which the electron pairs are actually 109.5° apart:

In this diagram the wedged line seems to protrude from the page whereas the dashed line recedes. Two examples showing representations of this arrangement, known as **tetrahedral structure**, are illustrated in Figure 11.11. When four pairs of electrons surround an atom they should be placed 109.5° apart to give them a tetrahedral structure.

tetrahedral structure

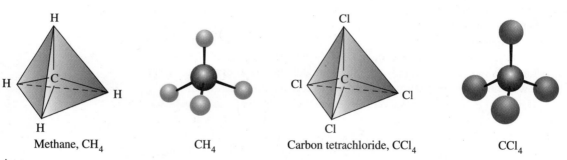

FIGURE 11.11 ▲
Ball-and-stick models of methane and carbon tetrachloride. Methane and carbon tetrachloride are nonpolar molecules because their polar bonds cancel each other in the tetrahedral arrangement of their atoms. The carbon atoms are located in the centers of the tetrahedrons.

The VSEPR model is based on the premise that we are counting electron pairs. It is quite possible that one or more of these electron pairs may be nonbonding or lone pairs. What happens to the molecular structure in these cases? Consider the ammonia molecule. First draw the Lewis structure to determine the number of electron pairs around the central atom:

$$H \!:\! \overset{\cdot\cdot}{\underset{H}{N}} \!:\! H$$

Since there are four pairs of electrons the arrangement of electrons around the central atom will be tetrahedral. However, only three of the pairs are bonded to another atom, so the shape of the molecule itself is pyramidal. It is important to understand that the placement of the electron pairs determines the structure but the name of the shape of the molecule is determined by the position of the atoms themselves. Therefore, ammonia has a pyramidal shape, not a tetrahedral shape. See Figure 11.12.

Now consider the effect of two unbonded pairs of electrons in the water molecule. The Lewis structure for water is

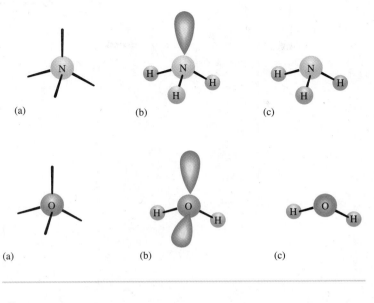

◀ **FIGURE 11.12**
(a) The tetrahedral arrangement of electron pairs around the **N** atom in the NH₃ molecule. (b) Three pairs are shared and one is unshared. (c) The **NH₃** molecule has a trigonal pyramidal structure.

◀ **FIGURE 11.13**
(a) The tetrahedral arrangement of the four electron pairs around oxygen in the **H₂O** molecule. (b) Two of the pairs are shared and two are unshared. (c) The **H₂O** molecule has a bent molecular structure.

$$H-\ddot{O}: \atop | \atop H$$

The four electron pairs indicate a tetrahedral arrangement is necessary (see Figure 11.11). The molecule is not tetrahedral because two of the electron pairs are unbonded pairs. The atoms in the water molecule form a "bent" shape as shown in Figure 11.13. Using the VSEPR model helps to explain some of the unique properties of the water molecule. Because it is bent and not linear we can see that the molecule is polar.

The properties of water that cause it to be involved in so many interesting and important roles are largely a function of its shape and polarity. We will consider water in greater detail in Chapter 13.

Predict the molecular structure for each molecule: H_2S, CCl_4, AlF_3.

1. Draw the Lewis structure.
2. Count the electron pairs and determine the arrangement that will minimize repulsions.
3. Determine the positions of the atoms and name the structure.

Example 11.14

Solution

Molecule	Lewis structure	Number of electron pairs	Electron pair arrangement	Molecular structure
H_2S	H:S̈:H	4	Tetrahedral	Bent
CCl_4	:C̈l:C:C̈l: with :C̈l: above and :C̈l: below	4	Tetrahedral	Tetrahedral
AlF_3	:F̈:Al:F̈: with :F̈: above	3	Trigonal planar	Trigonal planar

Liquid Crystals

What do color-changing pens or toy cars, bullet-resistant vests, wristwatches, calculators, and color-changing thermometers have in common? The chemicals that make each of them work are liquid crystals. In a normal crystal the molecules have an orderly arrangement. In a liquid crystal the molecules can flow and maintain an orderly arrangement at the same time.

The molecules in all types of liquid crystals are linear and polar. The atoms of linear molecules tend to lie in a relatively straight line and the molecules are generally much longer than they are wide. Polar molecules are attracted to each other, and certain linear ones are able to line up in an orderly fashion, without solidifying, to form liquid crystals. Such a substance can be used as a liquid crystal display (LCD), to change color with changing temperature, or to make a super-strong synthetic fiber.

The key to these products that change color with temperature is in the twisted arrangement of the molecules. The linear molecules form a generally flat surface. In the liquid crystal the molecules lie side by side in a nearly flat layer. The next layer is similar but at an angle to the one below. These closely packed flat layers have a special effect on light. As the light strikes the surface, some of it is reflected from the top layer and more from lower layers. When the same wavelength is reflected from many layers a color is observed. This is similar to the rainbow of colors formed by oil in a puddle on the street or the film of a soap bubble. As the temperature increases the molecules move faster, causing a change in the angle and the space between the layers. These changes result in a color change in the reflected light. Different compounds change color within different temperature ranges, allowing a variety of practical and amusing applications.

In liquid crystal displays (LCDs) in watches and calculators the process is sim-

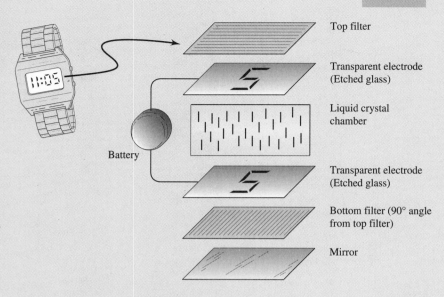

Adapted from *Chem Matters Magazine*.

ilar. Normally the LCD acts as a mirror reflecting the light that enters it. The display is created by a series of layers, consisting of a filter, a glass etched with tiny lines, a liquid crystal chamber, a second etched glass, a bottom filter, and a mirror (see diagram). The molecules at the top of the liquid crystal chamber align with the lines on the top layer of glass whereas those at the bottom align with the grooves on that glass (90-degree turn from the top). In between the molecules line up as closely as possible with neighboring molecules to form a twisted spiral. To display a number a tiny current is sent to the proper SiO_2 segments on the etched piece of glass. The plates become charged, and the polar molecules of the liquid crystal are attracted to the charged segments, thus destroying the carefully arranged spirals. The pattern of reflected light is changed and a numeral appears.

Another type of liquid crystal (nematic) contains molecules that all point in the same direction. These liquid crystals are used to manufacture very strong synthetic fibers. The molecules in nematic crystals all line up in the same direction but are free to slide past one another.

Perhaps the best example of these liquid crystals is the manufacture of Kevlar, a synthetic fiber used in bullet-resistant vests, canoes, and parts of the space shuttle. Kevlar is a synthetic polymer, like nylon or polyester, that gains strength by passing through a liquid crystal state during its manufacture. In a typical polymer the long molecular chains are jumbled together, somewhat like spaghetti. The strength of the material is limited by the disorderly arrangement.

The trick is to get the molecules to line up parallel to each other. Once the giant molecules have been synthesized they are dissolved in sulfuric acid. At the proper concentration the molecules align, and the solution is forced through tiny holes in a nozzle and further aligned. The sulfuric acid is removed in a water bath forming solid fibers in near perfect alignment. One strand of Kevlar is stronger than an equal-sized strand of steel. It has a much lower density as well, making it a material of choice in bullet-resistant vests.

Image labels: Top filter; Transparent electrode (Etched glass); Liquid crystal chamber; Transparent electrode (Etched glass); Bottom filter (90° angle from top filter); Mirror; Battery

Practice 11.6

Predict the shape for CF_4, NF_3, and BeF_2.

Concepts in Review

1. Describe how atomic radii vary (a) from left to right in a period, and (b) from top to bottom in a group.

2. Describe how the ionization energies of the elements vary with respect to (a) the position in the periodic table and (b) the removal of successive electrons.

3. Write Lewis structures for the representative elements from their position in the periodic table.

4. Describe (a) the formation of ions by electron transfer and (b) the nature of the chemical bond formed by electron transfer.

5. Show by means of Lewis structures the formation of an ionic compound from atoms.

6. Describe a crystal of sodium chloride.

7. Predict the relative sizes of an atom and a monatomic ion for a given element.

8. Describe the covalent bond and predict whether a given covalent bond will be polar or nonpolar.

9. Draw Lewis structures for the diatomic elements.

10. Identify single, double, and triple covalent bonds.

11. Describe the changes in electronegativity in (1) moving across a period and (2) moving down a group in the periodic table.

12. Predict formulas of simple compounds formed between the representative (Group A) elements using the periodic table.

13. Describe the effect of electronegativity on the type of chemical bonds in a compound.

14. Draw Lewis structures for (a) the molecules of covalent compounds and (b) polyatomic ions.

15. Describe the difference between polar and nonpolar bonds.

16. Distinguish clearly between ionic and molecular substances.

17. Predict whether the bonding in a compound will be primarily ionic or covalent.

18. Describe the VSEPR model for molecular shape.

19. Use the VSEPR model to determine molecular structure from Lewis structure for given compounds.

Key Terms

The terms listed here have been defined within this chapter. Section numbers are referenced in parenthesis for each term. More detailed definitions are given in the Glossary.

covalent bond (11.5)   ionization energy (11.1)   polar covalent bond (11.5)
dipole (11.6)   Lewis structure (11.2)   resonance structure (11.8)
electronegativity (11.6)   linear structure (11.11)   tetrahedral structure (11.11)
ionic bond (11.3)   nonpolar bond (11.6)   trigonal planar (11.11)

Questions

Questions refer to tables, figures, and key words and concepts defined within the chapter. A particularly challenging question or exercise is indicated with an asterisk.

1. Rank the following five elements according to the radii of their atoms, from smallest to largest: Na, Mg, Cl, K, and Rb. (Figure 11.2)

2. Explain why much more ionization energy is required to remove the first electron from neon than from sodium. (Table 11.1)

3. Explain the large increase in ionization energy needed to remove the third electron from beryllium compared with that needed for the second electron. (Table 11.1)

4. Does the first ionization energy increase or decrease from top to bottom in the periodic table for the alkali metal family? Explain. (Figure 11.3)

5. Does the first ionization energy increase or decrease from top to bottom in the periodic table for the noble gas family? Explain. (Figure 11.3)

6. Why does barium (Ba) have a lower ionization energy than beryllium (Be)? (Figure 11.3)

7. Why is there such a large increase in the ionization energy required to remove the second electron from a sodium atom as opposed to the first? (Table 11.1)

8. Which element in each of the following pairs has the larger atomic radius? (Figure 11.2)
(a) Na or K   (c) O or F   (e) Ti or Zr
(b) Na or Mg   (d) Br or I

9. Which element in each of Groups IA–VIIA has the smallest atomic radius? (Figure 11.2)

10. Why does the atomic size increase in going down any family of the periodic table?

11. All the atoms within each Group A family of elements can be represented by the same Lewis structure. Complete the table below, expressing the Lewis structure for each group. (Use E to represent the elements.) (Figure 11.4)

Group	IA	IIA	IIIA	IVA	VA	VIA	VIIA
$E\cdot$							

12. Draw the Lewis structure for Cs, Ba, Tl, Pb, Po, At, and Rn. How do these structures correlate with the group in which each element occurs?

13. In which general areas of the periodic table are the elements with (a) the highest and (b) the lowest electronegativities found?

14. What are valence electrons?

15. Explain why potassium usually forms a K^+ ion but not a K^{2+} ion.

16. Why does an aluminum ion have a +3 charge?

17. Which of these statements are correct? (Try to answer this exercise using only the periodic table.) Rewrite each incorrect statement to make it correct.
(a) If the formula for calcium iodide is CaI_2, then the formula for cesium iodide is CsI_2.
(b) Metallic elements tend to have relatively low electronegativities.
(c) If the formula for aluminum oxide is Al_2O_3, then the formula for gallium oxide is Ga_2O_3.
(d) Sodium and chlorine react to form molecules of NaCl.
(e) A chlorine atom has fewer electrons than a chloride ion.
(f) The noble gases have a tendency to lose one electron to become positively charged ions.
(g) The chemical bonds in a water molecule are ionic.
(h) The chemical bonds in a water molecule are polar.

(i) Valence electrons are those electrons in the outermost shell of an atom.

(j) An atom with eight electrons in its outer shell has all its s and p orbitals filled.

(k) Fluorine has the lowest electronegativity of all the elements.

(l) Oxygen has a greater electronegativity than carbon.

(m) A cation is larger than its corresponding neutral atom.

(n) Cl_2 is more ionic in character than HCl.

(o) A neutral atom with eight electrons in its valence shell will likely be an atom of a noble gas.

(p) A nitrogen atom has four valence electrons.

(q) An aluminum atom must lose three electrons to become an aluminum ion, Al^{3+}.

(r) A stable group of atoms that has either a positive or a negative charge and behaves as a single unit in many chemical reactions is called a *polyatomic ion.*

(s) Sodium sulfate, Na_2SO_4, has covalent bonds between sulfur and the oxygen atoms, and ionic bonds between the sodium ions and the sulfate ion.

(t) The water molecule is a dipole.

(u) In an ethylene molecule, C_2H_4,

$$\underset{H}{\overset{H}{\diagdown}}C=C\underset{H}{\overset{H}{\diagup}}$$

two pairs of electrons are shared between the carbon atoms.

(v) When electrons are transferred from one atom to another, the resulting compound contains ionic bonds.

(w) A phosphorus atom, $\cdot\overset{\cdot}{P}\cdot$, needs three additional electrons to attain a noble-gas electron structure.

(x) The simplest compound between oxygen, $\cdot\overset{\cdot\cdot}{O}\cdot$, and fluorine, $\overset{\cdot\cdot}{\underset{\cdot\cdot}{F}}\cdot$, atoms is FO_2.

(y) The $H{:}\overset{\cdot\cdot}{\underset{\cdot\cdot}{Cl}}{:}$ molecule has three unshared pairs of electrons.

18. Which of these statements are correct? (Try to answer this exercise using only the periodic table.) Rewrite each incorrect statement to make it correct.

(a) The smaller the difference in electronegativity between two atoms, the more ionic the bond between them will be.

(b) Lewis structures are mainly useful for the representative elements.

(c) The correct Lewis structure for NH_3 is

$$H{:}\overset{\cdot\cdot}{\underset{H}{N}}{:}H$$

(d) The correct Lewis structure for CO_2 is

$${:}\overset{\cdot\cdot}{\underset{\cdot\cdot}{O}}{:}C{:}\overset{\cdot\cdot}{\underset{\cdot\cdot}{O}}{:}$$

(e) The Lewis structure for the noble gas helium is

$${:}\overset{\cdot\cdot}{He}{:}$$

(f) In Period 5 of the periodic table, the element having the lowest ionization energy is Xe.

(g) An atom having an electron structure of $1s^2 2s^2 2p^6 3s^2 3p^2$ has four valence electrons.

(h) When an atom of bromine becomes a bromide ion, its size increases.

(i) The structures that show that H_2O is a dipole and that CO_2 is not a dipole are

$$H{-}\overset{\cdot\cdot}{\underset{\cdot\cdot}{O}}{-}H \quad \text{and} \quad {:}\overset{\cdot\cdot}{O}{=}C{=}\overset{\cdot\cdot}{O}{:}$$

(j) The Cl^- and S^{2-} ions have the same electron structure.

(k) A molecule with the central atom surrounded by two pairs of electrons has a linear shape.

(l) A molecule with a central atom having three bonding pairs and one nonbonding electron pair will have a tetrahedral electron structure and a tetrahedral shape.

(m) A molecule in which the central atom is surrounded by three bonding pairs of electrons will have a trigonal planar shape.

(n) A molecule with two bonding and two nonbonding pairs of electrons will have a bent shape.

(o) The Lewis structure for nitrogen is $:\overset{\cdot}{N}\cdot\,$.

(p) The Lewis structure for potassium is $P\cdot$.

Paired Exercises

These exercises are paired. Each odd-numbered exercise is followed by a similar even-numbered exercise. Answers to the even-numbered exercises are given in Appendix V.

19. Which is larger, a magnesium atom or a magnesium ion? Explain.

20. Which is smaller, a bromine atom or a bromide ion? Explain.

21. Using the table of electronegativity values (Table 11.5), indicate which element is more positive and which is more negative in the following compounds:

  (a) H_2O (c) NH_3 (e) NO

  (b) NaF   (d) PbS   (f) CH_4

22. Using the table of electronegativity values (Table 11.5), indicate which element is more positive and which is **more** negative in the following compounds:

  (a) HCl   (c) CCl_4 (e) $MGgH_2$

  (b) LiH   (d) IBr   (f) OF_2

23. Classify the bond between the following pairs of elements as principally ionic or principally covalent (use Table 11.5):
    (a) sodium and chlorine
    (b) carbon and hydrogen
    (c) chlorine and carbon
    (d) calcium and oxygen

24. Classify the bond between the following pairs of elements as principally ionic or principally covalent (use Table 11.5):
    (a) hydrogen and sulfur
    (b) barium and oxygen
    (c) fluorine and fluorine
    (d) potassium and fluorine

25. Explain what happens to the electron structures of Mg and Cl atoms when they react to form $MgCl_2$.

26. Write an equation representing
    (a) the change of a fluorine atom to a fluoride ion and
    (b) the change of a calcium atom to a calcium ion

27. Use Lewis structures to show the electron transfer for the formation of the following ionic compounds from the atoms:
    (a) MgF_2 (b) K_2O

28. Use Lewis structures to show the electron transfer for the formation of the following ionic compounds from the atoms:
    (a) CaO   (b) NaBr

29. How many valence electrons are in each of these atoms? H, K, Mg, He, Al

30. How many valence electrons are in each of these atoms? Si, N, P, O, Cl

31. How many electrons must be gained or lost for each of the following to achieve a noble-gas electron structure?
    (a) a calcium atom
    (b) a sulfur atom
    (c) a helium atom

32. How many electrons must be gained or lost for each of the following to achieve a noble-gas electron structure?
    (a) a chloride ion
    (b) a nitrogen atom
    (c) a potassium atom

33. Which is larger? Explain.
    (a) a potassium atom or a potassium ion
    (b) a bromine atom or a bromide ion

34. Which is larger? Explain.
    (a) a magnesium ion or an aluminum ion
    (b) Fe^{2+} or Fe^{3+}

35. Let E be any representative element. Following the pattern in the table, write the formulas for the hydrogen and oxygen compounds of
    (a) Na   (c) Al
    (b) Ca   (d) Sn

36. Let E be any representative element. Following the pattern in the table, write the formulas for the hydrogen and oxygen compounds of
    (a) Sb   (c) Cl
    (b) Se   (d) C

| Group | | | | | | |
IA	IIA	IIIA	IVA	VA	VIA	VIIA
EH	EH_2	EH_3	EH_4	EH_3	H_2E	HE
E_2O	EO	E_2O_3	EO_2	E_2O_5	EO_3	E_2O_7

| Group | | | | | | |
IA	IIA	IIIA	IVA	VA	VIA	VIIA
EH	EH_2	EH_3	EH_4	EH_3	H_2E	HE
E_2O	EO	E_2O_3	EO_2	E_2O_5	EO_3	E_2O_7

37. The formula for sodium sulfate is Na_2SO_4. Write the names and formulas for the other alkaline earth metal sulfates.

38. The formula for calcium bromide is $CaBr_2$. Write the names and formulas for the other alkaline earth metal bromides.

39. Write Lewis structures for
    (a) Na   (b) Br^- (c) O^{2-}

40. Write Lewis structures for
    (a) Ga   (b) Ga^{3+} (c) Ca^{2+}

41. Classify the bonding in each compound as ionic or covalent:
    (a) H_2O (c) MgO
    (b) NaCl   (d) Br_2

42. Classify the bonding in each compound as ionic or covalent:
    (a) HCl   (c) NH_3
    (b) $BaCl_2$ (d) SO_2

43. Predict the type of bond that would be formed between each of the following pairs of atoms:
    (a) Na and N
    (b) N and S
    (c) Br and I

44. Predict the type of bond that would be formed between each of the following pairs of atoms:
    (a) H and Si
    (b) O and F
    (c) Ca and I

45. Draw Lewis structures for
    (a) H_2 (b) N_2 (c) Cl_2

46. Draw Lewis structures for
    (a) O_2 (b) Br_2 (c) I_2

47. Draw Lewis structures for
   (a) NCl_3 **(c)** C_2H_6
   (b) H_2CO_3 **(d)** $NaNO_3$

49. Draw Lewis structures for
   (a) Ba^{2+} **(d)** CN^-
   (b) Al^{3+} **(e)** HCO_3^-
   (c) SO_3^{2-}

51. Classify the following molecules as polar or nonpolar:
   (a) H_2O
   (b) HBr
   (c) CF_4

53. Give the number and arrangement of the electron pairs around the central atom:
   (a) C in CCl_4
   (b) S in H_2S
   (c) Al in AlH_3

55. Use VSEPR theory to predict the structure of the following polyatomic ions:
   (a) sulfate ion
   (b) chlorate ion
   (c) periodate ion

57. Use VSEPR theory to predict the shape of the following molecules:
   (a) SiH_4
   (b) PH_3
   (c) SeF_2

59. Identify this element, from the following description: Element X reacts with sodium to form the compound Na_2X and is in the second period on the periodic table.

48. Draw Lewis structures for
   (a) H_2S **(c)** NH_3
   (b) CS_2 **(d)** NH_4Cl

50. Draw Lewis structures for
   (a) I^- **(d)** ClO_3^-
   (b) S^{2-} **(e)** NO_3^-
   (c) CO_3^{2-}

52. Classify the following molecules as polar or nonpolar:
   (a) F_2
   (b) CO_2
   (c) NH_3

54. Give the number and arrangement of the electron pairs around the central atom:
   (a) Be in BeF_2
   (b) N in NF_3
   (c) Cl in HCl

56. Use VSEPR theory to predict the structure of the following polyatomic ions:
   (a) ammonium ion
   (b) sulfite ion
   (c) phosphate ion

58. Use VSEPR theory to predict the shape of the following molecules:
   (a) SiF_4
   (b) OF_2
   (c) Cl_2O

60. Identify this element from the following description: Element Y reacts with oxygen to form the compound Y_2O and has the lowest ionization energy of any fourth-period element on the periodic table.

Additional Exercises

These exercises are not paired or labeled by topic and provide additional practice on concepts covered in this chapter.

61. Identify the element on the periodic table that satisfies each of the following descriptions:
   (a) the transition metal with the largest atomic radius
   (b) the alkaline earth metal with the greatest ionization energy
   (c) the least dense member of nitrogen's family
   (d) the alkali metal with the greatest ratio of neutrons to protons
   (e) the most electronegative transition metal

62. Choose the element that fits each of the following descriptions:
   (a) the lower electronegativity   As or Zn
   (b) the lower chemical reactivity   Ba or Be
   (c) the fewer valence electrons   N or Ne

63. Identify two reasons for the fact that fluorine has a much higher electronegativity than neon.

64. When one electron is removed from an atom of Li, it has two left. Helium atoms also have two electrons. Why is more energy required to remove the second electron from Li than to remove the first from He?

65. Group IB elements (see the periodic table on the inside cover of your book) have one electron in their outer shell, as do Group IA elements. Would you expect them to form compounds such as CuCl, AgCl, and AuCl? Explain.

66. The formula for lead(II) bromide is $PbBr_2$: predict formulas for tin(II) and germanium(II) bromides.

67. Why is it not proper to speak of sodium chloride molecules?

68. What is a covalent bond? How does it differ from an ionic bond?

69. Briefly comment on the structure $Na:\ddot{O}:Na$ for the compound Na_2O.

70. What are the four most electronegative elements?

71. Rank these elements from highest electronegativity to lowest: Mg, S, F, H, O, Cs.

72. Is it possible for a molecule to be nonpolar even though it contains polar covalent bonds? Explain.

73. Why is CO_2 a nonpolar molecule, whereas CO is a polar molecule?

74. Estimate the bond angle between atoms in each of these molecules.
    (a) H_2S
    (b) NH_3
    (c) NH_4^+
    (d) $SiCl_4$

75. What is meant by the term *superconductor*?

76. What is the important relationship between 77 K and superconductivity?

77. What was the relationship on the periodic table that enabled Chu to find a high temperature superconductor?

78. Indicate the limitations of current superconducting material.

79. Indicate three uses for liquid crystals.

80. Explain how a liquid crystal display works.

81. What is Kevlar? How does it attain the property of super-strength?

82. Consider the two molecules BF_3 and NF_3. Compare and contrast these two in terms of
    (a) the valence level orbitals on the central atom that are used for bonding
    (b) the shape of the molecule
    (c) the number of lone electron pairs on the central atom
    (d) the type and number of bonds found in the molecule

83. With respect to electronegativity, why is fluorine such an important atom? What combination of atoms on the periodic table results in the most ionic bond?

84. Why does the Lewis structure of each element in a given group of representative elements on the periodic table have the same number of dots?

85. A sample of an air pollutant composed of sulfur and oxygen was found to contain 1.40 g sulfur and 2.10 g oxygen. What is the empirical formula for this compound? Draw a Lewis structure to represent it.

86. A dry-cleaning fluid composed of carbon and chlorine was found to have the composition 14.5% carbon and 85.5% chlorine. Its known molar mass is 166 g/mol. Draw a Lewis structure to represent it.

Answers to Practice Exercises

11.1 (a) $:\dot{\ddot{N}}\cdot$ (b) $:\dot{Al}$ (c) $Sr:$ (d) $:\ddot{\ddot{Br}}\cdot$

11.2 (a) Na_2S (b) Pb_2O

11.3 (a) Mg_3N_2 (b) Ba_3As_2

11.4 (a) $:\ddot{Br}:$ (b) H (c) $H:\ddot{F}:$ (d) $\ddot{O}:$
     $:\ddot{P}:\ddot{Br}:$ $:\ddot{Cl}:\ddot{C}:\ddot{Cl}:$ $H:\ddot{C}:H$
     $:\ddot{Br}:$ $:\ddot{Cl}:$

11.5 (a) $H:\ddot{\ddot{N}}:H$ (b) $\left[H:\ddot{O}:H\right]^+$
         H H

    (c) $\left[\begin{array}{c} H \\ H:\ddot{N}:H \\ H \end{array}\right]^+$ (d) $\left[H:\ddot{O}:\ddot{C}::\ddot{O}:\right]^-$ with $:\ddot{O}:$

11.6 CF_4 tetrahedral, NF_3 pyramidal, BeF_2 linear

The Gaseous State of Matter

Kinetic-Molecular Theory (KMT)

◀ **Chapter Opening Photo: The colorful art of hot air ballooning illustrates the multiple properties of gases.**

Our atmosphere is composed of a mixture of gases, including nitrogen, oxygen, carbon dioxide, ozone, and trace amounts of other gases. But there are cautions to be observed and respected when dealing with our atmosphere. For example, carbon dioxide is valuable as it is taken in by plants and converted to carbohydrates, but it also is associated with the potentially hazardous greenhouse effect. Ozone surrounds the earth at high altitudes and protects us from harmful ultraviolet rays, but it also destroys rubber and plastics. We require air to live, yet scuba divers must be concerned about oxygen poisoning, and the "bends."

In chemistry, the study of the behavior of gases allows us to lay a foundation for understanding our atmosphere and the effects that gases have on our lives.

12.1   General Properties

In Chapter 3, solids, liquids, and gases were described briefly. In this chapter we shall consider the behavior of gases in greater detail.

Gases are the least dense and most mobile of the three states of matter. A solid has a rigid structure, and its particles remain in essentially fixed positions. When a solid absorbs sufficient heat, it melts and changes into a liquid. Melting occurs because the molecules (or ions) have absorbed enough energy to break out of the rigid crystal lattice structure of the solid. The molecules or ions in the liquid are more energetic than they were in the solid, as indicated by their increased mobility. Molecules in the liquid state cling to one another. When the liquid absorbs additional heat, the more energetic molecules break away from the liquid surface and go into the gaseous state—the most mobile state of matter. Gas molecules move at very high velocities and have high kinetic energy (KE). The average velocity of hydrogen molecules at 0°C is over 1600 meters (1 mile) per second. Mixtures of gases are uniformly distributed within the container in which they are confined.

A quantity of a substance occupies a much greater volume as a gas than it does as a liquid or a solid. For example, 1 mol of water (18.02 g) has a volume of 18 mL at 4°C. This same amount of water would occupy about 22,400 mL in the gaseous state—more than a 1200-fold increase in volume. We may assume from this difference in volume that (1) gas molecules are relatively far apart, (2) gases can be greatly compressed, and (3) the volume occupied by a gas is mostly empty space.

12.2   The Kinetic-Molecular Theory

Careful scientific studies of the behavior and properties of gases were begun in the 17th century by Robert Boyle (1627–1691). This work was carried forward by many investigators, and the accumulated data were used in the second half of the 19th century to formulate a general theory to explain the behavior and properties of gases. This theory is called the **Kinetic-Molecular Theory (KMT)**. The KMT has since been extended to cover, in part, the behavior of liquids and solids. It ranks today with the atomic theory as one of the greatest generalizations of modern science.

The KMT is based on the motion of particles, particularly gas molecules. A gas that behaves exactly as outlined by the theory is known as an **ideal gas**. Actually no ideal gases exist, but under certain conditions of temperature and pressure, gases approach ideal behavior, or at least show only small deviations from it. Under extreme conditions, such as very high pressure and low temperature, real gases may deviate greatly from ideal behavior. For example, at low temperature and high pressure many gases become liquids.

ideal gas

The principal assumptions of the Kinetic-Molecular Theory are:

1. Gases consist of tiny (submicroscopic) particles.
2. The distance between particles is large compared with the size of the particles themselves. The volume occupied by a gas consists mostly of empty space.
3. Gas particles have no attraction for one another.
4. Gas particles move in straight lines in all directions, colliding frequently with one another and with the walls of the container.
5. No energy is lost by the collision of a gas particle with another gas particle or with the walls of the container. All collisions are perfectly elastic.
6. The average kinetic energy for particles is the same for all gases at the same temperature, and its value is directly proportional to the Kelvin temperature.

The kinetic energy (KE) of a particle is one-half its mass times its velocity squared. It is expressed by the equation

$$KE = \frac{1}{2}mv^2$$

where m is the mass and v is the velocity of the particle.

All gases have the same kinetic energy at the same temperature. Therefore, from the kinetic energy equation we can see that, if we compare the velocities of the molecules of two gases, the lighter molecules will have a greater velocity than the heavier ones. For example, calculations show that the velocity of a hydrogen molecule is four times the velocity of an oxygen molecule.

Due to their molecular motion, gases have the property of **diffusion**, the ability of two or more gases to mix spontaneously until they form a uniform mixture. The diffusion of gases may be illustrated by use of the apparatus shown in Figure 12.1. Two large flasks, one containing reddish brown bromine vapors and the other dry air, are connected by a side tube. When the stopcock between the flasks is opened, the bromine and air will diffuse into each other. After standing awhile, both flasks will contain bromine and air.

diffusion

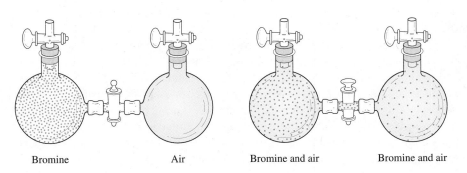

◀ **FIGURE 12.1**
Diffusion of gases. When the stopcock between the two flasks is opened, colored bromine molecules can be seen diffusing into the flask containing air.

Bromine   Air   Bromine and air   Bromine and air effusion

If we put a pinhole in a balloon, the gas inside will effuse or flow out of the balloon. **Effusion** is a process by which gas molecules pass through a very small orifice (opening) from a container at higher pressure to one at lower pressure.

Graham's law of effusion

Thomas Graham (1805–1869), a Scottish chemist, observed that the rate of effusion was dependent on the density of a gas. This observation led to **Graham's law of effusion**.

> **The rates of effusion of two gases at the same temperature and pressure are inversely proportional to the square roots of their densities or molar masses:**
>
> $$\frac{\text{rate of effusion of gas } A}{\text{rate of effusion of gas } B} = \sqrt{\frac{d\text{B}}{d\text{A}}} = \sqrt{\frac{\text{molar mass } B}{\text{molar mass } A}}$$

A major application of Graham's law occurred during World War II with the separation of the isotopes of uranium-235 (U-235) and uranium-238 (U-238). Naturally occurring uranium consists of 0.7% U-235, 99.3% U-238, and a trace of U-234. However, only U-235 is useful as fuel for nuclear reactors and atomic bombs, so the concentration of U-235 in the mixture of isotopes had to be increased.

Uranium was first changed to uranium hexafluoride, UF_6, a white solid that readily goes into the gaseous state. The gaseous mixture of $^{235}UF_6$ and $^{238}UF_6$ was then allowed to effuse through porous walls. Although the effusion rate of the lighter gas is only slightly faster than that of the heavier one,

$$\frac{\text{effusion rate } ^{235}UF_6}{\text{effusion rate } ^{238}UF_6} = \sqrt{\frac{\text{molar mass } ^{238}UF_6}{\text{molar mass } ^{235}UF_6}} = \sqrt{\frac{352}{349}} = 1.0043$$

the separation and enrichment of U-235 was accomplished by subjecting the gaseous mixture to several thousand stages of effusion.

12.3 Measurement of Pressure of Gases

pressure

Pressure is defined as force per unit area. When a rubber balloon is inflated with air, it stretches and maintains its larger size because the pressure on the inside is greater than that on the outside. Pressure results from the collisions of gas molecules with the walls of the balloon (see Figure 12.2). When the gas is released, the force or pressure of the air escaping from the small neck propels the balloon in a rapid, irregular flight. If the balloon is inflated until it bursts, the gas escaping all at once causes an explosive noise.

The effects of pressure are also observed in the mixture of gases surrounding the earth—our atmosphere. It is composed of about 78% nitrogen, 21% oxygen, and 1% argon, and other minor constituents by volume (see Table 12.1). The outer boundary of the atmosphere is not known precisely, but more than 99% of the atmosphere is below an altitude of 20 miles (32 km). Thus, the concentration of gas molecules in the atmosphere decreases with altitude, and at about 4 miles the amount of oxygen is insufficient to sustain human life. The gases in the atmosphere exert a pressure known as **atmospheric pressure**. The pressure exerted by a gas

atmospheric pressure

depends on the number of molecules of gas present, the temperature, and the volume in which the gas is confined. Gravitational forces hold the atmosphere relatively close to the earth and prevent air molecules from flying off into outer space. Thus, the atmospheric pressure at any point is due to the mass of the atmosphere pressing downward at that point.

The pressure of the gases in the atmosphere can be measured with a **barometer**. A mercury barometer may be prepared by completely filling a long tube with pure, dry mercury and inverting the open end into an open dish of mercury. If the tube is longer than 760 mm, the mercury level will drop to a point at which the column of mercury in the tube is just supported by the pressure of the atmosphere. If the tube is properly prepared, a vacuum will exist above the mercury column. The weight of mercury, per unit area, is equal to the pressure of the atmosphere. The column of mercury is supported by the pressure of the atmosphere, and the height of the column is a measure of this pressure (see Figure 12.3). The mercury barometer was invented in 1643 by the Italian physicist E. Torricelli (1608–1647), for whom the unit of pressure *torr* was named.

Air pressure is measured and expressed in many units. The standard atmospheric pressure, or simply **1 atmosphere** (atm), is the pressure exerted by a column of mercury 760 mm high at a temperature of 0°C. The normal pressure of the atmosphere at sea level is 1 atm or 760 torr or 760 mm Hg. The SI unit for pressure is the pascal (Pa), where 1 atm = 101,325 Pa or 101.3 kPa. Other units for expressing

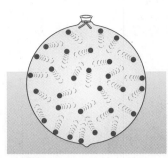

FIGURE 12.2
Here we see the pressure resulting from the collisions of gas molecules with the walls of the balloon. This pressure keeps the balloon inflated.

barometer
1 atmosphere

TABLE 12.1 Average Composition of Normal Dry Air

Gas	Percent by volume	Gas	Percent by volume
N_2	78.08	He	0.0005
O_2	20.95	CH_4	0.0002
Ar	0.93	Kr	0.0001
CO_2	0.033	Xe, H_2,	Trace
Ne	0.0018	and N_2O	

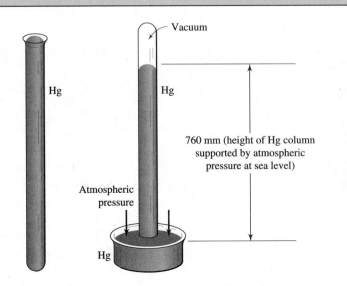

FIGURE 12.3
Preparation of a mercury barometer. The full tube of mercury at the left is inverted and placed in a dish of mercury.

TABLE 12.2 Pressure Units Equivalent to 1 Atmosphere

1 atm
760 torr
760 mm Hg
76 cm Hg
101,325 kPa
1013 mbar
29.9 in. Hg
14.7 lb/in.^{2}

pressure are inches of mercury, centimeters of mercury, the millibar (mbar), and pounds per square inch (lb/in.^{2} or psi). The meteorologist uses inches of mercury in reporting atmospheric pressure. The values of these units equivalent to 1 atm are summarized in Table 12.2 (1 atm $\equiv$ 760 torr $\equiv$ 760 mm Hg $\equiv$ 76 cm Hg $\equiv$ 101,325 Pa $\equiv$ 1013 mbar $\equiv$ 29.9 in. Hg $\equiv$ 14.7 lb/in.^{2}). (The symbol $\equiv$ means *identical with*.)

Atmospheric pressure varies with altitude. The average pressure at Denver, Colorado, 1.61 km (1 mile) above sea level, is 630 torr (0.83 atm). Atmospheric pressure is 0.5 atm at about 5.5 km (3.4 miles) altitude.

Pressure is often measured by reading the heights of mercury columns in millimeters on barometers. Thus pressure may be recorded as mm Hg. But in many applications the torr is superseding mm Hg as a unit of pressure. In problems dealing with gases it is necessary to make interconversions among the various pressure units. Since atm, torr, and mm Hg are common pressure units, we use illustrative problems involving all three of these units.

$$1 \text{ atm} = 760 \text{ torr} = 760 \text{ mm Hg}$$

Example 12.1

The average atmospheric pressure at Walnut, California, is 740. mm Hg. Calculate this pressure in (a) torr and (b) atmospheres.

Solution

This problem can be solved using conversion factors, relating one unit of pressure to another.
(a) To convert mm Hg to torr, use the conversion factor 760 torr/760 mm Hg (1 torr/1 mm Hg):

$$740. \ \cancel{\text{mm Hg}} \times \frac{1 \text{ torr}}{1 \ \cancel{\text{mm Hg}}} = 740. \text{ torr}$$

(b) To convert mm Hg to atm, use the conversion factor 1 atm/760. mm Hg:

$$740. \ \cancel{\text{mm Hg}} \times \frac{1 \text{ atm}}{760. \ \cancel{\text{mm Hg}}} = 0.974 \text{ atm}$$

Practice 12.1

A barometer reads 1.12 atm. Calculate the corresponding pressure in (a) torr and (b) mm Hg.

12.4 Dependence of Pressure on Number of Molecules and Temperature

Pressure is produced by gas molecules colliding with the walls of a container. At a specific temperature and volume the number of collisions depends on the number of gas molecules present. The number of collisions can be increased by increasing the number of gas molecules present. If we double the number of molecules, the frequency of collisions and the pressure should double. We find, for an ideal gas,

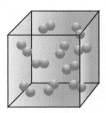

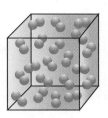

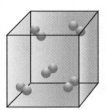

1 mol H_2 2 mol H_2 0.5 mol H_2
$P = 1$ atm $P = 2$ atm $P = 0.5$ atm

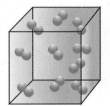

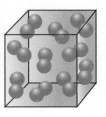

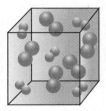

6.022×10^{23} molecules H_2 1 mol O_2 0.5 mol H_2 + 0.5 mol O_2
$P = 1$ atm $P = 1$ atm $P = 1$ atm

▲ FIGURE 12.4
The pressure exerted by a gas is directly proportional to the number of molecules present. In each case shown, the volume is 22.4 L and the temperature is 0°C.

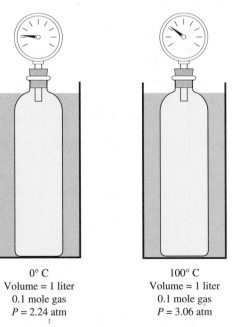

0° C   100° C
Volume = 1 liter   Volume = 1 liter
0.1 mole gas   0.1 mole gas
$P = 2.24$ atm $P = 3.06$ atm

◄ FIGURE 12.5
The pressure of a gas in a fixed volume increases with increasing temperature. The increased pressure is due to more frequent and more energetic collisions of the gas molecules with the walls of the container at the higher temperature.

kept constant, the pressure is directly proportional to the number of moles or molecules of gas present. Figure 12.4 illustrates this concept.

A good example of this molecule–pressure relationship may be observed in an ordinary cylinder of compressed gas that is equipped with a pressure gauge. When the valve is opened, gas escapes from the cylinder. The volume of the cylinder is constant, and the decrease in quantity (moles) of gas is registered by a drop in pressure indicated on the gauge.

The pressure of a gas in a fixed volume also varies with temperature. When the temperature is increased, the kinetic energy of the molecules increases, causing more frequent and more energetic collisions of the molecules with the walls of the container. This increase in collision frequency and energy results in a pressure increase (see Figure 12.5).

Robert Boyle (1627–1691).

Boyle's law

12.5   Boyle's Law

Through a series of experiments, Robert Boyle determined the relationship between the pressure (P) and volume (V) of a particular quantity of a gas. This relationship of P and V is known as **Boyle's law**.

> **At constant temperature (T), the volume (V) of a fixed mass of a gas is inversely proportional to the pressure (P), which may be expressed as:**
>
> $$V \propto \frac{1}{P} \quad \text{(mass and temperature are constant)}$$

This equation says that the volume varies ($\propto$) inversely with the pressure, at constant mass and temperature. When the pressure on a gas is increased, its volume will decrease, and vice versa. The inverse relationship of pressure and volume is shown graphically in Figure 12.6.

Boyle demonstrated that, when he doubled the pressure on a specific quantity of a gas, keeping the temperature constant, the volume was reduced to one-half the original volume; when he tripled the pressure on the system, the new volume was one-third the original volume; and so on. His demonstration showed that the product of volume and pressure is constant if the temperature is not changed:

$$PV = \text{constant} \quad \text{or} \quad PV = k \quad \text{(mass and temperature are constant)}$$

Let us demonstrate this law by using a cylinder with a movable piston so that the volume of gas inside the cylinder may be varied by changing the external pressure (see Figure 12.7). Assume that the temperature and the number of gas molecules do not change. Let us start with a volume of 1000 mL and a pressure of 1 atm. When we change the pressure to 2 atm, the gas molecules are crowded closer together, and the volume is reduced to 500 mL. When we increase the pressure to 4 atm, the volume becomes 250 mL.

Note that the product of the pressure times the volume is the same number in each case, substantiating Boyle's law. We may then say that

$$P_1V_1 = P_2V_2$$

where P_1V_1 is the pressure–volume product at one set of conditions, and P_2V_2 is the product at another set of conditions. In each case the new volume may be calculated by multiplying the starting volume by a ratio of the two pressures involved. Of course, the ratio of pressures used must reflect the direction in which the volume should change. When the pressure is changed from 1 atm to 2 atm, the ratio to be used is 1 atm/2 atm. Now we can verify the results given in Figure 12.7:

(a) Starting volume, 1000 mL; pressure change, 1 atm $\longrightarrow$ 2 atm

$$1000 \text{ mL} \times \frac{1 \text{ atm}}{2 \text{ atm}} = 500 \text{ mL}$$

(b) Starting volume, 1000 mL; pressure change, 1 atm $\longrightarrow$ 4 atm

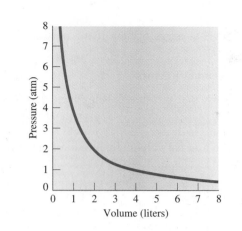

◀ **FIGURE 12.6**
Graph of pressure versus volume showing the inverse *PV* relationship of an ideal gas.

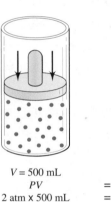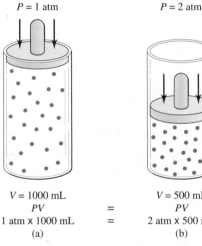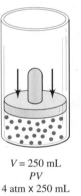

$P = 1$ atm		$P = 2$ atm		$P = 4$ atm
$V = 1000$ mL		$V = 500$ mL		$V = 250$ mL
PV	$=$	PV	$=$	PV
1 atm × 1000 mL	$=$	2 atm × 500 mL	$=$	4 atm × 250 mL
(a)		(b)		(c)

◀ **FIGURE 12.7**
The effect of pressure on the volume of a gas.

$$1000 \text{ mL} \times \frac{1 \text{ atm}}{4 \text{ atm}} = 250 \text{ mL}$$

(c) Starting volume, 500 mL; pressure change, 2 atm ⟶ 4 atm

$$500 \text{ mL} \times \frac{2 \text{ atm}}{4 \text{ atm}} = 250 \text{ mL}$$

In summary, a change in the volume of a gas due to a change in pressure can be calculated by multiplying the original volume by a ratio of the two pressures. If the pressure is increased, the ratio should have the smaller pressure in the numerator and the larger pressure in the denominator. If the pressure is decreased, the larger pressure should be in the numerator and the smaller pressure in the denominator.

new volume = original volume × ratio of pressures

The following examples are problems based on Boyle's law. If no mention is made of temperature, assume that it remains constant.

Example 12.2   What volume will 2.50 L of a gas occupy if the pressure is changed from 760. mm Hg to 630. mm Hg?

Solution   **Method A: Conversion Factors**

Step 1 Determine whether pressure is being increased or decreased:

pressure decreases $\longrightarrow$ volume increases

Step 2 Multiply the original volume by a ratio of pressures that will result in an increase in volume:

$$V = 2.50 \text{ L} \times \frac{760. \text{ mm Hg}}{630. \text{ mm Hg}} = 3.02 \text{ L (new volume)}$$

Decide which method is the best for you and stick with it.

Method B: Algebraic Equation

Step 1 Organize the given information:

$$P_1 = 760. \text{ mm Hg} \qquad V_1 = 2.50 \text{ L}$$

$$P_2 = 630. \text{ mm Hg} \qquad V_2 = ?$$

Step 2 Write and solve this equation for the unknown:

$$P_1 V_1 = P_2 V_2 \qquad V_2 = \frac{P_1 V_1}{P_2}$$

Step 3 Put the given information into this equation and calculate:

$$V_2 = \frac{760. \text{ mm Hg} \times 2.50 \text{ L}}{630. \text{ mm Hg}} = 3.02 \text{ L}$$

Example 12.3   A given mass of hydrogen occupies 40.0 L at 700. torr. What volume will it occupy at 5.00 atm pressure?

Solution   **Method A: Conversion Factors**

Step 1 Determine whether the pressure is being increased or decreased. Notice that in order to compare the values the units must be the same.

$$700. \text{ torr} \times \frac{1 \text{ atm}}{760 \text{ torr}} = 0.921 \text{ atm}$$

pressure increases $\longrightarrow$ volume decreases

Step 2 Multiply the original volume by a ratio of pressures that will result in a decrease in volume:

$$V = 40.0 \text{ L} \times \frac{0.921 \text{ atm}}{5.00 \text{ atm}} = 7.37 \text{ L}$$

Method B: Algebraic Equation

Step 1 Organize the given information. Remember to make the pressure units the same.

$$P_1 = 700. \text{ torr} = 0.921 \text{ atm} \qquad V_1 = 40.0 \text{ L}$$

$$P_2 = 5.00 \text{ atm} \qquad\qquad V_2 = ?$$

Step 2 Write and solve this equation for the unknown:

$$P_1V_1 = P_2V_2 \qquad V_2 = \frac{P_1V_1}{P_2}$$

Step 3 Put the given information into this equation and calculate:

$$V_2 = \frac{0.921 \text{ atm} \times 40.0 \text{ L}}{5.00 \text{ atm}} = 7.37 \text{ L}$$

A gas occupies a volume of 200. mL at 400. torr pressure. To what pressure must   **Example 12.4**
the gas be subjected in order to change the volume to 75.0 mL?

Method A: Conversion Factors

Step 1 Determine whether volume is being increased or decreased:

volume decreases $\longrightarrow$ pressure increases

Step 2 Multiply the original pressure by a ratio of volumes that will result in
an increase in pressure:

new pressure = original pressure $\times$ ratio of volumes

$$P = 400. \text{ torr} \times \frac{200. \text{ mL}}{75.0 \text{ mL}} = 1067 \text{ torr} \quad \text{or} \quad 1.07 \times 10^3 \text{ torr} \text{ (new pressure)}$$

Method B: Algebraic Equation

Step 1 Organize the given information. Remember to make units the same.

$$P_1 = 400. \text{ torr} \qquad V_1 = 200. \text{ mL}$$

$$P_2 = ? \qquad\qquad V_2 = 75.0 \text{ mL}$$

Step 2 Write and solve this equation for the unknown:

$$P_1V_1 = P_2V_2 \qquad P_2 = \frac{P_1V_1}{V_2}$$

Step 3 Put the given information into the equation and calculate:

$$P_2 = \frac{P_1V_1}{V_2} = \frac{400. \text{ torr} \times 200. \text{ mL}}{75.0 \text{ mL}} = 1.07 \times 10^3 \text{ torr}$$

Practice 12.2

A gas occupies a volume of 3.86 L at 0.750 atm. At what pressure will the
volume be 4.86 L?

12.6 Charles' Law

The effect of temperature on the volume of a gas was observed in about 1787 by the French physicist J. A. C. Charles (1746–1823). Charles found that various gases expanded by the same fractional amount when they underwent the same change in temperature. Later it was found that if a given volume of any gas initially at 0°C was cooled by 1°C, the volume decreased by $\frac{1}{273}$; if cooled by 2°C, it decreased by $\frac{2}{273}$; if cooled by 20°C, by $\frac{20}{273}$; and so on. Since each degree of cooling reduced the volume by $\frac{1}{273}$, it was apparent that any quantity of any gas would have zero volume if it could be cooled to −273°C. Of course, no real gas can be cooled to −273°C for the simple reason that it would liquefy before that temperature is reached. However, −273°C (more precisely −273.15°C) is referred to as **absolute zero**; this temperature is the zero point on the Kelvin (absolute) temperature scale—the temperature at which the volume of an ideal, or perfect, gas would become zero.

absolute zero

The volume–temperature relationship for methane is shown graphically in Figure 12.8. Experimental data show the graph to be a straight line that, when extrapolated, crosses the temperature axis at −273.15°C, or absolute zero. This is characteristic for all gases.

Charles' law

In modern form, **Charles' law** is as follows:

> At *constant pressure* the volume of a fixed mass of any gas is directly proportional to the absolute temperature, which may be expressed as:
>
> $$V \propto T \quad (P \text{ is constant})$$

A capital T is usually used for absolute temperature, K, and a small t for °C.

Mathematically this states that the volume of a gas varies directly with the absolute temperature when the pressure remains constant. In equation form Charles' law may be written as

$$V = kT \quad \text{or} \quad \frac{V}{T} = k \quad (\text{at constant pressure})$$

where k is a constant for a fixed mass of the gas. If the absolute temperature of a gas is doubled, the volume will double.

To illustrate, let us return to the gas cylinder with the movable or free-floating piston (see Figure 12.9). Assume that the cylinder labeled (a) contains a quantity of gas and the pressure on it is 1 atm. When the gas is heated, the molecules move faster, and their kinetic energy increases. This action should increase the number of collisions per unit of time and therefore increase the pressure. However, the increased internal pressure will cause the piston to rise to a level at which the internal and external pressures again equal 1 atm, as we see in cylinder (b). The net result is an increase in volume due to an increase in temperature.

Don't forget to change temperature to Kelvin when doing gas law calculations.

Another equation relating the volume of a gas at two different temperatures is

$$\frac{V_1}{T_1} = \frac{V_2}{T_2} \quad (\text{constant } P)$$

where V_1 and T_1 are one set of conditions and V_2 and T_2 are another set of conditions.

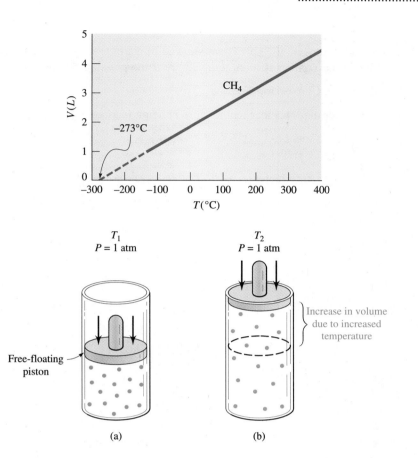

Volume–temperature relationship of methane (CH₄). Extrapolated portion of the graph is shown by the broken line.

◀ **FIGURE 12.9**
The effect of temperature on the volume of a gas. The gas in cylinder (a) is heated from T_1 to T_2. With the external pressure constant at 1 atm, the free-floating piston rises, resulting in an increased volume, as shown in cylinder (b).

A simple experiment showing the variation of the volume of a gas with temperature is illustrated in Figure 12.10. A balloon is placed in a beaker and liquid N_2 is poured over it. The volume is reduced, as shown by the collapse of the balloon; when the balloon is removed from the liquid N_2, the gas expands and the balloon increases in size.

(a)   (b)   (c)

▲
FIGURE 12.10
The air-filled balloons in (a) are placed in liquid nitrogen (b). The volume of the air decreases tremendously at this temperature. In (c) the balloons are removed from the beaker and are beginning to return to their original volume as they warm back to room temperature.

Example 12.5 Three liters of hydrogen at $-20.°C$ are allowed to warm to a room temperature of $27°C$. What is the volume at room temperature if the pressure remains constant?

Solution **Method A: Conversion Factors**

Remember temperature must be changed to Kelvin in gas problems. Note that we use 273 to convert instead of 273.15 since our original measurements are to the nearest degree.

Step 1 Determine whether temperature is being increased or decreased.

$$-20.°C + 273 = 253 \text{ K}$$
$$27°C + 273 = 300. \text{ K}$$

temperature increases ⟶ volume increases

Step 2 Multiply the original volume by a ratio of temperatures that will result in an increase in volume.

$$V = 3.00 \text{ L} \times \frac{300. \text{ K}}{253 \text{ K}} = 3.56 \text{ L}   \text{(new volume)}$$

Method B: Algebraic Equation

Step 1 Organize the given information. Remember to make units the same.

$$V_1 = 3.00 \text{ L} \qquad T_1 = 20.°C = 253 \text{ K}$$
$$V_2 = ? \qquad T_2 = 27°C = 300. \text{ K}$$

Step 2 Write and solve the equation for the unknown:

$$\frac{V_1}{T_1} = \frac{V_2}{T_2} \qquad V_2 = \frac{V_1 T_2}{T_1}$$

Step 3 Put the given information into the equation and calculate:

$$V_2 = \frac{V_1 T_2}{T_1} = \frac{3.00 \text{ L} \times 300. \text{ K}}{253 \text{ K}} = 3.56 \text{ L}$$

Example 12.6 If 20.0 L of oxygen are cooled from $100°C$ to $0°C$, what is the new volume?

Solution Since no mention is made of pressure, assume that pressure does not change.

Method A: Conversion Factors

Step 1 Change $°C$ to K:

$$100°C + 273 = 373 \text{ K}$$
$$0°C + 273 = 273 \text{ K}$$

Step 2 The ratio of temperature to be used is 273 K/373 K, because the final volume should be smaller than the original volume. The calculation is

$$V = 20.0 \text{ L} \times \frac{273 \text{ K}}{373 \text{ K}} = 14.6 \text{ L}   \text{(new volume)}$$

Method B: Algebraic Equation

Step 1 Organize the given information. Remember to make units coincide.

$$V_1 = 20.0 \text{ L} \qquad T_1 = 100°C = 373 \text{ K}$$
$$V_2 = ? \qquad T_2 = 0°C = 273 \text{ K}$$

Step 2 Write and solve equation for the unknown:

$$\frac{V_1}{T_1} = \frac{V_2}{T_2} \qquad V_2 = \frac{V_1 T_2}{T_1}$$

Step 3 Put the given information into the equation and calculate:

$$V_2 = \frac{V_1 T_2}{T_1} = \frac{20.0\ \text{L} \times 273\ \text{K}}{373\ \text{K}} = 14.6\ \text{L}$$

Practice 12.3

A 4.50-L container of nitrogen at 28.0°C is heated to 56.0°C. Assuming the volume of the container can vary, what is the new volume of the gas?

Joseph Louis Gay-Lussac (1778–1850).

12.7 Gay-Lussac's Law

J. L. Gay-Lussac was a French chemist involved in the study of volume relationships of gases. The three variables (pressure, P; volume, V; and temperature, T) are needed to describe a fixed amount of a gas. Boyle's law, $PV = k$, relates pressure and volume at constant temperature; Charles' law, $V = kT$, relates volume and temperature at constant pressure. A third relationship involving pressure and temperature at constant volume is a modification of Charles' law and is sometimes called **Gay-Lussac's law**:

Gay-Lussac's law

> **The pressure of a fixed mass of a gas, at constant volume, is directly proportional to the Kelvin temperature:**
>
> $$P = kT \quad \text{(at constant volume)} \qquad \frac{P_1}{T_1} = \frac{P_2}{T_2}$$

Example 12.7

The pressure of a container of helium is 650. torr at 25°C. If the sealed container is cooled to 0°C, what will the pressure be?

Solution

Method A: Conversion Factors

Step 1 Determine whether temperature is being increased or decreased.

temperature decreases $\longrightarrow$ pressure decreases

Step 2 Multiply the original pressure by a ratio of Kelvin temperatures that will result in a decrease in pressure:

$$650.\ \text{torr} \times \frac{273\ \text{K}}{298\ \text{K}} = 595\ \text{torr}$$

Method B: Algebraic Equation

Step 1 Organize the given information. Remember to make units the same.

$$P_1 = 650. \text{ torr} \qquad T_1 = 25°C = 298 \text{ K}$$

$$P_2 = ? \qquad T_2 = 0°C = 273 \text{ K}$$

Step 2 Write and solve equation for the unknown:

$$\frac{P_1}{T_1} = \frac{P_2}{T_2} \qquad P_2 = \frac{P_1 T_2}{T_1}$$

Step 3 Put given information into equation and calculate:

$$P_2 = \frac{650. \text{ torr} \times 273 \text{ K}}{298 \text{ K}} = 595 \text{ torr}$$

Practice 12.4

A gas cylinder contains 40.0 L of gas at 45.0°C and has a pressure of 650. torr. What will the pressure be if the temperature is changed to 100.°C?

We may summarize the effects of changes in pressure, temperature, and quantity of a gas as follows:

1. In the case of a constant volume,
   (a) when the temperature is increased, the pressure increases.
   (b) when the quantity of a gas is increased, the pressure increases (*T* remaining constant).
2. In the case of a variable volume,
   (a) when the external pressure is increased, the volume decreases (*T* remaining constant).
   (b) when the temperature of a gas is increased, the volume increases (*P* remaining constant).
   (c) when the quantity of a gas is increased, the volume increases (*P* and *T* remaining constant).

12.8   Standard Temperature and Pressure

In order to compare volumes of gases, common reference points of temperature and pressure were selected and called **standard conditions** or **standard temperature and pressure** (abbreviated **STP**). Standard temperature is 273.15 K (0°C), and standard pressure is 1 atm or 760 torr or 760 mm Hg or 101.325 kPa. For purposes of comparison, volumes of gases are usually changed to STP conditions.

standard conditions

standard temperature and pressure (STP)

> **standard temperature = 273.15 K or 0°C**
> **standard pressure = 1 atm or 760 torr or 760 mm Hg or**
> **101.325 kPa**

In this text we'll use 273 K for temperature conversions and calculations. Check with your instructor for rules in your class.

Combined Gas Laws: Simultaneous Changes in Pressure, Volume, and Temperature

12.9

When temperature and pressure change at the same time, the new volume may be calculated by multiplying the initial volume by the correct ratios of both pressure and temperature, as follows:

$$\text{final volume} = \text{initial volume} \times \left(\begin{array}{c}\text{ratio of}\\\text{pressures}\end{array}\right) \times \left(\begin{array}{c}\text{ratio of}\\\text{temperatures}\end{array}\right)$$

This equation combines Boyle's and Charles' laws, and the same considerations for the pressure and temperature ratios should be used in the calculation. The four possible variations are as follows:

1. Both T and P cause an increase in volume.
2. Both T and P cause a decrease in volume.
3. T causes an increase and P causes a decrease in volume.
4. T causes a decrease and P causes an increase in volume.

The P, V, and T relationships for a given mass of any gas, in fact, may be expressed as a single equation, $PV/T = k$. For problem solving this equation is usually written

$$\frac{P_1 V_1}{T_1} = \frac{P_2 V_2}{T_2}$$

where P_1, V_1, and T_1 are the initial conditions and P_2, V_2, and T_2 are the final conditions.

This equation can be solved for any one of the six variables and is useful in dealing with the pressure–volume–temperature relationships of gases. Note that when T is constant ($T_1 = T_2$), Boyle's law is represented; when P is constant ($P_1 = P_2$), Charles' law is represented; and when V is constant ($V_1 = V_2$), Gay-Lussac's law is represented.

Note, in the examples below, the use of 273 K does not change the number of significant figures in the temperature.
The converted temperature is expressed to the same precision as the original measurement.

Given 20.0 L of ammonia gas at 5°C and 730. torr, calculate the volume at 50.°C and 800. torr.

Example 12.8

Step 1 Organize the given information, putting temperatures in Kelvin:

Solution

$P_1 = 730.\text{ torr}$ $P_2 = 800.\text{ torr}$

$V_1 = 20.0\text{ L}$ $V_2 = ?$

$T_1 = 5°C = 278\text{ K}$ $T_2 = 50.°C = 323\text{ K}$

Method A: **Conversion Factors**

Step 2 Set up ratios of T and P:

$$T \text{ ratio} = \frac{323 \text{ K}}{278 \text{ K}} \text{ (increase in } T \text{ should increase } V\text{)}$$

$$P \text{ ratio} = \frac{730. \text{ torr}}{800. \text{ torr}} \text{ (increase in } P \text{ should decrease } V\text{)}$$

Step 3 Multiply the original pressure by the ratios:

$$V_2 = 20.0 \text{ L} \times \frac{730. \text{ torr}}{800. \text{ torr}} \times \frac{323 \text{ K}}{278 \text{ K}} = 21.2 \text{ L}$$

Method B: **Algebraic Equation**

Step 2 Write and solve the equation for the unknown. Solve

$$\frac{P_1 V_1}{T_1} = \frac{P_2 V_2}{T_2}$$

for V_2 by multiplying both sides of the equation by T_2/P_2 and rearranging to obtain

$$V_2 = \frac{V_1 P_1 T_2}{P_2 T_1}$$

Step 3 Put the given information into the equation and calculate:

$$V_2 = \frac{20.0 \text{ L} \times 730. \text{ torr} \times 323 \text{ K}}{800. \text{ torr} \times 278 \text{ K}} = 21.2 \text{ L}$$

Example 12.9 To what temperature (°C) must 10.0 L of nitrogen at 25°C and 700. torr be heated in order to have a volume of 15.0 L and a pressure of 760. torr?

Solution **Step 1** Organize the given information, putting temperatures in Kelvin:

$$P_1 = 700. \text{ torr} \qquad P_2 = 760. \text{ torr}$$

$$V_1 = 10.0 \text{ L} \qquad V_2 = 15.0 \text{ L}$$

$$T_1 = 25°C = 298 \text{ K} \qquad T_2 = ?$$

Method A: **Conversion Factors**

Step 2 Set up ratios of V and P.

$$P \text{ ratio} = \frac{760. \text{ torr}}{700. \text{ torr}} \text{ (increase in } P \text{ should increase } T\text{)}$$

$$V \text{ ratio} = \frac{15.0 \text{ L}}{10.0 \text{ L}} \text{ (increase in } V \text{ should increase } T\text{)}$$

Step 3 Multiply the original temperature by the ratios:

$$T_2 = 298 \text{ K} \times \frac{760. \text{ torr} \times 15.0 \text{ L}}{700. \text{ torr} \times 10.0 \text{ L}} = 485 \text{ K}$$

Method B: Algebraic Equation

Step 2 Write and solve the equation for the unknown:

$$\frac{P_1 V_1}{T_1} = \frac{P_2 V_2}{T_2} \qquad T_2 = \frac{T_1 P_2 V_2}{P_1 V_1}$$

Step 3 Put the given information into the equation and calculate:

$$T_2 = \frac{298\ \text{K} \times 760.\ \text{torr} \times 15.0\ \text{L}}{700.\ \text{torr} \times 10.0\ \text{L}} = 485\ \text{K}$$

In either method, since the problem asks for °C, we must subtract 273 from the Kelvin answer:

$$485\ \text{K} - 273 = 212°\text{C}$$

The volume of a gas-filled balloon is 50.0 L at 20.°C and 742 torr. What volume will it occupy at standard temperature and pressure (STP)?   **Example 12.10**

Step 1 Organize the given information, putting temperatures in Kelvin.   **Solution**

$P_1 = 742$ torr $\qquad\qquad P_2 = 760.$ torr (standard pressure)

$V_1 = 50.0$ L $\qquad\qquad V_2 = ?$

$T_1 = 20.°\text{C} = 293$ K $\qquad T_2 = 273$ K (standard temperature)

Method A: Conversion Factors

Step 2 Set up ratios of T and P:

$$T\ \text{ratio} = \frac{273\ \text{K}}{293\ \text{K}}\ \text{(decrease in } T \text{ should decrease } V\text{)}$$

$$P\ \text{ratio} = \frac{742\ \text{torr}}{760.\ \text{torr}}\ \text{(increase in } P \text{ should decrease } V\text{)}$$

Step 3 Multiply the original volume by the ratios:

$$V_2 = 50.0\ \text{L} \times \frac{273\ \text{K}}{293\ \text{K}} \times \frac{742\ \text{torr}}{760.\ \text{torr}} = 45.5\ \text{L}$$

Method B: Algebraic Equation

Step 2 Write and solve the equation for the unknown:

$$\frac{P_1 V_1}{T_1} = \frac{P_2 V_2}{T_2} \qquad V_2 = \frac{P_1 V_1 T_2}{P_2 T_1}$$

Step 3 Put the given information into the equation and calculate:

$$V_2 = \frac{742\ \text{torr} \times 50.0\ \text{L} \times 273\ \text{K}}{760.\ \text{torr} \times 293\ \text{K}} = 45.5\ \text{L}$$

Messenger Molecules

Nitrogen monoxide, NO, and carbon monoxide, CO, are small, gaseous molecules, often considered toxic to the body and the environment because of their contribution to photochemical smog and acid rain. But now, biochemists have found that these molecules are important players in a newly discovered group of biological messengers called *neurotransmitter molecules.*

Traditional neurotransmitter molecules, such as acetylcholine, norepinephrine, and epinephrine, are biogenic amines. In the 1950s, scientists discovered that amino acids could signal nerve cells, and in the 1970s researchers produced another class of transmitter molecules known as peptides—short strings of amino acids commonly called *endorphins* and *enkephalins* (associated with "runner's high"). In the late 1980s, nitrogen monoxide was discovered to be the first gas that could act as a neurotransmitter.

Until recently, neurotransmitter molecules were thought to be specific. That is, each neurotransmitter would fit into the next cell like a key in a lock. These transmitters were thought to be stored in tiny pouches in the cells where they were man-

Incoming synaptic transmission ignites a neuron on impulse neural pathway.

ufactured and released through transporters or channels only when needed. Nitrogen monoxide and carbon monoxide gases break the rules for explaining neurotransmission because gases are volatile and nonspecific. A gas freely diffuses into any nearby cell and must be made only when needed. The actions of gas molecules depend on their chemical properties rather than their molecular shape, as the lock-and-key model for neurotransmitters had suggested.

Nitrogen monoxide has been used for nearly a century to dilate blood vessels and increase blood flow, thereby lowering blood pressure.

In the late 1980s, scientists discovered that biologically derived NO is an important signaling molecule in both the central and peripheral nervous system. Nitrogen monoxide was found to be a mediator of certain neurons that do not respond to the normal chemical neurotransmitters (acetylcholine or norepinephrine). These NO-sensitive neurons are in the cardiovascular, respiratory, digestive, and urogenital systems. One study suggests that nitrogen monoxide may be a way to treat impotence by mediating the relaxation of the smooth muscle of the major erectile tissue in the penis, causing erection. Current research indicates that nitrogen monoxide plays a role in regulating blood pressure, blood clotting, and neurotransmission, and it may play a role in the ability of the immune system to kill tumor cells.

Researchers at Johns Hopkins Medical School reasoned that if one gas could act as a transmitter molecule, others might also be found. They began searching for other gases that might exhibit this biological activity. Carbon monoxide was proposed as a possible transmitter because the en-

zyme used to make it is localized within specific parts of the brain, such as the olfactory neurons and those cells thought to play a role in long-term memory. The enzyme used to make an intracellular messenger was found in exactly the same locations. Researchers then showed that nerve cells made the intracellular messenger when stimulated with carbon monoxide. An inhibitor for carbon monoxide blocked the production of the same messenger molecule.

Dr. Charles Stevens, from Salk Institute in La Jolla, California, thinks carbon monoxide might be an important factor in understanding long-term memory. Researchers think that in order to lay down a memory, a nerve cell is signaled repeatedly in a process called *long-term potentiation.* An important feature of the potentiation process is that the cell receiving the signal sends a message back to the cell that first signaled it. This message stimulates the originating cell to release its neurotransmitter molecule more easily. Investigators have never been able to find the message that the receiving cell sends to the originating cell.

When nitrogen monoxide was discovered to be a neurotransmitter, scientists thought that it might be the return signal. But the enzymes necessary to produce nitrogen monoxide were not found in the cells where long-term potentiation occurs. The enzyme that makes carbon monoxide, however, is abundant in these cells, and so Stevens began testing his theory. He found that inhibitors for carbon monoxide prevented long-term potentiation, and when CO was blocked from these cells, memories that were already there were erased. Further research may uncover still other roles for these gaseous molecules.

Practice 12.5

15.00 L of gas at 45.0°C and 800. torr is heated to 400.°C, and the pressure changed to 300. torr. What is the new volume?

Practice 12.6

To what temperature must 5.00 L of oxygen at 50.°C and 600. torr be heated in order to have a volume of 10.0 L and a pressure of 800. torr?

12.10   Dalton's Law of Partial Pressures

If gases behave according to the Kinetic-Molecular Theory, there should be no difference in the pressure–volume–temperature relationships whether the gas molecules are all the same or different. This similarity in the behavior of gases is the basis for an understanding of **Dalton's law of partial pressures**:

Dalton's law of partial pressures

> **The total pressure of a mixture of gases is the sum of the partial pressures exerted by each of the gases in the mixture.**

Each gas in the mixture exerts a pressure that is independent of the other gases present. These pressures are called **partial pressures**. Thus, if we have a mixture of three gases, A, B, and C, exerting partial pressures of 50 torr, 150 torr, and 400 torr, respectively, the total pressure will be 600 torr:

partial pressure

$$P_{Total} = p_A + p_B + p_C$$

$$P_{Total} = 50 \, torr + 150 \, torr + 400 \, torr = 600 \, torr$$

We can see an application of Dalton's law in the collection of insoluble gases over water. When prepared in the laboratory, oxygen is commonly collected by the downward displacement of water. Thus, the oxygen is not pure but is mixed with water vapor (see Figure 12.11). When the water levels are adjusted to the same height inside and outside the bottle, the pressure of the oxygen plus water vapor inside the bottle is equal to the atmospheric pressure:

$$P_{atm} = p_{O_2} + p_{H_2O}$$

To determine the amount of O_2 or any other gas collected over water, we must subtract the pressure of the water vapor from the total pressure of the gas. The vapor pressure of water at various temperatures is tabulated in Appendix II.

$$p_{O_2} = P_{atm} - p_{H_2O}$$

FIGURE 12.11
Oxygen collected over water. ▶

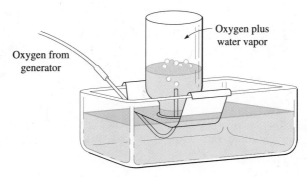

Oxygen from generator

Oxygen plus water vapor

Example 12.11   A 500.-mL sample of oxygen was collected over water at 23°C and 760. torr. What volume will the dry O_2 occupy at 23°C and 760. torr? The vapor pressure of water at 23°C is 21.2 torr.

Solution   To solve this problem, we must first determine the pressure of the oxygen alone, by subtracting the pressure of the water vapor present.

Step 1 Determine the pressure of dry O_2:

$$P_{Total} = 760. \text{ torr} = p_{O_2} + p_{H_2O}$$

$$p_{O_2} = 760. \text{ torr} - 21.2 \text{ torr} = 739 \text{ torr} \quad (\text{dry } O_2)$$

Step 2 Organize the given information:

$$P_1 = 739 \text{ torr} \qquad P_2 = 760. \text{ torr}$$

$$V_1 = 500. \text{ mL} \qquad V_2 = ?$$

T is constant

Step 3 Solve as a Boyle's law problem:

$$V = \frac{500. \text{ mL} \times 739 \text{ torr}}{760. \text{ torr}} = 486 \text{ mL dry } O_2$$

Practice 12.7

Hydrogen gas was collected by downward displacement of water. A volume of 600.0 mL of gas was collected at 25.0°C and 740.0 torr. What volume will the dry hydrogen occupy at STP?

12.11   Avogadro's Law

Early in the 19th century Gay-Lussac studied the volume relationships of reacting gases. His results, published in 1809, were summarized in a statement known as **Gay-Lussac's law of combining volumes**:

Gay-Lussac's law of combining volumes

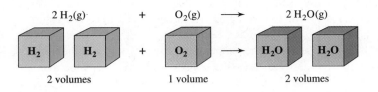

$2 H_2(g)$ + $O_2(g)$ $\longrightarrow$ $2 H_2O(g)$

| H_2 | H_2 | + | O_2 | $\longrightarrow$ | H_2O | H_2O |

2 volumes   1 volume   2 volumes

> **When measured at the same temperature and pressure, the ratios of the volumes of reacting gases are small whole numbers.**

Thus, H_2 and O_2 combine to form water vapor in a volume ratio of 2:1 (Figure 12.12); H_2 and Cl_2 react to form HCl in a volume ratio of 1:1; and H_2 and N_2 react to form NH_3 in a volume ratio of 3:1.

Two years later, in 1811, Amedeo Avogadro used the law of combining volumes of gases to make a simple but significant and far-reaching generalization concerning gases. **Avogadro's law** states:

Avogadro's law

> **Equal volumes of different gases at the same temperature and pressure contain the same number of molecules.**

This law was a real breakthrough in understanding the nature of gases.

1. It offered a rational explanation of Gay-Lussac's law of combining volumes of gases and indicated the diatomic nature of such elemental gases as hydrogen, chlorine, and oxygen.
2. It provided a method for determining the molar masses of gases and for comparing the densities of gases of known molar mass (see Sections 12.12 and 12.13).
3. It afforded a firm foundation for the development of the Kinetic-Molecular Theory.

By Avogadro's law, equal volumes of hydrogen and chlorine contain the same number of molecules. On a volume basis, hydrogen and chlorine react thus:

hydrogen + chlorine $\longrightarrow$ hydrogen chloride

1 volume   1 volume   2 volumes

Therefore, hydrogen molecules react with chlorine molecules in a 1:1 ratio. Since two volumes of hydrogen chloride are produced, one molecule of hydrogen and one molecule of chlorine must produce two molecules of hydrogen chloride. Therefore, each hydrogen molecule and each chlorine molecule must be made up of two atoms. The coefficients of the balanced equation for the reaction give the correct ratios for volumes, molecules, and moles of reactants and products:

$$H_2 \quad + \quad Cl_2 \quad \longrightarrow \quad 2 HCl$$

1 volume	1 volume	2 volumes
1 molecule	1 molecule	2 molecules
1 mol	1 mol	2 mol

By like reasoning, oxygen molecules also must contain at least two atoms because one volume of oxygen reacts with two volumes of hydrogen to produce two volumes of water vapor.

The volume of a gas depends on the temperature, the pressure, and the number of gas molecules. Different gases at the same temperature have the same average kinetic energy. Hence, if two different gases are at the same temperature, occupy equal volumes, and exhibit equal pressures, each gas must contain the same number of molecules. This statement is true because systems with identical *PVT* properties can be produced only by equal numbers of molecules having the same average kinetic energy.

12.12 Mole–Mass–Volume Relationships of Gases

molar volume

As with many constants, the molar volume is known more exactly to be 22.414 L. We use 22.4 L in our calculations since the extra figures do not often affect the result, given the other measurements in the calculation.

Because a mole contains 6.022×10^{23} molecules (Avogadro's number), a mole of any gas will have the same volume as a mole of any other gas at the same temperature and pressure. It has been experimentally determined that the volume occupied by a mole of any gas is 22.4 L at STP. This volume, 22.4 L, is known as the **molar volume** of a gas. The molar volume is a cube about 28.2 cm (11.1 in.) on a side. The molar masses of several gases, each occupying 22.4 L at STP, are shown in Figure 12.13.

> **One mole of a gas occupies 22.4 L at STP.**

The molar volume is useful for determining the molar mass of a gas or of substances that can be easily vaporized. If the mass and the volume of a gas at STP are known, we can calculate its molar mass. For example, 1 L of pure oxygen at STP has a mass of 1.429 g. The molar mass of oxygen may be calculated by

FIGURE 12.13 ▶
One mole of a gas occupies 22.4 L at STP. The mass given for each gas is the mass of 1 mol.

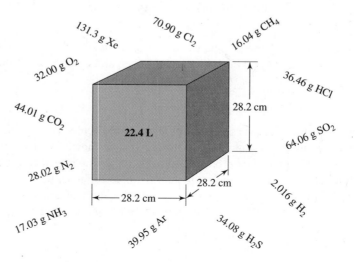

multiplying the mass of 1 L by 22.4 L/mol:

$$\frac{1.429 \text{ g}}{1 \text{ L}} \times \frac{22.4 \text{ L}}{1 \text{ mol}} = 32.00 \text{ g/mol} \quad \text{(molar mass)}$$

If the mass and volume are at other than standard conditions, we change the volume to STP and then calculate the molar mass.

The molar volume, 22.4 L/mol, is used as a conversion factor to convert grams per liter to grams per mole (molar mass) and also to convert liters to moles. The two conversion factors are

$$\frac{22.4 \text{ L}}{1 \text{ mol}} \quad \text{and} \quad \frac{1 \text{ mol}}{22.4 \text{ L}}$$

These conversions must be done at STP except under certain special circumstances. Examples follow.

Standard conditions apply only to pressure, temperature, and volume. Mass is not affected.

If 2.00 L of a gas measured at STP have a mass of 3.23 g, what is the molar mass of the gas?

Example 12.12

The unit of molar mass is g/mol; the conversion is from

Solution

$$\frac{\text{g}}{\text{L}} \longrightarrow \frac{\text{g}}{\text{mol}}$$

The starting amount is $\dfrac{3.23 \text{ g}}{2.00 \text{ L}}$. The conversion factor is $\dfrac{22.4 \text{ L}}{1 \text{ mol}}$.

The calculation is $\dfrac{3.23 \text{ g}}{2.00 \text{ L}} \times \dfrac{22.4 \text{ L}}{1 \text{ mol}} = 36.2 \text{ g/mol} \quad \text{(molar mass)}$

Measured at 40°C and 630. torr, the mass of 691 mL of ethyl ether is 1.65 g. Calculate the molar mass of ethyl ether.

Example 12.13

Step 1 Organize the given information, converting temperatures to Kelvin. Note that we must change to STP in order to determine molar mass.

Solution

$P_1 = 630. \text{ torr}$ $P_2 = 760. \text{ torr}$

$V_1 = 691 \text{ mL}$ $V_2 = ?$

$T_1 = 313 \text{ K } (40.°\text{C})$ $T_2 = 273 \text{ K}$

Step 2 Use either the conversion factor method or the algebraic method and the combined gas law to correct the volume (V_2) to STP:

$$V_2 = \frac{691 \text{ mL} \times 273 \text{ K} \times 630. \text{ torr}}{313 \text{ K} \times 760. \text{ torr}} = 500. \text{ mL} = 0.500 \text{ L} \quad \text{(at STP)}$$

Step 3 In the example, V_2 is the volume for 1.65 g of the gas, so we can now find the molar mass by converting g/L to g/mol:

$$\frac{1.65 \text{ g}}{0.500 \text{ L}} \times \frac{22.4 \text{ L}}{\text{mol}} = 73.9 \text{ g/mol}$$

Practice 12.8

A gas with a mass of 86 g occupies 5.00 L at 25°C and 3.00 atm pressure. What is the molar mass of the gas?

12.13 Density of Gases

The density, d, of a gas is its mass per unit volume, which is generally expressed in grams per liter as follows:

$$d = \frac{\text{mass}}{\text{volume}} = \frac{\text{g}}{\text{L}}$$

Because the volume of a gas depends on temperature and pressure, both should be given when stating the density of a gas. The volume of a solid or liquid is hardly affected by changes in pressure and is changed only slightly when the temperature is varied. Increasing the temperature from 0°C to 50°C will reduce the density of a gas by about 18% if the gas is allowed to expand, whereas a 50°C rise in the temperature of water (0°C ⟶ 50°C) will change its density by less than 0.2%.

The density of a gas at any temperature and pressure can be determined by calculating the mass of gas present in 1 L. At STP, in particular, the density can be calculated by multiplying the molar mass of the gas by 1 mol/22.4 L:

$$d \text{ (at STP)} = \text{molar mass} \times \frac{1 \text{ mol}}{22.4 \text{ L}}$$

$$\text{molar mass} = d \text{ (at STP)} \times \frac{22.4 \text{ L}}{1 \text{ mol}}$$

Table 12.3 lists the densities of some common gases.

TABLE 12.3 Density of Common Gases at STP

Gas	Molar mass (g/mol)	Density (g/L at STP)	Gas	Molar mass (g/mol)	Density (g/L at STP)
H_2	2.016	0.0900	H_2S		
CH_4	16.04	0.716	HCl	34.08	1.52
NH_3	17.03	0.760	F_2	36.45	1.63
C_2H_2	26.04	1.16	CO_2	38.00	1.70
HCN	27.03	1.21	C_3H_8	44.01	1.96
CO	28.01	1.25	O_3	44.09	1.97
N_2	28.02	1.25	SO_2	48.00	2.14
air	**(28.9)**	**(1.29)**	Cl_2	64.06	2.86
O_2	32.00	1.43		70.90	3.17

Physiological Effects of Pressure Changes

The human body has a variety of methods for coping with the changes in atmospheric pressure. As we travel to the mountains, fly in an airplane, or take a high-speed elevator to the top of a skyscraper, the pressure around us decreases. Our ears are sensitive to this because the eardrum (tympanic membrane) has air on both sides of it. The difference in pressure is relieved by yawning or moving the jaw to open the (Eustacian) tubes that connect the middle ear and throat and allow the pressure inside the eardrum to equalize with the outside.

Divers must also contend with the effects of pressure, most notably in body cavities containing air, such as the lungs, ears, and sinuses. Scuba divers do not experience a crushing effect of pressure at increased depths because the tank regulators deliver air at the same pressure as that of the surroundings. The diver must always breathe out regularly while ascending to the surface. Failure to do so may cause the lungs to expand, thus rupturing some of the alveoli, and resulting in loss of consciousness, brain damage, or heart attack. This is a clear application of Boyle's law.

Divers are also affected by consequences of **Henry's law,** which states that the amount of gas that will dissolve in a liquid varies directly with the pressure above the liquid. This means that during a dive the gases entering the lungs are absorbed into the blood to a greater extent than at the water's surface. If the diver returns too rapidly to the surface, the swift pressure reduction may cause dissolved gases to produce bubbles in the blood, resulting in a condition known as *decompression sickness* or "the bends." The only successful method of treatment for this involves the use of a decompression chamber to increase the pressure once again and slowly decompress the diver back to normal pressure.

In the medical field, *hyperbaric units* are used to treat patients who have cells starved for oxygen. In these units the whole room may be placed at high pressure (2 or 3 atm), and the entire staff as well as the patient undergo gradual compression and, following treatment, decompression. These units are widely used to treat carbon monoxide poisoning. Oxygen is dissolved directly into the plasma giving the tissues temporary relief from oxygen deprivation. Hyperbaric units are also effective in treating other problems such as skin grafts, severe thermal burns, and radiation tissue damage.

If a diver returns too quickly to the surface, the pressure reduction may produce bubbles in the blood—a condition known as "the bends."

···

Calculate the density of Cl_2 at STP.

Example 12.14

First calculate the molar mass of Cl_2. It is 70.90 g/mol. Since $d = $ g/L, the conversion is

Solution

$$\frac{g}{mol} \longrightarrow \frac{g}{L}$$

The conversion factor is $\frac{1 \text{ mol}}{22.4 \text{ L}}$:

$$d = \frac{70.90 \text{ g}}{1 \text{ mol}} \times \frac{1 \text{ mol}}{22.1 \text{ L}} = 3.165 \text{ g/L}$$

> **Practice 12.9**
>
> The molar mass of a gas is 20. g/mol. Calculate the density of the gas at STP.

12.14   Ideal Gas Equation

We have used four variables in calculations involving gases: the volume, V; the pressure, P; the absolute temperature, T; and the number of molecules or moles, (abbreviated n). Combining these variables into a single expression, we obtain

$$V \propto \frac{nT}{P} \quad \text{or} \quad V = \frac{nRT}{P}$$

where R is a proportionality constant known as the *ideal gas constant*. The equation is commonly written as

$$PV = nRT$$

ideal gas equation

and is known as the **ideal gas equation**. This equation states in a single expression what we have considered in our earlier discussions: The volume of a gas varies directly with the number of gas molecules and the absolute temperature, and varies inversely with the pressure. The value and units of R depend on the units of P, V, and T. We can calculate one value of R by taking 1 mol of a gas at STP conditions. Solve the equation for R:

$$R = \frac{PV}{nT} = \frac{1 \text{ atm} \times 22.4 \text{ L}}{1 \text{ mol} \times 273 \text{ K}} = 0.0821 \frac{\text{L-atm}}{\text{mol-K}}$$

The units of R in this case are liter-atmospheres (L-atm) per mole Kelvin (mol-K). When the value of $R = 0.0821$ L-atm/mol-K, P is in atmospheres, n is in moles, V is in liters, and T is in Kelvin.

The ideal gas equation can be used to calculate any one of the four variables when the other three are known.

Example 12.15   What pressure will be exerted by 0.400 mol of a gas in a 5.00-L container at 17.0°C?

Solution   **Step 1** Organize the given information, putting temperatures in Kelvin:

$$P = ?$$

$$V = 5.00 \text{ L}$$

$$T = 290. \text{ K}$$

$$n = 0.400 \text{ mol}$$

Step 2 Write and solve equation for the unknown:

$$PV = nRT \quad \text{or} \quad P = \frac{nRT}{V}$$

Step 3 Put given information into the equation and calculate:

$$P = \frac{0.400 \text{ mol} \times 0.0821 \text{ L-atm/mol-K} \times 290. \text{ K}}{5.00 \text{ L}} = 1.90 \text{ atm}$$

How many moles of oxygen gas are in a 50.0-L tank at 22.0°C if the pressure gauge reads 2000. lb/in.²?   **Example 12.16**

Step 1 Organize the given information, putting the temperature in Kelvin and changing pressure to atmospheres: **Solution**

$$P = \frac{2000. \text{ lb}}{\text{in.}^2} \times \frac{1 \text{ atm}}{14.7 \text{ lb/in.}^2} = 136.1 \text{ atm}$$

$$V = 50.0 \text{ L}$$

$$T = 295 \text{ K}$$

$$n = ?$$

Step 2 Write and solve the equation for the unknown:

$$PV = nRT \qquad \text{or} \qquad n = \frac{PV}{RT}$$

Step 3 Put given information into the equation and calculate:

$$n = \frac{136.1 \text{ atm} \times 50.0 \text{ L}}{(0.0821 \text{ L-atm/mol-K}) \times 295 \text{ K}} = 281 \text{ mol O}_2$$

Practice 12.10

A 23.8-L cylinder contains oxygen gas at 20.0°C and 732 torr. How many moles of oxygen are in the cylinder?

The molar mass of a gaseous substance can be determined using the ideal gas equation. Since molar mass = g/mol, then mol = g/molar mass. Using M for molar mass, we can substitute g/M for n (moles) in the ideal gas equation to get

$$PV = \frac{g}{M} RT \qquad \text{or} \qquad M = \frac{gRT}{PV} \quad \text{(modified ideal gas equation)}$$

This form of the gas equation is most useful in problems containing mass instead of moles.

which will allow us to calculate the molar mass, M, for any substance in the gaseous state.

Calculate the molar mass of butane gas, if 3.69 g occupy 1.53 L at 20.0°C and 1.00 atm.   **Example 12.17**

Change 20°C to 293 K and substitute the data into the modified ideal gas equation: **Solution**

$$M = \frac{gRT}{PV} = \frac{3.69 \text{ g} \times 0.0821 \text{ L-atm/mol-K} \times 293 \text{ K}}{1.00 \text{ atm} \times 1.53 \text{ L}} = 58.0 \text{ g/mol}$$

Practice 12.11

A sample of 0.286 g of a certain gas occupies 50.0 mL at standard temperature and 76.0 cm Hg. Determine the molar mass of the gas.

12.15 Stoichiometry Involving Gases

Mole–Volume and Mass–Volume Calculations

Stoichiometric problems involving gas volumes can be solved by the general mole-ratio method outlined in Chapter 9. The factors 1 mol/22.4 L and 22.4 L/1 mol are used for converting volume to moles and moles to volume, respectively. (See Figure 12.14.) These conversion factors are used under the assumption that the gases are at STP and that they behave as ideal gases. In actual practice, gases are measured at other than STP conditions, and the volumes are converted to STP for stoichiometric calculations.

In a balanced equation, the number preceding the formula of a gaseous substance represents the number of moles or molar volumes (22.4 L at STP) of that substance.

The following are examples of typical problems involving gases and chemical equations.

Example 12.18   What volume of oxygen (at STP) can be formed from 0.500 mol of potassium chlorate?

Solution   **Step 1** Write the balanced equation:

$$2 \text{ KClO}_3 \longrightarrow 2 \text{ KCl} + 3 \text{ O}_2(g)$$

   2 mol   3 mol

Step 2 The starting amount is 0.500 mol $KClO_3$. The conversion is from

   moles $KClO_3 \longrightarrow$ moles $O_2 \longrightarrow$ liters O_2

Step 3 Calculate the moles of O_2, using the mole-ratio method:

$$0.500 \text{ mol KClO}_3 \times \frac{3 \text{ mol O}_2}{2 \text{ mol KClO}_3} = 0.750 \text{ mol O}_2$$

Step 4 Convert moles of O_2 to liters of O_2. The moles of a gas at STP are converted to liters by multiplying by the molar volume, 22.4 L/mol:

$$0.750 \text{ mol O}_2 \times \frac{22.4 \text{ L}}{1 \text{ mol}} = 16.8 \text{ L O}_2$$

Setting up a continuous calculation, we obtain

$$0.500 \text{ mol KClO}_3 \times \frac{3 \text{ mol O}_2}{2 \text{ mol KClO}_3} \times \frac{22.4 \text{ L}}{1 \text{ mol}} = 16.8 \text{ L O}_2$$

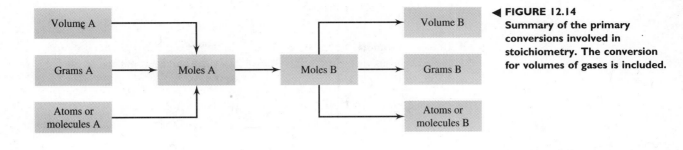

◀ **FIGURE 12.14**
**Summary of the primary
conversions involved in
stoichiometry. The conversion
for volumes of gases is included.**

How many grams of aluminum must react with sulfuric acid to produce 1.25 L of **Example 12.19**
hydrogen gas at STP?

Step 1 The balanced equation is   **Solution**

$$2\ Al(s) + 3\ H_2SO_4(aq) \longrightarrow Al_2(SO_4)_3(aq) + 3\ H_2(g)$$

$\quad$ 2 mol $\qquad\qquad\qquad\qquad\qquad\qquad\qquad$ 3 mol

Step 2 We first convert liters of H_2 to moles of H_2. Then the familiar stoichiometric calculation from the equation is used. The conversion is

$$L\ H_2 \longrightarrow mol\ H_2 \longrightarrow mol\ Al \longrightarrow g\ Al$$

$$1.25\ \cancel{L\ H_2} \times \frac{1\ \cancel{mol}}{22.4\ \cancel{L}} \times \frac{2\ \cancel{mol\ Al}}{3\ \cancel{mol\ H_2}} \times \frac{26.98\ g\ Al}{1\ \cancel{mol\ Al}} = 1.00\ g\ Al$$

What volume of hydrogen, collected at 30°C and 700. torr, will be formed by **Example 12.20**
reacting 50.0 g of aluminum with hydrochloric acid?

$$2\ Al(s) + 6\ HCl(aq) \longrightarrow 2\ AlCl_3(aq) + 3\ H_2(g)$$

$\quad$ 2 mol $\qquad\qquad\qquad\qquad\qquad\qquad$ 3 mol

In this problem the conditions are not at STP, so we cannot use the method shown   **Solution**
in Example 12.18. Either we need to calculate the volume at STP from the equation
and then convert this volume to the conditions given in the problem, or we can use
the ideal gas equation. Let's use the ideal gas equation.

First calculate the moles of H_2 obtained from 50.0 g of Al. Then, using the
ideal gas equation, calculate the volume of H_2 at the conditions given in the problem.

Step 1 Moles of H_2: The conversion is

$$grams\ Al \longrightarrow moles\ Al \longrightarrow moles\ H_2$$

$$50.0\ \cancel{g\ Al} \times \frac{1\ \cancel{mol\ Al}}{26.98\ \cancel{g\ Al}} \times \frac{3\ mol\ H_2}{2\ \cancel{mol\ Al}} = 2.78\ mol\ H_2$$

Step 2 Liters of H_2: Solve $PV = nRT$ for V and substitute the data into the
equation.
Convert °C to K: 30°C + 273 = 303 K.
Convert torr to atm: 700. $\cancel{torr} \times$ 1 atm/760. $\cancel{torr}$ = 0.921 atm.

$$V = \frac{nRT}{P} = \frac{2.78\ \cancel{mol}\ H_2 \times 0.0821\ L\text{-}\cancel{atm} \times 303\ \cancel{K}}{0.921\ \cancel{atm} \times \cancel{mol\text{-}K}} = 75.1\ L\ H_2$$

Note: The volume at STP is 62.3 L H_2.

Practice 12.12

If 10.0 g of sodium peroxide, Na_2O_2, react with water to produce sodium hydroxide and oxygen, how many liters of oxygen will be produced at 20°C and 750. torr?

$$2\ Na_2O_2(s) + 2\ H_2O(l) \longrightarrow 4\ NaOH(aq) + O_2(g)$$

Volume–Volume Calculations

When all substances in a reaction are in the gaseous state, simplifications in the calculation can be made. These are based on Avogadro's law, which states that gases under identical conditions of temperature and pressure contain the same number of molecules and occupy the same volume. Using this same law, we can also state that, under the same conditions of temperature and pressure, the volumes of gases reacting are proportional to the numbers of moles of the gases in the balanced equation. Consider the reaction:

$$H_2(g) + Cl_2(g) \longrightarrow 2\ HCl(g)$$

1 mol	1 mol	2 mol
22.4 L	22.4 L	2 × 22.4 L
1 volume	1 volume	2 volumes
Y volume	Y volume	2 Y volumes

In this reaction 22.4 L of hydrogen will react with 22.4 L of chlorine to give $2 \times 22.4 = 44.8$ L of hydrogen chloride gas. This statement is true because these volumes are equivalent to the number of reacting moles in the equation. Therefore, Y volume of H_2 will combine with Y volume of Cl_2 to give 2 Y volumes of HCl. For example, 100 L of H_2 react with 100 L of Cl_2 to give 200 L of HCl; if the 100 L of H_2 and of Cl_2 are at 50°C, they will give 200 L of HCl at 50°C. When the temperature and pressure before and after a reaction are the same, volumes can be calculated without changing the volumes to STP.

> **For reacting gases at constant temperature and pressure: Volume–volume relationships are the same as mole–mole relationships.**

Example 12.21   What volume of oxygen will react with 150 L of hydrogen to form water vapor? What volume of water vapor will be formed?

Solution   Assume that both reactants and products are measured at the same conditions. Calculate by using reacting volumes:

$$2\ H_2(g) + O_2(g) \longrightarrow 2\ H_2O(g)$$

2 mol	1 mol	2 mol
2 × 22.4 L	22.4 L	2 × 22.4 L
2 volumes	1 volume	2 volumes
150 L	75 L	150 L

For every two volumes of H_2 that react, one volume of O_2 reacts and two volumes of $H_2O(g)$ are produced:

$$150 \text{ L } H_2 \times \frac{1 \text{ volume } O_2}{2 \text{ volumes } H_2} = 75 \text{ L } O_2$$

$$150 \text{ L } H_2 \times \frac{2 \text{ volume } H_2O}{2 \text{ volume } H_2} = 150 \text{ L } H_2O$$

The equation for the preparation of ammonia is

Example 12.22

$$3H_2(g) + N_2(g) \xrightarrow{400°C} 2 \text{ NH}_3(g)$$

Assuming that the reaction goes to completion.

(a) What volume of H_2 will react with 50.0 L of N_2?
(b) What volume of NH_3 will be formed from 50.0 L of N_2?
(c) What volume of N_2 will react with 100. mL of H_2?
(d) What volume of NH_3 will be produced from 100. mL of H_2?
(e) If 600. mL of H_2 and 400. mL of N_2 are sealed in a flask and allowed to react, what amounts of H_2, N_2, and NH_3 are in the flask at the end of the reaction?

The answers to parts (a)–(d) are shown in the boxes and can be determined from the equation by inspection, using the principle of reacting volumes:

Solution

$$3 \text{ H}_2(g) \quad + \quad N_2(g) \quad \longrightarrow \quad 2 \text{ NH}_3(g)$$

3 volumes   1 volume   2 volumes

(a) $\boxed{150. \text{ L}}$ 50.0 L
(b)   50.0 L $\boxed{100. \text{ L}}$
(c) 100. mL $\boxed{33.3 \text{ mL}}$
(d) 100. mL $\boxed{66.7 \text{ mL}}$

(e) Volume ratio from the equation $= \dfrac{3 \text{ volumes } H_2}{1 \text{ volume } N_2}$

Volume ratio used $= \dfrac{600. \text{ mL } H_2}{400. \text{ mL } N_2} = \dfrac{3 \text{ volumes } H_2}{2 \text{ volumes } N_2}$

Comparing these two ratios, we see that an excess of N_2 is present in the gas mixture. Therefore, the reactant limiting the amount of NH_3 that can be formed is H_2:

$$3 \text{ H}_2(g) \quad + \quad N_2(g) \quad \longrightarrow \quad 2 \text{ NH}_3(g)$$

600 mL   200 mL   400 mL

To have a 3:1 ratio of volumes reacting, 600 mL of H_2 will react with 200 mL of N_2 to produce 400 mL of NH_3, leaving 200 mL of N_2 unreacted. At the end of the reaction the flask will contain 400 mL of NH_3 and 200 mL of N_2.

Practice 12.13

What volume of oxygen will react with 15.0 L of propane (C_3H_8) to form carbon dioxide and water? How much carbon dioxide will be formed? How much water?

$$C_3H_8(g) + 5 \text{ O}_2(g) \longrightarrow 3 \text{ CO}_2(g) + 4 \text{ H}_2O(g)$$

12.16 Real Gases

All the gas laws are based on the behavior of an ideal gas—that is, a gas with a behavior that is described exactly by the gas laws for all possible values of P, V, and T. Most real gases actually do behave very nearly as predicted by the gas laws over a fairly wide range of temperatures and pressures. However, when conditions are such that the gas molecules are crowded closely together (high pressure and/or low temperature), they show marked deviations from ideal behavior. Deviations occur because molecules have finite volumes and also have intermolecular attractions, which result in less compressibility at high pressures and greater compressibility at low temperatures than predicted by the gas laws. Many gases become liquids at high pressure and low temperature.

12.17 Air Pollution

Chemical reactions occur among the gases that are emitted into our atmosphere. In recent years, there has been growing concern over the effects these reactions have on our environment and our lives.

The outer portion (stratosphere) of the atmosphere plays a significant role in determining the conditions for life at the surface of the earth. This stratosphere protects the surface from the intense radiation and particles bombarding our planet. Some of the high-energy radiation from the sun acts upon oxygen molecules in the stratosphere, converting them into ozone, O_3. Different molecular forms of an element are called **allotropes** of that element. Thus oxygen and ozone are allotropic forms of oxygen:

allotrope

$$O_2 \xrightarrow{\text{sunlight}} O + O$$
$$\text{oxygen atoms}$$

$$O_2 + O \longrightarrow O_3$$

Ultraviolet radiation from the sun is highly damaging to living tissues of plants and animals. The ozone layer, however, shields the earth by absorbing ultraviolet radiation and thus prevents most of this lethal radiation from reaching the earth's surface. The reaction that occurs is the reverse of the preceding one:

$$O_3 \xrightarrow[\text{radiation}]{\text{ultraviolet}} O_2 + O + \text{heat}$$

Scientists have become concerned about a growing hazard to the ozone layer. Chlorofluorocarbon propellants, such as the Freons, CCl_3F and CCl_2F_2, which were used in aerosol spray cans and are used in refrigeration and air-conditioning units, are stable compounds and remain unchanged in the lower atmosphere. But when these chlorofluorocarbons are carried by convection currents to the stratosphere, they absorb ultraviolet radiation and produce chlorine atoms (chlorine free radicals), which in turn react with ozone. The following reaction sequence involving free

A free radical is a species containing an odd number of electrons. Free radicals are highly reactive.

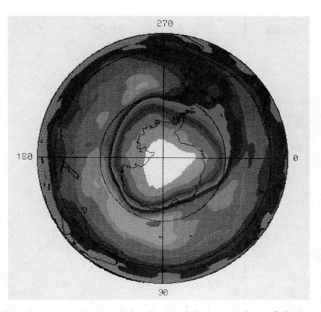

◀ **FIGURE 12.15**
Satellite map showing a severe depletion or "hole" in the ozone layer over Antarctica in October, 1990. The hole is believed to be due to pollution of the atmosphere by chlorofluorocarbons used in aerosols and refrigerants.

radicals has been proposed to explain the partial destruction of the ozone layer by chlorofluorocarbons.

$$CCl_3F \xrightarrow[\text{radiation}]{\text{ultraviolet}} \cdot CCl_2F \ + \ Cl\cdot \tag{1}$$

fluorocarbon   fluorocarbon   chlorine free
molecule   free radical   radical (atom)

$$Cl\cdot \ + \ O_3 \longrightarrow ClO\cdot \ + \ O_2 \tag{2}$$

$$ClO\cdot \ + \ O \longrightarrow O_2 \ + \ Cl\cdot \tag{3}$$

Because a chlorine atom is generated for each ozone molecule that is destroyed (reactions 2 and 3 can proceed repeatedly), a single chlorofluorocarbon molecule can be responsible for the destruction of many ozone molecules. During the past decade, scientists have discovered an annual thinning in the ozone layer over Antarctica. This is what we call the "hole" in the ozone layer. If this hole were to occur over populated regions of the world, severe effects would result, including a rise in the cancer rate, increased climatic temperatures, and vision problems. See Figure 12.15.

Ozone can be prepared by passing air or oxygen through an electrical discharge:

$$3\,O_2(g) \ + \ 286\,kJ \xrightarrow[\text{discharge}]{\text{electrical}} 2\,O_3(g)$$

The characteristic pungent odor of ozone is noticeable in the vicinity of electrical machines and power transmission lines. Ozone is formed in the atmosphere during electrical storms and by the photochemical action of ultraviolet radiation on a mixture of nitrogen dioxide and oxygen. Areas with high air pollution are subject to high atmospheric ozone concentrations.

Ozone is not a desirable low-altitude constituent of the atmosphere because it is known to cause extensive plant damage, cracking of rubber, and the formation of eye-irritating substances. Concentrations of ozone greater than 0.1 part per million (ppm) of air cause coughing, choking, headache, fatigue, and reduced resistance to respiratory infection. Concentrations between 10 and 20 ppm are fatal to humans.

A smoggy day in Los Angeles, ▶ California. High concentrations of ozone near the surface of the earth can cause plant damage, headaches, and deterioration of rubber, among other things.

In addition to ozone, the air in urban areas contains nitrogen oxides, which are components of smog. The term *smog* refers to air pollution in urban environments. Often the chemical reactions occur as part of a *photochemical process*. Nitrogen monoxide (NO) is oxidized in the air or in automobile engines to produce nitrogen dioxide (NO_2). In the presence of light,

$$NO_2 \xrightarrow{\text{light}} NO + O$$

In addition to nitrogen oxides, combustion of fossil fuels releases CO_2, CO, and sulfur oxides. Incomplete combustion releases unburned and partially burned hydrocarbons.

Society is continually attempting to discover, understand, and control emissions that contribute to this sort of atmospheric chemistry. It is a problem that each one of us faces as we look to the future if we want to continue to support life as we know it on our planet.

Concepts in Review

1. State the principal assumptions of the Kinetic-Molecular Theory.
2. Estimate the relative rates of effusion of two gases of known molar mass.
3. Sketch and explain the operation of a mercury barometer.
4. List two factors that determine gas pressure in a vessel of fixed volume.
5. State Boyle's, Charles' and Gay-Lussac's laws. Use all of them in problems.
6. State the combined gas law. Indicate when it is used.
7. Use Dalton's law of partial pressures and the combined gas law to determine the dry STP volume of a gas collected over water.

8. State Avogadro's law.

9. Understand the mole–mass–volume relationship of gases.

10. Determine the density of any gas at STP.

11. Determine the molar mass of a gas from its density at a known temperature and pressure.

12. Solve problems involving the ideal gas equation.

13. Make mole–volume, mass–volume, and volume–volume stoichiometric calculations from balanced chemical equations.

14. State two reasons why real gases may deviate from the behavior predicted for an ideal gas.

Key Terms

The terms listed here have been defined within this chapter. Section numbers are referenced in parenthesis for each term. More detailed definitions are given in the Glossary.

absolute zero (12.6)
allotrope (12.17)
1 (one) atmosphere (12.3)
atmospheric pressure (12.3)
Avogadro's law (12.11)
barometer (12.3)
Boyle's law (12.5)
Charles' law (12.6)
Dalton's law of partial pressures (12.10)
diffusion (12.2)
effusion (12.2)
Gay Lussac's law (12.7)

Gay Lussac's law of combining volumes (12.11)
Graham's law of effusion (12.2)
Henry's law (CIA)
ideal gas (12.2)
ideal gas equation (12.14)
Kinetic-Molecular Theory (KMT) (12.2)
molar volume (12.12)
partial pressure (12.10)
pressure (12.3)
standard conditions (12.8)
standard temperature and pressure (STP) (12.8)

Questions

Questions refer to tables, figures, and key words and concepts defined within the chapter. A particularly challenging question or exercise is indicated with an asterisk.

1. What evidence is used to show diffusion in Figure 12.1? If H_2 and O_2 were in the two flasks, how could you prove that diffusion had taken place?

2. How does the air pressure inside the balloon shown in Figure 12.2 compare with the air pressure outside the balloon? Explain.

3. According to Table 12.1, what two gases are the major constituents of dry air?

4. How does the pressure represented by 1 torr compare in magnitude to the pressure represented by 1 mm Hg? See Table 12.2.

5. In which container illustrated in Figure 12.5 are the molecules of gas moving faster? Assume both gases to be hydrogen.

6. In Figure 12.6, what gas pressure corresponds to a volume of 4 L?

7. How do the data illustrated in Figure 12.6 substantiate Boyle's law?

8. What effect would you observe in Figure 12.9 if T_2 were lower than T_1?

9. In the diagram shown in Figure 12.11, is the pressure of the oxygen plus water vapor inside the bottle equal to, greater than, or less than the atmospheric pressure outside the bottle? Explain.

10. List five gases in Table 12.3 that are more dense than air. Explain the basis for your selection.

11. What are the basic assumptions of the Kinetic-Molecular Theory?

12. Arrange the following gases, all at standard temperature, in order of increasing relative molecular velocities: H_2, CH_4, Rn, N_2, F_2, He. What is your basis for determining the order?

13. List, in descending order, the average kinetic energies of the molecules in Question 12.

14. What are the four parameters used to describe the behavior of a gas?

15. What are the characteristics of an ideal gas?

16. Under what condition of temperature, high or low, is a gas least likely to exhibit ideal behavior? Explain.

17. Under what condition of pressure, high or low, is a gas least likely to exhibit ideal behavior? Explain.

18. Compare, at the same temperature and pressure, equal volumes of H_2 and O_2 as to
    (a) number of molecules
    (b) mass
    (c) number of moles
    (d) average kinetic energy of the molecules
    (e) rate of effusion
    (f) density

19. How does the Kinetic-Molecular Theory account for the behavior of gases as described by
    (a) Boyle's law?
    (b) Charles' law?
    (c) Dalton's law of partial pressures?

20. Explain how the reaction

    $N_2(g) + O_2(g) \xrightarrow{\Delta} 2\ NO(g)$ proves that nitrogen and oxygen are diatomic molecules.

21. What is the reason for referring gases to STP?

22. Is the conversion of oxygen to ozone an exothermic or endothermic reaction? How do you know?

23. Write formulas for an oxygen atom, an oxygen molecule, and an ozone molecule. How many electrons are in an oxygen molecule?

24. When constant pressure is maintained, what effect does heating a mole of N_2 gas have on
    (a) its density?
    (b) its mass?
    (c) the average kinetic energy of its molecules?
    (d) the average velocity of its molecules?
    (e) the number of N_2 molecules in the sample?

25. State Henry's law and indicate its significance to a scuba diver.

26. Assuming ideal gas behavior, which of the following statements are correct? (Try to answer without referring to your text.) Rewrite each incorrect statement to make it correct.
    (a) The pressure exerted by a gas at constant volume is independent of the temperature of the gas.
    (b) At constant temperature, increasing the pressure exerted on a gas sample will cause a decrease in the volume of the gas sample.
    (c) At constant pressure, the volume of a gas is inversely proportional to the absolute temperature.
    (d) At constant temperature, doubling the pressure on a gas sample will cause the volume of the gas sample to decrease to one-half its original volume.
    (e) Compressing a gas at constant temperature will cause its density and mass to increase.
    (f) Equal volumes of CO_2 and CH_4 gases at the same temperature and pressure contain
        (1) the same number of molecules
        (2) the same mass
        (3) the same densities
        (4) the same number of moles
        (5) the same number of atoms
    (g) At constant temperature, the average kinetic energy of O_2 molecules at 200 atm pressure is greater than the average kinetic energy of O_2 molecules at 100 atm pressure.
    (h) According to Charles' law, the volume of a gas becomes zero at $-273°C$.
    (i) One liter of O_2 gas at STP has the same mass as 1 L of O_2 gas at $273°C$ and 2 atm pressure.
    (j) The volume occupied by a gas depends only on its temperature and pressure.
    (k) In a mixture containing O_2 molecules and N_2 molecules, the O_2 molecules, on the average, are moving faster than the N_2 molecules.
    (l) $PV = k$ is a statement of Charles' law.
    (m) If the temperature of a sample of gas is increased from $25°C$ to $50°C$, the volume of the gas will increase by 100%.
    (n) One mole of chlorine, Cl_2, at $20°C$ and 600 torr pressure contains 6.022×10^{23} molecules.
    (o) One mole of H_2 plus 1 mol of O_2 in an 11.2-L container exert a pressure of 4 atm at $0°C$.
    (p) When the pressure on a sample of gas is halved, with the temperature remaining constant, the density of the gas is also halved.
    (q) When the temperature of a sample of gas is increased at constant pressure, the density of the gas will decrease.
    (r) According to the equation

    $2\ KClO_3(s) \xrightarrow{\Delta} 2\ KCl(s) + 3\ O_2(g)$

    1 mol of $KClO_3$ will produce 67.2 L of O_2 at STP.
    (s) $PV = nRT$ is a statement of Avogadro's law.
    (t) STP conditions are 1 atm and $0°C$.

Paired Exercises

These exercises are paired. Each odd-numbered exercise is followed by a similar even-numbered exercise. Answers to the even-numbered exercises are given in Appendix V.

Pressure Units

27. The barometer reads 715 mm Hg. Calculate the corresponding pressure in
(a) atmospheres
(b) inches of Hg
(c) lb/in.^{2}

28. The barometer reads 715 mm Hg. Calculate the corresponding pressure in
(a) torrs
(b) millibars
(c) kilopascals

29. Express the following pressures in atmospheres:
(a) 28 mm Hg
(b) 6000. cm Hg
(c) 795 torr
(d) 5.00 kPa

30. Express the following pressures in atmospheres:
(a) 62 mm Hg
(b) 4250. cm Hg
(c) 225 torr
(d) 0.67 kPa

Boyle's and Charles' laws

31. A gas occupies a volume of 400. mL at 500. mm Hg pressure. What will be its volume, at constant temperature, if the pressure is changed to (a) 760 mm Hg? (b) 250 torr?

32. A gas occupies a volume of 400. mL at 500. mm Hg pressure. What will be its volume, at constant temperature, if the pressure is changed to (a) 2.00 atm? (b) 325 torr?

33. A 500.-mL sample of a gas is at a pressure of 640. mm Hg. What must be the pressure, at constant temperature, if the volume is changed to 855 mL?

34. A 500.-mL sample of a gas is at a pressure of 640. mm Hg. What must be the pressure, at constant temperature, if the volume is changed to 450. mL?

35. Given 6.00 L of N_2 gas at $-25°C$, what volume will the nitrogen occupy at (a) 0.0°C? (b) 100. K? (Assume constant pressure.)

36. Given 6.00 L of N_2 gas at $-25°C$, what volume will the nitrogen occupy at (a) 0.0°F? (b) 345. K? (Assume constant pressure.)

Combined Gas Laws

37. A gas occupies a volume of 410 mL at 27°C and 740 mm Hg pressure. Calculate the volume the gas would occupy at STP.

38. A gas occupies a volume of 410 mL at 27°C and 740 mm Hg pressure. Calculate the volume the gas would occupy at 250.°C and 680 mm Hg pressure.

39. An expandable balloon contains 1400. L of He at 0.950 atm pressure and 18°C. At an altitude of 22 miles (temperature 2.0°C and pressure 4.0 torr), what will be the volume of the balloon?

40. A gas occupies 22.4 L at 2.50 atm and 27°C. What will be its volume at 1.50 atm and $-5.00°C$?

Dalton's Law of Partial Pressures

41. What would be the partial pressure of N_2 gas collected over water at 20°C and 720. torr pressure? (Check Appendix II for the vapor pressure of water.)

42. What would be the partial pressure of N_2 gas collected over water at 25°C and 705. torr pressure? (Check Appendix II for the vapor pressure of water.)

43. A mixture contains H_2 at 600. torr pressure, N_2 at 200. torr pressure, and O_2 at 300. torr pressure. What is the total pressure of the gases in the system?

44. A mixture contains H_2 at 325. torr pressure, N_2 at 475. torr pressure, and O_2 at 650. torr pressure. What is the total pressure of the gases in the system?

45. A sample of methane gas, CH_4, was collected over water at 25.0°C and 720. torr. The volume of the wet gas is 2.50 L. What will be the volume of the dry methane at standard pressure?

46. A sample of propane gas, C_3H_8, was collected over water at 22.5°C and 745 torr. The volume of the wet gas is 1.25 L. What will be the volume of the dry propane at standard pressure?

Mole–Mass–Volume Relationships

47. What volume will 2.5 mol of Cl_2 occupy at STP?

48. What volume will 1.25 mol of N_2 occupy at STP?

49. How many grams of CO_2 are present in 2500 mL of CO_2 at STP?

50. How many grams of NH_3 are present in 1.75 L of NH_3 at STP?

51. What volume will each of the following occupy at STP?
    (a) 1.0 mol of NO_2
    (b) 17.05 g of NO_2
    (c) 1.20×10^{24} molecules of NO_2

52. What volume will each of the following occupy at STP?
    (a) 0.50 mol of H_2S
    (b) 22.41 g of H_2S
    (c) 8.55×10^{23} molecules of H_2S

53. How many molecules of NH_3 gas are present in a 1.00-L flask of NH_3 gas at STP?

54. How many molecules of CH_4 gas are present in a 1.00-L flask of CH_4 gas at STP?

Density of Gases

55. Calculate the density of the following gases at STP:
    (a) Kr
    (b) SO_3

56. Calculate the density of the following gases at STP:
    (a) He
    (b) C_4H_8

57. Calculate the density of:
    (a) F_2 gas at STP
    (b) F_2 gas at 27°C and 1.00 atm pressure

58. Calculate the density of:
    (a) Cl_2 gas at STP
    (b) Cl_2 gas at 22°C and 0.500 atm pressure

Ideal Gas Equation and Stoichiometry

59. At 27°C and 750 torr pressure, what will be the volume of 2.3 mol of Ne?

60. At 25°C and 725 torr pressure, what will be the volume of 0.75 mol of Kr?

61. What volume will a mixture of 5.00 mol of H_2 and 0.500 mol of CO_2 occupy at STP?

62. What volume will a mixture of 2.50 mol of N_2 and 0.750 mol of HCl occupy at STP?

63. Given the equation:

    $$4\ NH_3(g) + 5\ O_2(g) \longrightarrow 4\ NO(g) + 6\ H_2O(g)$$

    (a) How many moles of NH_3 are required to produce 5.5 mol of NO?
    (b) How many liters of NO can be made from 12 L of O_2 and 10. L of NH_3 at STP?
    (c) At constant temperature and pressure, what is the maximum volume, in liters, of NO that can be made from 3.0 L of NH_3 and 3.0 L of O_2?

64. Given the equation:

    $$4\ NH_3(g) + 5\ O_2(g) \longrightarrow 4\ NO(g) + 6\ H_2O(g)$$

    (a) How many moles of NH_3 will react with 7.0 mol of O_2?
    (b) At constant temperature and pressure, how many liters of NO can be made by the reaction of 800. mL of O_2?
    (c) How many grams of O_2 must react to produce 60. L of NO measured at STP?

65. Given the equation:

    $$4\ FeS(s) + 7\ O_2(g) \xrightarrow{\Delta} 2\ Fe_2O_3(s) + 4\ SO_2(g)$$

    how many liters of O_2, measured at STP, will react with 0.600 kg of FeS?

66. Given the equation:

    $$4\ FeS(s) + 7\ O_2(g) \xrightarrow{\Delta} 2\ Fe_2O_3(s) + 4\ SO_2(g)$$

    how many liters of SO_2, measured at STP, will be produced from 0.600 kg of FeS?

Additional Exercises

These exercises are not paired or labeled by topic and provide additional practice on concepts covered in this chapter.

67. Sketch a graph to show each of the following relationships:
  (a) P vs V at constant temperature and number of moles
  (b) T vs V at constant pressure and number of moles
  (c) T vs P at constant volume and number of moles
  (d) n vs V at constant temperature and pressure

68. Why is it dangerous to incinerate an aerosol can?

69. What volume does 1 mol of an ideal gas occupy at standard conditions?

70. Which of these occupies the greatest volume?
  (a) 0.2 mol of chlorine gas at 48°C and 80 cm Hg
  (b) 4.2 g of ammonia at 0.65 atm and -112°C
  (c) 21 g of sulfur trioxide at room temperature and 110 kPa

71. Which of these has the greatest density?
  (a) SF_6 at STP
  (b) C_2H_6 at room conditions
  (c) He at -80°C and 2.15 atm

72. A chemist carried out a chemical reaction that produced a gas. It was found that the gas contained 80.0% carbon and 20.0% hydrogen. It was also noticed that 1500 mL of the gas at STP had a mass of 2.01 g.
  (a) What is the empirical formula of the compound?
  (b) What is the molecular formula of the compound?
  (c) What Lewis structure fits this compound?

* 73. Three gases were added to the same 2.0-L container. The total pressure of the gases was 790 torr at room temperature (25.0°C). If the mixture contained 0.65 g of oxygen gas, 0.58 g of carbon dioxide, and an unknown amount of nitrogen gas, determine:
  (a) the total number of moles of gas in the container
  (b) the number of grams of nitrogen in the container
  (c) the partial pressure of each gas in the mixture

* 74. When carbon monoxide and oxygen gas react, carbon dioxide results. If 500. mL of O_2 at 1.8 atm and 15°C are mixed with 500. mL of CO at 800 mm Hg and 60°C, how many milliliters of CO_2 at STP could possibly result?

75. One of the methods for estimating the temperature at the center of the sun is based on the ideal gas equation. If the center is assumed to be a mixture of gases whose average molar mass is 2.0 g/mol, and if the density and pressure are 1.4 g/cm³ and 1.3×10^9 atm, respectively, find the temperature.

76. A soccer ball of constant volume 2.24 L is pumped up with air to a gauge pressure of 13 lb/in.² at 20.0°C. The molar mass of air is about 29 g/mol.
  (a) How many moles of air are in the ball?
  (b) What mass of air is in the ball?
  (c) During the game, the temperature rises to 30.0°C. What mass of air must be allowed to escape to bring the gauge pressure back to its original value?

77. A balloon will burst at a volume of 2.00 L. If it is partially filled at a temperature of 20.0°C and a pressure of 65 cm Hg to occupy 1.75 L, at what temperature will it burst if the pressure is exactly 1 atm at the time that it breaks?

78. At constant temperature, what pressure would be required to compress 2500 L of hydrogen gas at 1.0 atm pressure into a 25-L tank?

79. Given a sample of a gas at 27°C, at what temperature would the volume of the gas sample be doubled, the pressure remaining constant?

80. A gas sample at 22°C and 740 torr pressure is heated until its volume is doubled. What pressure would restore the sample to its original volume?

81. A gas occupies 250 mL at 700. torr and 22°C. When the pressure is changed to 500. torr, what temperature (°C) is needed to maintain the same volume?

82. Hydrogen stored in a metal cylinder has a pressure of 252 atm at 25°C. What will be the pressure in the cylinder when the cylinder is lowered into liquid nitrogen at -196°C?

83. The tires on an automobile were filled with air to 30. psi at 71.0°F. When driving at high speeds, the tires become hot. If the tires have a bursting pressure of 44 psi, at what temperature (°F) will the tires "blow out"?

84. What volume would 5.30 L of H_2 gas at STP occupy at 70°C and 830 torr pressure?

85. What pressure will 800. mL of a gas at STP exert when its volume is 250. mL at 30°C?

86. How many gas molecules are present in 600. mL of N_2O at 40°C and 400. torr pressure? How many atoms are present? What would be the volume of the sample at STP?

87. 5.00 L of CO_2 at 500. torr and 3.00 L of CH_4 at 400. torr are put into a 10.0-L container. What is the pressure exerted by the gases in the container?

88. A steel cylinder contains 60.0 mol of H_2 at a pressure of 1500 lb/in.².
  (a) How many moles of H_2 are in the cylinder when the pressure reads 850 lb/in.²?
  (b) How many grams of H_2 were initially in the cylinder?

89. At STP, 560. mL of a gas have a mass of 1.08 g. What is the molar mass of the gas?

*90. How many moles of Cl_2 are in one cubic meter ($1.00 \, m^3$) of Cl_2 gas at STP?

91. A gas has a density at STP of 1.78 g/L. What is its molar mass?

*92. At what temperature (°C) will the density of methane, CH_4, be 1.0 g/L at 1.0 atm pressure?

93. Using the ideal gas equation, $PV = nRT$, calculate:
  (a) the volume of 0.510 mol of H_2 at 47°C and 1.6 atm pressure
  (b) the number of grams in 16.0 L of CH_4 at 27°C and 600. torr pressure
  *(c) the density of CO_2 at 4.00 atm pressure and $-20.0°C$
  *(d) the molar mass of a gas having a density of 2.58 g/L at 27°C and 1.00 atm pressure.

94. What is the molar mass of a gas if 1.15 g occupy 0.215 L at 0.813 atm and 30.0°C?

95. What is the Kelvin temperature of a system in which 4.50 mol of a gas occupies 0.250 L at 4.15 atm?

96. How many moles of N_2 gas occupy 5.20 L at 250 K and 0.500 atm?

97. What volume of hydrogen at STP can be produced by reacting 8.30 mol of Al with sulfuric acid? The equation is

$$2 \, Al(s) + 3 \, H_2SO_4(aq) \longrightarrow Al_2(SO_4)_3(aq) + 3 \, H_2(g)$$

*98. Acetylene, C_2H_2, and hydrogen fluoride, HF, react to give difluoroethane.

$$C_2H_2(g) + 2 \, HF(g) \longrightarrow C_2H_4F_2(g)$$

When 1.0 mol of C_2H_2 and 5.0 mol of HF are reacted in a 10.0-L flask, what will be the pressure in the flask at 0°C when the reaction is complete?

99. What are the relative rates of effusion of N_2 and He?

*100. **(a)** What are the relative rates of effusion of CH_4 and He?
  (b) If these two gases are simultaneously introduced into opposite ends of a 100.-cm tube and allowed to diffuse toward each other, at what distance from the helium end will molecules of the two gases meet?

*101. A gas has a percent composition by mass of 85.7% carbon and 14.3% hydrogen. At STP the density of the gas is 2.50 g/L. What is the molecular formula of the gas?

*102. Assume that the reaction

$$2 \, CO(g) + O_2(g) \longrightarrow 2 \, CO_2(g)$$

goes to completion. When 10. mol of CO and 8.0 mol of O_2 react in a closed 10.-L vessel,
  (a) how many moles of CO, O_2, and CO_2 are present at the end of the reaction?
  (b) what will be the total pressure in the flask at 0°C?

*103. 250 mL of O_2, measured at STP, were obtained by the decomposition of the $KClO_3$ in a 1.20-g mixture of KCl and $KClO_3$:

$$2 \, KClO_3(s) \longrightarrow 2 \, KCl(s) + 3 \, O_2(g)$$

What is the percent by mass of $KClO_3$ in the mixture?

*104. Look at the apparatus below. When a small amount of water is squirted into the flask containing ammonia gas (by squeezing the bulb of the medicine dropper), water from the beaker fills the flask through the long glass tubing. Explain this phenomenon. (Remember that ammonia dissolves in water.)

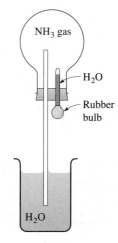

*105. Determine the pressure of the gas in each of the figures below:

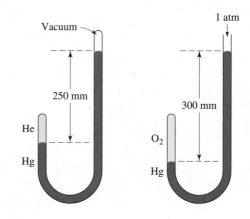

*106. Consider the arrangement of gases shown below. If the valve between the gases is opened and the temperature is held constant:
(a) determine the pressure of each gas
(b) determine the total pressure in the system

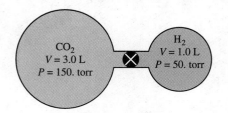

*107. Air has a density of 1.29 g/L at STP. Calculate the density of air on Pikes Peak, where the pressure is 450 torr and the temperature is 17°C.

*108. A room is 16 ft × 12 ft × 12 ft. Would air enter or leave the room and how much, if the temperature in the room changed from 27°C to −3°C with the pressure remaining constant? Show evidence.

*109. A steel cylinder contained 50.0 L of oxygen gas under a pressure of 40.0 atm and a temperature of 25°C. What was the pressure in the cylinder during a storeroom fire that caused the temperature to rise 152°C? (Be careful!)

Answers to Practice Exercises

12.1 (a) 851 torr, (b) 851 mm Hg
12.2 0.596 atm
12.3 4.92 L
12.4 762 torr
12.5 84.7 L
12.6 861 K (588°C)
12.7 518 mL

12.8 1.4×10^2 g/mol
12.9 0.89 g/L
12.10 0.953 mol
12.11 128 g/mol
12.12 1.56 L O_2
12.13 75.0 L O_2, 45.0 L CO_2, 60.0 L H_2O

Water and the Properties of Liquids

Planet earth, that magnificent blue sphere we enjoy viewing from the space shuttle, is spectacular. Over 75% of the earth is covered with water. We are born from it, drink it, bathe in it, cook with it, enjoy its beauty in waterfalls and rainbows, and stand in awe of the majesty of icebergs. Water supports and enhances life.

In chemistry, water provides the medium for numerous reactions. The shape of the water molecule is the basis for hydrogen bonds. These bonds determine the unique properties and reactions of water. The tiny water molecule holds the answers to many of the mysteries of chemical reactions.

13.1 Liquids and Solids

In the last chapter, we found that a gas can be a substance containing particles that are far apart, in rapid random motion, and independent of each other. The Kinetic-Molecular Theory, along with the ideal gas equation, summarizes the behavior of most gases at relatively high temperatures and low pressures.

Solids are obviously very different from gases. A solid contains particles that are very close together, has a high density, compresses negligibly, and maintains its shape regardless of container. These characteristics indicate large attractive forces between particles. The model for solids is very different from the one for gases.

Liquids, on the other hand, lie somewhere in between the extremes of gases and solids. A liquid contains particles that are close together, is essentially incompressible, and has a definite volume. These properties are very similar to solids. But, a liquid also takes the shape of its container, which is closer to the model of a gas.

Although liquids and solids show similar properties, they differ tremendously from gases. No simple mathematical relationship, like the ideal gas equation, works well for liquids or solids. Instead, these models are directly related to the forces of attraction between molecules. With these general statements in mind, let us consider some of the specific properties of liquids.

13.2 Evaporation

When beakers of water, ethyl ether, and ethyl alcohol are allowed to stand uncovered in an open room, their volumes gradually decrease. The process by which this change takes place is called *evaporation.*

Attractive forces exist between molecules in the liquid state. Not all of these molecules, however, have the same kinetic energy. Molecules that have greater than average kinetic energy can overcome the attractive forces and break away from the surface of the liquid to become a gas. **Evaporation** or **vaporization** is the escape of molecules from the liquid state to the gas or vapor state.

In evaporation, molecules of higher-than-average kinetic energy escape from a liquid, leaving it cooler than it was before they escaped. For this reason, evaporation of perspiration is one way the human body cools itself and keeps its temperature

evaporation
vaporization

◀ **Chapter Opening Photo: Water on planet earth as viewed from an Apollo spacecraft.**

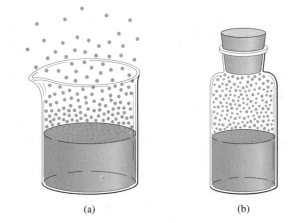

(a)   (b)

constant. When volatile liquids such as ethyl chloride, C_2H_5Cl, are sprayed on the skin, they evaporate rapidly, cooling the area by removing heat. The numbing effect of the low temperature produced by evaporation of ethyl chloride allows it to be used as a local anesthetic for minor surgery.

Solids such as iodine, camphor, naphthalene (moth balls), and, to a small extent, even ice will go directly from the solid to the gaseous state, bypassing the liquid state. This change is a form of evaporation and is called **sublimation**:

sublimation

$$\text{liquid} \xrightarrow{\text{evaporation}} \text{vapor}$$

$$\text{solid} \xrightarrow{\text{sublimation}} \text{vapor}$$

13.3   Vapor Pressure

When a liquid vaporizes in a closed system as shown in Figure 13.1, part (b), some of the molecules in the vapor or gaseous state strike the surface and return to the liquid state by the process of **condensation**. The rate of condensation increases until it is equal to the rate of vaporization. At this point, the space above the liquid is said to be saturated with vapor, and an equilibrium, or steady state, exists between the liquid and the vapor. The equilibrium equation is

condensation

$$\text{liquid} \underset{\text{condensation}}{\overset{\text{vaporization}}{\rightleftarrows}} \text{vapor}$$

This equilibrium is dynamic; both processes—vaporization and condensation—are taking place, even though one cannot see or measure a change. The number of molecules leaving the liquid in a given time interval is equal to the number of molecules returning to the liquid.

At equilibrium the molecules in the vapor exert a pressure like any other gas. The pressure exerted by a vapor in equilibrium with its liquid is known as the **vapor pressure** of the liquid. The vapor pressure may be thought of as a measure of the "escaping" tendency of molecules to go from the liquid to the vapor state. The

vapor pressure

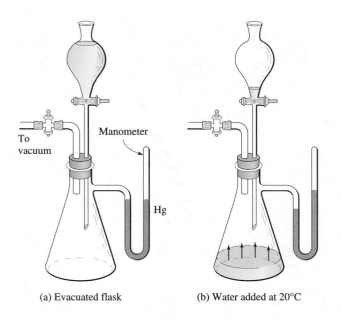
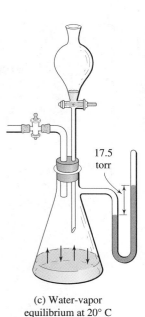
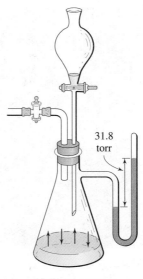

(a) Evacuated flask

(b) Water added at 20°C

(c) Water-vapor
equilibrium at 20° C

(d) Water-vapor
equilibrium at 30° C

vapor pressure of a liquid is independent of the amount of liquid and vapor present, but it increases as the temperature rises. Figure 13.2 illustrates a liquid–vapor equilibrium and the measurement of vapor pressure.

When equal volumes of water, ethyl ether, and ethyl alcohol are placed in separate beakers and allowed to evaporate at the same temperature, we observe that the ether evaporates faster than the alcohol, which evaporates faster than the water. This order of evaporation is consistent with the fact that ether has a higher vapor pressure at any particular temperature than ethyl alcohol or water. One reason for this higher vapor pressure is that the attraction is less between ether molecules than between alcohol molecules or between water molecules. The vapor pressures of these three compounds at various temperatures are compared in Table 13.1.

Substances that evaporate readily are said to be **volatile**. A volatile liquid has a relatively high vapor pressure at room temperature. Ethyl ether is a very volatile liquid, water is not too volatile, and mercury, which has a vapor pressure of 0.0012 torr at 20°C, is essentially a nonvolatile liquid. Most substances that are normally in a solid state are nonvolatile (solids that sublime are exceptions).

▲
FIGURE 13.2
Measurement of the vapor pressure of water at 20°C and 30°C. In flask (a) the system is evacuated. The mercury manometer attached to the flask shows equal pressure in both legs. In (b) water has been added to the flask and begins to evaporate, exerting pressure as indicated by the manometer. In (c), when equilibrium is established, the pressure inside the flask remains constant at 17.5 torr. In (d) the temperature is changed to 30°C, and equilibrium is reestablished with the vapor pressure at 31.8 torr.

volatile

13.4   Surface Tension

Have you ever observed water and mercury in the form of small drops? These liquids occur as drops due to *surface tension* of liquids. A droplet of liquid that is not falling or under the influence of gravity (as on the space shuttle) will form a sphere. Minimum surface area is found in the geometrical form of the sphere. The molecules within the liquid are attracted to the surrounding liquid molecules. However, at the surface of the liquid, the attraction is nearly all inward, pulling the surface into a spherical shape. The resistance of a liquid to an increase in its surface

Water beading on a newly waxed car is an example of surface tension.

TABLE 13.1 The Vapor Pressure of Water, Ethyl Alcohol, and Ethyl Ether at Various Temperatures

Temperature (°C)	Vapor pressure (torr)		
	Water	Ethyl alcohol	Ethyl ether*
0	4.6	12.2	185.3
10	9.2	23.6	291.7
20	17.5	43.9	442.2
30	31.8	78.8	647.3
40	55.3	135.3	921.3
50	92.5	222.2	1276.8
60	152.9	352.7	1729.0
70	233.7	542.5	2296.0
80	355.1	812.6	2993.6
90	525.8	1187.1	3841.0
100	760.0	1693.3	4859.4
110	1074.6	2361.3	6070.1

* Note that the vapor pressure of ethyl ether at temperatures of 40°C and higher exceeds standard pressure, 760 torr, which indicates that the substance has a low boiling point and therefore should be stored in a cool place in a tightly sealed container.

surface tension

area is called the **surface tension** of the liquid. Substances with large attractive forces between molecules have high surface tensions. The effect of surface tension in water is illustrated by the phenomenon of floating a needle on the surface of still water. Other examples include the movement of the water strider insect across a calm pond, or the beading of water on a freshly waxed car.

capillary action

Liquids also exhibit a phenomenon known as **capillary action**, which is the spontaneous rising of a liquid in a narrow tube. This action results from the *cohesive forces* within the liquid, and the *adhesive forces* between the liquid and the walls of the container. If the forces between the liquid and the container are greater than those within the liquid itself, the liquid will climb the walls of the container. For example, consider the California sequoia tree, which can be over 200 feet in height. Under atmospheric pressure, water will only rise 33 feet in a glass tube, but capillary action will cause water to rise from the roots to all parts of the tree.

meniscus

The meniscus in liquids is further evidence of these cohesive and adhesive forces. When a liquid is placed in a glass cylinder the surface of the liquid shows a curve called the **meniscus** (see Figure 13.3). The concave shape of the meniscus of water shows that the adhesive forces between the glass and the liquid are stronger than the cohesive forces within the liquid. In a nonpolar substance such as mercury, the meniscus is convex, indicating that the cohesive forces within the mercury are greater than the adhesive forces between the glass wall and the mercury.

13.5 Boiling Point

The boiling temperature of a liquid is associated with its vapor pressure. We have seen that the vapor pressure increases as the temperature increases. When the internal or vapor pressure of a liquid becomes equal to the external pressure, the liquid boils. (By external pressure we mean the pressure of the atmosphere above the

TABLE 13.2 Physical Properties of Ethyl Chloride, Ethyl Ether, Ethyl Alcohol, and Water

Substance	Boiling point (°C)	Melting point (°C)	Heat of vaporization J/g (cal/g)	Heat of fusion J/g (cal/g)
Ethyl chloride	13	−139	387 (92.5)	—
Ethyl ether	34.6	−116	351 (83.9)	—
Ethyl alcohol	78.4	−112	855 (204.3)	104 (24.9)
Water	100.0	0	2259 (540)	335 (80)

FIGURE 13.3
The meniscus is the characteristic curve of the surface of a liquid in a narrow capillary tube. Here we see the meniscus of mercury (left) and water (right)

liquid.) The boiling temperature of a pure liquid remains constant as long as the external pressure does not vary.

The boiling point (bp) of water is 100°C at 1 atm pressure. Table 13.1 shows that the vapor pressure of water at 100°C is 760 torr, a figure we have seen many times before. The significant fact here is that the boiling point is the temperature at which the vapor pressure of the water or other liquid is equal to standard, or atmospheric, pressure at sea level. These relationships lead to the following definition: The **boiling point** is the temperature at which the vapor pressure of a liquid is equal to the external pressure above the liquid.

boiling point

We can readily see that a liquid has an infinite number of boiling points. When we give the boiling point of a liquid, we should also state the pressure. When we express the boiling point without stating the pressure, we mean it to be the **normal boiling point** at standard pressure (760 torr). Using Table 13.1 again, we see that the normal boiling point of ethyl ether is between 30°C and 40°C, and for ethyl alcohol it is between 70°C and 80°C, because, for each compound, 760 torr pressure lies within these stated temperature ranges. At the normal boiling point, 1 g of a liquid changing to a vapor (gas) absorbs an amount of energy equal to its heat of vaporization (see Table 13.2).

normal boiling point

The boiling point at various pressures may be evaluated by plotting the data of Table 13.2 on the graph in Figure 13.4, where temperature is plotted horizontally along the x axis and vapor pressure is plotted vertically along the y axis. The resulting curves are known as **vapor-pressure curves**. Any point on these curves represents a vapor–liquid equilibrium at a particular temperature and pressure. We may find the boiling point at any pressure by tracing a horizontal line from the designated pressure to a point on the vapor-pressure curve. From this point we draw a vertical line to obtain the boiling point on the temperature axis. Four such points are shown in Figure 13.4; they represent the normal boiling points of the four compounds at 760 torr pressure. By reversing this process, you can ascertain at what pressure a substance will boil at a specific temperature. The boiling point is one of the most commonly used physical properties for characterizing and identifying substances.

vapor-pressure curves

Practice 13.1

Use the graph in Figure 13.4 to determine the boiling points of ethyl chloride, ethyl ether, ethyl alcohol, and water at 600 torr.

FIGURE 13.4 ▶
**Vapor-pressure–temperature
curves for ethyl chloride, ethyl
ether, ethyl alcohol, and water.**

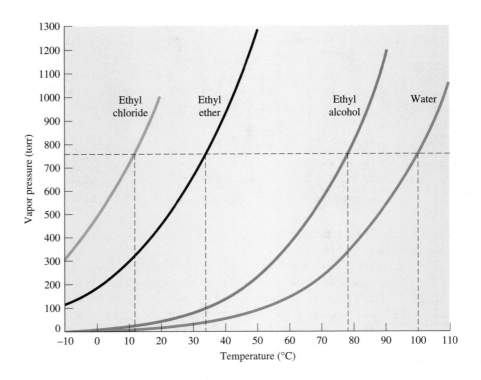

Practice 13.2
────────────
The average atmospheric pressure in Denver is 0.83 atm. What is the boiling point of water in Denver?

13.6   Freezing Point or Melting Point

As heat is removed from a liquid, the liquid becomes colder and colder, until a temperature is reached at which it begins to solidify. A liquid changing into a solid is said to be *freezing,* or *solidifying.* When a solid is heated continuously, a temperature is reached at which the solid begins to liquefy. A solid that is changing into a liquid is said to be *melting.* The temperature at which the solid phase of a substance is in equilibrium with its liquid phase is known as the **freezing point** or **melting point** of that substance. The equilibrium equation is

freezing or melting point

$$\text{solid} \underset{\text{freezing}}{\overset{\text{melting}}{\rightleftarrows}} \text{liquid}$$

When a solid is slowly and carefully heated so that a solid–liquid equilibrium is achieved and then maintained, the temperature will remain constant as long as both phases are present. The energy is used solely to change the solid to the liquid. The melting point is another physical property that is commonly used for characterizing substances.

The most common example of a solid–liquid equilibrium is ice and water. In a well-stirred system of ice and water, the temperature remains at 0°C as long as both phases are present. The melting point changes with pressure but only slightly unless the pressure change is very large.

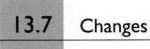

13.7   Changes of State

The majority of solids undergo two changes of state upon heating. A solid changes to a liquid at its melting point, and a liquid changes to a gas at its boiling point. This warming process can be represented by a graph called a *heating curve* (Figure 13.5). This figure shows ice being heated at a constant rate. As energy flows into the ice, the vibrations within the crystal increase and the temperature rises ($A \longrightarrow B$). Eventually, the molecules begin to break free from the crystal and melting occurs ($B \longrightarrow C$). During the melting process all energy goes into breaking down the crystal structure; the temperature remains constant.

The energy required to change 1 g of a solid at its melting point into a liquid is called the **heat of fusion**. When the solid has completely melted, the temperature once again rises ($C \longrightarrow D$); the energy input is increasing the molecular motion within the water. At 100°C, the water reaches its boiling point; the temperature remains constant while the added energy is used to vaporize the water to steam ($D \longrightarrow E$). The **heat of vaporization** is the energy required to change 1 g of liquid to vapor at its normal boiling point. The attractive forces between the liquid molecules are overcome during vaporization. Beyond this temperature, all the water exists as steam and is being heated further ($E \longrightarrow F$).

heat of fusion

heat of vaporization

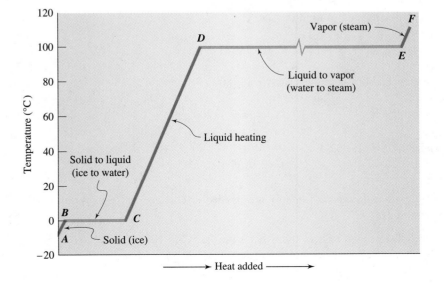

◀ **FIGURE 13.5**
Heating curve for a pure substance—the absorption of heat by a substance from the solid state to the vapor state. Using water as an example, the *AB* interval represents the ice phase; *BC* interval, the melting of ice to water; *CD* interval, the elevation of the temperature of water from 0°C to 100°C; *DE* interval, the boiling of water to steam; and *EF* interval, the heating of steam.

Comet Power

When a comet is near the sun, some of the ice on its surface sublimates into the vacuum of space. This rapidly moving water vapor drags comet dust along with it, creating two of the characteristic features of a comet, its tail and its dusty shroud (usually called a coma). Scientists understand that the production of a tail on a comet traveling near the sun is the result of the rapid conversion of ice to water vapor from the heat of the sun. But why would a comet traveling far from the sun have a tail when there is not enough heat to transform ice to water vapor? This is a mystery. How can these comets produce such celestial wonders?

New observations from the University of Hawaii, Honolulu, suggest that the release of carbon monoxide provides the

spectacle. Astronomers Matthew C. Senay and David Jewitt used two short-wavelength radio telescopes to detect the emissions of gases from Comet Schwassmann-Wachmann 1. This particular comet never gets closer to the sun than Jupiter. Their telescopes revealed that the comet releases a startling amount of carbon monoxide. Carbon monoxide sublimes at a temperature of 25 K. The carbon monoxide coming from the comet has about the same speed as the comet but moves in general toward the sun. Senay and Jewitt suggest that the gas comes from the surface of the comet most illuminated by the sun. The sun sublimes the carbon monoxide at a rate of nearly 2000 kg/sec, producing the tail of gas headed toward the sun.

Example 13.1   How many joules of energy are needed to change 10.0 g of ice at 0.00°C to water at 20.0°C?

Solution   Ice will absorb 335 J/g (heat of fusion) in going from a solid at 0°C to a liquid at 0°C. An additional 4.184 J/g°C (specific heat of water) are needed to raise the temperature of the water for each 1°C.

Joules needed to melt the ice:

$$10.0 \ g \times \frac{335 \ J}{1 \ g} = 3.35 \times 10^3 \ J \ (801 \ cal)$$

Joules needed to heat the water from 0.00°C to 20.0°C:

$$10.0 \ g \times \frac{4.184 \ J}{1 \ g°C} \times 20.0°C = 837 \ J \ (200. \ cal)$$

Thus, 3350 J + 837 J = 4.19×10^3 J (1.00×10^3 cal) are needed.

Example 13.2   How many kilojoules of energy are needed to change 20.0 g of water at 20.°C to steam at 100.°C?

Solution   Kilojoules needed to heat the water from 20.°C to 100.°C:

$$20.0 \ g \times \frac{4.184 \ J}{1 \ g°C} \times \frac{1 \ kJ}{1000 \ J} \times 80.°C = 6.7 \ kJ$$

Kilojoules needed to change water at 100.°C to steam at 100.°C:

$$20.0 \text{ g} \times \frac{2.26 \text{ kJ}}{1 \text{ g}} = 45.2 \text{ kJ}$$

Thus, 6.7 kJ + 45.2 kJ = 51.9 kJ are needed.

Practice 13.3

How many kilojoules of energy are required to change 50.0 g of ethyl alcohol from 60.0°C to vapor at 78.4°C? The specific heat of ethyl alcohol is 2.138 J/g°C.

13.8   Occurrence of Water

Water is our most common natural resource. It covers about 75% of the earth's surface. Not only is it found in the oceans and seas, in lakes, rivers, streams, and in glacial ice deposits, it is also always present in the atmosphere and in cloud formations.

About 97% of the earth's water is in the oceans. This *saline* water contains vast amounts of dissolved minerals. More than 70 elements have been detected in the mineral content of seawater. Only four of these—chlorine, sodium, magnesium, and bromine—are now commercially obtained from the sea. The world's *fresh* water comprises the other 3%, of which about two-thirds is locked up in polar ice caps and glaciers. The remaining fresh water is found in groundwater, lakes, and the atmosphere.

Water is an essential constituent of all living matter. It is the most abundant compound in the human body, making up about 70% of total body mass. About 92% of blood plasma is water; about 80% of muscle tissue is water; and about 60% of a red blood cell is water. Water is more important than food in the sense that a person can survive much longer without food than without water.

13.9   Physical Properties of Water

Water is a colorless, odorless, tasteless liquid with a melting point of 0°C and a boiling point of 100°C at 1 atm. The heat of fusion of water is 335 J/g (80 cal/g). The heat of vaporization of water is 2.26 kJ/g (540 cal/g). The values for water for both the heat of fusion and the heat of vaporization are high compared with those for other substances; these high values indicate that strong attractive forces are acting between the molecules.

Ice and water exist together in equilibrium at 0°C, as shown in Figure 13.6. When ice at 0°C melts, it absorbs 335 J/g (80 cal/g) in changing into a liquid; the temperature remains at 0°C. In order to refreeze the water we have to remove 335 J/g (80 cal/g) from the liquid at 0°C.

In Figure 13.6 both boiling water and steam are shown to have a temperature of 100°C. It takes 418 J (100 cal) to heat 1 g of water from 0°C to 100°C, but water at its boiling point absorbs 2.26 kJ/g (540 cal/g) in changing to steam. Although boiling water and steam are both at the same temperature, steam contains considerably more heat per gram and can cause more severe burns than hot water. In Table 13.3 the physical properties of water are tabulated and compared with those of other hydrogen compounds of Group VIA elements.

The maximum density of water is 1.000 g/mL at 4°C. Water has the unusual property of contracting in volume as it is cooled to 4°C and then expanding when cooled from 4°C to 0°C. Therefore 1 g of water occupies a volume greater than 1 mL at all temperatures except 4°C. Although most liquids contract in volume all the way down to the point at which they solidify, a large increase (about 9%) in volume occurs when water changes from a liquid at 0°C to a solid (ice) at 0°C. The density of ice at 0°C is 0.917 g/mL, which means that ice, being less dense than water, will float in water.

TABLE 13.3 Physical Properties of Water and Other Hydrogen Compounds of Group VIA Elements

Formula	Color	Molar mass (g/mol)	Melting point (°C)	Boiling point, 1 atm (°C)	Heat of fusion J/g (cal/g)	Heat of vaporization J/g (cal/g)
H_2O	Colorless	18.02	0.00	100.0	335 (80.0)	2.26×10^3 (540)
H_2S	Colorless	34.08	−85.5	−60.3	69.9 (16.7)	548 (131)
H_2Se	Colorless	80.98	−65.7	−41.3	31 (7.4)	238 (57.0)
H_2Te	Colorless	129.6	−49	−2	—	179 (42.8)

13.10 Structure of the Water Molecule

A single water molecule consists of two hydrogen atoms and one oxygen atom. Each hydrogen atom is attached to the oxygen atom by a single covalent bond. This bond is formed by the overlap of the $1s$ orbital of hydrogen with an unpaired $2p$ orbital of oxygen. The average distance between the two nuclei is known as the *bond length*. The O—H bond length in water is 0.096 nm. The water molecule is nonlinear and has a bent structure with an angle of about 105 degrees between the two bonds (see Figure 13.7).

Oxygen is the second most electronegative element. As a result, the two covalent OH bonds in water are polar. If the three atoms in a water molecule were aligned in a linear structure, such as H+——⟶O⟵——+H, the two polar bonds would be acting in equal and opposite directions and the molecule would be nonpolar. However, water is a highly polar molecule. Therefore, it does not have a linear structure. When atoms are bonded together in a nonlinear fashion, the angle formed by the bonds is called the *bond angle*. In water, the HOH bond angle is 105°. The two polar covalent bonds and the bent structure result in a partial negative charge on the oxygen atom and a partial positive charge on each hydrogen atom. The polar nature of water is responsible for many of its properties, including its behavior as a solvent.

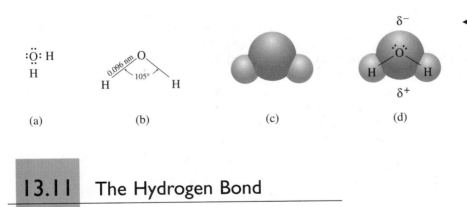

(a)   (b)   (c)   (d)

◀ **FIGURE 13.7**
Diagrams of a water molecule: (a) electron distribution, (b) bond angle and O—H bond length, (c) molecular orbital structure, and (d) dipole representation.

13.11 The Hydrogen Bond

Table 13.3 compares the physical properties of H_2O, H_2S, H_2Se, and H_2Te. From this comparison it is apparent that four physical properties of water—melting point, boiling point, heat of fusion, and heat of vaporization—are extremely high and do not fit the trend relative to the molar masses of the four compounds. If the properties of water followed the progression shown by the other three compounds, we would expect the melting point of water to be below $-85°C$ and the boiling point to be below $-60°C$.

Why does water have these anomalous physical properties? It is because liquid water molecules are held together more strongly than other molecules in the same family. The intermolecular force acting between water molecules is called a **hydrogen bond**, which acts like a very weak bond between two polar molecules. A

hydrogen bond

hydrogen bond is formed between polar molecules that contain hydrogen covalently bonded to a small, highly electronegative atom such as fluorine, oxygen, or nitrogen (F—H, O—H, N—H). A hydrogen bond is actually the dipole–dipole attraction between polar molecules containing these three types of polar bonds.

> **Elements that have significant hydrogen-bonding ability are F, O, and N.**

Because a hydrogen atom has only one electron, it can form only one covalent bond. When it is attached to a strong electronegative atom such as oxygen, a hydrogen atom will also be attracted to an oxygen atom of another molecule, forming a dipole–dipole attraction (H-bond) between the two molecules. Water has two types of bonds: covalent bonds that exist between hydrogen and oxygen atoms within a molecule and hydrogen bonds that exist between hydrogen and oxygen atoms in *different* water molecules.

Hydrogen bonds are *intermolecular* bonds; that is, they are formed between atoms in different molecules. They are somewhat ionic in character because they are formed by electrostatic attraction. Hydrogen bonds are much weaker than the ionic or covalent bonds that unite atoms to form compounds. Despite their weakness, they are of great chemical importance.

The oxygen atom in water can form two hydrogen bonds—one through each of the unbonded pairs of electrons. Figure 13.8 shows (a) two water molecules linked by a hydrogen bond and (b) six water molecules linked by hydrogen bonds. A dash (—) is used for the covalent bond and a dotted line (····) for the hydrogen bond. In water each molecule is linked to others through hydrogen bonds to form a three-dimensional aggregate of water molecules. This intermolecular hydrogen bonding effectively gives water the properties of a much larger, heavier molecule, explaining in part its relatively high melting point, boiling point, heat of fusion, and heat of vaporization. As water is heated and energy is absorbed, hydrogen bonds are continually being broken until at 100°C, with the absorption of an additional 2.26 kJ/g (540 cal/g), water separates into individual molecules, going into the gaseous state. Sulfur, selenium, and tellurium are not sufficiently electronegative for their hydrogen compounds to behave like water. The lack of hydrogen bonding is one reason why H_2S is a gas and not a liquid at room temperature.

FIGURE 13.8 ▶
Hydrogen bonding. Water in the liquid and solid states exists as aggregates in which the water molecules are linked together by hydrogen bonds.

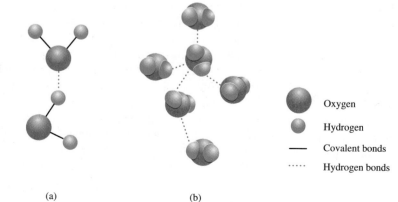

Oxygen

Hydrogen

—— Covalent bonds

····· Hydrogen bonds

(a)   (b)

Fluorine, the most electronegative element, forms the strongest hydrogen bonds. This bonding is strong enough to link hydrogen fluoride molecules together as *dimers*, H_2F_2, or as larger $(HF)_n$ molecular units. The dimer structure may be represented in this way:

Hydrogen bonding can occur between two different atoms that are capable of forming H-bonds. Thus we may have an O····H—N or O—H····N linkage in which the hydrogen atom forming the H-bond is between an oxygen and a nitrogen atom. This form of the H-bond exists in certain types of protein molecules and many biologically active substances.

Would you expect hydrogen bonding to occur between molecules of the following substances?

Example 14.3

(a) ethyl alcohol   (b) dimethyl ether

(a) Hydrogen bonding should occur in ethyl alcohol because one hydrogen atom is bonded to an oxygen atom:

Solution

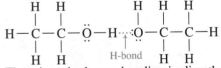

(b) There is no hydrogen bonding in dimethyl ether because all the hydrogen atoms are bonded only to carbon atoms.

Both ethyl alcohol and dimethyl ether have the same molar mass (46.07). Although both compounds have the same molecular formula, C_2H_6O, ethyl alcohol has a much higher boiling point (78.4°C) than dimethyl ether (−23.7°C) because of hydrogen bonding between the alcohol molecules.

Practice 13.4

Would you expect hydrogen bonding to occur between molecules of the following substances?

(a)   (b)   (c)

How Sweet It Is!

Intermolecular forces provide the chemical basis for the sweet taste of many consumer products. The artificial sweetener industry was built on the sweet taste of chemicals discovered quite by accident in the chemistry laboratory. In 1878, Ira Remsen was working late in his laboratory and glanced at the clock to discover he was about to miss a dinner with friends. In his haste to leave the lab he forgot to wash his hands. Later that evening at dinner he broke a piece of bread and tasted it only to discover that it was very sweet. He realized the sweet taste was the result of the chemical he had been working with in the lab and rushed back to the lab where he isolated saccharin—the first of the artificial sweeteners.

In 1937, while working in his laboratory, Michael Sveda was smoking a cigarette (a very dangerous practice to say the least!). He touched the cigarette to his lips and was surprised by the exceedingly sweet taste. When he purified the chemical he had on his hands it was found to be cyclamate, a staple of the artificial sweetener industry for many years.

In 1965, James Schlatter was researching anti-ulcer drugs for the pharmaceutical firm G. D. Searle. In the course of his work, he accidentally ingested a small amount of his preparation and found to his surprise that it had an extremely sweet taste. When purified, the sweet-tasting substance turned out to be aspartame, a molecule consisting of two amino acids joined together. Since only very small quantities of aspartame are necessary to produce sweetness, it proved to be an excellent low-calorie artificial sweetener. Today under the trade names of "Equal" and "Nutrasweet" aspartame is one of the cornerstones of the artificial sweetener industry.

There are more than 50 different molecules that have a sweet taste and all of them have similar molecular shapes. The triangle of sweetness theory, developed

N—H or —OH group
(seeking to H-bond with
O or N on a taste bud)

O or N atom
(seeking to H-bond with
polar H on a taste bud)

Any hydrophobic group
(e.g., CH_3, C_6H_5)

Triangle of sweetness

by Lamont Kier (Massachusetts College of Pharmacy) indicates that these molecules contain three sites that produce the proper structure to attach to the taste bud and trigger the response that registers "sweet" in our brains.

Taste buds are composed of proteins that can form hydrogen bonds with other molecules. The proteins contain —N—H and —OH groups (with hydrogen available to bond) as well as C=O groups (providing oxygen for hydrogen bonding). Molecules that are sweet also contain H-bonding groups including —OH, $—NH_2$, and O or N. The molecules must not only have the proper atoms to form hydrogen bonds, but must also contain a region that is hydrophobic (repels H_2O). The triangle in the diagram shows the three necessary sites. The molecule must contain the three sites located at just the proper distances.

The search for new and better sweeteners is continuing. The perfect sweetener would have the following qualities: (1) be as sweet or sweeter than sucrose (table sugar), (2) be nontoxic, (3) be quick to register sweet on the taste buds, (4) be easy to release so the taste doesn't linger, (5) have no calories, (6) be stable when cooked or dissolved, and (7) of course, be inexpensive. Scientists are continuing to search for different molecules that possess these qualities.

13.12 Formation and Chemical Properties of Water

Water is very stable to heat; it decomposes to the extent of only about 1% at temperatures up to 2000°C. Pure water is a nonconductor of electricity. But when a small amount of sulfuric acid or sodium hydroxide is added, the solution is readily decomposed into hydrogen and oxygen by an electric current. Two volumes of hydrogen are produced for each volume of oxygen:

$$2\ H_2O(l) \xrightarrow[\text{H}_2\text{SO}_4 \text{ or NaOH}]{\text{electrical energy}} 2\ H_2(g)\ +\ O_2(g)$$

Formation

Water is formed when hydrogen burns in air. Pure hydrogen burns very smoothly in air, but mixtures of hydrogen and air or oxygen explode when ignited. The reaction is strongly exothermic:

$$2\ H_2(g)\ +\ O_2(g) \longrightarrow 2\ H_2O(g)\ +\ 484\ \text{kJ}$$

Water is produced by a variety of other reactions, especially by (1) acid–base neutralizations, (2) combustion of hydrogen-containing materials, and (3) metabolic oxidation in living cells:

1. $HCl(aq)\ +\ NaOH(aq) \longrightarrow NaCl(aq)\ +\ H_2O(l)$
2. $2\ C_2H_2(g)\ +\ 5\ O_2(g) \longrightarrow 4\ CO_2(g)\ +\ 2\ H_2O(g)\ +\ 1212\ \text{kJ}$
   acetylene

   $CH_4(g)\ +\ 2\ O_2(g) \longrightarrow CO_2(g)\ +\ 2\ H_2O(g)\ +\ 803\ \text{kJ}$
   methane
3. $C_6H_{12}O_6(aq)\ +\ 6\ O_2(g) \xrightarrow{\text{enzymes}} 6\ CO_2(g)\ +\ 6\ H_2O(l)\ +\ 2519\ \text{kJ}$
   glucose

The combustion of acetylene shown in (2) is strongly exothermic and is capable of producing very high temperatures. It is used in oxygen–acetylene torches to cut and weld steel and other metals. Methane is known as natural gas and is commonly used as fuel for heating and cooking. The reaction of glucose with oxygen shown in (3) is the reverse of photosynthesis. It is the overall reaction by which living cells obtain needed energy by metabolizing glucose to carbon dioxide and water.

Reactions of Water with Metals and Nonmetals

The reactions of metals with water at different temperatures show that these elements vary greatly in their reactivity. Metals such as sodium, potassium, and calcium react with cold water to produce hydrogen and a metal hydroxide. A small piece of sodium added to water melts from the heat produced by the reaction, forming a silvery metal ball, which rapidly flits back and forth on the surface of the water. Caution must be used when experimenting with this reaction, because the hydrogen produced is frequently ignited by the sparking of the sodium, and it will explode, spattering sodium. Potassium reacts even more vigorously than sodium. Calcium sinks in water and liberates a gentle stream of hydrogen. The equations for these reactions are

$$2\,Na(s) + 2\,H_2O(l) \longrightarrow H_2(g) + 2\,NaOH(aq)$$

$$2\,K(s) + 2\,H_2O(l) \longrightarrow H_2(g) + 2\,KOH(aq)$$

$$Ca(s) + 2\,H_2O(l) \longrightarrow H_2(g) + Ca(OH)_2(aq)$$

Zinc, aluminum, and iron do not react with cold water but will react with steam at high temperatures, forming hydrogen and a metallic oxide. The equations are

$$Zn(s) + H_2O(g) \longrightarrow H_2(g) + ZnO(s)$$

$$2\,Al(s) + 3\,H_2O(g) \longrightarrow 3\,H_2(g) + Al_2O_3(s)$$

$$3\,Fe(s) + 4\,H_2O(g) \longrightarrow 4\,H_2(g) + Fe_3O_4(s)$$

Copper, silver, and mercury are examples of metals that do not react with cold water or steam to produce hydrogen. We conclude that sodium, potassium, and calcium are chemically more reactive than zinc, aluminum, and iron, which are more reactive than copper, silver, and mercury.

Certain nonmetals react with water under various conditions. For example, fluorine reacts violently with cold water, producing hydrogen fluoride and free oxygen. The reactions of chlorine and bromine are much milder, producing what is commonly known as "chlorine water" and "bromine water," respectively. Chlorine water contains HCl, HOCl, and dissolved Cl_2; the free chlorine gives it a yellow-green color. Bromine water contains HBr, HOBr, and dissolved Br_2; the free bromine gives it a reddish-brown color. Steam passed over hot coke (carbon) produces a mixture of carbon monoxide and hydrogen that is known as "water gas." Since water gas is combustible, it is useful as a fuel. It is also the starting material for the commercial production of several alcohols. The equations for these reactions are

$$2\,F_2(g) + 2\,H_2O(l) \longrightarrow 4\,HF(aq) + O_2(g)$$

$$Cl_2(g) + H_2O(l) \longrightarrow HCl(aq) + HOCl(aq)$$

$$Br_2(l) + H_2O(l) \longrightarrow HBr(aq) + HOBr(aq)$$

$$C(s) + H_2O(g) \xrightarrow{\text{1000°C}} CO(g) + H_2(g)$$

Reactions of Water with Metal and Nonmetal Oxides

Metal oxides that react with water to form hydroxides are known as **basic anhydrides**. Examples are

basic anhydride

$$CaO(s) + H_2O(l) \longrightarrow Ca(OH)_2(aq)$$
calcium hydroxide

$$Na_2O(s) + H_2O(l) \longrightarrow 2\,NaOH(aq)$$
sodium hydroxide

Certain metal oxides, such as CuO and Al_2O_3, do not form solutions containing OH^- ions because the oxides are insoluble in water.

Nonmetal oxides that react with water to form acids are known as **acid anhydrides**. Examples are

acid anhydride

$$CO_2(g) + H_2O(l) \rightleftharpoons H_2CO_3(aq)$$
carbonic acid

$$SO_2(g) + H_2O(l) \rightleftharpoons H_2SO_3(aq)$$
sulfurous acid

$$N_2O_5(s) + H_2O(l) \longrightarrow 2\,HNO_3(aq)$$
nitric acid

The word *anhydrous* means "without water." An anhydride is a metal oxide or a nonmetal oxide derived from a base or an oxy-acid by the removal of water. To determine the formula of an anhydride, the elements of water are removed from an acid or base formula until all the hydrogen is removed. Sometimes more than one formula unit is needed to remove all the hydrogen as water. The formula of the anhydride then consists of the remaining metal or nonmetal and the remaining oxygen atoms. In calcium hydroxide, removal of water as indicated leaves CaO as the anhydride:

$$Ca\!\!\begin{array}{c} O\boxed{H} \\ \boxed{OH} \end{array} \xrightarrow{\Delta} CaO + H_2O$$

In sodium hydroxide, H_2O cannot be removed from one formula unit, so two formula units of NaOH must be used, leaving Na_2O as the formula of the anhydride:

$$\begin{array}{c} NaO\boxed{H} \\ Na\boxed{OH} \end{array} \xrightarrow{\Delta} Na_2O + H_2O$$

The removal of H_2O from H_2SO_4 gives the acid anhydride SO_3:

$$H_2SO_4 \xrightarrow{\Delta} SO_3 + H_2O$$

The foregoing are examples of typical reactions of water but are by no means a complete list of the known reactions of water.

13.13 Hydrates

When certain solutions containing ionic compounds are allowed to evaporate, some water molecules remain as part of the crystalline compound that is left after evaporation is complete. Solids that contain water molecules as part of their crystalline structure are known as **hydrates**. Water in a hydrate is known as **water of hydration**, or **water of crystallization**.

Formulas for hydrates are expressed by first writing the usual anhydrous (without water) formula for the compound and then adding a dot followed by the number of water molecules present. An example is $BaCl_2 \cdot 2\,H_2O$. This formula tells us that each formula unit of this compound contains one barium ion, two chloride ions, and two water molecules. A crystal of the compound contains many of these units in its crystalline lattice.

hydrate
water of hydration
water of crystallization

As a solution of CuSO₄ evaporates beautiful blue crystals of CuSO₄ · 5 H₂O are formed.

In naming hydrates, we first name the compound exclusive of the water and then add the term *hydrate,* with the proper prefix representing the number of water molecules in the formula. For example, $BaCl_2 \cdot 2\ H_2O$ is called *barium chloride dihydrate.* Hydrates are true compounds and follow the Law of Definite Composition. The molar mass of $BaCl_2 \cdot 2\ H_2O$ is 244.2 g/mol; it contains 56.22% barium, 29.03% chlorine, and 14.76% water.

Water molecules in hydrates are bonded by electrostatic forces between polar water molecules and the positive or negative ions of the compound. These forces are not as strong as covalent or ionic chemical bonds. As a result, water of crystallization can be removed by moderate heating of the compound. A partially dehydrated or completely anhydrous compound may result. When $BaCl_2 \cdot 2\ H_2O$ is heated, it loses its water at about 100°C:

$$BaCl_2 \cdot 2\ H_2O(s) \xrightarrow{\ 100°C\ } BaCl_2(s)\ +\ 2\ H_2O(g)$$

When a solution of copper(II) sulfate ($CuSO_4$) is allowed to evaporate, beautiful blue crystals containing 5 moles of water per mole of $CuSO_4$ are formed. The formula for this hydrate is $CuSO_4 \cdot 5\ H_2O$; it is called copper(II) sulfate pentahydrate. When $CuSO_4 \cdot 5\ H_2O$ is heated, water is lost, and a pale green-white powder, anhydrous $CuSO_4$, is formed:

$$CuSO_4 \cdot 5\ H_2O(s) \xrightarrow{\ 250°C\ } CuSO_4(s)\ +\ 5\ H_2O(g)$$

When water is added to anhydrous copper(II) sulfate, the foregoing reaction is reversed, and the compound turns blue again. Because of this outstanding color change, anhydrous copper(II) sulfate has been used as an indicator to detect small amounts of water. The formation of the hydrate is noticeably exothermic.

The formula for plaster of paris is $(CaSO_4)_2 \cdot H_2O$. When mixed with the proper quantity of water, plaster of paris forms a dihydrate and sets to a hard mass. It is, therefore, useful for making patterns for the reproduction of art objects, molds, and surgical casts. The chemical reaction is

$$(CaSO_4)_2 \cdot H_2O(s)\ +\ 3\ H_2O(l) \longrightarrow 2\ CaSO_4 \cdot 2\ H_2O(s)$$

Table 13.4 lists a number of common hydrates.

13.14   Hygroscopic Substances

hygroscopic substance

Many anhydrous compounds and other substances readily absorb water from the atmosphere. Such substances are said to be **hygroscopic**. This property can be observed in the following simple experiment: Spread a 10–20 g sample of anhydrous copper(II) sulfate on a watch glass and set it aside so that the compound is exposed to the air. Then determine the mass of the sample periodically for 24 hours, noting the increase in mass and the change in color. Over time, water is absorbed from the atmosphere, forming the blue pentahydrate $CuSO_4 \cdot 5\ H_2O$.

deliquescence

Some compounds continue to absorb water beyond the hydrate stage to form solutions. A substance that absorbs water from the air until it forms a solution is said to be **deliquescent**. A few granules of anhydrous calcium chloride or pellets of sodium hydroxide exposed to the air will appear moist in a few minutes, and within an hour will absorb enough water to form a puddle of solution. Diphosphorus

TABLE 13.4 Selected Hydrates

Hydrate	Name	Hydrate	Name
$CaCl_2 \cdot 2\,H_2O$	Calcium chloride dihydrate	$Na_2CO_3 \cdot 10\,H_2O$	Sodium carbonate decahydrate
$Ba(OH)_2 \cdot 8\,H_2O$	Barium hydroxide octahydrate	$(NH_4)_2C_2O_4 \cdot H_2O$	Ammonium oxalate monohydrate
$MgSO_4 \cdot 7\,H_2O$	Magnesium sulfate heptahydrate	$NaC_2H_3O_2 \cdot 3\,H_2O$	Sodium acetate trihydrate
$SnCl_2 \cdot 2\,H_2O$	Tin(II) chloride dihydrate	$Na_2B_4O_7 \cdot 10\,H_2O$	Sodium tetraborate decahydrate
$CoCl_2 \cdot 6\,H_2O$	Cobalt(II) chloride hexahydrate	$Na_2S_2O_3 \cdot 5\,H_2O$	Sodium thiosulfate pentahydrate

pentoxide (P_2O_5) picks up water so rapidly that its mass cannot be determined accurately except in an anhydrous atmosphere.

Compounds that absorb water are useful as drying agents (desiccants). Refrigeration systems must be kept dry with such agents or the moisture will freeze and clog the tiny orifices in the mechanism. Bags of drying agents are often enclosed in packages containing iron or steel parts to absorb moisture and prevent rusting. Anhydrous calcium chloride, magnesium sulfate, sodium sulfate, calcium sulfate, silica gel, and diphosphorus pentoxide are some of the compounds commonly used for drying liquids and gases that contain small amounts of moisture.

This common hygroscopic substance is packaged with many consumer products containing metals.

13.15   Natural Waters

Natural fresh waters are not pure, but contain dissolved minerals, suspended matter, and sometimes harmful bacteria. The water supplies of large cities are usually drawn from rivers or lakes. Such water is generally unsafe to drink without treatment. To make such water safe to drink it is treated by some or all of the following processes (See Figure 13.9).

1. *Screening.* Removal of relatively large objects, such as trash, fish, and so on.
2. *Flocculation and sedimentation.* Chemicals, usually lime, CaO, and alum (aluminum sulfate), $Al_2(SO_4)_3$, are added to form a flocculent jelly-like precipitate of aluminum hydroxide. This precipitate traps most of the fine suspended matter in the water and carries it to the bottom of the sedimentation basin.
3. *Sand filtration.* Water is drawn from the top of the sedimentation basin and passed downward through fine sand filters. Nearly all the remaining suspended matter and bacteria are removed by the sand filters.
4. *Aeration.* Water is drawn from the bottom of the sand filters and is aerated by spraying. The purpose of this process is to remove objectionable odors and tastes.
5. *Disinfection.* In the final stage chlorine gas is injected into the water to kill harmful bacteria before the water is distributed to the public. Ozone is also used in some countries to disinfect water. In emergencies water may be disinfected by simply boiling it for a few minutes.

Moisturizers have long been used to protect and rehydrate the skin. These products contain humectants and emollients that increase the water content of the skin in different ways. The emollients cover the skin with a layer of material that is immiscible with water thus preventing water from within the skin from evaporating. In contrast, humectants add water to the skin by attracting water vapor from the air.

The most common humectants are sorbitol, glycerin, and polypropylene glycol. Each molecule is polar, containing multiple —OH groups (see formulas).

The oxygen atom in each —OH group is considerably more electronegative than the hydrogen atom. This electronegativity difference results in a partial negative charge on the oxygen, whereas the hydrogen carries a partial positive charge. This polarity is the basis for the attraction between the humectant and the water molecules (also polar).

Emollients are composed of hydrophobic (water-insoluble) molecules. These products are made of nonpolar molecules. There is a great diversity of compounds in this category, including animal oils, vegetable oils, exotic oils (such as jojoba and aloe vera), and synthetic oils. In each case the molecules form a water-insoluble layer on the skin, which traps the skin's own moisture and feels smooth to the touch.

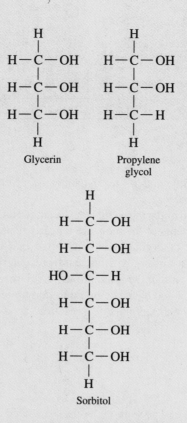

Glycerin   Propylene glycol

Sorbitol

Many consumers believe that alcohols can dry the skin and should not be used in skin care products. Is this concern justified? What is the purpose of alcohols in skin care products? The problem is not so simple as we might expect. There are two different categories of alcohol with distinctly different properties. Fatty acid alcohols, such as caprylic, isocetyl, oleyl, and stearyl, are made from long-chain fatty acids. These molecules are essentially nonpolar and behave like emollients. They form a water-insoluble layer on the skin that traps the moisture. The second type of alcohol is the simple alcohol, which includes ethyl, isopropyl, and methyl alcohols. These substances act as solvents in the skin care product and can be drying. They absorb excess oil, dissolve one ingredient into another and make products evaporate. Some also act to keep products from spoiling and separating. The problem with alcohol in skin care products results from overuse or use by people who don't need them.

One of the best tests for whether a skin care product will tend to dry the skin is to examine the texture. Liquids are typically the most drying formulations and are best used on oily skin. Gels containing lightweight emollients are best for skin with varying degrees of oiliness. Creams tend to contain heavier moisturizers and are best for normal to dry skin. Ointments are very heavy, creamy products that form a barrier to the skin, acting in the same manner as a humectant. Ointments are for use on severely dry or damaged skin. The key to selecting the proper moisturizer lies in understanding the way in which the product functions.

If the drinking water of children contains an optimum amount of fluoride ion, their teeth will be more resistant to decay. Therefore, in many communities NaF or Na_2SiF_6 is added to the water supply to bring the fluoride ion concentration up to the optimum level of about 1.0 ppm. Excessively high concentrations of fluoride ion can cause brown mottling of the teeth.

Water that contains dissolved calcium and magnesium salts is called *hard water.* One drawback of hard water is that ordinary soap does not lather well in it; the soap reacts with the calcium and magnesium ions to form an insoluble greasy scum. However, synthetic soaps, known as detergents, are available that have excellent

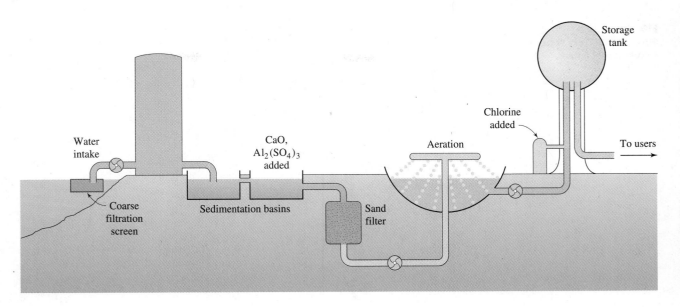

▲
FIGURE 13.9
Typical municipal water
treatment plant.

cleaning qualities and do not form precipitates with hard water. Hard water is also undesirable because it causes "scale" to form on the walls of water heaters, teakettles, coffee pots, and steam irons, which greatly reduces their efficiency.

Four techniques used to "soften" hard water are:

1. **Distillation** The water is boiled, and the steam formed is condensed into a liquid again, leaving the minerals behind in the distilling vessel. Figure 13.10 illustrates a simple laboratory distillation apparatus. Commercial stills are capable of producing hundreds of liters of distilled water per hour.

2. **Calcium precipitation** Calcium and magnesium ions are precipitated from hard water by adding sodium carbonate and lime. Insoluble calcium carbonate and magnesium hydroxide are precipitated and are removed by filtration or sedimentation.

3. **Ion-exchange** Hard water is effectively softened as it is passed through a bed or tank of zeolite—a complex sodium aluminum silicate. In this process, sodium ions replace objectionable calcium and magnesium ions, and the water is thereby softened:

$$Na_2(\text{zeolite})(s) \;+\; Ca^{2+}(aq) \longrightarrow Ca(\text{zeolite})(s) \;+\; 2\,Na^+(aq)$$

The zeolite is regenerated by back-flushing with concentrated sodium chloride solution, reversing the foregoing reaction.

4. **Demineralization** Both cations and anions are removed by a two-stage ion-exchange system. Special synthetic organic resins are used in the ion-exchange beds. In the first stage metal cations are replaced by hydrogen ions. In the second stage anions are replaced by hydroxide ions. The hydrogen and hydroxide ions react, and essentially pure, mineral-free water leaves the second stage.

The oceans are an enormous source of water, but seawater contains about 3.5 lb of salts per 100 lb of water. This 35,000 ppm of dissolved salts makes seawater unfit for agricultural and domestic uses. Water that contains less than 1000 ppm of

The sodium ions present in water softened either by chemical precipitation or by the zeolite process are not objectionable to most users of soft water.

FIGURE 13.10 ▶
Simple laboratory setup for distillation of liquids.

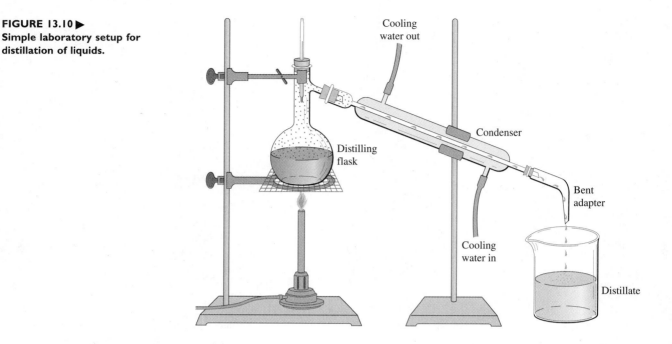

salts is considered reasonably good for drinking, and safe drinking water is already being obtained from the sea in many parts of the world. Continuous research is being done in an effort to make usable water from the oceans more abundant and economical. See Figure 13.11.

13.16   Water Pollution

Polluted water was formerly thought of as water that was unclear, had a bad odor or taste, and contained disease-causing bacteria. However, such factors as increased population, industrial requirements for water, atmospheric pollution, toxic waste dumps, and use of pesticides have greatly expanded the problem of water pollution.

Many of the newer pollutants are not removed or destroyed by the usual water-treatment processes. For example, among the 66 organic compounds found in the drinking water of a major city on the Mississippi River, 3 are labeled slightly toxic, 17 moderately toxic, 15 very toxic, 1 extremely toxic, and 1 supertoxic. Two are known carcinogens (cancer-producing agents), 11 are suspect, and 3 are metabolized to carcinogens. The U.S. Public Health Service classifies water pollutants under eight broad categories. These categories are shown in Table 13.5.

Many outbreaks of disease or poisoning, such as typhoid, dysentery, and cholera have been attributed directly to drinking water. Rivers and streams are an easy means for municipalities to dispose of their domestic and industrial waste products. Much of this water is used again by people downstream, and then discharged back into the water source. Then another community still farther downstream draws the same water and discharges its own wastes. Thus, along waterways such as the

◄ **FIGURE 13.11**
Catalina Island, California (left), gets most of its water from a desalinization plant. Below is a series of filtration units used to treat the sea water, which is then pumped into the main reservoir.

TABLE 13.5 Classification of Water Pollutants

Type of pollutant	Examples
Oxygen-demanding wastes	Decomposable organic wastes from domestic sewage and industrial wastes of plant and animal origin
Infectious agents	Bacteria, viruses, and other organisms from domestic sewage, animal wastes, and animal process wastes
Plant nutrients	Principally compounds of nitrogen and phosphorus
Organic chemicals	Large numbers of chemicals synthesized by industry, pesticides, chlorinated organic compounds
Other minerals and chemicals	Inorganic chemicals from industrial operations, mining, oil field operations, and agriculture
Radioactive substances	Waste products from mining and processing radioactive materials, airborne radioactive fallout, increased use of radioactive materials in hospitals and research
Heat from industry	Large quantities of heated water returned to water bodies from power plants and manufacturing facilities after use for cooling
Sediment from land erosion	Solid matter washed into streams and oceans by erosion, rain, and water runoff

Mississippi and Delaware rivers, water is withdrawn and discharged many times. If this water is not properly treated, harmful pollutants build up, causing epidemics of various diseases.

The disposal of hazardous waste products adds to the water pollution problem. These products are unavoidable in the manufacture of many products that we use in everyday life. One common way to dispose of these wastes is to place them in toxic waste dumps. What has been found after many years of disposing of wastes in this manner is that toxic substances have seeped into the groundwater deposits. As a result many people have become ill, and water wells have been closed until satisfactory methods of detoxifying this water are found. This problem is serious, because one-half the United States population gets its drinking water from groundwater. To clean up the thousands of industrial dumps and to find and implement new and safe methods of disposing of wastes is ongoing and costly.

Many major water pollutants have been recognized and steps have been taken to eliminate them. Three that pose serious problems are lead, detergents, and chlorine-containing organic compounds. Lead poisoning, for example, has been responsible for many deaths in past years. One major toxic action of lead in the body is the inhibition of the enzyme necessary for the production of hemoglobin in the blood. The usual intake of lead into the body is through food. However, extraordinary amounts of lead can be ingested from water running through lead pipes and by using lead-containing ceramic containers for storage of food and beverages.

It has been clearly demonstrated that waterways rendered so polluted that the water is neither fit for human use nor able to sustain marine life can be successfully restored. However, keeping our lakes and rivers free from pollution is a very costly and complicated process.

Concepts in Review

1. List the common properties of liquids and solids. Explain how they are different from gases.

2. Explain the process of evaporation from the standpoint of kinetic energy.

3. Relate vapor-pressure data or vapor-pressure curves of different substances to their relative rates of evaporation and to their relative boiling points.

4. Explain the forces involved in surface tension of a liquid. Give two examples.

5. Explain why a meniscus forms on the surface of liquids in a container.

6. Explain what is occurring throughout the heating curve for water.

7. Describe a water molecule with respect to the Lewis structure, bond angle, and polarity.

8. Make sketches showing hydrogen bonding (a) between water molecules, (b) between hydrogen fluoride molecules, and (c) between ammonia molecules.

9. Explain the effect of hydrogen bonding on the physical properties of water.

10. Determine whether a compound will or will not form hydrogen bonds.

11. Identify metal oxides as basic anhydrides and write balanced equations for their reactions with water.

12. Identify nonmetal oxides as acid anhydrides and write balanced equations for their reactions with water.

13. Deduce the formula of the acid anhydride or basic anhydride when given the formula of the corresponding acid or base.

14. Identify the product, name each reactant and product, and write equations for the complete dehydration of hydrates.

15. Outline the processes necessary to prepare safe drinking water from a contaminated river source.

16. Describe how water may be softened by distillation, chemical precipitation, ion exchange, and demineralization.

17. Complete and balance equations for (a) the reactions of water with Na, K, and Ca; (b) the reaction of steam with Zn, Al, Fe, and C; and (c) the reaction of water with halogens.

Key Terms

The terms listed here have been defined within this chapter. Section numbers are referenced in parenthesis for each term. More detailed definitions are given in the Glossary.

acid anhydride (13.12)
basic anhydride (13.12)
boiling point (13.15)
capillary action (13.4)
condensation (13.3)
deliquescence (13.14)
evaporation (13.2)
freezing or melting point (13.6)

heat of fusion (13.7)
heat of vaporization (13.7)
hydrate (13.13)
hydrogen bond (13.11)
hygroscopic substance (13.14)
meniscus (13.4)
nonvolatile (13.3)
normal boiling point (13.5)

sublimation (13.2)
surface tension (13.4)
vapor pressure (13.3)
vapor pressure curves (13.5)
vaporization (13.2)
volatile (13.3)
water of crystallization (13.13)
water of hydration (13.13)

Questions

Questions refer to tables, figures, and key words and concepts defined within the chapter. A particularly challenging question or exercise is indicated with an asterisk.

1. Compare the potential energy of the two states of water shown in Figure 13.6.

2. In what state (solid, liquid, or gas) would H_2S, H_2Se, and H_2Te be at 0°C? (See Table 13.3.)

3. The two thermometers in the flask on the hotplate (Figure 13.6) read 100°C. What is the pressure of the atmosphere?

4. Draw a diagram of a water molecule and point out the areas that are the negative and positive ends of the dipole.

5. If the water molecule were linear, with all three atoms in a straight line rather than in the shape of a V, as shown in Figure 13.7, what effect would this have on the physical properties of water?

6. Based on Table 13.4, how do we specify 1, 2, 3, 4, 5, 6, 7, and 8 molecules of water in the formulas of hydrates?

7. Would the distillation setup in Figure 13.10 be satisfactory for separating salt and water? Ethyl alcohol and water? Explain.

8. If the liquid in the flask in Figure 13.10 is ethyl alcohol and the atmospheric pressure is 543 torr, what temperature will show on the thermometer? (Use Figure 13.4.)

9. If water were placed in both containers in Figure 13.1, would both have the same vapor pressure at the same temperature? Explain.

10. In Figure 13.1, in which case, (a) or (b), will the atmosphere above the liquid reach a point of saturation?

11. Suppose that a solution of ethyl ether and ethyl alcohol were placed in the closed bottle in Figure 13.1. Use Figure 13.4 for information on the substances.
    (a) Would both substances be present in the vapor?
    (b) If the answer to part (a) is yes, which would have more molecules in the vapor?

12. In Figure 13.2, if 50% more water had been added in part (b), what equilibrium vapor pressure would have been observed in (c)?

13. At approximately what temperature would each of the substances listed in Table 13.2 boil when the pressure is 30 torr? (See Figure 13.4.)

14. Use the graph in Figure 13.4 to find the following:
    (a) The boiling point of water at 500 torr
    (b) The normal boiling point of ethyl alcohol
    (c) The boiling point of ethyl ether at 0.50 atm

15. Consider Figure 13.5.
    (a) Why is line BC horizontal? What is happening in this interval?
    (b) What phases are present in the interval BC?
    (c) When heating is continued after point C, another horizontal line, DE, is reached at a higher temperature. What does this line represent?

16. List six physical properties of water.

17. What condition is necessary for water to have its maximum density? What is its maximum density?

18. Account for the fact that an ice–water mixture remains at 0°C until all the ice is melted, even though heat is applied to it.

19. Which contains less heat, ice at 0°C or water at 0°C? Explain.

20. Why does ice float in water? Would ice float in ethyl alcohol ($d = 0.789$ g/mL)? Explain.

21. If water molecules were linear instead of bent, would the heat of vaporization be higher or lower? Explain.

22. The heat of vaporization for ethyl ether is 351 J/g (83.9 cal/g) and that for ethyl alcohol is 855 J/g (204.3 cal/g). Which of these compounds has hydrogen bonding? Explain.

23. Would there be more or less H-bonding if water molecules were linear instead of bent? Explain.

24. Which would show hydrogen bonding, ammonia, NH_3, or methane, CH_4? Explain.

25. In which condition are there fewer hydrogen bonds between molecules: water at 40°C or water at 80°C?

26. Which compound,

    $H_2NCH_2CH_2NH_2$ or $CH_3CH_2CH_2NH_2$,

    would you expect to have the higher boiling point? Explain your answer. (Both compounds have similar molar masses.)

27. Explain why rubbing alcohol which has been warmed to body temperature still feels cold when applied to your skin.

28. The vapor pressure at 20°C is given for the following compounds:

methyl alcohol	96 torr
acetic acid	11.7 torr
benzene	74.7 torr
bromine	173 torr
water	17.5 torr
carbon tetrachloride	91 torr
mercury	0.0012 torr
toluene	23 torr

    (a) Arrange these compounds in their order of increasing rate of evaporation.
    (b) Which substance listed would have the highest boiling point? The lowest?

29. Suggest a method whereby water could be made to boil at 50°C.

30. Explain why a higher temperature is obtained in a pressure cooker than in an ordinary cooking pot.

31. What is the relationship between vapor pressure and boiling point?

32. On the basis of the Kinetic-Molecular Theory, explain why vapor pressure increases with temperature.

33. Why does water have such a relatively high boiling point?

34. The boiling point of ammonia, NH_3, is -33.4°C and that of sulfur dioxide, SO_2, is -10.0°C. Which has the higher vapor pressure at -40°C?

35. Explain what is occurring physically when a substance is boiling.

36. Explain why HF (bp $= 19.4$°C) has a higher boiling point than HCl (bp $= -85$°C), whereas F_2 (bp $= -188$°C) has a lower boiling point than Cl_2 (bp $= -34$°C).

37. Why does a boiling liquid maintain a constant temperature when heat is continuously being added?

38. At what specific temperature will ethyl ether have a vapor pressure of 760 torr?

39. Why does a lake freeze from the top down?

40. What water temperature would you theoretically expect to find at the bottom of a very deep lake? Explain.

41. Write equations to show how the following metals react with water: aluminum, calcium, iron, sodium, zinc. State the conditions for each reaction.

42. Is the formation of hydrogen and oxygen from water an exothermic or an endothermic reaction? How do you know?

43. (a) What is an anhydride?
    (b) What type of compound will be an acid anhydride?
    (c) What type of compound will be a basic anhydride?

44. Which of the following statements are correct? Rewrite each incorrect statement to make it correct.
    (a) The process of a substance changing directly from a solid to a gas is called sublimation.
    (b) When water is decomposed, the volume ratio of H_2 to O_2 is $2:1$, but the mass ratio of H_2 to O_2 is $1:8$.
    (c) Hydrogen sulfide is a larger molecule than water.
    (d) The changing of ice into water is an exothermic process.

(e) Water and hydrogen fluoride are both nonpolar molecules.

(f) Hydrogen bonding is stronger in H_2O than in H_2S because oxygen is more electronegative than sulfur.

(g) $H_2O_2 \longrightarrow 2 H_2O + O_2$ represents a balanced equation for the decomposition of hydrogen peroxide.

(h) Steam at 100°C can cause more severe burns than liquid water at 100°C.

(i) The density of water is independent of temperature.

(j) Liquid A boils at a lower temperature than liquid B. This fact indicates that liquid A has a lower vapor pressure than liquid B at any particular temperature.

(k) Water boils at a higher temperature in the mountains than at sea level.

(l) No matter how much heat you put under an open pot of pure water on a stove, you cannot heat the water above its boiling point.

(m) The vapor pressure of a liquid at its boiling point is equal to the prevailing atmospheric pressure.

(n) The normal boiling temperature of water is 273°C.

(o) The pressure exerted by a vapor in equilibrium with its liquid is known as the vapor pressure of the liquid.

(p) Sodium, potassium, and calcium each react with water to form hydrogen gas and a metal hydroxide.

(q) Calcium oxide reacts with water to form calcium hydroxide and hydrogen gas.

(r) Carbon dioxide is the hydride of carbonic acid.

(s) Water in a hydrate is known as water of hydration or water of crystallization.

(t) A substance that absorbs water from the air until it forms a solution is deliquescent.

(u) Distillation is effective for softening water because the minerals boil away, leaving soft water behind.

(v) Disposal of toxic industrial wastes in toxic waste dumps has been found to be a very satisfactory long-term solution to the problem of what to do with these wastes.

(w) The amount of heat needed to change 1 mol of ice at 0°C to a liquid at 0°C is 6.02 kJ (1.44 kcal).

(x) $BaCl_2 \cdot 2 H_2O$ has a higher percentage of water than does $CaCl_2 \cdot 2 H_2O$.

Paired Exercises

These exercises are paired. Each odd-numbered exercise is followed by a similar even-numbered exercise. Answers to the even-numbered exercises are given in Appendix V.

45. Write the formulas for the anhydrides of the following acids:
H_2SO_3, H_2SO_4, HNO_3

46. Write the formulas for the anhydrides of the following acids:
$HClO_4$, H_2CO_3, H_3PO_4

47. Write the formulas for the anhydrides of the following bases:
LiOH, NaOH, $Mg(OH)_2$

48. Write the formulas for the anhydrides of the following bases:
KOH, $Ba(OH)_2$, $Ca(OH)_2$

49. Complete and balance the following equations:
(a) $Ba(OH)_2 \xrightarrow{\Delta}$
(b) $CH_3OH + O_2 \longrightarrow$
  methyl alcohol
(c) $Rb + H_2O \longrightarrow$
(d) $SnCl_2 \cdot 2 H_2O \xrightarrow{\Delta}$
(e) $HNO_3 + NaOH \longrightarrow$
(f) $CO_2 + H_2O \longrightarrow$

50. Complete and balance the following equations:
(a) $Li_2O + H_2O \longrightarrow$
(b) $KOH \xrightarrow{\Delta}$
(c) $Ba + H_2O \longrightarrow$
(d) $Cl_2 + H_2O \longrightarrow$
(e) $SO_3 + H_2O \longrightarrow$
(f) $H_2SO_3 + KOH \longrightarrow$

51. Name each of the following hydrates:
(a) $BaBr_2 \cdot 2 H_2O$
(b) $AlCl_3 \cdot 6 H_2O$
(c) $FePO_4 \cdot 4 H_2O$

52. Name each of the following hydrates:
(a) $MgNH_4PO_4 \cdot 6 H_2O$
(b) $FeSO_4 \cdot 7 H_2O$
(c) $SnCl_4 \cdot 5 H_2O$

53. Distinguish between deionized water and
(a) hard water
(b) soft water

54. Distinguish between deionized water and
(a) distilled water
(b) natural water

55. How many moles of compound are in 100. g of $CoCl_2 \cdot 6 H_2O$?

56. How many moles of compound are in 100. g of $FeI_2 \cdot 4 H_2O$?

57. How many moles of water can be obtained from 100. g of $CoCl_2 \cdot 6 H_2O$?

58. How many moles of water can be obtained from 100. g of $FeI_2 \cdot 4 H_2O$?

59. When a person purchases epsom salts, $MgSO_4 \cdot 7 H_2O$, what percent of the compound is water?

60. Calculate the mass percent of water in the hydrate $Al_2(SO_4)_3 \cdot 18 H_2O$.

61. Sugar of lead, a hydrate of lead acetate, $Pb(C_2H_3O_2)_2$, contains 14.2% H_2O. What is the empirical formula for the hydrate?

62. A 25.0-g sample of a hydrate of $FePO_4$ was heated until no more water was driven off. The mass of anhydrous sample is 16.9 g. What is the empirical formula of the hydrate?

63. How many joules are needed to change 120. g of water at 20.°C to steam at 100.°C?

64. How many joules of energy must be removed from 126 g of water at 24°C to form ice at 0°C?

*65. Suppose 100. g of ice at 0°C are added to 300. g of water at 25°C. Is this sufficient ice to lower the temperature of the system to 0°C and still have ice remaining? Show evidence for your answer.

*66. Suppose 35.0 g of steam at 100.°C are added to 300. g of water at 25°C. Is this sufficient steam to heat all the water to 100.°C and still have steam remaining? Show evidence for your answer.

*67. If 75 g of ice at 0.0°C were added to 1.5 L of water at 75°C, what would be the final temperature of the mixture?

*68. If 9560 J of energy were absorbed by 500. g of ice at 0.0°C, what would be the final temperature?

69. How many grams of water will react with each of the following?
    (a) 1.00 mol K
    (b) 1.00 mol Ca
    (c) 1.00 mol SO_3

70. How many grams of water will react with each of the following?
    (a) 1.00 g Na
    (b) 1.00 g MgO
    (c) 1.00 g N_2O_5

Additional Exercises

These exercises are not paired or labeled by topic and provide additional practice on concepts covered in this chapter.

71. Which would cause a more severe burn, liquid water at 100°C or steam at 100°C? Why?

72. Imagine a shallow dish of alcohol set into a tray of water. If one were to blow across the tray, the alcohol would evaporate, while the water would cool significantly and eventually freeze. Explain why.

73. Regardless of how warm the outside temperature may be, one always feels cool when stepping out of a swimming pool, the ocean, or a shower. Why is this so?

74. Sketch a heating curve for a substance X whose melting point is 40°C and whose boiling point is 65°C.
    (a) Describe what you will observe as a 60. g sample of X is warmed from 0°C to 100°C.
    (b) If the heat of fusion of X is 80. J/g, the heat of vaporization is 190. J/g, and if 3.5 J are required to warm 1 g of X each degree, how much energy will be needed to accomplish the change in (a)?

75. Why does the vapor pressure of a liquid increase as the temperature of it is increased?

76. At the top of Mount Everest, which is just about 29,000 feet above sea level, the atmospheric pressure is about 270 torr. Using Figure 13.4 determine the approximate boiling temperature of water on Mt. Everest.

77. Explain how anhydrous copper(II) sulfate, $CuSO_4$, can act as an indicator for moisture.

78. Write formulas of magnesium sulfate heptahydrate and disodium hydrogen phosphate dodecahydrate.

79. How can soap function to make soft water from hard water? What objections are there to using soap for this purpose?

80. What substance is commonly used to destroy bacteria in water?

81. What chemical, other than chlorine or chlorine compounds, can be used to disinfect water for domestic use?

82. Some organic pollutants in water can be oxidized by dissolved molecular oxygen. What harmful effect can result from this depletion of oxygen in the water?

83. Why should you not drink liquids that are stored in ceramic containers, especially unglazed ones?

84. Write the chemical equation showing how magnesium ions are removed by a zeolite water softener.

85. Write an equation to show how hard water containing calcium chloride, $CaCl_2$, is softened by using sodium carbonate, Na_2CO_3.

86. How is the chemical structure of a humectant different from that of an emollient?

87. Explain how humectants and emollients help to moisturize the skin.

88. Explain the triangle theory of sweetness.

89. What are the characteristics of a good sweetener?

90. How many calories are required to change 225 g of ice at 0°C to steam at 100.°C?

91. The molar heat of vaporization is the number of joules required to change one mole of a substance from liquid to vapor at its boiling point. What is the molar heat of vaporization of water?

*92. The specific heat of zinc is 0.096 cal/g°C. Determine the energy required to raise the temperature of 250. g of zinc from room temperature (20.0°C) to 150.°C.

*93. Suppose 150. g of ice at 0.0°C is added to 0.120 L of water at 45°C. If the mixture is stirred and allowed to cool to 0.0°C, how many grams of ice remain?

94. How many joules of energy would be liberated by condensing 50.0 mol of steam at 100.0°C and allowing the liquid to cool to 30.0°C?

95. How many kilojoules of energy are needed to convert 100. g of ice at −10.0°C to water at 20.0°C. (The specific heat of ice at −10.0°C is 2.01 J/g°C.)

96. What mass of water must be decomposed to produce 25.0 L of oxygen at STP?

97. Compare the volume occupied by 1.00 mol of liquid water at 0°C and 1.00 mol of water vapor at STP.

*98. Suppose 1.00 mol of water evaporates in 1.00 day. How many water molecules, on the average, leave the liquid each second?

*99. A quantity of sulfuric acid is added to 100. mL of water. The final volume of the solution is 122 mL and has a density of 1.26 g/mL. What mass of acid was added? Assume the density of the water is 1.00 g/mL.

100. A mixture of 80.0 mL of hydrogen and 60.0 mL of oxygen is ignited by a spark to form water.
    (a) Does any gas remain unreacted? Which one, H_2 or O_2?
    (b) What volume of which gas (if any) remains unreacted? (Assume the same conditions before and after the reaction.)

101. A student (with slow reflexes) puts his hand in a stream of steam at 100.°C until 1.5 g of water have condensed. If the water then cools to room temperature (20.0°C), how many joules have been absorbed by the student's hand?

Answers to Practice Exercises

13.1 8.5°C, 28°C, 73°C, 93°C
13.2 approximately 93°C
13.3 44.7 kJ
13.4 (a) yes, (b) yes, (c) no

Most of the substances we encounter in our daily lives are mixtures. Often they are homogeneous mixtures, which are called *solutions*. When you think of a solution, juices, blood plasma, shampoo, soft drinks, or wine may come to mind. These solutions all have water as a main component, but many common items, such as air, gasoline, and steel, are also solutions that do not contain water. What are the necessary components of a solution? Why do some substances mix while others do not? What effect does a dissolved substance have on the properties of the solution? Answering these questions is the first step in understanding the solutions we encounter in our daily lives.

14.1   General Properties of Solutions

The term **solution** is used in chemistry to describe a system in which one or more substances are homogeneously mixed or dissolved in another substance. A simple solution has two components, a solute and a solvent. The **solute** is the component that is dissolved or is the least abundant component in the solution. The **solvent** is the dissolving agent or the most abundant component in the solution. For example, when salt is dissolved in water to form a solution, salt is the solute and water is the solvent. Complex solutions containing more than one solute and/or more than one solvent are common.

From the three states of matter—solid, liquid, and gas—it is possible to have nine different types of solutions: solid dissolved in solid, solid dissolved in liquid, solid dissolved in gas, liquid dissolved in liquid, and so on. Of these, the most common solutions are solid dissolved in liquid, liquid dissolved in liquid, gas dissolved in liquid, and gas dissolved in gas. Some common types of solutions are listed in Table 14.1.

A true solution is one in which the particles of dissolved solute are molecular or ionic in size, generally in the range of 0.1 to 1 nm (10^{-8} to 10^{-7} cm). The properties of a true solution are as follows:

1. It is a homogeneous mixture of two or more components—solute and solvent—and has a variable composition; that is, the ratio of solute to solvent may be varied.
2. The dissolved solute is molecular or ionic in size.
3. It may be either colored or colorless but it is usually transparent.
4. The solute remains uniformly distributed throughout the solution and will not settle out with time.
5. The solute generally can be separated from the solvent by purely physical means (e.g., by evaporation).

These properties are illustrated by water solutions of sugar and of potassium permanganate. Suppose that we prepare two sugar solutions, the first containing 10 g of sugar added to 100 mL of water and the second containing 20 g of sugar added to 100 mL of water. Each solution is stirred until all the solute dissolves, demonstrating that we can vary the composition of a solution. Every portion of the

solution
solute
solvent

◀ **Chapter Opening Photo: The ocean is a salt solution covering the majority of the earth's surface.**

325

Note the beautiful purple trails of KMnO₄ as the crystals dissolve.

TABLE 14.1 Common Types of Solutions

Phase of solution	Solute	Solvent	Example
Gas	Gas	Gas	Air
Liquid	Gas	Liquid	Soft drinks
Liquid	Liquid	Liquid	Antifreeze
Liquid	Solid	Liquid	Salt water
Solid	Gas	Solid	H₂ in Pt
Solid	Solid	Solid	Brass

solution has the same sweet taste because the sugar molecules are uniformly distributed throughout. If confined so that no solvent is lost, the solution will taste and appear the same a week or a month later. The properties of the solution are unaltered after the solution is passed through filter paper. But by carefully evaporating the water, we can recover the sugar from the solution.

To observe the dissolving of potassium permanganate, $KMnO_4$, we affix a few crystals of it to paraffin wax or rubber cement at the end of a glass rod and submerge the entire rod, with the wax-permanganate end up, in a cylinder of water. Almost at once the beautiful purple color of dissolved permanganate ions, MnO_4^-, appears at the top of the rod and streams to the bottom of the cylinder as the crystals dissolve. The purple color at first is mostly at the bottom of the cylinder because $KMnO_4$ is denser than water. But after a while the purple color disperses until it is evenly distributed throughout the solution. This dispersal demonstrates that molecules and ions move about freely and spontaneously (diffuse) in a liquid or solution.

Solution permanency is explained in terms of the Kinetic-Molecular Theory (see Section 12.2). According to the KMT both the solute and solvent particles (molecules and/or ions) are in constant random motion. This motion is energetic enough to prevent the solute particles from settling out under the influence of gravity.

14.2   Solubility

solubility

The term **solubility** describes the amount of one substance (solute) that will dissolve in a specified amount of another substance (solvent) under stated conditions. For example, 36.0 g of sodium chloride will dissolve in 100 g of water at 20°C. We say, then, that the solubility of NaCl in water is 36.0 g/100 g H_2O at 20°C.

Solubility is often used in a relative way. For instance, we say that a substance is very soluble, moderately soluble, slightly soluble, or insoluble. Although these terms do not accurately indicate how much solute will dissolve, they are frequently used to describe the solubility of a substance qualitatively.

miscible
immiscible

Two other terms often used to describe solubility are *miscible* and *immiscible*. Liquids that are capable of mixing and forming a solution are **miscible**; those that do not form solutions or are generally insoluble in each other are **immiscible**. Methyl alcohol and water are miscible in each other in all proportions. Oil and water are immiscible, forming two separate layers when they are mixed, as illustrated in Figure 14.1.

The general guidelines for the solubility of common ionic compounds (salts) are given in Figure 14.2. These guidelines have some exceptions but they provide a solid foundation for the compounds considered in this course. The solubilities of over 200 compounds are given in the Solubility Table in Appendix IV. Solubility data for thousands of compounds can be found by consulting standard reference sources.*

The quantitative expression of the amount of dissolved solute in a particular quantity of solvent is known as the **concentration of a solution**. Several methods of expressing concentration will be described in Section 14.6.

The term *salt* is used interchangeably with *ionic compound* by many chemists.

concentration of a solution

14.3 Factors Related to Solubility

Predicting solubilities is complex and difficult. Many variables, such as size of ions, charge on ions, interaction between ions, interaction between solute and solvent, and temperature, complicate the problem. Because of the factors involved, the general rules of solubility given in Figure 14.2 have many exceptions. However, the rules are very useful, because they do apply to many of the more common compounds that we encounter in the study of chemistry. Keep in mind that these are rules, not laws, and are therefore subject to exceptions. Fortunately the solubility of a solute is relatively easy to determine experimentally. We will now discuss factors related to solubility.

FIGURE 14.1
An immiscible mixture of oil and water.

Soluble		Insoluble
Salts of Na^+, K^+, NH_4^+		
Nitrates, NO_3^- Acetates, $C_2H_3O_2^-$		
Chlorides, Cl^- Bromides Br^- Iodides, I^-	except →	Ag^+, Hg_2^{2+}, Pb^{2+}
Sulfates, SO_4^{2-} Ag^+, Ca^{2+} are slightly soluble	except →	Ba^{2+}, Sr^{2+}, Pb^{2+}
NH_4^+ alkali metal cations, alkaline earth metal cations	← except	Carbonates, CO_3^{2-} Phosphates, PO_4^{3-} Hydroxides, OH^- Sulfides, S^{2-}

◄ **FIGURE 14.2**
The solubility of various common ions. Substances containing the ions on the left are generally soluble in cold water, while those substances containing the ions on the right are insoluble in cold water. The arrows point to the exceptions.

---

*Two commonly used handbooks are *Lange's Handbook of Chemistry*, 13th ed. (New York: McGraw-Hill, 1985), and *Handbook of Chemistry and Physics*, 76th ed. (Cleveland: Chemical Rubber Co., 1996).

= Water

= Na^+

= Cl^-

FIGURE 14.3
Dissolution of sodium chloride in water. Polar water molecules are attracted to Na^+ and Cl^- ions in the salt crystal, weakening the attraction between the ions. As the attraction between the ions weakens, the ions move apart and become surrounded by water dipoles. The hydrated ions slowly diffuse away from the crystal to become dissolved in solution.

The Nature of the Solute and Solvent

The old adage "like dissolves like" has merit, in a general way. Polar or ionic substances tend to be more miscible, or soluble, with other polar substances. Nonpolar substances tend to be miscible with other nonpolar substances and less miscible with polar substances. Thus, ionic compounds, which are polar, tend to be much more soluble in water, which is polar, than in solvents such as ether, hexane, or benzene, which are essentially nonpolar. Sodium chloride, an ionic substance, is soluble in water, slightly soluble in ethyl alcohol (less polar than water), and insoluble in ether and benzene. Pentane, C_5H_{12}, a nonpolar substance, is only slightly soluble in water but is very soluble in benzene and ether.

At the molecular level the formation of a solution from two nonpolar substances, such as hexane and benzene, can be visualized as a process of simple mixing. The nonpolar molecules, having little tendency to either attract or repel one another, easily intermingle to form a homogeneous mixture.

Solution formation between polar substances is much more complex. See for example, the process by which sodium chloride dissolves in water (Figure 14.3). Water molecules are very polar and are attracted to other polar molecules or ions. When salt crystals are put into water, polar water molecules become attracted to the sodium and chloride ions on the crystal surfaces and weaken the attraction between Na^+ and Cl^- ions. The positive end of the water dipole is attracted to the Cl^- ions, and the negative end of the water dipole to the Na^+ ions. The weakened attraction permits the ions to move apart, making room for more water dipoles. Thus, the surface ions are surrounded by water molecules, becoming hydrated ions, $Na^+(aq)$ and $Cl^-(aq)$, and slowly diffuse away from the crystals and dissolve in solution:

$$NaCl(crystal) \xrightarrow{H_2O} Na^+(aq) + Cl^-(aq)$$

Examination of the data in Table 14.2 reveals some of the complex questions relating to solubility. For example: Why are lithium halides, except for lithium fluoride, more soluble than sodium and potassium halides? Why are the solubilities of lithium fluoride and sodium fluoride so low in comparison with those of the other metal halides? Why doesn't the solubility of LiF, NaF, and NaCl increase proportionately with temperature, as do the solubilities of the other metal halides? Sodium chloride is appreciably soluble in water but is insoluble in concentrated hydrochloric acid, HCl, solution. On the other hand, LiF and NaF are not very soluble in water but are quite soluble in hydrofluoric acid, HF, solution—Why? These questions will not be answered directly here, but are meant to arouse your curiosity to the point that you will do some reading and research on the properties of solutions.

The Effect of Temperature on Solubility

Temperature has major effects on the solubility of most substances. Most solutes have a limited solubility in a specific solvent at a fixed temperature. For most solids dissolved in a liquid, an increase in temperature results in increased solubility (see Figure 14.4). However, no completely valid general rule governs the solubility of solids in liquids with change in temperature. Some solids increase in solubility only slightly with increasing temperature (see NaCl in Figure 14.4); other solids decrease in solubility with increasing temperature (see Li_2SO_4 in Figure 14.4).

On the other hand, the solubility of a gas in water usually decreases with

TABLE 14.2 Solubility of Alkali Metal Halides in Water

Salt	Solubility (g salt/100 g H₂O) 0°C	Solubility (g salt/100 g H₂O) 100°C
LiF	0.12	0.14 (at 35°C)
LiCl	67	127.5
LiBr	143	266
LiI	151	481
NaF	4	5
NaCl	35.7	39.8
NaBr	79.5	121
NaI	158.7	302
KF	92.3 (at 18°C)	Very soluble
KCl	27.6	57.6
KBr	53.5	104
KI	127.5	208

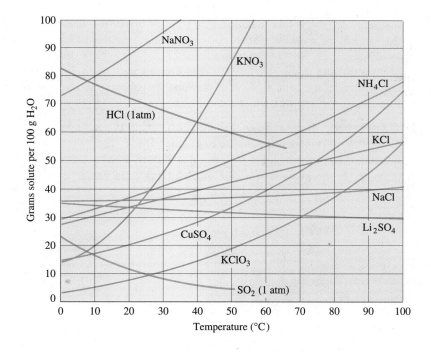

◄ FIGURE 14.4
Solubility of various compounds in water. Solids are shown in red and gases are shown in blue.

increasing temperature (see HCl and SO₂ in Figure 14.4). The tiny bubbles that form when water is heated are due to the decreased solubility of air at higher temperatures. The decreased solubility of gases at higher temperatures is explained in terms of the KMT by assuming that, in order to dissolve, the gas molecules must form bonds of some sort with the molecules of the liquid. An increase in temperature decreases the solubility of the gas because it increases the kinetic energy (speed) of the gas molecules and thereby decreases their ability to form "bonds" with the liquid molecules.

Pouring root beer into a glass illustrates the effect of pressure on solubility. The escaping CO_2 produces the foam.

The Effect of Pressure on Solubility

Small changes in pressure have little effect on the solubility of solids in liquids or liquids in liquids but have a marked effect on the solubility of gases in liquids. The solubility of a gas in a liquid is directly proportional to the pressure of that gas above the solution. Thus, the amount of a gas that is dissolved in solution will double if the pressure of that gas over the solution is doubled. For example, carbonated beverages contain dissolved carbon dioxide under pressures greater than atmospheric pressure. When a bottle of carbonated soda is opened, the pressure is immediately reduced to the atmospheric pressure, and the excess dissolved carbon dioxide bubbles out of the solution.

14.4   Rate of Dissolving Solids

The rate at which a solid dissolves is governed by (1) the size of the solute particles, (2) the temperature, (3) the concentration of the solution, and (4) agitation or stirring. Let's look at each of these conditions.

1. *Particle Size.*   A solid can dissolve only at the surface that is in contact with the solvent. Because the surface-to-volume ratio increases as size decreases, smaller crystals dissolve faster than large ones. For example, if a salt crystal 1 cm on a side (6-cm^2 surface area) is divided into 1000 cubes, each 0.1 cm on a side, the total surface of the smaller cubes is 60 cm^2—a tenfold increase in surface area (see Figure 14.5).

2. *Temperature.*   In most cases the rate of dissolving of a solid increases with temperature. This increase is due to kinetic effects. The solvent molecules move more rapidly at higher temperatures and strike the solid surfaces more often and harder, causing the rate of dissolving to increase.

FIGURE 14.5 ▶
Surface area of crystals. A crystal 1 cm on a side has a surface area of 6 cm^2. Subdivided into 1000 smaller crystals, each 0.1 cm on a side, the total surface area is increased to 60 cm^2.

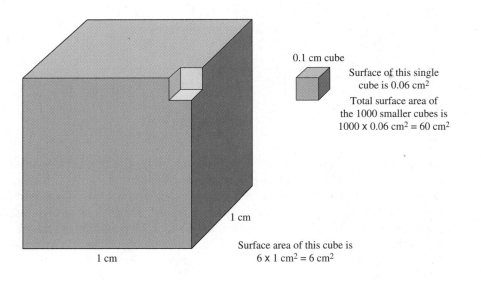

0.1 cm cube

Surface of this single cube is 0.06 cm^2

Total surface area of the 1000 smaller cubes is 1000 x 0.06 cm^2 = 60 cm^2

1 cm

1 cm

Surface area of this cube is 6 x 1 cm^2 = 6 cm^2

Killer Lakes

In a tiny African nation called Cameroon, two towns border on lakes that are people killers. The townspeople do not die by drowning in the lakes, or by drinking contaminated water. Instead they die from carbon dioxide asphyxiation. The lakes give off clouds of carbon dioxide at irregular intervals. Thirty-seven people died near Lake Monoun in August, 1984. Just two years later 1700 people died at nearby Lake Nyos.

Scientists studying these volcanic crater lakes have found that CO_2 percolates upward from groundwater into the bottom of these lakes. The CO_2 accumulates to dangerous levels because the water is naturally stratified into layers that do not mix. A boundary called a *chemocline* separates the layers, keeping fresh water at the surface of the lake. The lower layers of the lake contain dissolved minerals and gases (including CO_2).

The disasters occur when something disturbs the layers. An earthquake, a landslide, or even winds can trigger the phenomenon. As waves form and move across the lake, the layers within the lake are mixed. When the deep water containing the CO_2 rises near the top of the lake, the dissolved CO_2 is released from the solution (similar to the bubbles released on opening a can of soda). At Lake Nyos, where 1700 people died, the cloud of CO_2 spilled over the edge of the crater and traveled down a river valley. Since CO_2 is denser than air, the cloud stayed near the ground. It traveled at an amazing speed of 45 mph and killed people as far away as 25 miles.

Although scientists do not know precisely what causes the water layers to turn-over, they have succeeded in measuring the rate at which gas seeps into the lake bottoms. The rate is so fast that some scientists think the bottom waters of Lake Nyos could be saturated in less than 20 years, and Lake Monoun could be saturated in less than 10 years.

Scientists and engineers are working to lower gas concentrations in both lakes. In Lake Monoun, water is being pumped through pipes from the lake bottom to the surface to release the gas slowly. Lake Nyos is a larger lake and represents a more difficult problem. One end of the lake is supported by a weak natural dam. If the dam were to break, the water from the lake would spill into a valley with about 10,000 residents and could trigger a CO_2 release. Water could be pumped out of the lake to release the CO_2 and lower the level of the lake to accommodate the dam, but funding for these projects is uncertain. Until money is found to relieve the buildup of gases, these lakes will remain disasters waiting to happen.

3. *Concentration of the Solution.* When the solute and solvent are first mixed, the rate of dissolving is at its maximum. As the concentration of the solution increases and the solution becomes more nearly saturated with the solute, the rate of dissolving decreases greatly. The rate of dissolving is pictured graphically in Figure 14.6. Note that about 17 g dissolve in the first 5-minute interval, but only about 1 g dissolves in the fourth 5-minute interval. Although

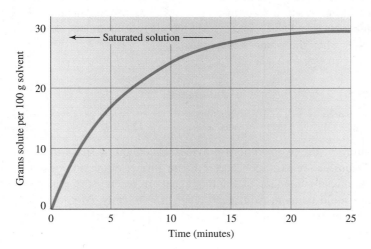

◀ FIGURE 14.6
Rate of dissolution of a solid solute in a solvent. The rate is maximum at the beginning and decreases as the concentration approaches saturation.

different solutes show different rates, the rate of dissolving always becomes very slow as the concentration approaches the saturation point.

4. *Agitation or Stirring.* The effect of agitation or stirring is kinetic. When a solid is first put into water, the only solvent with which it comes in contact is in the immediate vicinity. As the solid dissolves, the amount of dissolved solute around the solid becomes more and more concentrated, and the rate of dissolving slows down. If the mixture is not stirred, the dissolved solute diffuses very slowly through the solution; weeks may pass before the solid is entirely dissolved. Stirring distributes the dissolved solute rapidly through the solution, and more solvent is brought into contact with the solid, causing it to dissolve more rapidly.

14.5   Solutions: A Reaction Medium

Many solids must be put into solution in order to undergo appreciable chemical reaction. We can write the equation for the double displacement reaction between sodium chloride and silver nitrate:

$$NaCl + AgNO_3 \longrightarrow AgCl + NaNO_3$$

But suppose we mix solid $NaCl$ and solid $AgNO_3$ and look for a chemical change. If any reaction occurs, it is slow and virtually undetectable. In fact, the crystalline structures of $NaCl$ and $AgNO_3$ are so different that we could separate them by tediously picking out each kind of crystal from the mixture. But if we dissolve the $NaCl$ and $AgNO_3$ separately in water and mix the two solutions, we observe the immediate formation of a white, curd-like precipitate of silver chloride.

Molecules or ions must come into intimate contact or collide with one another in order to react. In the foregoing example, the two solids did not react because the ions were securely locked within their crystal structures. But when the $NaCl$ and $AgNO_3$ are dissolved, their crystal lattices are broken down and the ions become mobile. When the two solutions are mixed, the mobile Ag^+ and Cl^- ions come into contact and react to form insoluble $AgCl$, which precipitates out of solution. The soluble Na^+ and NO_3^- ions remain mobile in solution but form the crystalline salt $NaNO_3$ when the water is evaporated:

$$NaCl(aq) + AgNO_3(aq) \longrightarrow AgCl(s) + NaNO_3(aq)$$

$$Na^+(aq) + Cl^-(aq) + Ag^+(aq) + NO_3^-(aq) \longrightarrow AgCl(s) + Na^+(aq) + NO_3^-(aq)$$

sodium chloride solution	silver nitrate solution	silver chloride	sodium nitrate in solution

The mixture of the two solutions provides a medium or space in which the Ag^+ and Cl^- ions can react. (See Chapter 15 for further discussion of ionic reactions.)

Solutions also function as diluting agents in reactions in which the undiluted reactants would combine with each other too violently. Moreover, a solution of known concentration provides a convenient method for delivering specific amounts of reactants.

14.6 Concentration of Solutions

The concentration of a solution expresses the amount of solute dissolved in a given quantity of solvent or solution. Because reactions are often conducted in solution, it is important to understand the methods of expressing concentration and to know how to prepare solutions of particular concentrations. The concentration of a solution may be expressed qualitatively or quantitatively. Let's begin with a look at the qualitative methods of expressing concentration.

Dilute and Concentrated Solutions

When we say that a solution is *dilute* or *concentrated*, we are expressing, in a relative way, the amount of solute present. One gram of a compound and 2 g of a compound in solution are both dilute solutions when compared with the same volume of a solution containing 20 g of a compound. Ordinary concentrated hydrochloric acid contains 12 mol of HCl per liter of solution. In some laboratories the dilute acid is made by mixing equal volumes of water and the concentrated acid. In other laboratories the concentrated acid is diluted with two or three volumes of water, depending on its use. The term **dilute solution**, then, describes a solution that contains a relatively small amount of dissolved solute. Conversely, a **concentrated solution** contains a relatively large amount of dissolved solute.

dilute solution
concentrated solution

Saturated, Unsaturated, and Supersaturated Solutions

At a specific temperature there is a limit to the amount of solute that will dissolve in a given amount of solvent. When this limit is reached, the resulting solution is said to be *saturated*. For example, when we put 40.0 g of KCl into 100 g of H_2O at 20°C, we find that 34.0 g of KCl dissolves and 6.0 g of KCl remains undissolved. The solution formed is a saturated solution of KCl.

Two processes are occurring simultaneously in a saturated solution. The solid is dissolving into solution and, at the same time, the dissolved solute is crystallizing out of solution. This may be expressed as

solute (undissolved) $\rightleftharpoons$ solute (dissolved)

When these two opposing processes are occurring at the same rate, the amount of solute in solution is constant, and a condition of equilibrium is established between dissolved and undissolved solute. Therefore, a **saturated solution** contains dissolved solute in equilibrium with undissolved solute.

saturated solution

It is important to state the temperature of a saturated solution, because a solution that is saturated at one temperature may not be saturated at another. If the temperature of a saturated solution is changed, the equilibrium is disturbed, and the amount of dissolved solute will change to reestablish equilibrium.

A saturated solution may be either dilute or concentrated, depending on the solubility of the solute. A saturated solution can be conveniently prepared by dissolving a little more than the saturated amount of solute at a temperature somewhat higher than room temperature. Then the amount of solute in solution will be in

TABLE 14.3 Saturated Solutions at 20°C and 50°C		
	Solubility (g solute/100 g H_2O)	
Solute	20°C	50°C
NaCl	36.0	37.0
KCl	34.0	42.6
$NaNO_3$	88.0	114.0
$KClO_3$	7.4	19.3
$AgNO_3$	222.0	455.0
$C_{12}H_{22}O_{11}$	203.9	260.4

excess of its solubility at room temperature, and, when the solution cools, the excess solute will crystallize, leaving the solution saturated. (In this case, the solute must be more soluble at higher temperatures and must not form a supersaturated solution.) Examples expressing the solubility of saturated solutions at two different temperatures are given in Table 14.3.

unsaturated solution

An **unsaturated solution** contains less solute per unit of volume than does its corresponding saturated solution. In other words, additional solute can be dissolved in an unsaturated solution without altering any other conditions. Consider a solution made by adding 40 g of KCl to 100 g of H_2O at 20°C (see Table 14.3). The solution formed will be saturated and will contain about 6 g of undissolved salt, because the maximum amount of KCl that can dissolve in 100 g of H_2O at 20°C is 34 g. If the solution is now heated and maintained at 50°C, all the salt will dissolve and, in fact, even more can be dissolved. Thus the solution at 50°C is unsaturated.

In some circumstances, solutions can be prepared that contain more solute than that needed for a saturated solution at a particular temperature. These solutions are said to be **supersaturated**. However, we must qualify this definition by noting that a supersaturated solution is unstable. Disturbances, such as jarring, stirring, scratching the walls of the container, or dropping in a "seed" crystal, cause the supersaturation to return to saturation. When a supersaturated solution is disturbed, the excess solute crystallizes out rapidly, returning the solution to a saturated state.

supersaturated solution

Supersaturated solutions are not easy to prepare but may be made from certain substances by dissolving, in warm solvent, an amount of solute greater than that needed for a saturated solution at room temperature. The warm solution is then allowed to cool very slowly. With the proper solute and careful work, a supersaturated solution will result.

Example 14.1   Will a solution made by adding 2.5 g of $CuSO_4$ to 10 g of H_2O be saturated or unsaturated at 20°C?

Solution   To answer this question we first need to know the solubility of $CuSO_4$ at 20°C. From Figure 14.4 we see that the solubility of $CuSO_4$ at 20°C is about 21 g per 100 g of H_2O. This amount is equivalent to 2.1 g of $CuSO_4$ per 10 g of H_2O.

Since 2.5 g per 10 g of H_2O is greater than 2.1 g per 10 g of H_2O, the solution will be saturated and 0.4 g of $CuSO_4$ will be undissolved.

Practice 14.1

Will a solution made by adding 9.0 g NH_4Cl to 20 g of H_2O be saturated or unsaturated at 50°C?

Mass Percent Solution

The mass percent method expresses concentration of the solution as the percent of solute in a given mass of solution. It says that for a given mass of solution a certain percent of that mass is solute. Suppose that we take a bottle from the reagent shelf that reads "sodium hydroxide, NaOH, 10%." This statement means that for every 100 g of this solution, 10 g will be NaOH and 90 g will be water. (Note that this amount of solution is 100 g and not 100 mL.) We could also make this same concentration of solution by dissolving 2.0 g of NaOH in 18 g of water. Mass percent concentrations are most generally used for solids dissolved in liquids:

$$\text{mass percent} = \frac{\text{g solute}}{\text{g solute} + \text{g solvent}} \times 100 = \frac{\text{g solute}}{\text{g solution}} \times 100$$

As instrumentation advances are made in chemistry, our ability to measure the concentration of dilute solutions is increasing as well. Instead of mass percent, chemists now commonly use **parts per million (ppm)**:

$$\text{parts per million} = \frac{\text{g solute}}{\text{g solute} + \text{g solvent}} \times 1,000,000$$

Currently, air and water contaminants, drugs in the human body, and pesticide residues are some substances measured in parts per million.

The heat released in this hot pack results from the crystallization of a supersaturated solution of sodium acetate.

parts per million (ppm)

Note that mass percent is independent of the formula for the solute.

What is the mass percent of sodium hydroxide in a solution that is made by dissolving 8.00 g of NaOH in 50.0 g of H_2O?

Example 14.2

grams of solute (NaOH) = 8.00 g

Solution

grams of solvent (H_2O) = 50.0 g

$$\frac{8.00 \text{ g NaOH}}{8.00 \text{ g NaOH} + 50.0 \text{ g } H_2O} \times 100 = 13.8\% \text{ NaOH solution}$$

What masses of potassium chloride and water are needed to make 250. g of 5.00% solution?

Example 14.3

The percent expresses the mass of the solute:

Solution

250. g = total mass of solution

5.00% of 250. g = 0.0500 × 250. g = 12.5 g KCl (solute)

250. g − 12.5 g = 237.5 H_2O

Dissolving 12.5 g of KCl in 237.5 g of H_2O gives a 5.00% KCl solution.

Example 14.4   A 34.0% sulfuric acid solution has a density of 1.25 g/mL. How many grams of H_2SO_4 are contained in 1.00 L of this solution?

Solution   Since H_2SO_4 is the solute, we first solve the mass percent equation for grams of solute:

$$\text{mass percent} = \frac{\text{g solute}}{\text{g solution}} \times 100$$

$$\text{g solute} = \frac{\text{mass percent} \times \text{g solution}}{100}$$

The mass percent is given in the problem. We need to determine the grams of solution. The mass of the solution can be calculated from the density data. Convert density (g/mL) to grams:

$$1.00 \text{ L} = 1.00 \times 10^3 \text{ mL}$$

$$\frac{1.25 \text{ g}}{\text{mL}} \times 1.00 \times 10^3 \text{ mL} = 1250 \text{ g} \quad \text{(mass of solution)}$$

Now we have all the figures to calculate the grams of solute:

$$\text{g solute} = \frac{34.0 \text{ g} \times 1250 \text{ g}}{100} = 425 \text{ g } H_2SO_4$$

Thus, 1.00 L of 34.0% H_2SO_4 solution contains 425 g of H_2SO_4.

> **Practice 14.2**
>
> What is the mass percent of Na_2SO_4 in a solution that is made by dissolving 25.0 g of Na_2SO_4 in 225.0 g of H_2O?

Mass/Volume Percent (m/v)

This method expresses concentration as grams of solute per 100 mL of solution. With this system, a 10.0% (m/v) glucose solution is made by dissolving 10.0 g of glucose in water, diluting to 100 mL, and mixing. The 10.0% (m/v) solution could also be made by diluting 20.0 g to 200 mL, 50.0 g to 500 mL, and so on. Of course, any other appropriate dilution ratio may be used:

$$\text{mass/volume percent} = \frac{\text{g solute}}{\text{mL solution}} \times 100$$

Volume Percent

Solutions that are formulated from two liquids are often expressed as *volume percent* with respect to the solute. The volume percent is the volume of a liquid in 100 mL of solution. The label on a bottle of ordinary rubbing alcohol reads "isopropyl alcohol, 70% by volume." Such a solution could be made by mixing 70 mL of alcohol with water to make a total volume of 100 mL, but we cannot use 30 mL of water because the two volumes are not necessarily additive:

$$\text{volume percent} = \frac{\text{volume of liquid in question}}{\text{total volume of solution}} \times 100$$

Volume percent is used to express the concentration of alcohol in beverages. Wines generally contain 12% alcohol by volume. This translates into 12 mL of alcohol in each 100 mL of wine. The beverage industry also uses the concentration unit of *proof* (twice the volume percent). Pure alcohol is 100%, therefore 200 proof. Scotch whiskey is 86 proof or 43% alcohol.

Molarity

Mass percent solutions do not equate or express the molar masses of the solute in solution. For example, 1000. g of 10.0% NaOH solution contains 100. g of NaOH; 1000. g 10.0% KOH solution contains 100. g of KOH. In terms of moles of NaOH and KOH, these solutions contain

$$\text{mol NaOH} = 100. \text{ g NaOH} \times \frac{1 \text{ mol NaOH}}{40.00 \text{ g NaOH}} = 2.50 \text{ mol NaOH}$$

$$\text{mol KOH} = 100. \text{ g KOH} \times \frac{1 \text{ mol KOH}}{56.11 \text{ g KOH}} = 1.78 \text{ mol KOH}$$

From these figures we see that the two 10.0% solutions do not contain the same number of moles of NaOH and KOH. Yet 1 mol of each of these bases will neutralize the same amount of acid. As a result we find that a 10.0% NaOH solution has more reactive alkali than a 10.0% KOH solution.

We need a method of expressing concentration that will easily indicate how many moles of solute are present per unit volume of solution. For this purpose the molar method of expressing concentration is used.

A 1 molar solution contains 1 mol of solute per liter of solution. For example, to make a 1 molar solution of sodium hydroxide, NaOH, we dissolve 40.00 g of NaOH (1 mol) in water and dilute the solution with more water to a volume of 1 L. The solution contains 1 mol of the solute in 1 L of solution and is said to be 1 molar in concentration. Figure 14.7 illustrates the preparation of a 1 molar solution. Note that the volume of the solute and the solvent together is 1 L.

The concentration of a solution can, of course, be varied by using more or less solute or solvent; but in any case the **molarity** of a solution is the number of moles of solute per liter of solution. The abbreviation for molarity is *M*. The units of molarity are moles per liter. The expression "2.0 *M* NaOH" means a 2.0 molar solution of NaOH (2.0 mol, or 80.00 g, of NaOH dissolved in 1 L of solution).

molarity (M)

$$\textbf{molarity} = M = \frac{\text{number of moles of solute}}{\text{liter of solution}} = \frac{\text{moles}}{\text{liter}}$$

Flasks that are calibrated to contain specific volumes at a particular temperature are used to prepare solutions of a desired concentration. These *volumetric flasks* have a calibration mark on the neck to indicate accurately the measured volume. Molarity is based on a specific volume of solution and therefore will vary slightly

FIGURE 14.7
Preparation of a 1 *M* solution. ▶

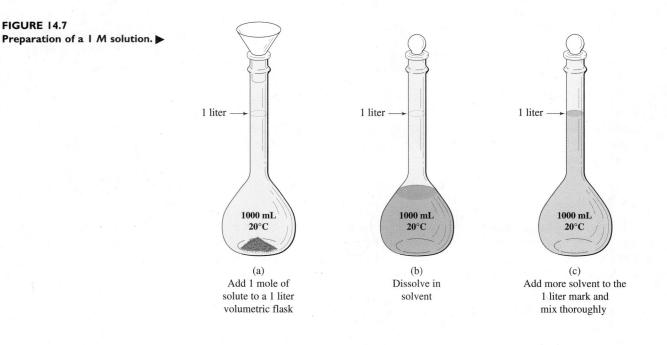

(a)	(b)	(c)
Add 1 mole of solute to a 1 liter volumetric flask	Dissolve in solvent	Add more solvent to the 1 liter mark and mix thoroughly

with temperature because volume varies with temperature (1000 mL of H_2O at 20°C = 1001 mL at 25°C).

Suppose we want to make 500 mL of 1 *M* solution. This solution can be prepared by determining the mass of 0.5 mol of the solute and diluting with water in a 500-mL (0.5-L) volumetric flask. The molarity will be

$$M = \frac{0.5 \text{ mol solute}}{0.5 \text{ L solution}} = 1 \text{ molar}$$

Thus you can see that it is not necessary to have a liter of solution to express molarity. All we need to know is the number of moles of dissolved solute and the volume of solution. Thus 0.001 mol of NaOH in 10 mL of solution is 0.1 *M*:

$$\frac{0.001 \text{ mol}}{10 \text{ mL}} \times \frac{1000 \text{ mL}}{1 \text{ L}} = 0.1 \; M$$

When we stop to think that a balance is not calibrated in moles but in grams, we can incorporate grams into the molarity formula. We do so by using the relationship:

$$\text{moles} = \frac{\text{grams of solute}}{\text{molar mass}}$$

Substituting this relationship into our expression for molarity, we get

$$M = \frac{\text{mol}}{\text{L}} = \frac{\text{g solute}}{\text{molar mass solute} \times \text{L solution}}$$

$$= \frac{\text{g}}{\text{molar mass} \times \text{L}}$$

We can now determine the mass of any amount of a solute that has a known formula, dilute it to any volume, and calculate the molarity of the solution using this formula.

The molarities of the concentrated acids commonly used in the laboratory are

HCl   12 *M*
$HC_2H_3O_2$ 17 *M*
HNO_3 16 *M*
H_2SO_4 18 *M*

What is the molarity of a solution containing 1.4 mol of acetic acid, $HC_2H_3O_2$, in 250. mL of solution?   **Example 14.5**

Substitute the data, 1.4 mol and 250. mL (0.250 L), directly into the equation for molarity: **Solution**

$$M = \frac{mol}{L} = \frac{1.4\ mol}{0.250\ L} = \frac{5.6\ mol}{1\ L} = 5.6\ M$$

By the unit conversion method we note that the concentration given in the problem statement is 1.4 mol per 250. mL (mol/mL). Since molarity = mol/L, the needed conversion is

$$\frac{mol}{mL} \longrightarrow \frac{mol}{L} = M$$

$$\frac{1.4\ mol}{250.\ mL} \times \frac{1000\ mL}{1\ L} = \frac{5.6\ mol}{1\ L} = 5.6\ M$$

What is the molarity of a solution made by dissolving 2.00 g of potassium chlorate in enough water to make 150. mL of solution?   **Example 14.6**

This problem can be solved using the unit conversion method. The steps in the conversions must lead to units of moles/liter: **Solution**

$$\frac{g\ KClO_3}{mL} \longrightarrow \frac{g\ KClO_3}{L} \longrightarrow \frac{mol\ KClO_3}{L} = M$$

The data are

$$g = 2.00\ g \qquad molar\ mass\ KClO_3 = 122.6\ g/mol \qquad volume = 150.\ mL$$

$$\frac{2.00\ g\ KClO_3}{150.\ mL} \times \frac{1000\ mL}{1\ L} \times \frac{1\ mol\ KClO_3}{122.6\ g\ KClO_3} = \frac{0.109\ mol}{1\ L} = 0.109\ M$$

How many grams of potassium hydroxide are required to prepare 600. mL of 0.450 M KOH solution?   **Example 14.7**

The conversion is   **Solution**

$$milliliters \longrightarrow liters \longrightarrow moles \longrightarrow grams$$

The data are

$$volume = 600.\ mL \quad M = \frac{0.450\ mol}{L} \quad molar\ mass\ KOH = \frac{56.11\ g\ KOH}{mol}$$

The calculation is

$$600.\ mL \times \frac{1\ L}{1000\ mL} \times \frac{0.450\ mol}{L} \times \frac{56.11\ g\ KOH}{mol} = 15.1\ g\ KOH$$

Practice 14.3

What is the molarity of a solution made by dissolving 7.50 g of magnesium nitrate, $Mg(NO_3)_2$, in enough water to make 25.0 mL of solution?

Practice 14.4

How many grams of sodium chloride are needed to prepare 125 mL of a 0.037 M NaCl solution?

Example 14.8 How many milliliters of 2.00 M HCl will react with 28.0 g of NaOH?

Solution

Step 1 Write and balance the equation for the reaction:

$$HCl(aq) + NaOH(aq) \longrightarrow NaCl(aq) + H_2O(aq)$$

The equation states that 1 mol of HCl reacts with 1 mol of NaOH.

Step 2 Find the number of moles of NaOH in 28.0 g of NaOH:

g NaOH $\longrightarrow$ mol NaOH

$$28.0 \text{ g NaOH} \times \frac{1 \text{ mol}}{40.00 \text{ g}} = 0.700 \text{ mol NaOH}$$

28.0 g NaOH = 0.700 mol NaOH

Step 3 Solve for moles and volume of HCl needed. From Steps 1 and 2 we see that 0.700 mol of HCl will react with 0.700 mol of NaOH, because the ratio of moles reacting is 1:1. We know that 2.00 M HCl contains 2.00 mol of HCl per liter, and so the volume that contains 0.700 mol of HCl will be less than 1 L:

mol NaOH $\longrightarrow$ mol HCl $\longrightarrow$ L HCl $\longrightarrow$ mL HCl

$$0.700 \text{ mol NaOH} \times \frac{1 \text{ mol HCl}}{1 \text{ mol NaOH}} \times \frac{1 \text{ L HCl}}{2.00 \text{ mol HCl}} = 0.350 \text{ L HCl}$$

$$0.350 \text{ L HCl} \times \frac{1000 \text{ mL}}{1 \text{ L}} = 350. \text{ mL HCl}$$

Therefore, 350. mL of 2.00 M HCl contains 0.700 mol HCl and will react with 0.700 mol, or 28.0 g, of NaOH.

Example 14.9 What volume of 0.250 M solution can be prepared from 16.0 g of potassium carbonate?

Solution

We start with 16.0 g of K_2CO_3 and need to find the volume of 0.250 M solution that can be prepared from this K_2CO_3. The conversion, therefore, is

g K_2CO_3 $\longrightarrow$ mol K_2CO_3 $\longrightarrow$ L solution

The data are

$$16.0 \text{ g K}_2\text{CO}_3 \quad M = \frac{0.250 \text{ mol}}{1 \text{ L}} \quad \text{molar mass K}_2\text{CO}_3 = \frac{138.2 \text{ g K}_2\text{CO}_3}{1 \text{ mol}}$$

$$16.0 \text{ g K}_2\text{CO}_3 \times \frac{1 \text{ mol K}_2\text{CO}_3}{138.2 \text{ g K}_2\text{CO}_3} \times \frac{1 \text{ L}}{0.250 \text{ mol K}_2\text{CO}_3} = 0.463 \text{ L (463 mL)}$$

Thus, 463 mL of 0.250 M solution can be made from 16.0 g K_2CO_3.

Calculate the number of moles of nitric acid in 325 mL of 16 M HNO$_3$ solution. **Example 14.10**

Solution

Use the equation

$$\text{moles} = \text{liters} \times M$$

Substitute the data given in the problem and solve:

$$\text{moles} = 0.325 \text{ L} \times \frac{16 \text{ mol HNO}_3}{1 \text{ L}} = 5.2 \text{ mol HNO}_3$$

Practice 14.5

What volume of 0.035 M AgNO$_3$ can be made from 5.0 g of AgNO$_3$?

Practice 14.6

How many milliliters of 0.50 M NaOH are required to react completely with 25.00 mL of 1.5 M HCl?

Dilution Problems

Chemists often find it necessary to dilute solutions from one concentration to another by adding more solvent to the solution. If a solution is diluted by adding pure solvent, the volume of the solution increases, but the number of moles of solute in the solution remains the same. Thus, the moles/liter (molarity) of the solution decreases. It is important to read a problem carefully to distinguish between (1) how much solvent must be added to dilute a solution to a particular concentration and (2) to what volume a solution must be diluted to prepare a solution of a particular concentration.

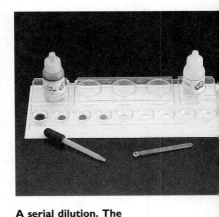

A serial dilution. The concentration of food coloring in cup 1 is 1 part per 10 (by weight), cup 2 is 1 part per 100; cup 3 is 1 part per 1000, and so on. The concentration in cup 6 is 1 part per million (ppm).

Calculate the molarity of a sodium hydroxide solution that is prepared by mixing 100. mL of 0.20 M NaOH with 150. mL of water. Assume the volumes are additive. **Example 14.11**

Solution

This problem is a dilution problem. If we double the volume of a solution by adding water, we cut the concentration in half. Therefore, the concentration of the above solution should be less than 0.10 M. In the dilution, the moles of NaOH remain constant; the molarity and volume change. The final volume is (100. mL + 150. mL) or 250. mL.

To solve this problem, (1) calculate the moles of NaOH in the original solution, and (2) divide the moles of NaOH by the final volume of the solution to obtain the new molarity.

Step 1 Calculate the moles of NaOH in the original solution:

$$M = \frac{\text{mol}}{\text{L}} \qquad \text{mol} = \text{L} \times M$$

$$0.100 \; \cancel{L} \times \frac{0.20 \text{ mol NaOH}}{1 \; \cancel{L}} = 0.020 \text{ mol NaOH}$$

Step 2 Solve for the new molarity, taking into account that the total volume of the solution after dilution is 250. mL (0.250 L):

$$M = \frac{0.020 \text{ mol NaOH}}{0.250 \text{ L}} = 0.080 \; M \text{ NaOH}$$

Alternative Solution

When the moles of solute in a solution before and after dilution are the same, then the moles before and after dilution may be set equal to each other:

$$\text{mol}_1 = \text{mol}_2$$

where mol_1 = moles before dilution, and mol_2 = moles after dilution. Then

$$\text{mol}_1 = \text{L}_1 \times M_1 \qquad \text{mol}_2 = \text{L}_2 \times M_2$$

$$\text{L}_1 \times M_1 = \text{L}_2 \times M_2$$

When both volumes are in the same units, a more general statement can be made:

$$V_1 \times M_1 = V_2 \times M_2$$

For this problem

$$V_1 = 100. \text{ mL} \qquad M_1 = 0.20 \; M$$
$$V_2 = 250. \text{ mL} \qquad M_2 = \text{(unknown)}$$

Then

$$100. \text{ mL} \times 0.20 \; M = 250. \text{ mL} \times M_2$$

Solving for M_2, we get

$$M_2 = \frac{100. \; \cancel{mL} \times 0.20 \; M}{250. \; \cancel{mL}} = 0.080 \; M \text{ NaOH}$$

Practice 14.7

Calculate the molarity of a solution prepared by diluting 125 mL of 0.400 M $K_2Cr_2O_7$ with 875 mL of water.

How many grams of silver chloride will be precipitated by adding sufficient silver nitrate to react with 1500. mL of 0.400 M barium chloride solution?   **Example 14.12**

$$2\,AgNO_3(aq) \;+\; BaCl_2(aq) \longrightarrow 2\,AgCl(s) \;+\; Ba(NO_3)_2(aq)$$

                        1 mol   2 mol

This problem is a stoichiometry problem. The fact that $BaCl_2$ is in solution means that we need to consider the volume and concentration of the solution in order to determine the number of moles of $BaCl_2$ reacting.   **Solution**

Step 1 Determine the number of moles of $BaCl_2$ in 1500. mL of 0.400 M solution:

$$M = \frac{mol}{L} \qquad mol = L \times M \qquad 1500.\ mL = 1.500\ L$$

$$1.500\ \cancel{L} \times \frac{0.400\ mol\ BaCl_2}{\cancel{L}} = 0.600\ mol\ BaCl_2$$

Step 2 Use the mole-ratio method to calculate the moles and grams of AgCl:

$$mol\ BaCl_2 \longrightarrow mol\ AgCl \longrightarrow g\ AgCl$$

$$0.600\ \cancel{mol\ BaCl_2} \times \frac{2\ \cancel{mol\ AgCl}}{1\ \cancel{mol\ BaCl_2}} \times \frac{143.4\ g\ AgCl}{\cancel{mol\ AgCl}} = 172\ g\ AgCl$$

Practice 14.8

How many grams of lead(II) iodide will be precipitated by adding sufficient $Pb(NO_3)_2$ to react with 750 mL of 0.250 M KI solution?

$$2\,KI(aq) \;+\; Pb(NO_3)_2(aq) \longrightarrow PbI_2(s) \;+\; 2\,KNO_3(aq)$$

Normality

Normality is another way of expressing the concentration of a solution. It is based on an alternative chemical unit of mass called the *equivalent mass*. The **normality** of a solution is the concentration expressed as the number of equivalent masses (equivalents, abbreviated equiv) of solute per liter of solution. A 1 normal (1 N) solution contains 1 equivalent mass of solute per liter of solution. Normality is widely used in analytical chemistry because it simplifies many of the calculations involving solution concentration: normality

$$normality = N = \frac{number\ of\ equivalents\ of\ solute}{1\ liter\ of\ solution} = \frac{equivalents}{liter}$$

where

$$number\ of\ equivalents\ of\ solute = \frac{grams\ of\ solute}{equivalent\ mass\ of\ solute}$$

Every substance may be assigned an equivalent mass. The equivalent mass may be equal either to the molar mass of the substance or to an integral fraction of the molar mass (i.e., the molar mass divided by 2, 3, 4, and so on). To gain an understanding of the meaning of equivalent mass, let us start by considering these two reactions:

$$HCl(aq) + NaOH(aq) \longrightarrow NaCl(aq) + H_2O(l)$$

1 mole   1 mol
(36.46 g)   (40.00 g)

$$H_2SO_4(aq) + 2\,NaOH(aq) \longrightarrow Na_2SO_4(aq) + 2\,H_2O(l)$$

1 mole   2 mol
(98.08 g)   (80.00 g)

We note first that 1 mol of hydrochloric acid reacts with 1 mol of sodium hydroxide and 1 mol of sulfuric acid reacts with 2 mol of NaOH. If we make 1 M solutions of these substances, 1 L of 1 M HCl will react with 1 L of 1 M NaOH, and 1 L of 1 M H_2SO_4 will react with 2 L of 1 M NaOH. From this reaction, we can see that H_2SO_4 has twice the chemical capacity of HCl when reacting with NaOH. We can, however, adjust these acid solutions to be equivalent in reactivity by dissolving only 0.5 mol of H_2SO_4 per liter of solution. By doing so, we find that we are required to use 49.04 g of H_2SO_4/L (instead of 98.08 g of H_2SO_4/L) to make a solution that is equivalent to one made from 36.46 g of HCl/L. These masses, 49.04 g of H_2SO_4 and 36.46 g of HCl, are chemically equivalent and are known as the equivalent masses of these substances, because each will react with the same amount of NaOH (40.00 g). The equivalent mass of HCl is equal to its molar mass, but that of H_2SO_4 is one-half its molar mass.

Thus, 1 L of solution containing 36.46 g of HCl would be 1 N, and 1 L of solution containing 49.04 g of H_2SO_4 would also be 1 N. A solution containing 98.08 g of H_2SO_4 (1 mol per liter) would be 2 N when reacting with NaOH in the given equation.

equivalent mass

The **equivalent mass** is the mass of a substance that will react with, combine with, contain, replace, or in any other way be equivalent to 1 mol of hydrogen atoms or hydrogen ions.

Normality and molarity can be interconverted in the following manner:

$$N = \frac{equiv}{L} \qquad M = \frac{mol}{L}$$

$$N = M \times \frac{equiv}{mol} = \frac{mol}{L} \times \frac{equiv}{mol} = \frac{equiv}{L}$$

$$M = N \times \frac{mol}{equiv} = \frac{equiv}{L} \times \frac{mol}{equiv} = \frac{mol}{L}$$

Thus a 2.0 N H_2SO_4 solution is 1.0 M.

One application of normality and equivalents is in acid–base neutralization reactions. An equivalent of an acid is that mass of the acid that will furnish 1 mol of H^+ ions. An equivalent of a base is that mass of base that will furnish 1 mol of OH^- ions. Using concentrations in normality, 1 equivalent of acid (A) will react with 1 equivalent of base (B):

$$N_A = \frac{equiv_A}{L_A} \qquad and \qquad N_B = \frac{equiv_B}{L_B}$$

$$\text{equiv}_A = L_A \times N_A \quad \text{and} \quad \text{equiv}_B = L_B \times N_B$$

Since $\text{equiv}_A = \text{equiv}_B$,

$$L_A \times N_A = L_B \times N_B$$

When both volumes are in the same units, we can write a more general equation:

$$V_A N_A = V_B N_B$$

which states that the volume of acid times the normality of the acid equals the volume of base times the normality of the base.

(a) What is the normality of an H_2SO_4 solution if 25.00 mL of the solution requires 22.48 mL of 0.2018 N NaOH for complete neutralization? (b) What is the molarity of the H_2SO_4 solution?   **Example 14.13**

(a) Solve for N_A by substituting the data into   **Solution**

$$V_A N_A = V_B N_B$$

$$25.00 \text{ mL} \times N_A = 22.48 \text{ mL} \times 0.2018 \ N$$

$$N_A = \frac{22.48 \text{ mL} \times 0.2018 \ N}{25.00 \text{ mL}} = 0.1815 \ N \ H_2SO_4$$

(b) When H_2SO_4 is completely neutralized it furnishes 2 equivalents of H^+ ions per mole of H_2SO_4. The conversion from N to M is

$$\frac{\text{equiv}}{L} \longrightarrow \frac{\text{mol}}{L}$$

$$H_2SO_4 = 0.1815 \ N$$

$$\frac{0.1815 \text{ equiv}}{1 \text{ L}} \times \frac{1 \text{ mol}}{2 \text{ equiv}} = 0.09075 \text{ mol/L}$$

The H_2SO_4 solution is 0.09075 M.

Practice 14.9

What is the normality of a NaOH solution if 50.0 mL of the solution requires 23.72 mL of 0.0250 N H_2SO_4? What is the molarity of the NaOH?

The equivalent mass of a substance may be variable; its value is dependent on the reaction that the substance is undergoing. Consider the reactions represented by these equations:

$$NaOH + H_2SO_4 \longrightarrow NaHSO_4 + H_2O$$

$$2 \ NaOH + H_2SO_4 \longrightarrow Na_2SO_4 + 2 \ H_2O$$

In the first reaction 1 mol of H_2SO_4 furnishes 1 mol of hydrogen atoms. Therefore the equivalent mass of H_2SO_4 is the molar mass, namely 98.08 g. But in the

Microencapsulation

Producing chemical reactions that will occur at precisely the correct moment is one of the tasks facing the chemist in industry. For this to happen, one or more of the reactants must be stored separately and released under controlled conditions precisely when the reaction is desired. A technique developed to accomplish this is microencapsulation, in which reactive chemicals—solids, liquids, or gases—are sealed inside tiny capsules. The material forming the wall of the capsule is carefully chosen so that the encapsulated chemicals can be released, at the appropriate time, by one of several methods. This release can be accomplished in a variety of ways, which include dissolving the capsules, diffusion through the capsule walls, and by mechanical, thermal, electrical, or chemical disruption of the capsules.

In one type of microencapsulation, water diffuses into the capsule and forms a solution which then diffuses out into the surroundings at a constant rate. Some types of capsules contain materials that dissolve at a certain level of acidity and form pores in the capsule through which the encapsulated materials escape. Still other types of capsules dissolve completely over a given period of time, releasing their contents into the system.

Applications of microencapsulation are found everywhere. Carbonless paper, often used in receipts makes use of pressure-sensitive microcapsules containing colorless dye precursors. Another reactive substance is present and converts the precursor to the colored form when pressure is applied by a pen or printer.

Adhesives are frequently encapsulated to prevent them from becoming tacky too soon. The active surfaces on pressure-sensitive labels and certain self-sealing envelopes are coated with encapsulated adhesives that are released by pressure. Heat-sensitive encapsulation is used for

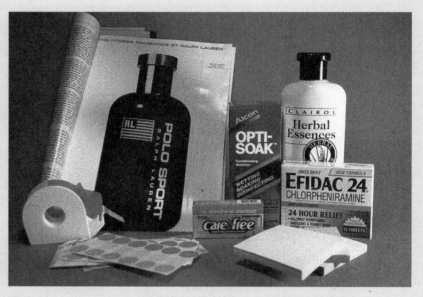

Many items we come across in our daily lives are microencapsulated. These are just a few examples.

the adhesives in iron-on patches for clothing.

Many microencapsulated products are to be found in the kitchen. Flavorings are encapsulated to make them easier to store in a powdered state, cut evaporation, and reduce reactions with the air. These advantages increase the shelf life of products. Flavoring microcapsules may also be heat sensitive and release their contents during cooking or pressure sensitive (as in chewing gum) and release their contents upon chewing.

Still other encapsulated products are found in our bathrooms. Time-release microencapsulation is used in deodorants, moisturizers, colognes, and perfumes. The encapsulation process prevents evaporation, decomposition, and unwanted reactions with the air and other ingredients. Drugs and medications are frequently encapsulated to dissolve slowly over a long

period of time in the body. These medications generally work in the intestinal tract, but some may also be given by injection to work within other tissues.

Fragrances have undergone microencapsulation in such products as cosmetics, health care products, detergents, and even foods. Gas companies use encapsulated propyl mercaptan, $CH_3CH_2CH_2SH$, to teach children how to detect a gas leak (natural gas without this substance is odorless). Encapsulated fragrances are responsible for the ever-present scratch-and-sniff labels found in children's books and fashion magazines. When the paper is scratched or pulled open, the fragrance is released into the air.

Other applications of microencapsulation include time-release pesticides and neutralizer for contact lenses, as well as special additives in detergents, cleaners, and paints.

TABLE 14.4 Concentration Units for Solutions

Units	Symbol	Definition
Mass percent	% m/m	$\dfrac{\text{Mass solute}}{\text{Mass solution}} \times 100$
Parts per million	ppm	$\dfrac{\text{Mass solute}}{\text{Mass solution}} \times 1{,}000{,}000$
Mass/volume percent	% m/v	$\dfrac{\text{Mass solute}}{\text{mL solution}} \times 100$
Volume percent	% v/v	$\dfrac{\text{mL solute}}{\text{mL solution}} \times 100$
Molarity	M	$\dfrac{\text{Moles solute}}{\text{L solution}}$
Normality	N	$\dfrac{\text{Equivalents solute}}{\text{L solution}}$
Molality	m	$\dfrac{\text{Moles solute}}{\text{kg solvent}}$

second reaction 1 mol of H_2SO_4 furnishes 2 mol of hydrogen atoms. Therefore, the equivalent mass of the H_2SO_4 is one-half the molar mass, or 49.04 g. A summary of the quantitative concentration units is found in Table 14.4.

14.7   Colligative Properties of Solutions

Two solutions—one containing 1 mol (60.06 g) of urea, NH_2CONH_2, and the other containing 1 mol (342.3 g) of sucrose, $C_{12}H_{22}O_{11}$, in 1 kg of water—both have a freezing point of $-1.86°C$, not $0°C$ as for pure water. Urea and sucrose are distinctly different substances, yet they lower the freezing point of the water by the same amount. The only thing apparently common to these two solutions is that each contains 1 mol (6.022×10^{23} molecules) of solute and 1 kg of solvent. In fact, when we dissolve 1 mol of any nonionizable solute in 1 kg of water, the freezing point of the resulting solution is $-1.86°C$.

These results lead us to conclude that the freezing-point depression for a solution containing 6.022×10^{23} solute molecules (particles) and 1 kg of water is a constant, namely, $1.86°C$. Freezing-point depression is a general property of solutions. Furthermore, the amount by which the freezing point is depressed is the same for all solutions made with a given solvent; that is, each solvent shows a characteristic *freezing-point depression constant*. Freezing-point depression constants for several solvents are given in Table 14.5.

The solution formed by the addition of a nonvolatile solute to a solvent has a lower freezing point, a higher boiling point, and a lower vapor pressure than that of the pure solvent. All these effects are related and are known as colligative

TABLE 14.5 Freezing-Point Depression and Boiling-Point Elevation Constants of Selected Solvents

Solvent	Freezing point of pure solvent (°C)	Freezing-point depression constant, K_f $\left(\dfrac{°C \text{ kg solvent}}{\text{mol solute}}\right)$	Boiling point of pure solvent (°C)	Boiling-point elevation constant, K_b $\left(\dfrac{°C \text{ kg solvent}}{\text{mol solute}}\right)$
Water	0.00	1.86	100.0	0.512
Acetic acid	16.6	3.90	118.5	3.07
Benzene	5.5	5.1	80.1	2.53
Camphor	178	40	208.2	5.95

colligative properties

properties. The **colligative properties** are properties that depend only on the number of solute particles in a solution and not on the nature of those particles. Freezing-point depression, boiling-point elevation, and vapor-pressure lowering are colligative properties of solutions.

The colligative properties of a solution can be considered in terms of vapor pressure. The vapor pressure of a pure liquid depends on the tendency of molecules to escape from its surface. Thus, if 10% of the molecules in a solution are nonvolatile solute molecules, the vapor pressure of the solution is 10% lower than that of the pure solvent. The vapor pressure is lower because the surface of the solution contains 10% nonvolatile molecules and 90% of the volatile solvent molecules. A liquid boils when its vapor pressure equals the pressure of the atmosphere. Thus, we can see that the solution just described as having a lower vapor pressure will have a higher boiling point than the pure solvent. The solution with a lowered vapor pressure does not boil until it has been heated above the boiling point of the solvent (see Figure 14.8(a)). Each solvent has its own characteristic boiling-point elevation constant (Table 14.5). The boiling-point elevation constant is based on a solution that contains 1 mol of solute particles per kilogram of solvent. For example, the boiling-point elevation constant for a solution containing 1 mol of solute particles per kilogram of water is 0.512°C, which means that this water solution will boil at 100.512°C.

The freezing behavior of a solution can also be considered in terms of lowered vapor pressure. Figure 14.8(b) shows the vapor-pressure relationships of ice, water, and a solution containing 1 mol of solute per kilogram of water. The freezing point of water is at the intersection of the water and ice vapor-pressure curves (i.e., at the point where water and ice have the same vapor pressure). Because the vapor pressure of water is lowered by the solute, the vapor-pressure curve of the solution does not intersect the vapor-pressure curve of ice until the solution has been cooled below the freezing point of pure water. Thus, it is necessary to cool the solution below 0°C in order for it to freeze.

The foregoing discussion dealing with freezing-point depressions is restricted to *un-ionized* substances. The discussion of boiling-point elevations is restricted to *nonvolatile* and un-ionized substances. The colligative properties of ionized substances are not under consideration at this point, but are discussed in Chapter 15.

Some practical applications involving colligative properties are (1) use of salt–ice mixtures to provide low freezing temperatures for homemade ice cream,

Engine coolant is an example of the use of colligative properties. The additon of coolant to the water in a radiator raises its boiling point and lowers its freezing point.

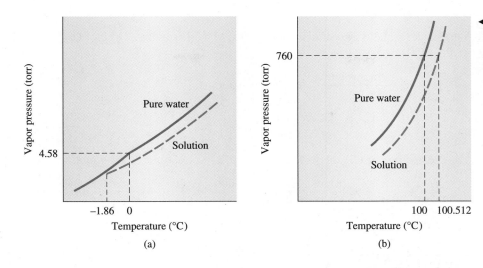

◀ FIGURE 14.8
Vapor-pressure curves of pure
water and water solutions,
showing (a) freezing-point
depression and (b) boiling-point
elevation effects (concentration:
1 mol solute/1 kg water).

(2) use of sodium chloride or calcium chloride to melt ice from streets, and (3) use of ethylene glycol and water mixtures as antifreeze in automobile radiators (ethylene glycol also raises the boiling point of radiator fluid, thus allowing the engine to operate at a higher temperature).

Both the freezing-point depression and the boiling-point elevation are directly proportional to the number of moles of solute per kilogram of solvent. When we deal with the colligative properties of solutions, another concentration expression, *molality*, is used. The **molality (*m*)** of a solute is the number of moles of solute per kilogram of solvent:

molality (*m*)

$$m = \frac{\text{mol solute}}{\text{kg solvent}}$$

Note that a lowercase *m* is used for molality concentrations and a capital *M* for molarity. The difference between molality and molarity is that molality refers to moles of solute *per kilogram of solvent*, whereas molarity refers to moles of solute *per liter of solution*. For un-ionized substances, the colligative properties of a solution are directly proportional to its molality.

Molality is independent of volume. It is a mass-to-mass relationship of solute to solvent and allows for experiments, such as freezing-point depression and boiling-point elevation, to be conducted at variable temperatures.

The following equations are used in calculations involving colligative properties and molality:

$$\Delta t_f = mK_f \qquad \Delta t_b = mK_b \qquad m = \frac{\text{mol solute}}{\text{kg solvent}}$$

m = molality; mol solute/kg solvent

Δt_f = freezing-point depression; °C

Δt_b = boiling-point elevation; °C

K_f = freezing-point depression constant; °C kg solvent/mol solute

K_b = boiling-point elevation constant; °C kg solvent/mol solute

Sodium chloride or calcium chloride is used to melt ice on snowy streets and highways.

Example 14.14 What is the molality (*m*) of a solution prepared by dissolving 2.70 g of CH_3OH in 25.0 g of H_2O?

Solution

Since $m = \dfrac{\text{mol solute}}{\text{kg solvent}}$, the conversion is

$$\frac{2.70 \text{ g } CH_3OH}{25.0 \text{ g } H_2O} \longrightarrow \frac{\text{mol } CH_3OH}{25.0 \text{ g } H_2O} \longrightarrow \frac{\text{mol } CH_3OH}{1 \text{ kg } H_2O}$$

The molar mass of CH_3OH is $(12.01 + 4.032 + 16.00)$ or 32.04 g/mol:

$$\frac{2.70 \text{ g } CH_3OH}{25.0 \text{ g } H_2O} \times \frac{1 \text{ mol } CH_3OH}{32.04 \text{ g } CH_3OH} \times \frac{1000 \text{ g } H_2O}{1 \text{ kg } H_2O} = \frac{3.37 \text{ mol } CH_3OH}{1 \text{ kg } H_2O}$$

The molality is 3.37 *m*.

Practice 14.10

What is the molality of a solution prepared by dissolving 150.0 g of $C_6H_{12}O_6$ in 600.0 g of H_2O?

Example 14.15 A solution is made by dissolving 100. g of ethylene glycol ($C_2H_6O_2$) in 200. g of water. What is the freezing point of this solution?

Solution To calculate the freezing point of the solution, we first need to calculate Δt_f, the change in freezing point. Use the equation

$$\Delta t_f = mK_f = \frac{\text{mol solute}}{\text{kg solvent}} \times K_f$$

K_f (for water): $\dfrac{1.86°C \text{ kg solvent}}{\text{mol solute}}$ (from Table 14.5)

mol solute: $100. \text{ g } C_2H_6O_2 \times \dfrac{1 \text{ mol } C_2H_6O_2}{62.07 \text{ g } C_2H_6O_2} = 1.61 \text{ mol } C_2H_6O_2$

kg solvent: $200. \text{ g } H_2O \times \dfrac{1 \text{ kg}}{1000 \text{ g}} = 0.200 \text{ kg } H_2O$

$$\Delta t_f = \frac{1.61 \text{ mol } C_2H_6O_2}{0.200 \text{ kg } H_2O} \times \frac{1.86°C \text{ kg } H_2O}{1 \text{ mol } C_2H_6O_2} = 15.0°C$$

The freezing-point depression, 15.0°C, must be subtracted from 0°C, the freezing point of the pure solvent (water):

freezing point of solution = freezing point of solvent $- \Delta t_f$

$$= 0.0°C - 15.0°C = -15.0°C$$

Therefore, the freezing point of the solution is $-15.0°C$. The calculation can also be done using the equation

$$\Delta t_f = K_f \times \frac{\text{g solute}}{\text{molar mass solute}} \times \frac{1}{\text{kg solvent}}$$

A solution made by dissolving 4.71 g of a compound of unknown molar mass in 100.0 g of water has a freezing point of $-1.46°C$. What is the molar mass of the compound?

Example 14.16

First substitute the data in $\Delta t_f = mK_f$ and solve for m:

$\Delta t_f = +1.46$ (since the solvent, water, freezes at $0°C$)

$$K_f = \frac{1.86°C \text{ kg } H_2O}{\text{mol solute}}$$

$$1.46°C = mK_f = m \times \frac{1.86°C \text{ kg } H_2O}{\text{mol solute}}$$

$$m = \frac{1.46°C \times \text{mol solute}}{1.86°C \times \text{kg } H_2O} = \frac{0.785 \text{ mol solute}}{\text{kg } H_2O}$$

Now convert the data, 4.71 g solute/100.0 g H_2O, to g/mol:

$$\frac{4.71 \text{ g solute}}{100.0 \text{ g } H_2O} \times \frac{1000 \text{ g } H_2O}{1 \text{ kg } H_2O} \times \frac{1 \text{ kg } H_2O}{0.785 \text{ mol solute}} = 60.0 \text{ g/mol}$$

The molar mass of the compound is 60.0 g/mol.

Practice 14.11

What is the freezing point of the solution in Practice exercise 14.10? What is the boiling point?

14.8 Osmosis and Osmotic Pressure

When red blood cells are put into distilled water, they gradually swell and, in time, may burst. If red blood cells are put in a 5% urea (or a 5% salt) solution, they gradually shrink and take on a wrinkled appearance. The cells behave in this fashion because they are enclosed in semipermeable membranes. A **semipermeable membrane** allows the passage of water (solvent) molecules through it in either direction but prevents the passage of larger solute molecules or ions. When two solutions of different concentrations (or water and a water solution) are separated by a semipermeable membrane, water diffuses through the membrane from the solution of lower concentration into the solution of higher concentration. The diffusion of water, either from a dilute solution or from pure water, through a semipermeable membrane into a solution of higher concentration is called **osmosis**.

semipermeable membrane

osmosis

A 0.90% (0.15 M) sodium chloride solution is known as a *physiological saline solution* because it is *isotonic* with blood plasma; that is, it has the same osmotic pressure as blood plasma. Because each mole of NaCl yields about 2 mol of ions when in solution, the solute particle concentration in physiological saline solution is nearly 0.30 M. Five percent glucose solution (0.28 M) is also approximately isotonic with blood plasma. Blood cells neither swell nor shrink in an isotonic solution. The cells described in the first paragraph of this section swell in water

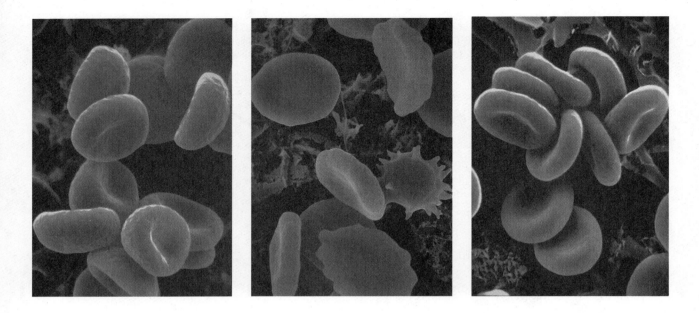

▲
Human red blood cells. *Left:* In a hypotonic solution (0.2% saline) the cells swell as water moves into the cell center. *Center:* In a hypertonic solution (1.6% saline) water leaves the cells causing them to crenate (shrink). *Right:* In an isotonic solution the concentration is the same inside and outside the cell (0.9% saline). Cells do not change in size. Magnification is 260,000 X.

because water is *hypotonic* to cell plasma. The cells shrink in 5% urea solution because the urea solution is *hypertonic* to the cell plasma. In order to prevent possible injury to blood cells by osmosis, fluids for intravenous use are usually made up at approximately isotonic concentration.

All solutions exhibit *osmotic pressure*, which is another colligative property. Osmotic pressure is dependent only on the concentration of the solute particles and is independent of their nature. The osmotic pressure of a solution can be measured by determining the amount of counterpressure needed to prevent osmosis; this pressure can be very large. The osmotic pressure of a solution containing 1 mol of solute particles in 1 kg of water is about 22.4 atm, which is about the same as the pressure exerted by 1 mol of a gas confined in a volume of 1 L at 0°C.

Osmosis has a role in many biological processes, and semipermeable membranes occur commonly in living organisms. An example is the roots of plants, which are covered with tiny structures called root hairs; soil water enters the plant by osmosis, passing through the semipermeable membranes covering the root hairs. Artificial or synthetic membranes can also be made.

Osmosis can be demonstrated with the simple laboratory setup shown in Figure 14.9. As a result of osmotic pressure, water passes through the cellophane membrane into the thistle tube, causing the solution level to rise. In osmosis the net transfer of water is always from a less concentrated to a more concentrated solution; that is, the effect is toward equalization of the concentration on both sides of the membrane. It should also be noted that the effective movement of water in osmosis is always from the region of *higher water concentration* to the region of *lower water concentration*.

Osmosis can be explained by assuming that a semipermeable membrane has passages that permit water molecules and other small molecules—to pass in either direction. Both sides of the membrane are constantly being struck by water molecules in random motion. The number of water molecules crossing the membrane is proportional to the number of water molecule-to-membrane impacts per unit of time.

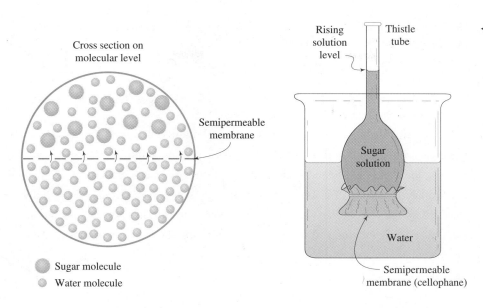

Cross section on
molecular level

Semipermeable
membrane

Rising
solution
level

Thistle
tube

Sugar
solution

Water

Semipermeable
membrane (cellophane)

Sugar molecule
Water molecule

◄ FIGURE 14.9
Laboratory demonstration of
osmosis: As a result of osmosis,
water passes through the
membrane causing the solution
to rise in the thistle tube.

Because the solute molecules or ions reduce the concentration of water, there are more water molecules and more water-molecule impacts on the side with the lower solute concentration (more dilute solution). The greater number of water molecule-to-membrane impacts on the dilute side thus causes a net transfer of water to the more concentrated solution. Again, note that the overall process involves the net transfer, by diffusion through the membrane, of water molecules from a region of higher water concentration (dilute solution) to one of lower water concentration (more concentrated solution).

This explanation is a simplified picture of osmosis. No one has ever seen the hypothetical passages that allow water molecules and other small molecules or ions to pass through them. Alternative explanations have been proposed, but our discussion has been confined to water solutions. Osmotic pressure is a general colligative property, however, and is known to occur in nonaqueous systems.

Concepts in Review

1. Describe the types of solutions.

2. List the general properties of solutions.

3. Describe and illustrate the process by which an ionic substance dissolves in water.

4. Indicate the effects of temperature and pressure on the solubility of solids and gases in liquids.

5. Identify and explain the factors affecting the rate at which a solid dissolves in a liquid.

6. Use a solubility table or graph to determine whether a solution is saturated, unsaturated, or supersaturated at a given temperature.

7. Calculate the mass percent or volume percent for a solution.

8. Calculate the amount of solute in a given quantity of a solution when given the mass percent or volume percent of a solution.

9. Calculate the molarity of a solution from the volume and the mass, or moles, of solute.

10. Calculate the mass of a substance necessary to prepare a solution of specified volume and molarity.

11. Determine the resulting molarity in a typical dilution problem.

12. Apply stoichiometry to chemical reactions involving solutions.

13. Use the concepts of equivalent mass and normality in calculations.

14. Explain the effect of a solute on the vapor pressure of a solvent.

15. Explain the effect of a solute on boiling point and freezing point of a solution.

16. Calculate the boiling and freezing points of a solution from concentration data.

17. Calculate molality and molar mass of a solute from boiling/freezing point data.

Key Terms

The terms listed here have been defined within this chapter. Section numbers are referenced in parenthesis for each term.

colligative properties (14.7)
concentrated solution (14.6)
concentration of a solution (14.2)
dilute solution (14.6)
equivalent mass (14.6)
immiscible (14.2)
miscible (14.2)
molality (*m*) (14.7)
molarity (*M*) (14.6)
normality (14.6)

osmosis (14.8)
parts per million (ppm) (14.6)
saturated solution (14.6)
semipermeable membrane (14.8)
solubility (14.2)
solute (14.1)
solution (14.1)
solvent (14.1)
supersaturated solution (14.6)
unsaturated solution (14.6)

Questions

Questions refer to tables, figures, and key words and concepts defined within the chapter. A particularly challenging question or exercise is indicated with an asterisk.

1. Make a sketch indicating the orientation of water molecules (a) about a single sodium ion and (b) about a single chloride ion in solution.

2. Estimate the number of grams of sodium fluoride that would dissolve in 100 g of water at 50°C. (See Table 14.2.)

3. What is the solubility at 25°C of each of the substances listed below? (See Figure 14.4.)
   (a) potassium chloride   (c) potassium nitrate
   (b) potassium chlorate

4. What is different in the solubility trend of the potassium halides compared with that of the lithium halides and the sodium halides? (See Table 14.2.)

5. What is the solubility, in grams of solute per 100 g of H_2O of (a) $KClO_3$ at 60°C, (b) HCl at 20°C, (c) Li_2SO_4 at 80°C, and (d) KNO_3 at 0°C? (See Figure 14.4.)

6. Which substance, KNO_3 or NH_4Cl, shows the greater increase in solubility with increased temperature? (See Figure 14.4.)

7. Does a 2 molal solution in benzene or a 1 molal solution in camphor show the greater freezing-point depression? (See Table 14.5.)

8. What would be the total surface area if the 1-cm cube in Figure 14.5 were cut into cubes 0.01 cm on a side?

9. At which temperatures—10°C, 20°C, 30°C, 40°C, or 50°C—would you expect a solution made from 63 g of ammonium chloride and 150 g of water to be unsaturated? (See Figure 14.4.)

10. Explain why the rate of dissolving decreases as shown in Figure 14.6.

11. Would the volumetric flasks in Figure 14.7 be satisfactory for preparing normal solutions? Explain.

12. Assume that the thistle tube in Figure 14.9 contains 1.0 M sugar solution and that the water in the beaker has just been replaced by a 2.0 M solution of urea. Would the solution level in the thistle tube continue to rise, remain constant, or fall? Explain.

13. Name and distinguish between the two components of a solution.

14. Is it always apparent in a solution which component is the solute, for example, in a solution of a liquid in a liquid?

15. Explain why the solute does not settle out of a solution.

16. Is it possible to have one solid dissolved in another? Explain.

17. An aqueous solution of KCl is colorless, $KMnO_4$ is purple, and $K_2Cr_2O_7$ is orange. What color would you expect of an aqueous solution of $Na_2Cr_2O_7$? Explain.

18. Explain why hexane will dissolve benzene but will not dissolve sodium chloride.

19. Some drinks like tea are consumed either hot or cold, whereas others like Coca Cola are drunk only cold. Why?

20. Why is air considered to be a solution?

21. In which will a teaspoonful of sugar dissolve more rapidly, 200 mL of iced tea or 200 mL of hot coffee? Explain in terms of the KMT.

22. What is the effect of pressure on the solubility of gases in liquids? Solids in liquids?

23. Why do smaller particles dissolve faster than large ones?

24. In a saturated solution containing undissolved solute, solute is continuously dissolving, but the concentration of the solution remains unchanged. Explain.

25. Explain why there is no apparent reaction when crystals of $AgNO_3$ and NaCl are mixed, but a reaction is apparent immediately when solutions of $AgNO_3$ and NaCl are mixed.

26. What do we mean when we say that concentrated nitric acid, HNO_3, is 16 molar?

27. Will 1 L of 1 M NaCl contain more chloride ions than 0.5 L of 1 M $MgCl_2$? Explain.

28. Champagne is usually cooled in a refrigerator prior to opening. It is also opened very carefully. What would happen if a warm bottle of champagne is shaken and opened quickly and forcefully?

29. Explain how a supersaturated solution of $NaC_2H_3O_2$ can be prepared and proven to be supersaturated.

30. Explain in terms of the KMT how a semipermeable membrane functions when placed between pure water and a 10% sugar solution.

31. Which has the higher osmotic pressure, a solution containing 100 g of urea, NH_2CONH_2, in 1 kg of H_2O or a solution containing 150 g of glucose, $C_6H_{12}O_6$, in 1 kg of H_2O?

32. Explain why a lettuce leaf in contact with salad dressing containing salt and vinegar soon becomes wilted and limp whereas another lettuce leaf in contact with plain water remains crisp.

33. A group of shipwreck survivors floated for several days on a life raft before being rescued. Those who had drunk some seawater were found to be suffering the most from dehydration. Explain.

34. Which of the following statements are correct? Rewrite the incorrect statements to make them correct.
    (a) A solution is a homogeneous mixture.
    (b) It is possible for the same substance to be the solvent in one solution and the solute in another.
    (c) A solute can be removed from a solution by filtration.
    (d) Saturated solutions are always concentrated solutions.
    (e) If a solution of sugar in water is allowed to stand undisturbed for a long time, the sugar will gradually settle to the bottom of the container.
    (f) It is not possible to prepare an aqueous 1.0 M AgCl solution.
    (g) Gases are generally more soluble in hot water than in cold water.
    (h) It is impossible to prepare a two-phase liquid mixture from two liquids that are miscible with each other in all proportions.
    (i) A solution that is 10% NaCl by mass always contains 10 g of NaCl.
    (j) Small changes in pressure have little effect on the solubility of solids in liquids but a marked effect on the solubility of gases in liquids.
    (k) How fast a solute dissolves depends mainly on the size of the solute particles, the temperature of the solvent, and the degree of agitation or stirring taking place.
    (l) In order to have a 1 M solution you must have 1 mol of solute dissolved in sufficient solvent to give 1 L of solution.
    (m) Dissolving 1 mol of NaCl in 1 L of water will give a 1 M solution.
    (n) One mole of solute in 1 L of solution has the same concentration as 0.1 mol of solute in 100 mL of solution.
    (o) When 100 mL of 0.200 M HCl is diluted to 200-mL volume by the addition of water, the resulting solution is 0.100 M and contains one-half the number of moles of HCl as were in the original solution.

(p) Fifty milliliters of 0.1 *M* H_2SO_4 will neutralize the same volume of 0.1 *M* NaOH as 100 mL of 0.1 *M* HCl.

(q) Fifty milliliters of 0.1 *N* H_2SO_4 will neutralize the same volume of 0.1 *M* NaOH as 100 mL of 0.1 *M* HCl.

(r) The molarity of a solution will vary slightly with temperature.

(s) The equivalent mass of $Ca(OH)_2$ is one-half its molar mass.

(t) Gram for gram, methyl alcohol, CH_3OH, is more effective than ethyl alcohol, C_2H_5OH, in lowering the freezing point of water.

(u) An aqueous solution that freezes below 0°C will have a normal boiling point below 100°C.

(v) The colligative properties of a solution depend on the number of solute particles dissolved in solution.

(w) A solution of 1.00 mol of a nonionizable solute and 1000 g of water will freeze at −1.86°C and will boil at 99.5°C at atmospheric pressure.

(x) Water will diffuse from a 0.1 *M* sugar solution to a 0.2 *M* sugar solution when these two solutions are separated by a semipermeable membrane.

(y) An isotonic salt solution has the same osmotic pressure as blood plasma.

(z) Red blood cells will neither swell nor shrink when placed into an isotonic salt solution.

35. What disadvantages are there in expressing the concentration of solutions as dilute or concentrated?

36. Explain how concentrated H_2SO_4 can be both 18 *M* and 36 *N* in concentration.

37. Describe how you would prepare 750 mL of 5 *M* NaCl solution.

38. Arrange the following bases (in descending order) according to the volume of each that will react with 1 L of 1 *M* HCl; (a) 1 *M* NaOH, (b) 1.5 *M* $Ca(OH)_2$, (c) 2 *M* KOH, and (d) 0.6 *M* $Ba(OH)_2$.

*39. Explain in terms of vapor pressure why the boiling point of a solution containing a nonvolatile solute is higher than that of the pure solvent.

40. Explain why the freezing point of a solution is lower than the freezing point of the pure solvent.

41. Which would be colder, a glass of water and crushed ice or a glass of Seven-Up and crushed ice? Explain.

42. When water and ice are mixed, the temperature of the mixture is 0°C. But, if methyl alcohol and ice are mixed, a temperature of −10°C is readily attained. Explain why the two mixtures show such different temperature behavior.

43. Which would be more effective in lowering the freezing point of 500. g of water?
(a) 100. g of sucrose ($C_{12}H_{22}O_{11}$) or 100. g of ethyl alcohol (C_2H_5OH)
(b) 100. g of sucrose or 20.0 g of ethyl alcohol
(c) 20.0 g of ethyl alcohol or 20.0 g of methyl alcohol (CH_3OH)

44. Is the molarity of a 5 molal aqueous solution of NaCl greater or less than 5 molar? Explain.

45. What is microencapsulation?

46. Explain how a scratch-and-sniff label works.

47. What are the purposes for timed-release microencapsulation?

48. State three types of microencapsulation systems and give a practical application of each.

Paired Exercises

These exercises are paired. Each odd-numbered exercise is followed by a similar even-numbered exercise. Answers to the even-numbered exercises are given in Appendix V.

49. Which of the substances listed below are reasonably soluble and which are insoluble in water? (See Figure 14.2 or Appendix IV.)
(a) KOH
(b) $NiCl_2$
(c) ZnS
(d) $AgC_2H_3O_2$
(e) Na_2CrO_4

50. Which of the substances listed below are reasonably soluble and which are insoluble in water? (See Figure 14.2 or Appendix IV.)
(a) PbI_2
(b) $MgCO_3$
(c) $CaCl_2$
(d) $Fe(NO_3)_3$
(e) $BaSO_4$

Percent Solutions

51. Calculate the mass percent of the following solutions:
    (a) 25.0 g NaBr + 100.0 g H_2O
    (b) 1.20 g K_2SO_4 + 10.0 g H_2O

53. How many grams of a solution that is 12.5% by mass $AgNO_3$ would contain 30.0 g of $AgNO_3$?

55. Calculate the mass percent of the following solutions:
    (a) 60.0 g NaCl + 200.0 g H_2O
    (b) 0.25 mol $HC_2H_3O_2$ + 3.0 mol H_2O

57. How much solute is present in 65 g of 5.0% KCl solution?

59. Calculate the mass/volume percent of a solution made by dissolving 22.0 g of CH_3OH (methanol) in C_2H_5OH (ethanol) to make 100. mL of solution.

61. What is the volume percent of 10.0 mL of CH_3OH (methanol) dissolved in water to a volume of 40.0 mL?

52. Calculate the mass percent of the following solutions.
    (a) 40.0 g $Mg(NO_3)_2$ + 500.0 g H_2O
    (b) 17.5 g $NaNO_3$ + 250.0 g H_2O

54. How many grams of a solution that is 12.5% by mass $AgNO_3$ would contain 0.400 mol of $AgNO_3$?

56. Calculate the mass percent of the following solutions:
    (a) 145.0 g NaOH in 1.5 kg H_2O
    (b) 1.0 m solution of $C_6H_{12}O_6$ in water

58. How much solute is present in 250. g of 15.0% K_2CrO_4 solution?

60. Calculate the mass/volume percent of a solution made by dissolving 4.20 g of NaCl in H_2O to make 12.5 mL of solution.

62. What is the volume percent of 2.0 mL of hexane, C_6H_{14}, dissolved in benzene, C_6H_6, to a volume of 9.0 mL?

Molarity Problems

63. Calculate the molarity of the following solutions:
    (a) 0.10 mol of solute in 250 mL of solution
    (b) 2.5 mol of NaCl in 0.650 L of solution
    (c) 53.0 g of Na_2CrO_4 in 1.00 L of solution
    (d) 260 g of $C_6H_{12}O_6$ in 800. mL of solution

65. Calculate the number of moles of solute in each of the following solutions:
    (a) 40.0 L of 1.0 M LiCl
    (b) 25.0 mL of 3.00 M H_2SO_4

67. Calculate the grams of solute in each of the following solutions:
    (a) 150 L of 1.0 M NaCl
    (b) 260 mL of 18 M H_2SO_4

69. How many milliliters of 0.256 M KCl solution will contain the following?
    (a) 0.430 mol of KCl
    (b) 20.0 g of KCl

64. Calculate the molarity of the following solutions:
    (a) 0.025 mol of HCl in 10. mL of solution
    (b) 0.35 mol $BaCl_2 \cdot H_2O$ in 593 mL of solution
    (c) 1.50 g of $Al_2(SO_4)_3$ in 2.00 L of solution
    (d) 0.0282 g of $Ca(NO_3)_2$ in 1.00 mL of solution

66. Calculate the number of moles of solute in each of the following solutions:
    (a) 349 mL of 0.0010 M NaOH
    (b) 5000. mL of 3.1 M $CoCl_2$

68. Calculate the grams of solute in each of the following solutions:
    (a) 0.035 L of 10.0 M HCl
    (b) 8.00 mL of 8.00 M $Na_2C_2O_4$

70. How many milliliters of 0.256 M KCl solution will contain the following?
    (a) 10.0 mol of KCl
    *(b) 71.0 g of chloride ion, Cl^-

Dilution Problems

71. What will be the molarity of the resulting solutions made by mixing the following? Assume volumes are additive.
    (a) 200. mL of 12 M HCl + 200.0 mL H_2O
    (b) 60.0 mL of 0.60 M $ZnSO_4$ + 500. mL H_2O

73. Calculate the volume of concentrated reagent required to prepare the diluted solutions indicated:
    (a) 12 M HCl to prepare 400. mL of 6.0 M HCl
    (b) 16 M HNO_3 to prepare 100. mL of 2.5 N HNO_3

75. What will be the molarity of each of the solutions made by mixing 250 mL of 0.75 M H_2SO_4 with (a) 150 mL of H_2O? (b) 250 mL of 0.70 M H_2SO_4?

72. What will be the molarity of the resulting solutions made by mixing the following? Assume volumes are additive.
    (a) 100. mL 1.0 M HCl + 150 mL 2.0 M HCl
    (b) 25.0 mL 12.5 M NaCl + 75.0 mL 2.00 M NaCl

74. Calculate the volume of concentrated reagent required to prepare the diluted solutions indicated:
    (a) 15 M NH_3 to prepare 50. mL of 6.0 M NH_3
    (b) 18 M H_2SO_4 to prepare 250 mL of 10.0 N H_2SO_4

76. What will be the molarity of each of the solutions made by mixing 250 mL of 0.75 M H_2SO_4 with (a) 400. mL of 2.50 M H_2SO_4? (b) 375 mL of H_2O?

Stoichiometry Problems

77. $BaCl_2(aq) + K_2CrO_4(aq) \rightarrow BaCrO_4(s) + 2\ KCl(aq)$
Using the above equation, calculate
  (a) The grams of $BaCrO_4$ that can be obtained from 100.0 mL of 0.300 M $BaCl_2$
  (b) The volume of 1.0 M $BaCl_2$ solution needed to react with 50.0 mL of 0.300 M K_2CrO_4 solution

79. Given the balanced equation
$6\ FeCl_2(aq) + K_2Cr_2O_7(aq) + 14\ HCl(aq) \rightarrow$
$6\ FeCl_3(aq) + 2\ CrCl_3(aq) + 2\ KCl(aq) + 7\ H_2O(l)$
  (a) How many moles of KCl will be produced from 2.0 mol of $FeCl_2$?
  (b) How many moles of $CrCl_3$ will be produced from 1.0 mol of $FeCl_2$?
  (c) How many moles of $FeCl_2$ will react with 0.050 mol of $K_2Cr_2O_7$?
  (d) How many milliliters of 0.060 M $K_2Cr_2O_7$ will react with 0.025 mol of $FeCl_2$?
  (e) How many milliliters of 6.0 M HCl will react with 15.0 mL of 6.0 M $FeCl_2$?

78. $3\ MgCl_2(aq) + 2\ Na_3PO_4(aq) \rightarrow$
$Mg_3(PO_4)_2(s) + 6\ NaCl(aq)$
Using the above equation, calculate:
  (a) The milliliters of 0.250 M Na_3PO_4 that will react with 50.0 mL of 0.250 M $MgCl_2$.
  (b) The grams of $Mg_3(PO_4)_2$ that will be formed from 50.0 mL of 0.250 M $MgCl_2$.

80. $2\ KMnO_4(aq) + 16\ HCl(aq) \rightarrow$
$2\ MnCl_2(aq) + 5\ Cl_2(g) + 8\ H_2O(l) + 2\ KCl(aq)$
Calculate the following using the above equation:
  (a) The moles of Cl_2 produced from 0.050 mol of $KMnO_4$
  (b) The moles of HCl required to react with 1.0 L of 2.0 M $KMnO_4$
  (c) The milliliters of 6.0 M HCl required to react with 200. mL of 0.50 M $KMnO_4$
  (d) The liters of Cl_2 gas at STP produced by the reaction of 75.0 mL of 6.0 M HCl

Equivalent Mass and Normality Problems

81. Calculate the equivalent mass of the acid and base in each of the following reactions:
  (a) $HCl + NaOH \rightarrow NaCl + H_2O$
  (b) $2\ HCl + Ba(OH)_2 \rightarrow 2\ H_2O + BaCl_2$
  (c) $H_2SO_4 + Ca(OH)_2 \rightarrow CaSO_4 + 2\ H_2O$

83. What is the normality of the following solutions? Assume complete neutralization.
  (a) 4.0 M HCl
  (b) 0.243 M HNO_3
  (c) 3.0 M H_2SO_4

85. What volume of 0.2550 N NaOH is required to neutralize
  (a) 20.22 mL of 0.1254 N HCl?
  (b) 14.86 mL of 0.1246 N H_2SO_4?

82. Calculate the equivalent mass of the acid and base in each of the following reactions:
  (a) $H_2SO_4 + KOH \rightarrow KHSO_4 + H_2O$
  (b) $H_3PO_4 + 2\ LiOH \rightarrow Li_2HPO_4 + 2\ H_2O$
  (c) $HNO_3 + NaOH \rightarrow NaNO_3 + H_2O$

84. What is the normality of the following solutions? Assume complete neutralization.
  (a) 1.85 M H_3PO_4
  (b) 0.250 M $HC_2H_3O_2$
  (c) 1.25 M NaOH

86. What volume of 0.2550 N NaOH is required to neutralize
  (a) 21.30 mL of 0.1430 M HCl?
  (b) 18.00 mL of 0.1430 M H_2SO_4?

Molality and Colligative Properties Problems

87. Calculate the molality of these solutions:
  (a) 14.0 g of CH_3OH in 100.0 g of H_2O
  (b) 2.50 mol of benzene (C_6H_6) in 250 g of hexane (C_6H_{14})

89. (a) What is the molality of a solution containing 100.0 g of ethylene glycol, $C_2H_6O_2$, in 150.0 g of water?
  (b) What is the boiling point of this solution?
  (c) What is the freezing point of this solution?

*91. The freezing point of a solution of 8.00 g of an unknown compound dissolved in 60.0 g of acetic acid is 13.2°C. Calculate the molar mass of the compound.

88. Calculate the molality of these solutions:
  (a) 1.0 g of $C_6H_{12}O_6$ in 1.0 g of H_2O
  (b) 0.250 mol of iodine in 1.0 kg of H_2O

90. What is (a) the molality, (b) the freezing point, and (c) the boiling point of a solution containing 2.68 g of naphthalene, $C_{10}H_8$, in 38.4 g of benzene, C_6H_6?

*92. What is the molar mass of a compound if 4.80 g of the compound dissolved in 22.0 g of H_2O gives a solution that freezes at -2.50°C?

Additional Exercises

These exercises are not paired or labeled by topic and provide additional practice on concepts covered in this chapter.

93. How many grams of solution, 10.0% NaOH by mass, are required to neutralize 150 mL of a 1.0 M HCl solution?

*94. How many grams of solution, 10.0% NaOH by mass, are required to neutralize 250.0 g of a 1.0 m solution of HCl?

*95. A sugar syrup solution contains 15.0% sugar, $C_{12}H_{22}O_{11}$, by mass and has a density of 1.06 g/mL.
  (a) How many grams of sugar are in 1.0 L of this syrup?
  (b) What is the molarity of this solution?
  (c) What is the molality of this solution?

*96. A solution of 3.84 g of C_4H_2N (empirical formula) in 250.0 g of benzene depresses the freezing point of benzene 0.614°C. What is the molecular formula for the compound?

*97. Hydrochloric acid, HCl, is sold as a concentrated aqueous solution (12.0 mol/L). If the density of the solution is 1.18 g/mL, determine the molality of the solution.

*98. How many grams of KNO_3 must be used to make 450 mL of a solution that is to contain 5.5 mg/mL of potassium ion? Calculate the molarity of the solution.

99. What mass of 5.50% solution can be prepared from 25.0 g KCl?

100. Physiological saline, NaCl, solutions used in intravenous injections have a concentration of 0.90% NaCl (mass/volume).
  (a) How many grams of NaCl are needed to prepare 500.0 mL of this solution?
  *(b) How much water must evaporate from this solution to give a solution that is 9.0% NaCl (mass/volume)?

*101. A solution is made from 50.0 g of KNO_3 and 175 g of H_2O. How many grams of water must evaporate to give a saturated solution of KNO_3 in water at 20°C? (See Figure 14.4.)

102. What volume of 70.0% rubbing alcohol can you prepare if you have only 150 mL of pure isopropyl alcohol on hand?

103. At 20°C an aqueous solution of HNO_3 that is 35.0% HNO_3 by mass has a density of 1.21 g/mL.
  (a) How many grams of HNO_3 are present in 1.00 L of this solution?
  (b) What volume of this solution will contain 500. g of HNO_3?

*104. What is the molarity of a nitric acid solution, if the solution is 35.0% HNO_3 by mass and has a density of 1.21 g/mL?

105. To what volume must a solution of 80.0 g of H_2SO_4 in 500.0 mL of solution be diluted to give a 0.10 M solution?

106. How many milliliters of water must be added to 300.0 mL of 1.40 M HCl to make a solution that is 0.500 M HCl?

107. A 10.0-mL sample of 16 M HNO_3 is diluted to 500.0 mL. What is the molarity of the final solution?

108. Given a 5.00 M KOH solution, how would you prepare 250.0 mL of 0.625 M KOH?

109. **(a)** How many moles of hydrogen will be liberated from 200.0 mL of 3.00 M HCl reacting with an excess of magnesium? The equation is

$$Mg(s) + 2\ HCl(aq) \rightarrow MgCl_2(aq) + H_2(g)$$

  (b) How many liters of hydrogen gas, H_2, measured at 27°C and 720 torr, will be obtained? (*Hint*: Use the ideal gas equation.)

*110. What is the molarity of an HCl solution, 150.0 mL of which, when treated with excess magnesium, liberates 3.50 L of H_2 gas measured at STP?

111. What is the normality of an H_2SO_4 solution if 36.26 mL are required to neutralize 2.50 g of $Ca(OH)_2$?

112. Which will be more effective in neutralizing stomach acid, HCl: a tablet containing 12.0 g of $Mg(OH)_2$ or a tablet containing 10.0 g of $Al(OH)_3$? Show evidence for your answer.

113. Which would be more effective as an antifreeze in an automobile radiator? A solution containing
  (a) 10 kg of methyl alcohol, CH_3OH, or 10 kg of ethyl alcohol, C_2H_5OH?
  (b) 10 m solution of methyl alcohol or 10 m solution of ethyl alcohol?

114. Automobile battery acid is 38% H_2SO_4 and has a density of 1.29 g/mL. Calculate the molality and the molarity of this solution.

*115. A sugar solution made to feed hummingbirds contains 1.00 lb of sugar, $C_{12}H_{22}O_{11}$, to 4.00 lb of water. Can this solution be put outside without freezing where the temperature falls to 20.0°F at night? Show evidence for your answer.

*116. What would be (a) the molality and (b) the boiling point of an aqueous sugar, $C_{12}H_{22}O_{11}$, solution that freezes at -5.4°C?

117. A solution of 6.20 g of $C_2H_6O_2$ in water has a freezing point of $-0.372°C$. How many grams of H_2O are in the solution?

118. What (a) mass and (b) volume of ethylene glycol ($C_2H_6O_2$, density = 1.11 g/mL) should be added to 12.0 L of water in an automobile radiator to protect it fom freezing at $-20°C$? (c) To what temperature Fahrenheit will the radiator be protected?

119. Can a saturated solution ever be a dilute solution? Explain.

120. What volume of 0.65 M HCl is needed to completely neutralize 12 g of NaOH?

*121. If 150 mL of 0.055 M HNO_3 are needed to completely neutralize 1.48 g of an *impure* sample of sodium hydrogen carbonate (baking soda), what percent of the sample is baking soda?

122. (a) How much water must be added to concentrated sulfuric acid, H_2SO_4 (17.8 M), to prepare 8.4 L of 1.5 M sulfuric acid solution?
(b) How many moles of H_2SO_4 would be in each milliliter of the original concentrate?
(c) How many moles would be in each milliliter of the diluted solution?

123. An aqueous solution freezes at $-3.6°C$. What is its boiling temperature?

*124. How would you prepare a 6.00 M HNO_3 solution if only 3.00 M and 12.0 M solutions of the acid were available for mixing?

*125. A 20.0-mL portion of an HBr solution of unknown strength was diluted to exactly 240 mL. Now if 100.0 mL of this diluted solution required 88.4 mL of 0.37 M NaOH to achieve complete neutralization, what was the strength of the original HBr solution?

126. When 80.5 mL of 0.642 M $Ba(NO_3)_2$ is mixed with 44.5 mL of 0.743 M KOH, a precipitate of $Ba(OH)_2$ forms. How many grams of $Ba(OH)_2$ do you expect?

127. Exactly three hundred grams of a 5.0% sucrose solution is to be prepared. How many grams of a 2.0% solution of sucrose would contain the same number of grams of sugar?

128. Lithium carbonate, Li_2CO_3, is a drug used to treat manic depression. A 0.25 M solution of it is prepared.
(a) How many moles of Li_2CO_3 are present in 45.8 mL of the solution?
(b) How many grams of Li_2CO_3 are in 750 mL of the same solution?
(c) How many milliliters of the solution would be needed to supply 6.0 g of the solute?
(d) If the solution has a density of 1.22 g/mL, what is its mass percent?

129. If a student accidentally mixed 400.0 mL of 0.35 M HCl with 1100 mL of 0.65 M HCl, what would be the final molarity of the hydrochloric acid solution?

130. Suppose you start with 100. mL of distilled water. Then suppose you add one drop (20 drops/mL) of concentrated acetic acid, $HC_2H_3O_2$ (17.8 M). What is the molarity of the resulting solution?

Answers to Practice Exercises

14.1 unsaturated
14.2 10.0% Na_2SO_4 solution
14.3 2.02 M
14.4 0.27 g NaCl
14.5 0.84 L (840 mL)
14.6 75 mL NaOH
14.7 5.00×10^{-2} M
14.8 43 g
14.9 1.19×10^{-2} N NaOH, 1.19×10^{-2} M NaOH
14.10 1.387 m
14.11 freezing point = $-2.58°C$, boiling point = $100.71°C$

Water is the medium for the reactions that provide the foundation for life on this planet. Many significant chemical reactions occur in aqueous solutions. Volcanos and hot springs yield acidic solutions formed from hydrochloric acid and sulfur dioxide. These acids mix with bases (such as calcium carbonate) dissolved from rocks and microscopic animals to form the salts of our oceans and produce the beauty of stalactites and stalagmites in limestone caves. Photosynthesis and respiration reactions cannot occur without an adequate balance between acids and bases. Even small deviations in this balance can be detrimental to living organisms. Human activity has resulted in changes in this delicate acid–base balance. For example, in recent years we have become very concerned about the effects of acid rain upon our environment. We are beginning to examine ways in which we can maintain the balance as we continue to change our planet.

15.1 Acids and Bases

The word *acid* is derived from the Latin *acidus,* meaning "sour" or "tart," and is also related to the Latin word *acetum,* meaning "vinegar." Vinegar has been known since antiquity as a product of the fermentation of wine and apple cider. The sour constituent of vinegar is acetic acid, $HC_2H_3O_2$. Some of the characteristic properties commonly associated with acids are the following:

1. sour taste
2. change the color of litmus, a vegetable dye, from blue to red
3. react with
   • metals such as zinc and magnesium to produce hydrogen gas
   • hydroxide bases to produce water and an ionic compound (salt)
   • carbonates to produce carbon dioxide

These properties are due to the hydrogen ions, H^+, released by acids in a water solution.

Classically, a *base* is a substance capable of liberating hydroxide ions, OH^-, in water solution. Hydroxides of the alkali metals (Group IA) and alkaline earth metal (Group IIA), such as LiOH, NaOH, KOH, $Ca(OH)_2$, and $Ba(OH)_2$, are the most common inorganic bases. Water solutions of bases are called *alkaline solutions* or *base solutions.* Some of the characteristic properties commonly associated with bases are the following:

1. bitter or caustic taste
2. a slippery, soapy feeling
3. the ability to change litmus from red to blue
4. the ability to interact with acids

Several theories have been proposed to answer the question "What is an acid and a base?" One of the earliest, most significant of these theories was advanced

◄ **Chapter Opening Photo:** As lava flows into the sea, a slightly acidic solution is formed.

in a doctoral thesis in 1884 by Svante Arrhenius (1859–1927), a Swedish scientist, who stated that "an acid is a hydrogen-containing substance that dissociates to produce hydrogen ions, and that a base is a hydroxide-containing substance that dissociates to produce hydroxide ions in aqueous solutions." Arrhenius postulated that the hydrogen ions were produced by the dissociation of acids in water, and that the hydroxide ions were produced by the dissociation of bases in water:

$$HA \longrightarrow H^+ + A^-$$
acid

$$MOH \longrightarrow M^+ + OH^-$$
base

Thus, an acid solution contains an excess of hydrogen ions and a base an excess of hydroxide ions.

In 1923, the Brønsted–Lowry proton transfer theory was introduced by J. N. Brønsted (1897–1947), a Danish chemist, and T. M. Lowry (1847–1936), an English chemist. This theory states that an acid is a proton donor and a base is a proton acceptor.

> **A Brønsted–Lowry acid is a proton (H^+) donor.**
> **A Brønsted–Lowry base is a proton (H^+) acceptor.**

Consider the reaction of hydrogen chloride gas with water to form hydrochloric acid:

$$HCl(g) + H_2O(l) \longrightarrow H_3O^+(aq) + Cl^-(aq) \tag{1}$$

In the course of the reaction, HCl donates, or gives up, a proton to form a Cl^- ion, and H_2O accepts a proton to form the H_3O^+ ion. Thus, HCl is an acid and H_2O is a base, according to the Brønsted–Lowry theory.

A hydrogen ion, H^+, is nothing more than a bare proton and does not exist by itself in an aqueous solution. In water a proton combines with a polar water molecule to form a hydrated hydrogen ion, H_3O^+, commonly called a **hydronium ion**. The proton is attracted to a polar water molecule, forming a bond with one of the two pairs of unshared electrons:

hydronium ion

$$H^+ + H\!:\!\ddot{O}\!:\ \longrightarrow\ \left[H\!:\!\ddot{O}\!:\!H\right]^+$$
$$\qquad\quad \overset{..}{H} \qquad\qquad\quad \overset{..}{H}$$
hydronium ion

Note the electron structure of the hydronium ion. For simplicity of expression in equations, we often use H^+ instead of H_3O^+, with the explicit understanding that H^+ is always hydrated in solution.

As illustrated in Equation (1) when a Brønsted–Lowry acid donates a proton, it forms the conjugate base of that acid. When a base accepts a proton, it forms the conjugate acid of that base. A conjugate acid and base are produced as products. The formulas of a conjugate acid–base pair differ by one proton (H^+). Consider what happens when HCl(g) is bubbled through water, as shown by the equation below:

conjugate acid–base pair

$$HCl(g) + H_2O(l) \longrightarrow Cl^-(aq) + H_3O^+(aq)$$

conjugate acid–base pair

acid   base   base   acid

The conjugate acid–base pairs are HCl—Cl^- and H_3O^+—H_2O. Cl^- is the conjugate base of HCl, and HCl is the conjugate acid of Cl^-. H_2O is the conjugate base of H_3O^+, and H_3O^+ is the conjugate acid of H_2O.

Another example of conjugate acid–base pairs is observed in the following equation:

$$NH_4^+ + H_2O \longrightarrow H_3O^+ + NH_3$$

acid   base   acid   base

In this equation the conjugate acid–base pairs are NH_4^+—NH_3 and H_3O^+—H_2O.

Example 15.1

Write the formula for (a) the conjugate base of H_2O and of HNO_3, and (b) the conjugate acid of SO_4^{2-} and of $C_2H_3O_2^-$.

Solution

(a) To write the conjugate base of an acid, remove one proton from the acid formula. Thus,

Remember, the difference between an acid or a base and its conjugate is one proton, H^+.

$$H_2O \xrightarrow{-H^+} OH^- \quad \text{(conjugate base)}$$

$$HNO_3 \xrightarrow{-H^+} NO_3^- \quad \text{(conjugate base)}$$

Note that, by removing an H^+, the conjugate base becomes more negative than the acid by one minus charge.

(b) To write the conjugate acid of a base, add one proton to the formula of the base. Thus,

$$SO_4^{2-} \xrightarrow{+H^+} HSO_4^- \quad \text{(conjugate acid)}$$

$$C_2H_3O_2^- \xrightarrow{+H^+} HC_2H_3O_2 \quad \text{(conjugate acid)}$$

In each case the conjugate acid becomes more positive than the base by one positive charge due to the addition of H^+.

Practice 15.1

Indicate the conjugate base for each of the following acids:
(a) H_2CO_3 (b) HNO_2 (c) $HC_2H_3O_2$

Practice 15.2

Indicate the conjugate acid for each of the following bases:
(a) HSO_4^- (b) NH_3 (c) OH^-

A more general concept of acids and bases was introduced by Gilbert N. Lewis. The Lewis theory deals with the way in which a substance with an unshared pair of electrons reacts in an acid–base type of reaction. According to this theory a base is any substance that has an unshared pair of electrons (electron-pair donor), and an acid is any substance that will attach itself to or accept a pair of electrons.

> **A Lewis acid is an electron pair acceptor.**
> **A Lewis base is an electron pair donor.**

In the reaction

$$H^+ \; + \; \overset{\displaystyle H}{\underset{\displaystyle H}{:\!\ddot{N}\!:\!H}} \; \longrightarrow \; \left[\overset{\displaystyle H}{\underset{\displaystyle H}{H\!:\!\ddot{N}\!:\!H}} \right]^+$$

$$\quad acid \qquad\qquad base$$

The H^+ is a Lewis acid and $:NH_3$ is a Lewis base. According to the Lewis theory, substances other than proton donors (e.g., BF_3) behave as acids:

$$\overset{\displaystyle F}{\underset{\displaystyle F}{F\!:\!\ddot{B}}} \; + \; \overset{\displaystyle H}{\underset{\displaystyle H}{:\!\ddot{N}\!:\!H}} \; \longrightarrow \; \overset{\displaystyle F\;H}{\underset{\displaystyle F\;H}{F\!:\!\ddot{B}\!:\!\ddot{N}\!:\!H}}$$

$$\quad acid \qquad\qquad base$$

The Lewis and Brønsted–Lowry bases are identical because, to accept a proton, a base must have an unshared pair of electrons.

The three theories are summarized in Table 15.1. These theories explain how acid–base reactions occur. We will generally use the theory that best explains the reaction that is under consideration. Most of our examples will refer to aqueous solutions. It is important to realize that in an aqueous acidic solution the H^+ ion concentration is always greater than OH^- ion concentration. And, vice versa, in an aqueous basic solution the OH^- ion concentration is always greater than the H^+ ion concentration. When the H^+ and OH^- ion concentrations in a solution are equal, the solution is neutral; that is, it is neither acidic nor basic.

TABLE 15.1 Summary of Acid–Base Definitions

Theory	Acid	Base
Arrhenius	A hydrogen-containing substance that produces hydrogen ions in aqueous solution	A hydroxide-containing substance that produces hydroxide ions in aqueous solution
Brønsted-Lowry	A proton (H^+) donor	A proton (H^+) acceptor
Lewis	Any species that will bond to an unshared pair of electrons (electron-pair acceptor)	Any species that has an unshared pair of electrons (electron-pair donor)

15.2 Reactions of Acids

In aqueous solutions the H^+ or H_3O^+ ions are responsible for the characteristic reactions of acids. The following reactions are in an aqueous medium.

Reaction with Metals Acids react with metals that lie above hydrogen in the activity series of elements to produce hydrogen and an ionic compound (salt) (see Section 17.5):

acid + metal $\longrightarrow$ hydrogen + ionic compound

$$2\ HCl(aq)\ +\ Ca(s)\ \longrightarrow\ H_2(g)\ +\ CaCl_2(aq)$$

$$H_2SO_4(aq)\ +\ Mg(s)\ \longrightarrow\ H_2(g)\ +\ MgSO_4(aq)$$

$$6\ HC_2H_3O_2(aq)\ +\ 2\ Al(s)\ \longrightarrow\ 3\ H_2(g)\ +\ 2\ Al(C_2H_3O_2)_3(aq)$$

Acids such as nitric acid (HNO_3) are oxidizing substances (see Chapter 17) and react with metals to produce water instead of hydrogen. For example:

$$3\ Zn(s)\ +\ 8\ HNO_3(dilute)\ \longrightarrow\ 3\ Zn(NO_3)_2(aq)\ +\ 2\ NO(g)\ +\ 4\ H_2O(l)$$

Reaction with Bases The interaction of an acid and a base is called a *neutralization reaction*. In aqueous solutions, the products of this reaction are a salt and water:

acid + base $\longrightarrow$ salt + water

$$HBr(aq)\ +\ KOH(aq)\ \longrightarrow\ KBr(aq)\ +\ H_2O(l)$$

$$2\ HNO_3(aq)\ +\ Ca(OH)_2(aq)\ \longrightarrow\ Ca(NO_3)_2(aq)\ +\ 2\ H_2O(l)$$

$$2\ H_3PO_4(aq)\ +\ 3\ Ba(OH)_2(aq)\ \longrightarrow\ Ba_3(PO_4)_2(s)\ +\ 6\ H_2O(l)$$

Reaction with Metal Oxides This reaction is closely related to that of an acid with a base. With an aqueous acid, the products are a salt and water:

acid + metal oxide $\longrightarrow$ salt + water

$$2\ HCl(aq)\ +\ Na_2O(s)\ \longrightarrow\ 2\ NaCl(aq)\ +\ H_2O(l)$$

$$H_2SO_4(aq)\ +\ MgO(s)\ \longrightarrow\ MgSO_4(aq)\ +\ H_2O(l)$$

$$6\ HCl(aq)\ +\ Fe_2O_3(s)\ \longrightarrow\ 2\ FeCl_3(aq)\ +\ 3\ H_2O(l)$$

Reaction with Carbonates Many acids react with carbonates to produce carbon dioxide, water, and an ionic compound:

$$H_2CO_3(aq)\ \longrightarrow\ CO_2(g)\ +\ H_2O(l)$$

Carbonic acid (H_2CO_3) is not the product, because it is unstable and decomposes into water and carbon dioxide.

acid + carbonate $\longrightarrow$ salt + water + carbon dioxide

$$2\ HCl(aq)\ +\ Na_2CO_3(aq)\ \longrightarrow\ 2\ NaCl(aq)\ +\ H_2O(l)\ +\ CO_2(g)$$

$$H_2SO_4(aq)\ +\ MgCO_3(s)\ \longrightarrow\ MgSO_4(aq)\ +\ H_2O(l)\ +\ CO_2(g)$$

15.3   Reactions of Bases

The OH^- ions are responsible for the characteristic reactions of bases. All the following reactions are in an aqueous medium.

Reaction with Acids Bases react with acids to produce a salt and water. See reaction of acids with bases in Section 15.2.

Amphoteric Hydroxides Hydroxides of certain metals, such as zinc, aluminum, and chromium, are **amphoteric**—that is, they are capable of reacting as either an acid or a base. When treated with a strong acid, they behave like bases; when reacted with a strong base, they behave like acids:

amphoteric

$$Zn(OH)_2(s) + 2\ HCl(aq) \longrightarrow ZnCl_2(aq) + 2\ H_2O(l)$$

$$Zn(OH)_2(s) + 2\ NaOH(aq) \longrightarrow Na_2Zn(OH)_4(aq)$$

Reaction of NaOH and KOH with Certain Metals Some amphoteric metals react directly with the strong bases sodium hydroxide and potassium hydroxide to produce hydrogen:

$$base + metal + water \longrightarrow salt + hydrogen$$

$$2\ NaOH(aq) + Zn(s) + 2\ H_2O(l) \longrightarrow Na_2Zn(OH)_4(aq) + H_2(g)$$

$$2\ KOH(aq) + 2\ Al(s) + 6\ H_2O(l) \longrightarrow 2\ KAl(OH)_4(aq) + 3\ H_2(g)$$

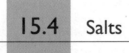

15.4   Salts

Salts are very abundant in nature. Most of the rocks and minerals of the earth's mantle are salts of one kind or another. Huge quantities of dissolved salts also exist in the oceans. Salts can be considered to be compounds that have been derived from acids and bases. They consist of positive metal or ammonium ions (H^+ excluded) combined with negative nonmetal ions (OH^- and O^{2-} excluded). The positive ion is the base counterpart and the nonmetal ion is the acid counterpart:

Chemists use the terms *ionic compound* and *salt* interchangeably.

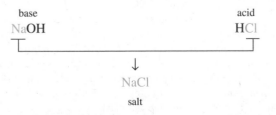

Salts are usually crystalline and have high melting and boiling points.

Pucker Power

The candy counter is filled with a variety of treats designed to cause your lips to pucker and to stimulate your tongue to send signals to your brain saying "SOUR." The human tongue has four types of taste receptors (known as "taste buds"). Sweet, bitter, salty, and sour taste buds are each concentrated in a different area of the tongue. These receptors are molecules that fit together with molecules in the candy (or other food), sending a signal to the brain, which is interpreted as a taste.

Substances that taste sour are acids. The substances that produce this sour taste in candy and other confections are often malic acid and/or citric acid.

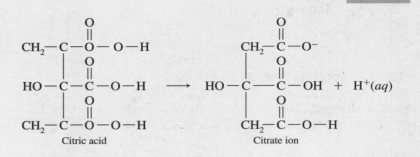

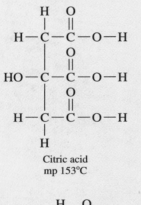

Citric acid
mp 153°C

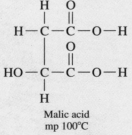

Malic acid
mp 100°C

Citric acid tastes more sour than malic acid. When these substances are mixed with other ingredients such as sugar, corn syrup, flavorings, and preservatives, the result is a candy both sweet and tart! The same acids can be mixed with synthetic rubber or chicle (dried latex from the sapo-

dilla tree) to produce bubble gum with real pucker power. The most sour of this type of gum is called Face Slammers.™ Try some on your favorite 12-year-old.

Gum manufacturers have gone a step further in incorporating acid–base chemistry into their products for an even greater surprise. In Mad Dawg™ gum the initial taste is sour. But after a couple of minutes of chewing, brightly colored foam begins to accumulate in your mouth and ooze out over your lips. What is going on here?

The foam is a mixture of sugar and saliva mixed with carbon dioxide bubbles released when several of the gum's ingredients are mixed in the watery environment of your mouth. The citric and malic acids dissociate to form hydrogen ions while sodium hydrogen carbonate (baking soda) dissolves into sodium and hydrogen carbonate ions:

$$NaHCO_3(s) \longrightarrow Na^+(aq) + HCO_3^-(aq)$$

The hydrogen ions from the acids mix with the hydrogen carbonate ions from the baking soda to produce water and carbon dioxide gas:

$$H^+(aq) + HCO_3^-(aq) \longrightarrow H_2O(l) + CO_2(g)$$

The acids stimulate production of saliva and the food coloring adds color to the foamy mess. The major problem for chemists in creating this treat was keeping the reactants apart until the consumer pops the gum into his or her mouth. As solids,

sodium hydrogen carbonate and citric acid do not react. But with the slightest amount of water present the process begins. When a tablet such as Alka Seltzer™ is manufactured, the ingredients (solid citric acid, sodium hydrogen carbonate, aspirin, and flavoring) are compressed into a tablet that is sealed into a dry foil packet. When opened and dropped into water the reaction (and the relief) begins and the bubbles are released.

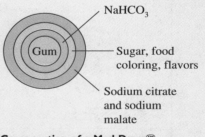

Cross-section of a Mad Dawg™ gum ball.

In Mad Dawg™ gum balls, the center core is moist rubber and the coating is applied in solution. Some of the early versions of these gum balls actually exploded as they were removed from the candy machine. To eliminate this problem multiple coatings are now used to keep the acid in one layer (on the outside to give the first sour taste) and the sodium hydrogen carbonate in an inner layer (see diagram). When the consumer crunches on the gum ball the layers begin to mix in the saliva and the fun begins!

From a single acid, such as hydrochloric acid (HCl), we can produce many chloride compounds by replacing the hydrogen with metal ions (e.g., NaCl, KCl, RbCl, $CaCl_2$, $NiCl_2$). Hence, the number of known salts greatly exceeds the number of known acids and bases. If the hydrogen atoms of a binary acid are replaced by a nonmetal, the resulting compound has covalent bonding and is therefore not considered to be ionic (e.g., PCl_3, S_2Cl_2, Cl_2O, NCl_3, ICl).

You may want to review Chapter 6 for nomenclature of acids, bases, and salts.

15.5   Electrolytes and Nonelectrolytes

We can show that solutions of certain substances are conductors of electricity by using a simple conductivity apparatus consisting of a pair of electrodes connected to a voltage source through a light bulb and switch (see Figure 15.1). If the medium between the electrodes is a conductor of electricity, the light bulb will glow when the switch is closed. When chemically pure water is placed in the beaker and the switch is closed, the light does not glow, indicating that water is a virtual nonconductor. When we dissolve a small amount of sugar in the water and test the solution, the light still does not glow, showing that a sugar solution is also a nonconductor. But, when a small amount of salt, NaCl, is dissolved in water and this solution is tested, the light glows brightly. Thus, the salt solution conducts electricity. A fundamental difference exists between the chemical bonding of sugar and that of salt. Sugar is a covalently bonded (molecular) substance; common salt is a substance with ionic bonds.

Substances whose aqueous solutions are conductors of electricity are called **electrolytes**. Substances whose solutions are nonconductors are known as **nonelectrolytes**. The classes of compounds that are electrolytes are acids, bases, and other ionic compounds. Solutions of certain oxides also are conductors because the oxides form an acid or a base when dissolved in water. One major difference between

electrolyte
nonelectrolyte

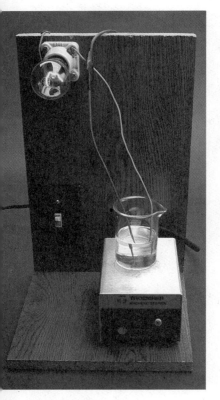

FIGURE 15.1
A simple conductivity apparatus for testing electrolytes and nonelectrolytes in solution. The light bulb glows because the solution contains an electrolyte (acetic acid solution).

electrolytes and nonelectrolytes is that electrolytes are capable of producing ions in solution, whereas nonelectrolytes do not have this property. Solutions that contain a sufficient number of ions will conduct an electric current. Although pure water is essentially a nonconductor, many city water supplies contain enough dissolved ionic matter to cause the light to glow dimly when the water is tested in a conductivity apparatus. Table 15.2 lists some common electrolytes and nonelectrolytes.

> **Acids, bases, and salts are electrolytes.**

15.6   Dissociation and Ionization of Electrolytes

Arrhenius received the 1903 Nobel Prize in chemistry for his work on electrolytes. He stated that a solution conducts electricity because the solution dissociates immediately upon dissolving into electrically charged particles (ions). The movement of these ions toward oppositely charged electrodes causes the solution to be a conductor. According to his theory, solutions that are relatively poor conductors contain electrolytes that are only partly dissociated. Arrhenius also believed that ions exist in solution whether or not an electric current is present. In other words, the electric current does not cause the formation of ions. Remember that positive ions are cations; negative ions are anions.

We have seen that sodium chloride crystals consist of sodium and chloride ions held together by ionic bonds. **Dissociation** is the process by which the ions of a salt separate as the salt dissolves. When placed in water, the sodium and chloride ions are attracted by the polar water molecules, which surround each ion as it dissolves. In water, the salt dissociates, forming hydrated sodium and chloride ions (see Figure 15.2). The sodium and chloride ions in solution are surrounded by a specific number of water dipoles and have less attraction for each other than they

dissociation   had in the crystalline state. The equation representing this dissociation is

$$NaCl(s) + (x + y)\ H_2O \longrightarrow Na^+(H_2O)_x + Cl^-(H_2O)_y$$

A simplified dissociation equation in which the water is omitted but understood to be present is

$$NaCl(s) \longrightarrow Na^+(aq) + Cl^-(aq)$$

TABLE 15.2 Representative Electrolytes and Nonelectrolytes

Electrolytes		Nonelectrolytes	
H_2SO_4	$HC_2H_3O_2$	$C_{12}H_{22}O_{11}$ (sugar)	CH_3OH (methyl alcohol)
HCl	NH_3	C_2H_5OH (ethyl alcohol)	$CO(NH_2)_2$ (urea)
HNO_3	K_2SO_4	$C_2H_4(OH)_2$ (ethylene glycol)	O_2
NaOH	$NaNO_3$	$C_3H_5(OH)_3$ (glycerol)	H_2O

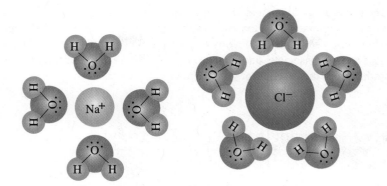

◄ FIGURE 15.2
Hydrated sodium and chloride ions. When sodium chloride dissolves in water, each Na⁺ and Cl⁻ ion becomes surrounded by water molecules. The negative end of the water dipole is attracted to the Na⁺ ion, and the positive end is attracted to the Cl⁻ ion.

It is important to remember that sodium chloride exists in an aqueous solution as hydrated ions and not as NaCl units, even though the formula NaCl or Na^+ + Cl^- is often used in equations.

The chemical reactions of salts in solution are the reactions of their ions. For example, when sodium chloride and silver nitrate react and form a precipitate of silver chloride, only the Ag^+ and Cl^- ions participate in the reaction. The Na^+ and NO_3^- remain as ions in solution:

$$Ag^+(aq) + Cl^-(aq) \longrightarrow AgCl(s)$$

Ionization is the formation of ions; it occurs as a result of a chemical reaction of certain substances with water. Glacial acetic acid (100% $HC_2H_3O_2$) is a liquid that behaves as a nonelectrolyte when tested by the method described in Section 15.5. But a water solution of acetic acid conducts an electric current (as indicated by the dull-glowing light of the conductivity apparatus). The equation for the reaction with water, which forms hydronium and acetate ions, is

ionization

$$\underset{\text{acid}}{HC_2H_3O_2} + \underset{\text{base}}{H_2O} \rightleftharpoons \underset{\text{acid}}{H_3O^+} + \underset{\text{base}}{C_2H_3O_2^-}$$

or, in the simplified equation,

$$HC_2H_3O_2 \rightleftharpoons H^+ + C_2H_3O_2^-$$

In this ionization reaction, water serves not only as a solvent but also as a base according to the Brønsted–Lowry theory.

Hydrogen chloride is predominantly covalently bonded, but when dissolved in water it reacts to form hydronium and chloride ions:

$$HCl(g) + H_2O(l) \longrightarrow H_3O^+(aq) + Cl^-(aq)$$

When a hydrogen chloride solution is tested for conductivity, the light glows brilliantly, indicating many ions in the solution.

Ionization occurs in each of the above two reactions with water, producing ions in solution. The necessity for water in the ionization process can be demonstrated by dissolving hydrogen chloride in a nonpolar solvent such as hexane, and testing the solution for conductivity. The solution fails to conduct electricity, indicating that no ions are produced.

The terms *dissociation* and *ionization* are often used interchangeably to describe processes taking place in water. But, strictly speaking, the two are different. In the dissociation of a salt, the salt already exists as ions; when it dissolves in water, the ions separate, or dissociate, and increase in mobility. In the ionization process, ions are produced by the reaction of a compound with water.

15.7   Strong and Weak Electrolytes

strong electrolyte
weak electrolyte

Electrolytes are classified as strong or weak depending on the degree, or extent, of dissociation or ionization. **Strong electrolytes** are essentially 100% ionized in solution; **weak electrolytes** are much less ionized (based on comparing 0.1 *M* solutions). Most electrolytes are either strong or weak, with a few classified as moderately strong or weak. Most salts are strong electrolytes. Acids and bases that are strong electrolytes (highly ionized) are called *strong acids* and *strong bases*. Acids and bases that are weak electrolytes (slightly ionized) are called *weak acids* and *weak bases*.

For equivalent concentrations, solutions of strong electrolytes contain many more ions than do solutions of weak electrolytes. As a result, solutions of strong electrolytes are better conductors of electricity. Consider the two solutions, 1 *M* HCl and 1 *M* $HC_2H_3O_2$. Hydrochloric acid is almost 100% ionized; acetic acid is about 1% ionized. Thus HCl is a strong acid, and $HC_2H_3O_2$ is a weak acid. Hydrochloric acid has about 100 times as many hydronium ions in solution as acetic acid, making the HCl solution much more acidic.

One can distinguish between strong and weak electrolytes experimentally using the apparatus described in Section 15.5. A 1 *M* HCl solution causes the light to glow brilliantly, but a 1 *M* $HC_2H_3O_2$ solution causes only a dim glow. In a similar fashion the strong base sodium hydroxide, NaOH, may be distinguished from the weak base ammonia, NH_3. The ionization of a weak electrolyte in water is represented by an equilibrium equation showing that both the un-ionized and ionized forms are present in solution. In the equilibrium equation of $HC_2H_3O_2$ and its ions, we say that the equilibrium lies "far to the left" because relatively few hydrogen and acetate ions are present in solution:

$$HC_2H_3O_2(aq) \rightleftharpoons H^+(aq) + C_2H_3O_2^-(aq)$$

We have previously used a double arrow in an equation to represent reversible processes in the equilibrium between dissolved and undissolved solute in a saturated solution. A double arrow ($\rightleftharpoons$) is also used in the ionization equation of soluble weak electrolytes to indicate that the solution contains a considerable amount of the un-ionized compound in equilibrium with its ions in solution. (See Section 16.1 for a discussion of reversible reactions.) A single arrow is used to indicate that the electrolyte is essentially all in the ionic form in the solution. For example, nitric acid is a strong acid; nitrous acid is a weak acid. Their ionization equations in water may be indicated as

$$HNO_3(aq) \xrightarrow{H_2O} H^+(aq) + NO_3^-(aq)$$

$$HNO_2(aq) \overset{H_2O}{\rightleftharpoons} H^+(aq) + NO_2^-(aq)$$

TABLE 15.3 Strong and Weak Electrolytes

Strong electrolytes		Weak electrolytes	
Most soluble salts	$HClO_4$	$HC_2H_3O_2$	$H_2C_2O_4$
H_2SO_4	NaOH	H_2CO_3	H_3BO_3
HNO_3	KOH	HNO_2	HClO
HCl	$Ca(OH)_2$	H_2SO_3	NH_3
HBr	$Ba(OH)_2$	H_2S	HF

Practically all soluble salts, acids (such as sulfuric, nitric, and hydrochloric acids), and bases (such as sodium, potassium, calcium, and barium hydroxides) are strong electrolytes. Weak electrolytes include numerous other acids and bases such as acetic acid, nitrous acid, carbonic acid, and ammonia. The terms *strong acid, strong base, weak acid,* and *weak base* refer to whether an acid or base is a strong or weak electrolyte. A brief list of strong and weak electrolytes is given in Table 15.3.

Electrolytes yield two or more ions per formula unit upon dissociation—the actual number being dependent on the compound. Dissociation is complete or nearly complete for nearly all soluble ionic compounds and for certain other strong electrolytes, such as those given in Table 15.3. The following are dissociation equations for several strong electrolytes. In all cases the ions are actually hydrated:

$$NaOH \xrightarrow{H_2O} Na^+(aq) + OH^-(aq) \qquad \text{2 ions in solution per formula unit}$$

$$Na_2SO_4 \xrightarrow{H_2O} 2\,Na^+(aq) + SO_4^{2-}(aq) \qquad \text{3 ions in solution per formula unit}$$

$$Fe_2(SO_4)_3 \xrightarrow{H_2O} 2\,Fe^{3+}(aq) + 3\,SO_4^{2-}(aq)$$
$$\text{5 ions in solution per formula unit}$$

One mole of NaCl will give 1 mol of Na^+ ions and 1 mol of Cl^- ions in solution, assuming complete dissociation of the salt. One mole of $CaCl_2$ will give 1 mol of Ca^{2+} ions and 2 mol of Cl^- ions in solution:

$$NaCl \xrightarrow{H_2O} Na^+(aq) + Cl^-(aq)$$
$$\text{1 mol} \qquad\;\; \text{1 mol} \qquad\quad\; \text{1 mol}$$

$$CaCl_2 \xrightarrow{H_2O} Ca^{2+}(aq) + 2\,Cl^-(aq)$$
$$\text{1 mol} \qquad\;\; \text{1 mol} \qquad\quad\; \text{2 mol}$$

What is the molarity of each ion in a solution of (a) 2.0 *M* NaCl, and (b) 0.40 *M* K_2SO_4? Assume complete dissociation.

Example 15.2

(a) According to the dissociation equation,

Solution

$$NaCl \xrightarrow{H_2O} Na^+(aq) + Cl^-(aq)$$
$$\text{1 mol} \qquad\;\; \text{1 mol} \qquad\quad\; \text{1 mol}$$

the concentration of Na^+ is equal to that of NaCl (1 mol NaCl $\longrightarrow$ 1 mol Na^+), and the concentration of Cl^- is also equal to that of NaCl. Therefore, the concentrations of the ions in 2.0 M NaCl are 2.0 M Na^+ and 2.0 M Cl^-.

(b) According to the dissociation equation,

$$K_2SO_4 \xrightarrow{H_2O} 2\ K^+(aq)\ +\ SO_4^{2-}(aq)$$

$$\underset{\text{1 mol}}{} \qquad \underset{\text{2 mol}}{} \qquad \underset{\text{1 mol}}{}$$

the concentration of K^+ is twice that of K_2SO_4 and the concentration of SO_4^{2-} is equal to that of K_2SO_4. Therefore, the concentrations of the ions in 0.40 M K_2SO_4 are 0.80 M K^+ and 0.40 M SO_4^{2-}.

Practice 15.3

What is the molarity of each ion in a solution of (a) 0.050 M $MgCl_2$, and (b) 0.070 M $AlCl_3$?

Colligative Properties of Electrolyte Solutions

We have learned that when 1 mol of sucrose, a nonelectrolyte, is dissolved in 1000 g of water, the solution freezes at $-1.86°C$. When 1 mol of NaCl is dissolved in 1000 g of water, the freezing point of the solution is not $-1.86°C$, as might be expected, but is closer to $-3.72°C$ (-1.86×2). The reason for the lower freezing point is that 1 mol of NaCl in solution produces 2 mol of particles ($2 \times 6.022 \times 10^{23}$ ions) in solution. Thus, the freezing-point depression produced by 1 mol of NaCl is essentially equivalent to that produced by 2 mol of a nonelectrolyte. An electrolyte such as $CaCl_2$, which yields three ions in water, gives a freezing-point depression of about three times that of a nonelectrolyte. These freezing-point data provide additional evidence that electrolytes dissociate when dissolved in water. The other colligative properties are similarly affected by substances that yield ions in aqueous solutions.

15.8 Ionization of Water

The more we study chemistry, the more intriguing the water molecule becomes. Two equations commonly used to show how water ionizes are

$$H_2O\ +\ H_2O \rightleftharpoons H_3O^+\ +\ OH^-$$
$$\underset{\text{acid}}{} \quad \underset{\text{base}}{} \qquad \underset{\text{acid}}{} \quad \underset{\text{base}}{}$$

and

$$H_2O \rightleftharpoons H^+\ +\ OH^-$$

The first equation represents the Brønsted–Lowry concept, with water reacting as both an acid and a base, forming a hydronium ion and a hydroxide ion. The second equation is a simplified version, indicating that water ionizes to give a hydrogen and a hydroxide ion. Actually, the proton, H^+, is hydrated and exists as a hydronium ion. In either case equal molar amounts of acid and base are produced so that water is neutral, having neither H^+ nor OH^- ions in excess. The ionization of water at 25°C produces an H^+ ion concentration of 1.0×10^{-7} mol/L and an OH^- ion concentration of 1.0×10^{-7} mol/L. Square brackets, [ ], indicate that the concentration is in moles per liter. Thus $[H^+]$ means the concentration of H^+ is in moles per liter. These concentrations are usually expressed as

$$[H^+] \text{ or } [H_3O^+] = 1.0 \times 10^{-7} \text{ mol/L}$$

$$[OH^-] = 1.0 \times 10^{-7} \text{ mol/L}$$

These figures mean that about two out of every billion water molecules are ionized. This amount of ionization, small as it is, is a significant factor in the behavior of water in many chemical reactions.

15.9   Introduction to pH

The acidity of an aqueous solution depends on the concentration of hydrogen or hydronium ions. The pH scale of acidity was devised to fill the need for a simple, convenient numerical way to state the acidity of a solution. Values on the pH scale are obtained by mathematical conversion of H^+ ion concentrations to pH by the expression:

$$pH = -\log[H^+]$$

pH

where $[H^+]$ = H^+ or H_3O^+ ion concentration in moles per liter. **pH** is defined as the *negative* logarithm of the H^+ or H_3O^+ concentration in moles per liter:

$$pH = -\log[H^+] = -\log(1 \times 10^{-7}) = -(-7) = 7$$

For example, the pH of pure water at 25°C is 7 and is said to be neutral; that is, it is neither acidic nor basic, because the concentrations of H^+ and OH^- are equal. Solutions that contain more H^+ ions than OH^- ions have pH values less than 7, and solutions that contain less H^+ ions than OH^- ions have values greater than 7.

> **pH < 7.00 is an acidic solution**
> **pH = 7.00 is a neutral solution**
> **pH > 7.00 is a basic solution**

When $[H^+] = 1 \times 10^{-5}$ mol/L, pH = 5 (acidic)

When $[H^+] = 1 \times 10^{-9}$ mol/L, pH = 9 (basic)

Instead of saying that the hydrogen ion concentration in the solution is 1×10^{-5} mol/L, it is customary to say that the pH of the solution is 5. The smaller the pH value, the more acidic the solution (see Figure 15.3).

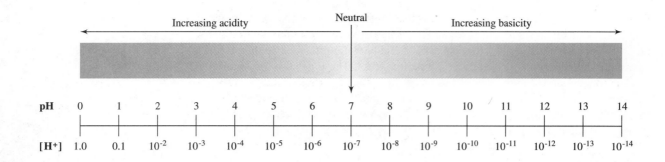

▲ **FIGURE 15.3**
The pH scale of acidity and basicity.

The pH scale, along with its interpretation, is given in Table 15.4. Table 15.5 lists the pH of some common solutions. Note that a change of only 1 pH unit means a tenfold increase or decrease in H^+ ion concentration. For example, a solution with a pH of 3.0 is ten times more acidic than a solution with a pH of 4.0. A simplified method of determining pH from $[H^+]$ follows:

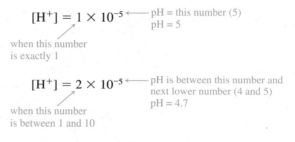

Calculation of the pH value corresponding to any H^+ ion concentration requires the use of logarithms, which are exponents. The **logarithm** (log) of a number is

logarithm

TABLE 15.4 The pH Scale for Expressing Acidity		
$[H^+]$ (mol/L)	pH	
1×10^{-14}	14	
1×10^{-13}	13	
1×10^{-12}	12	
1×10^{-11}	11	Increasing
1×10^{-10}	10	basicity
1×10^{-9}	9	
1×10^{-8}	8	
1×10^{-7}	7	Neutral
1×10^{-6}	6	
1×10^{-5}	5	
1×10^{-4}	4	
1×10^{-3}	3	Increasing
1×10^{-2}	2	acidity
1×10^{-1}	1	
1×10^{0}	0	

TABLE 15.5 The pH of Some Common Solutions	
Solution	pH
Gastric juice	1.0
0.1 *M* HCl	1.0
Lemon juice	2.3
Vinegar	2.8
0.1 *M* HC$_2$H$_3$O$_2$	2.9
Orange juice	3.7
Tomato juice	4.1
Coffee, black	5.0
Urine	6.0
Milk	6.6
Pure water (25°C)	7.0
Blood	7.4
Household ammonia	11.0
1 *M* NaOH	14.0

simply the power to which 10 must be raised to give that number. Thus the log of 100 is 2 $(100 = 10^2)$, and the log of 1000 is 3 $(1000 = 10^3)$. The log of 500 is 2.70, but you cannot easily determine this value without a scientific calculator.

Let us determine the pH of a solution with $[H^+] = 2 \times 10^{-5}$. The exponent (-5) indicates that the pH is between 4 and 5. Enter 2×10^{-5} into your calculator and press the log key. The number $-4.69 \ldots$ will be displayed. The pH is then

$$pH = -\log[H^+] = -(-4.69\ldots) = 4.7$$

Next we must determine the correct number of significant figures in the logarithm. The rules for logs are different from those we use in other math operations. The number of decimal places for a log must equal the number of significant figures in the original number. Since 2×10^{-5} has one significant figure, we should round the log to one decimal place $(4.69\ldots) = 4.7$.

Remember to change the sign on your calculator since $pH = -\log[H^+]$.

What is the pH of a solution with an $[H^+]$ of (a) 1.0×10^{-11}, and (b) 6.0×10^{-4}, and (c) 5.47×10^{-8}?

Example 15.3

Solution

(a) $[H^+] = 1.0 \times 10^{-11}$
   (2 significant figures)
   $pH = -\log(1.0 \times 10^{-11})$
   $pH = 11.00$
   (2 decimal places)

(b) $[H^+] = 6.0 \times 10^{-4}$
   (2 significant figures)
   $\log 6.0 \times 10^{-4} = -3.22$
   $pH = -\log[H^+]$
   $pH = -(-3.22) = 3.22$
   (2 decimal places)

(c) $[H^+] = 5.47 \times 10^{-8}$
   (3 significant figures)
   $\log 5.47 \times 10^{-8} = -7.262$
   $pH = -\log[H^+]$
   $pH = -(-7.262) = 7.262$
   (3 decimal places)

Practice 15.4

What is the pH of a solution with $[H^+]$ of (a) 3.9×10^{-12} M, (b) 1.3×10^{-3} M, and (c) 3.72×10^{-6} M?

The measurement and control of pH is extremely important in many fields of science and technology. The proper soil pH is necessary to grow certain types of plants successfully. The pH of certain foods is too acidic for some diets. Many biological processes are delicately controlled pH systems. The pH of human blood is regulated to very close tolerances through the uptake or release of H^+ by mineral ions, such as HCO_3^-, HPO_4^{2-}, and $H_2PO_4^-$. Changes in the pH of the blood by as little as 0.4 pH unit result in death.

Compounds with colors that change at particular pH values are used as indicators in acid–base reactions. For example, phenolphthalein, an organic compound, is colorless in acid solution and changes to pink at a pH of 8.3. When a solution of sodium hydroxide is added to a hydrochloric acid solution containing phenolphthalein, the change in color (from colorless to pink) indicates that all the acid is neutralized. Commercially available pH test paper, such as shown in Figure 15.4, contains chemical indicators. The indicator in the paper takes on different colors when wetted with solutions of different pH. Thus the pH of a solution can be estimated by placing a drop on the test paper and comparing the color of the test

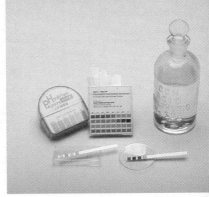

FIGURE 15.4
pH test paper for determining the approximate acidity of solutions.

Hair Care and pH—A Delicate Balance

Hair shampoo advertisements often proclaim the proper pH for their products, but does controlling the pH of hair care products really make hair cleaner, shiny, or stronger?

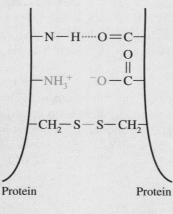

▲
Interaction between strands of hair protein.

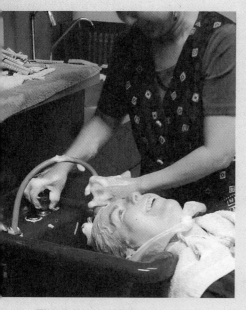

The chemistry of shampoo is ever improving to suit different hair conditions.

Each strand of hair is composed of many long chains of amino acid linked together as polymers called *proteins*. The individual chains can connect with other chains in one of three ways: (1) hydrogen bonds, red; (2) salt bridges (the result of acid–base interactions), green; and (3) disulfide bonds, blue. These interactions are shown in the diagram.

When hair is wet with water, the hydrogen bonds are broken. As the wet hair is shaped, set, or dried, the hydrogen bonds form at new positions and hold the hair in the style desired. If an acidic solution (pH 1.0–2.0) is used on the hair, the hydrogen bonds and the salt bridges are both broken, leaving only the disulfide bonds to hold the chains together. In a mildly alkaline solution (pH 8.5) some of the disulfide bonds are also broken. The outer surface of the hair becomes rough, and light does not reflect evenly from the surface, making the hair look dull. Using an alkaline shampoo will cause damage by continued breakage of the disulfide bonds, which results in "split ends." If the pH is increased further to approximately 12.0 the hair dissolves as all types of bonds break. This is the working basis for depilatories (hair removers), such as Neet™ and Nair™.

Hair has its maximum strength at pH 4.0–5.0. Shampooing tends to leave the hair slightly alkaline, so an acid rinse is sometimes used to bring the pH back into the normal range. Lemon juice or vinegar are common household products that are used for this purpose. The shampoo may also be "acid-balanced," containing a weak acid (such as citric acid) to counteract the alkalinity of the solution formed when the detergent interacts with water.

paper with a color chart calibrated at different pH values. Common applications of pH test indicators are the kits used to measure and adjust the pH of swimming pools and hot tubs. Electronic pH meters are used for making rapid and precise pH determinations.

15.10 Neutralization

neutralization

The reaction of an acid and a base to form a salt and water is known as **neutralization**. We have seen this reaction before, but now with our knowledge about ions and ionization, let us reexamine the process of neutralization.

Consider the reaction that occurs when solutions of sodium hydroxide and hydrochloric acid are mixed. The ions present initially are Na^+ and OH^- from the base and H^+ and Cl^- from the acid. The products, sodium chloride and water, exist as Na^+ and Cl^- ions and H_2O molecules. A chemical equation representing this reaction is

$$HCl(aq) + NaOH(aq) \longrightarrow NaCl(aq) + H_2O(l)$$

This equation, however, does not show that HCl, NaOH, and NaCl exist as ions in solution. The following total ionic equation gives a better representation of the reaction:

$$(H^+ + Cl^-) + (Na^+ + OH^-) \longrightarrow Na^+ + Cl^- + H_2O(l)$$

This equation shows that the Na^+ and Cl^- ions did not react. These ions are called **spectator ions** because they were present but did not take part in the reaction. The only reaction that occurred was that between the H^+ and OH^- ions. Therefore, the equation for the neutralization can be written as this net ionic equation:

 spectator ion

$$\underset{\text{acid}}{H^+(aq)} + \underset{\text{base}}{OH^-(aq)} \longrightarrow \underset{\text{water}}{H_2O(l)}$$

This simple net ionic equation represents not only the reaction of sodium hydroxide and hydrochloric acid, but also the reaction of any strong acid with any water-soluble hydroxide base in an aqueous solution. The driving force of a neutralization reaction is the ability of an H^+ ion and an OH^- ion to react and form a molecule of un-ionized water.

The amount of acid, base, or other species in a sample may be determined by **titration**, which is the process of measuring the volume of one reagent that is required to react with a measured mass or volume of another reagent.

 titration

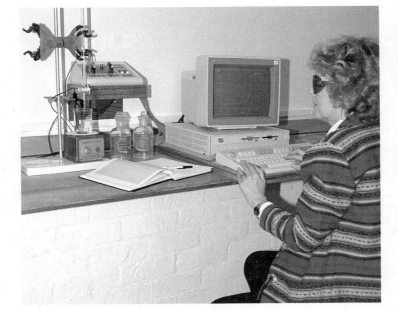

This pH meter measures the acidity and alkalinity of garden soil.

◀ **The progress of a titration can be monitored by computer graphing.**

Let us consider the titration of an acid with a base. A measured volume of acid of unknown concentration is placed in a flask, and a few drops of an indicator solution are added. Base solution of known concentration is slowly added from a buret to the acid until the indicator changes color The indicator selected is one that changes color when the stoichiometric quantity (according to the equation) of base has been added to the acid. At this point, known as the *end point of the titration*, the titration is complete, and the volume of base used to neutralize the acid is read from the buret. The concentration or amount of acid in solution can be calculated from the titration data and the chemical equation for the reaction. Let's look at some illustrative examples.

Example 15.4

Suppose that 42.00 mL of 0.150 *M* NaOH solution is required to neutralize 50.00 mL of hydrochloric acid solution. What is the molarity of the acid solution?

Solution

The equation for the reaction is

$$NaOH(aq) + HCl(aq) \longrightarrow NaCl(aq) + H_2O(l)$$

In this neutralization NaOH and HCl react in a 1:1 mole ratio. Therefore, the moles of HCl in solution are equal to the moles of NaOH required to react with it. First we calculate the moles of NaOH used, and from this value we determine the moles of HCl:

Data: 42.00 mL of 0.150 *M* NaOH   50.00 mL of HCl
        Molarity of acid = *M* (unknown)

Determine the moles of NaOH:

$$M = mol/L \qquad 42.00 \text{ mL} = 0.04200 \text{ L}$$

$$0.04200 \text{ L} \times \frac{0.150 \text{ mol NaOH}}{1 \text{ L}} = 0.00630 \text{ mol NaOH}$$

Since NaOH and HCl react in a 1:1 ratio, 0.00630 mol of HCl was present in the 50.00 mL of HCl solution. Therefore, the molarity of the HCl is

$$M = \frac{mol}{L} = \frac{0.00630 \text{ mol HCl}}{0.05000 \text{ L}} = 0.126 \ M \text{ HCl}$$

Example 15.5

Suppose that 42.00 mL of 0.150 *M* NaOH solution is required to neutralize 50.00 mL of H_2SO_4 solution. What is the molarity of the acid solution?

Solution

The equation for the reaction is

$$2 \text{ NaOH}(aq) + H_2SO_4(aq) \longrightarrow Na_2SO_4(aq) + 2 \text{ H}_2O(l)$$

The same amount of base (0.00630 mol of NaOH) is used in this titration as in Example 15.4, but the mole ratio of acid to base in the reaction is 1:2. The moles of H_2SO_4 reacted can be calculated by using the mole-ratio method:

Data: 42.00 mL of 0.150 *M* NaOH = 0.00630 mol NaOH

$$0.00630 \text{ mol NaOH} \times \frac{1 \text{ mol } H_2SO_4}{2 \text{ mol NaOH}} = 0.00315 \text{ mol } H_2SO_4$$

Therefore, 0.00315 mol of H_2SO_4 was present in 50.00 mL of H_2SO_4 solution. The molarity of the H_2SO_4 is

$$M = \frac{mol}{L} = \frac{0.00315 \text{ mol } H_2SO_4}{0.05000 \text{ L}} = 0.0630 \text{ M } H_2SO_4$$

A 25.00-mL sample of H_2SO_4 solution required 14.26 mL of 0.2240 N NaOH for complete neutralization. What is the normality and the molarity of the sulfuric acid? **Example 15.6**

The equation for the reaction is **Solution**

$$2 \text{ NaOH}(aq) + H_2SO_4(aq) \longrightarrow Na_2SO_4(aq) + 2 H_2O(l)$$

The normality of the acid can be calculated from

$$V_A N_A = V_B N_B$$

Substitute the data into the equation and solve for N_A:

$$25.00 \text{ mL} \times N_A = 14.26 \text{ mL} \times 0.2240 \text{ N}$$

$$N_A = \frac{14.26 \text{ mL} \times 0.2240 \text{ N}}{25.00 \text{ mL}} = 0.1278 \text{ N } H_2SO_4$$

The normality of the acid is 0.1278 N.

Because H_2SO_4 furnishes 2 equivalents of H^+ per mole, the conversion to molarity is

$$\frac{equiv}{L} \times \frac{mol}{equiv}$$

$$\frac{0.1278 \text{ equiv } H_2SO_4}{1 \text{ L}} \times \frac{1 \text{ mol } H_2SO_4}{2 \text{ equiv } H_2SO_4} = 0.06390 \text{ mol/L } H_2SO_4$$

The H_2SO_4 solution is 0.06390 M.

Practice 15.5

A 50.0-mL sample of HCl required 24.81 mL of 0.1250 M NaOH for neutralization. What is the molarity of the acid?

15.11 Acid Rain

Acid rain is defined as any atmospheric precipitation that is more acidic than normal. The increase in acidity might be from natural or man-made sources. Rain acidity varies throughout the world and across the United States. The pH of rain is generally lower in the eastern United States and higher in the west. Unpolluted rain has a pH of 5.6, and so is slightly acidic. This acidity results from the dissolution of carbon dioxide in the water producing carbonic acid:

$$CO_2(g) + H_2O(l) \longrightarrow H_2CO_3(aq) \rightleftharpoons H^+(aq) + HCO_3^-(aq)$$

Marble masterpieces, sculpted to last forever, are slowly disappearing as acid rain dissolves the calcium carbonate ($CaCO_3$). ▶

Although the details of acid-rain formation are not yet fully understood, chemists know the general process involves the following steps:

1. emission of nitrogen and sulfur oxides into the air
2. transportation of these oxides throughout the atmosphere
3. chemical reactions between the oxides and water forming sulfuric acid (H_2SO_4) and nitric acid (HNO_3)
4. rain or snow, which carries the acids to the surface

The oxides may also be deposited directly on a dry surface and become acidic when normal rain falls on them.

Acid rain is not a new phenomenon. Rain was probably acidic in the early days of our planet as volcanic eruptions, fires, and decomposition of organic matter released large volumes of nitrogen and sulfur oxides into the atmosphere. Use of fossil fuels, especially since the industrial revolution about 250 years ago, has made significant changes in the amounts of pollutants being released into the atmosphere. As increasing amounts of fossil fuels have been burned, more and more sulfur and nitrogen oxides have poured into the atmosphere, thus increasing the acidity of rain.

Acid rain affects a variety of factors in our environment. For example, fresh water plants and animals decline significantly when rain is acidic; large numbers of fish and plants die when acidic water from spring thaws enters the lakes. Aluminum is leached from the soil into lakes by acidic rain water where the aluminum compounds adversely affect the gills of fish. In addition to leaching aluminum from the soil, acid rain also causes other valuable minerals, such as magnesium and calcium, to dissolve and run into lakes and streams. It can also dissolve the waxy protective coat on plant leaves making them vulnerable to attack by bacteria and fungi.

In our cities, acid rain is responsible for extensive and continuing damage to buildings, monuments, and statues. It may also reduce the durability of paint and promote the deterioration of paper, leather, and cloth. In short, we are just beginning to explore the effects of acid rain on human beings and on our food chain.

15.12   Writing Net Ionic Equations

In Section 15.10 we wrote the reaction of hydrochloric acid and sodium hydroxide in three different equations:

(1) $HCl(aq) + NaOH(aq) \longrightarrow NaCl(aq) + H_2O(l)$

(2) $(H^+ + Cl^-) + (Na^+ + OH^-) \longrightarrow Na^+ + Cl^- + H_2O(l)$

(3) $H^+ + OH^- \longrightarrow H_2O$

In the **un-ionized equation** (1), compounds are written in their molecular, or normal, formula expressions. In the **total ionic equation** (2), compounds are written to show the form in which they are predominantly present: strong electrolytes as ions in solution; and nonelectrolytes, weak electrolytes, precipitates, and gases in their molecular (or un-ionized) forms. In the **net ionic equation** (3), only those molecules or ions that have changed are included in the equation; ions or molecules that do not change (spectators) are omitted.

un-ionized equation
total ionic equation

net ionic equation

Up to this point, when balancing an equation we have been concerned only with the atoms of the individual elements. Because ions are electrically charged, ionic equations often end up with a net electrical charge. A balanced equation must have the same net charge on each side, whether that charge is positive, negative, or zero. Therefore, when balancing an ionic equation, we must make sure that both the same number of each kind of atom and the same net electrical charge are present on each side.

Following is a list of rules for writing ionic equations:

1. Strong electrolytes in solution are written in their ionic form.
2. Weak electrolytes are written in their molecular (un-ionized) form.
3. Nonelectrolytes are written in their molecular form.
4. Insoluble substances, precipitates, and gases are written in their molecular forms.
5. The net ionic equation should include only substances that have undergone a chemical change. Spectator ions are omitted from the net ionic equation.
6. Equations must be balanced, both in atoms and in electrical charge.

Study the following examples. In each one the un-ionized equation is given. Write the total ionic equation and the net ionic equation for each.

$HNO_3(aq) + KOH(aq) \longrightarrow KNO_3(aq) + H_2O(l)$
un-ionized equation

Example 15.7

$(H^+ + NO_3^-) + (K^+ + OH^-) \longrightarrow (K^+ + NO_3^-) + H_2O$
total ionic equation

Solution

$H^+ + OH^- \longrightarrow H_2O$
net ionic equation

The HNO_3, KOH, and KNO_3 are soluble, strong electrolytes. The K^+ and NO_3^- ions are spectator ions, have not changed, and are not included in the net ionic equation. Water is a nonelectrolyte and is written in the molecular form.

Example 15.8 $2 \, AgNO_3(aq) + BaCl_2(aq) \longrightarrow 2 \, AgCl(s) + Ba(NO_3)_2(aq)$

un-ionized equation

$$(2 \, Ag^+ + 2 \, NO_3^-) + (Ba^{2+} + 2 \, Cl^-) \longrightarrow 2 \, AgCl(s) + (Ba^{2+} + 2 \, NO_3^-)$$

total ionic equation

$$Ag^+ + Cl^- \longrightarrow AgCl(s)$$

net ionic equation

Solution   Although AgCl is an ionic compound, it is written in the un-ionized form on the right side of the ionic equations because most of the Ag^+ and Cl^- ions are no longer in solution but have formed a precipitate of AgCl. The Ba^{2+} and NO_3^- ions are spectator ions.

Example 15.9 $Na_2CO_3(aq) + H_2SO_4(aq) \longrightarrow Na_2SO_4(aq) + H_2O(l) + CO_2(g)$

un-ionized equation

$$(2 \, Na^+ + CO_3^{2-}) + (2 \, H^+ + SO_4^{2-}) \longrightarrow (2 \, Na^+ + SO_4^{2-}) + H_2O(l) + CO_2(g)$$

total ionic equation

$$CO_3^{2-} + 2 \, H^+ \longrightarrow H_2O(l) + CO_2(g)$$

net ionic equation

Solution   Carbon dioxide, CO_2, is a gas and evolves from the solution; Na^+ and SO_4^{2-} are spectator ions.

Example 15.10 $HC_2H_3O_2(aq) + NaOH(aq) \longrightarrow NaC_2H_3O_2(aq) + H_2O(l)$

un-ionized equation

$$HC_2H_3O_2 + (Na^+ + OH^-) \longrightarrow (Na^+ + C_2H_3O_2^-) + H_2O$$

total ionic equation

$$HC_2H_3O_2 + OH^- \longrightarrow C_2H_3O_2^- + H_2O$$

net ionic equation

Solution   Acetic acid, $HC_2H_3O_2$, a weak acid, is written in the molecular form, but sodium acetate, $NaC_2H_3O_2$, a soluble salt, is written in the ionic form. The Na^+ ion is the only spectator ion in this reaction. Both sides of the net ionic equation have a -1 electrical charge.

Example 15.11 $Mg(s) + 2 \, HCl(aq) \longrightarrow MgCl_2(aq) + H_2(g)$

un-ionized equation

$$Mg + (2 \, H^+ + 2 \, Cl^-) \longrightarrow (Mg^{2+} + 2 \, Cl^-) + H_2(g)$$

total ionic equation

$$Mg + 2 \, H^+ \longrightarrow Mg^{2+} + H_2(g)$$

net ionic equation

Solution   The net electrical charge on both sides of the equation is $+2$.

$$H_2SO_4(aq) \ + \ Ba(OH)_2(aq) \longrightarrow BaSO_4(s) \ + \ 2 \ H_2O(l)$$

un-ionized equation

$$(2 \ H^+ \ + \ SO_4^{2-}) \ + \ (Ba^{2+} \ + \ 2 \ OH^-) \longrightarrow BaSO_4(s) \ + \ 2 \ H_2O(l)$$

total ionic equation

$$2 \ H^+ \ + \ SO_4^{2-} \ + \ Ba^{2+} \ + \ 2 \ OH^- \longrightarrow BaSO_4(s) \ + \ 2 \ H_2O(l)$$

net ionic equation

Example 15.12

Barium sulfate, $BaSO_4$, is a highly insoluble salt. If we conduct this reaction using the conductivity apparatus described in Section 15.5, the light glows brightly at first but goes out when the reaction is complete because almost no ions are left in solution. The $BaSO_4$ precipitates out of solution, and water is a nonconductor of electricity.

Solution

Practice 15.6

Write the net ionic equation for
$$3 \ H_2S(aq) \ + \ 2 \ Bi(NO_3)_3(aq) \longrightarrow Bi_2S_3(s) \ + \ 6 \ HNO_3(aq)$$

15.13   Colloids: An Introduction

When we add sugar to a flask of water and shake it, the sugar dissolves and forms a clear homogeneous *solution.* When we do the same experiment with very fine sand and water, the sand particles form a *suspension,* which settles when the shaking stops. When we repeat the experiment again using ordinary cornstarch, we find that the starch does not dissolve in cold water. But if the mixture is heated and stirred, the starch forms a cloudy, opalescent *dispersion.* This dispersion does not appear to be clear and homogeneous like the sugar solution, yet it is not obviously heterogeneous and does not settle like the sand suspension. In short, its properties are intermediate between those of the sugar solution and those of the sand suspension. The starch dispersion is actually a *colloid,* a name derived from the Greek *kolla,* meaning "glue," and was coined by the English scientist Thomas Graham in 1861. Graham classified solutes as crystalloids if they diffused through a parchment membrane and as colloids if they did not diffuse through the membrane.

As it is now used, the word **colloid** means a dispersion in which the dispersed particles are larger than the solute ions or molecules of a true solution and smaller than the particles of a mechanical suspension. The term does not imply a glue-like quality, although most glues are colloidal materials. The size of colloidal particles ranges from a lower limit of about 1 nm (10^{-7} cm) to an upper limit of about 1000 nm (10^{-4} cm).

colloid

The fundamental difference between a colloidal dispersion and a true solution is the size not the nature of the particles. The solute particles in a solution are usually single ions or molecules that may be hydrated to varying degrees. Colloidal particles are usually aggregations of ions or molecules. However, the molecules of

TABLE 15.6 Types of Colloidal Dispersions

Type	Name	Examples
Gas in liquid	Foam	Whipped cream, soap suds
Gas in solid	Solid foam	Styrofoam, foam rubber, pumice
Liquid in gas	Liquid aerosol	Fog, clouds
Liquid in liquid	Emulsion	Milk, vinegar in oil salad dressing, mayonnaise
Liquid in solid	Solid emulsion	Cheese, opals, jellies
Solid in gas	Solid aerosol	Smoke, dust in air
Solid in liquid	Sol	India ink, gold sol
Solid in solid	Solid Sol	Tire rubber, certain gems (e.g., rubies)

some polymers, such as proteins, are large enough to be classified as colloidal particles when in solution. To appreciate fully the differences in relative sizes, the volumes (not just the linear dimensions) of colloidal particles and solute particles must be compared. The difference in volumes can be approximated by assuming that the particles are spheres. A large colloidal particle has a diameter of about 500 nm, whereas a fair-sized ion or molecule has a diameter of about 0.5 nm. Thus, the diameter of the colloidal particle is about 1000 times that of the solute particle. Because the volumes of spheres are proportional to the cubes of their diameters, we can calculate that the volume of a colloidal particle can be up to a billion ($10^3 \times 10^3 \times 10^3 = 10^9$) times greater than that of a solution particle.

Colloids are mixtures in which one component, the *dispersal phase,* exists as discrete particles in the other component, the *dispersing phase* or *dispersing medium.* The components of a colloidal dispersion are also sometimes called the *discontinuous phase* and the *continuous phase.* Each component, or phase, can exist as a solid, a liquid, or a gas. The components cannot be mutually soluble, nor can both phases of the dispersion be gases, because such conditions would result in an ordinary solution. Hence, only eight types of colloidal dispersions, based on the possible physical states of the phases, are known. The eight types, with specific examples, are listed in Table 15.6.

15.14   Preparation of Colloids

Colloidal dispersions can be prepared by two methods: (1) *dispersion,* the breaking down of larger particles to colloidal size, and (2) *condensation,* the formation of colloidal particles from solutions.

Homogenized milk is a good example of a colloid prepared by dispersion. Milk, as drawn from the cow, is an unstable emulsion of fat in water. The fat globules are so large that they rise and form a cream layer in a few hours. To avoid separation of the cream, the milk is homogenized by pumping it through very small holes, or orifices, at high pressure. The violent shearing action of this treatment breaks the fat globules into particles well within the colloidal size range. The butterfat in homogenized milk remains dispersed indefinitely. Colloid mills, which reduce particles to colloidal size by grinding or shearing, are used in preparing many commercial products such as paints, cosmetics, and salad dressings.

The preparation of colloids by condensation frequently involves a precipitation reaction in a dilute solution. A colloidal dispersion in a liquid is called a **sol**. For example, a good colloidal sulfur sol can be made by bubbling hydrogen sulfide into a solution of sulfur dioxide. Solid sulfur is formed and dispersed as a colloid:

$$SO_2 + 2\ H_2S \longrightarrow \underset{\substack{\text{colloidal}\\\text{sulfur}}}{3\ S} + 2\ H_2O$$

sol

A colloidal dispersion is also easily made by adding iron(III) chloride solution to boiling water. The reddish-brown colloidal dispersion that is formed probably consists of iron(III) hydroxide and hydrated iron(III) oxide:

$$FeCl_3 + 3\ HOH \xrightarrow{H_2O} Fe(OH)_3 + 3\ HCl$$

$$Fe(OH)_3 \xrightarrow{H_2O} Fe_2O_3 \cdot xH_2O$$

A great many products for home use (insecticides, insect repellents, and deodorants, to name a few) are packaged as aerosols. The active ingredient, either a liquid or a solid, is dissolved in a liquefied gas and sealed under pressure in a container fitted with a release valve. When this valve is opened, the pressurized solution is ejected. The liquefied gas vaporizes, and the active ingredient is converted to a colloidal aerosol almost instantaneously.

15.15   Properties of Colloids

In 1827 Robert Brown (1773–1858), while observing a strongly illuminated aqueous suspension of pollen under a high-powered microscope, noted that the pollen grains appeared to have a trembling, erratic motion. He determined later that this erratic motion was not confined to pollen but was characteristic of colloidal particles in general. This random motion of colloidal particles is called **Brownian movement**. It can be readily observed by confining cigarette smoke in a small transparent chamber and illuminating it with a strong beam of light at right angles to the optical axis of the microscope. The smoke particles appear as tiny randomly moving lights, because the light is reflected from their surfaces. This motion is due to the continual bombardment of the smoke particles by air molecules. Since Brownian movement can be seen when colloidal particles are dispersed in either a gaseous or a liquid medium, it affords nearly direct visual proof that matter at the molecular level actually is moving randomly, as postulated by the Kinetic-Molecular Theory.

Brownian movement

When an intense beam of light is passed through an ordinary solution and is viewed at an angle, the beam passing through the solution is hardly visible. A beam of light, however, is clearly visible and sharply outlined when it is passed through a colloidal dispersion (see Figure 15.5). This phenomenon is known as the **Tyndall effect**. It was first described by Michael Faraday in 1857 and later amplified by John Tyndall. The Tyndall effect, like the Brownian movement, can be observed

Tyndall effect

FIGURE 15.5
The Tyndall effect. The beakers on the left and right each contain a true solution, whereas the beaker in the center contains a collodial starch solution. A laser beam, emitted by the laser on the far left, scatters in the colloid but is invisible in the true solution. ▶

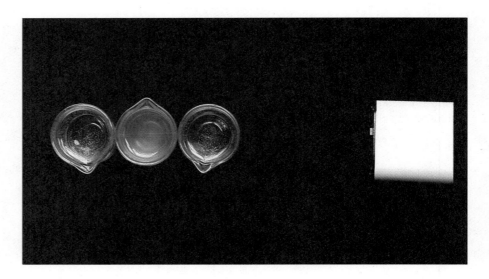

in nearly all colloidal dispersions. It occurs because the colloidal particles are large enough to scatter the rays of visible light. The ions or molecules of true solutions are too small to scatter light and, therefore, do not exhibit a noticeable Tyndall effect.

Another important characteristic of colloids is that the particles have relatively huge surface areas. We saw in Section 14.6 that the surface area is increased tenfold when a 1-cm cube is divided into 1000 cubes with sides of 0.1 cm. When a 1-cm cube is divided into colloidal-size cubes measuring 10^{-6} cm, the combined surface area of all the particles becomes a million times greater than that of the original cube.

Colloidal particles become electrically charged when they adsorb ions on their surfaces. *Adsorption* should not be confused with *absorption*. Adsorption refers to the adhesion of molecules or ions to a surface, whereas absorption refers to the taking in of one material by another material. Adsorption occurs because the atoms or ions at the surface of a particle are not completely surrounded by other atoms or ions as are those in the interior. Consequently these surface atoms or ions attract and adsorb ions or polar molecules from the dispersion medium onto the surfaces of the colloidal particles. This property is directly related to the large surface area presented by the many tiny particles.

The particles in a given dispersion tend to adsorb only ions having one kind of charge. For example, cations are primarily adsorbed on iron(III) hydroxide sol, resulting in positively charged colloidal particles. On the other hand, the particles of an arsenic(III) sulfide sol primarily adsorb anions, resulting in negatively charged colloidal particles. The properties of true solutions, colloidal dispersions, and mechanical suspensions are summarized and compared in Table 15.7.

15.16   Stability of Colloids

The stability of different dispersions varies with the properties of the dispersed and dispersing phases. We have noted that nonhomogenized milk is a colloid, yet it will separate after standing for a few hours. However, the particles of a good colloidal dispersion will remain in suspension indefinitely. As a case in point, a ruby-red

TABLE 15.7 Comparison of the Properties of True Solutions, Colloidal Dispersions, and Suspensions

	Particle size (nm)	Ability to pass through filter paper	Ability to pass through parchment	Exhibits Tyndall effect	Exhibits Brownian movement	Settles out on standing	Appearance
True solution	<1	Yes	Yes	No	No	No	Transparent, homogeneous
Colloidal dispersion	1–1000	Yes	No	Yes	Yes	Generally does not	Usually not transparent, but may appear to be homogeneous
Suspension	>1000	No	No	—	No	Yes	Not transparent, heterogeneous

gold sol has been kept in the British Museum for more than a century without noticeable settling. (This specimen is kept for historical interest; it was prepared by Michael Faraday.) The particles of a specific colloid remain dispersed for two reasons:

1. They are bombarded by the molecules of the dispersing phase, which keeps the particles in motion (Brownian movement) so that gravity does not cause them to settle out.
2. Since the colloidal particles have the same kind of electrical charge, they repel each other. This mutual repulsion prevents the dispersed particles from coalescing to larger particles, which would settle out of suspension.

In certain types of colloids, the presence of a material known as a protective colloid is necessary for stability. Egg yolk, for example, acts as a stabilizer, or protective colloid, in mayonnaise. The yolk adsorbs on the surfaces of the oil particles and prevents them from coalescing.

15.17 Applications of Colloidal Properties

Activated charcoal has an enormous surface area, approximately 1 million square centimeters per gram in some samples. Hence, charcoal is very effective in selectively adsorbing the polar molecules of some poisonous gases and is therefore used in gas masks. Charcoal can be used to adsorb impurities from liquids as well as from gases, and large amounts are used to remove substances that have objectionable tastes and odors from water supplies. In sugar refineries activated charcoal is used to adsorb colored impurities from the raw sugar solutions.

A process widely used for dust and smoke control in many urban and industrial areas was devised by an American, Frederick Cottrell (1877–1948). The Cottrell process takes advantage of the fact that the particulate matter in dust and smoke is electrically charged. Air to be cleaned of dust or smoke is passed between electrode plates charged with a high voltage. Positively charged particles are attracted to,

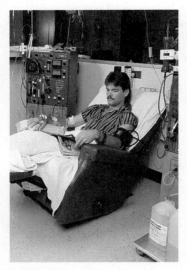

This dialysis machine acts as artificial kidneys by removing soluble waste products from the blood.

dialysis

neutralized, and thereby precipitated at the negative electrodes. Negatively charged particles are removed in the same fashion at the positive electrodes. Large Cottrell units are fitted with devices for automatic removal of precipitated material. In some installations, particularly at cement mills and smelters, the value of the dust collected may be sufficient to pay for the precipitation equipment. Small units, designed for removing dust and pollen from air in the home, are now on the market. Unfortunately, Cottrell units remove only particulate matter; they cannot remove gaseous pollutants such as carbon monoxide, sulfur dioxide, and nitrogen oxides.

Thomas Graham found that a parchment membrane would allow the passage of true solutions but would prevent the passage of colloidal dispersions. Dissolved solutes can be removed from colloidal dispersions through the use of such a membrane by a process called **dialysis**. The membrane itself is called a *dialyzing membrane*. Artificial membranes are made from such materials as parchment paper, collodion, or certain kinds of cellophane. Dialysis can be demonstrated by putting a colloidal starch dispersion and some copper(II) sulfate solution in a parchment paper bag and suspending it in running water. In a few hours the blue color of the copper(II) sulfate has disappeared, and only the starch dispersion remains in the bag.

A life-saving application of dialysis has been the development of artificial kidneys. The blood of a patient suffering from partial kidney failure is passed through the artificial kidney machine for several hours, during which time the soluble waste products are removed by dialysis.

Concepts in Review

1. State the general characteristics of acids and bases.
2. Define an acid and base in terms of the Arrhenius, Brønsted–Lowry, and Lewis theories.
3. Identify acid–base conjugate pairs in a reaction.
4. When given the reactants, complete and balance equations for the reactions of acids with bases, metals, metal oxides, and carbonates.
5. When given the reactants, complete and balance equations of the reaction of an amphoteric hydroxide with either a strong acid or a strong base.
6. Write balanced equations for the reaction of sodium hydroxide or potassium hydroxide with zinc and with aluminum.
7. Classify common compounds as electrolytes or nonelectrolytes.
8. Distinguish between strong and weak electrolytes.
9. Explain the process of dissociation and ionization. Indicate how they differ.
10. Write equations for the dissociation and/or ionization of acids, bases, and salts in water.
11. Describe and write equations for the ionization of water.
12. Explain how pH expresses hydrogen-ion concentration or hydronium-ion concentration.
13. Given pH as an integer, calculate the H^+ molarity, and vice versa.
14. Use a calculator to calculate pH values from corresponding H^+ molarities.

15. Explain the process of acid–base neutralization.

16. Calculate the molarity, normality, or volume of an acid or base solution from appropriate titration data.

17. Write un-ionized, total ionic, and net ionic equations for neutralization equations.

18. Discuss colloids and describe methods for their preparation.

19. Describe the characteristics that distinguish true solutions, colloidal dispersions, and mechanical suspensions.

20. Explain (a) how colloidal dispersions can be stabilized and (b) how they can be precipitated.

Key Terms

The terms listed here have been defined within this chapter. Section numbers are referenced in parenthesis for each term. More detailed definitions are given in the Glossary.

amphoteric (15.2)	nonelectrolyte (15.5)
Brownian movement (15.15)	pH (15.9)
colloid (15.13)	sol (15.14)
dialysis (15.17)	spectator ion (15.10)
dissociation (15.6)	strong electrolyte (15.7)
electrolyte (15.5)	titration (15.10)
hydronium ion (15.1)	total ionic equation (15.12)
ionization (15.6)	Tyndall effect (15.15)
logarithm (15.9)	un-ionized equation (15.12)
net ionic equation (15.12)	weak electrolyte (15.7)
neutralization (15.10)	

Questions

Questions refer to tables, figures, and key words and concepts defined within the chapter. A particularly challenging question or exercise is indicated with an asterisk.

1. Since a hydrogen ion and a proton are identical, what differences exist between the Arrhenius and Brønsted–Lowry definitions of an acid? (See Table 15.1.)

2. According to Figure 15.1, what type of substance must be in solution in order for the bulb to light?

3. Which of the following classes of compounds are electrolytes: acids, alcohols, bases, salts? (See Table 15.2.)

4. What two differences are apparent in the arrangement of water molecules about the hydrated ions as depicted in Figure 15.2?

5. The pH of a solution with a hydrogen ion concentration of 0.003 M is between what two whole numbers? (See Table 15.4.)

6. Which is the more acidic, tomato juice or blood? (See Table 15.5.)

7. Using each of the three acid–base theories (Arrhenius, Brønsted–Lowry, and Lewis), define an acid and a base.

8. For each of the acid–base theories referred to in Exercise 7, write an equation illustrating the neutralization of an acid with a base.

9. Write the Lewis structure for the (a) bromide ion, (b) hydroxide ion, and (c) cyanide ion. Why are these ions considered to be bases according to the Brønsted–Lowry and Lewis acid–base theories?

10. Into what three classes of compounds do electrolytes generally fall?

11. Name each compound listed in Table 15.3.

12. A solution of HCl in water conducts an electric current, but a solution of HCl in hexane does not. Explain this behavior in terms of ionization and chemical bonding.

13. How do ionic compounds exist in their crystalline structure? What occurs when they are dissolved in water?

14. An aqueous methyl alcohol, CH_3OH, solution does not conduct an electric current, but a solution of sodium hydroxide, NaOH, does. What does this information tell us about the OH group in the alcohol?

15. Why does molten NaCl conduct electricity?

16. Explain the difference between dissociation of ionic compounds and ionization of molecular compounds.

17. Distinguish between strong and weak electrolytes.

18. Explain why ions are hydrated in aqueous solutions.

19. What is the main distinction between water solutions of strong and weak electrolytes?

20. What are the relative concentrations of $H^+(aq)$ and OH^- (aq) in (a) a neutral solution, (b) an acid solution, and (c) a basic solution?

21. Write the net ionic equation for the reaction of a strong acid with a water-soluble hydroxide base in an aqueous solution.

22. The solubility of HCl gas in water, a polar solvent, is much greater than its solubility in hexane, a nonpolar solvent. How can you account for this difference?

23. Pure water, containing equal concentrations of both acid and base ions, is neutral. Why?

24. Which of the following statements are correct? Rewrite each incorrect statement to make it correct.
    (a) The Arrhenius theory of acids and bases is restricted to aqueous solutions.
    (b) The Brønsted–Lowry theory of acids and bases is restricted to solutions other than aqueous solutions.
    (c) All substances that are acids according to the Brønsted–Lowry theory will also be acids by the Lewis theory.
    (d) All substances that are acids according to the Lewis theory will also be acids by the Brønsted–Lowry theory.
    (e) An electron-pair donor is a Lewis acid.
    (f) All Arrhenius acid–base neutralization reactions can be represented by a single net ionic equation.
    (g) When an ionic compound dissolves in water, the ions separate; this process is called *ionization.*
    (h) In the auto-ionization of water,

    $$2\ H_2O \rightleftharpoons H_3O^+ + OH^-$$

    the H_3O^+ and the OH^- constitute a conjugate acid–base pair.
    (i) In the reaction in part (h), H_2O is both the acid and the base.
    (j) Most common Na^+, K^+, and NH_4^+ salts are soluble in water.
    (k) A solution of pH 3 is 100 times more acidic than a solution of pH 5.
    (l) In general, ionic substances when placed in water will give a solution capable of conducting an electric current.
    (m) The terms *dissociation* and *ionization* are synonymous.
    (n) A solution of $Mg(NO_3)_2$ contains three ions per formula unit in solution.
    (o) The terms *strong acid, strong base, weak acid,* and *weak base* refer to whether an acid or base solution is concentrated or dilute.
    (p) pH is defined as the negative logarithm of the molar concentration of H^+ ions (or H_3O^+ ions).
    (q) All reactions may be represented by net ionic equations.
    (r) One mole of $CaCl_2$ contains more anions than cations.
    (s) It is possible to boil seawater at a lower temperature than that required to boil pure water (both at the same pressure).
    (t) It is possible to have a neutral aqueous solution whose pH is not 7.
    (u) The size of colloidal particles ranges from 1 mm to 1000 mm.
    (v) The Tyndall effect is observable in both colloidal and true solutions.

25. Indicate the fundamental difference between a colloidal dispersion and a true solution.

26. List the two methods used to prepare colloidal dispersions and briefly explain how each is accomplished.

27. Explain the Tyndall effect and how it may be used to distinguish between a colloidal dispersion and a true solution.

28. Distinguish between adsorption and absorption.

29. Why do particles in a specific colloid remain dispersed?

30. Explain the process of dialysis, giving a practical application in society.

31. Diagram and label the types of interactions occurring between strands of protein in the hair.

32. What is the difference between a shampoo and a depilatory? Include a discussion of the types of bonds broken in your answer.

Paired Exercises

These exercises are paired. Each odd-numbered exercise is followed by a similar even-numbered exercise. Answers to the even-numbered exercises are given in Appendix V.

33. Identify the conjugate acid–base pairs in the following equations:
   (a) $HCl + NH_3 \longrightarrow NH_4^+ + Cl^-$
   (b) $HCO_3^- + OH^- \rightleftharpoons CO_3^{2-} + H_2O$
   (c) $HCO_3^- + H_3O^+ \rightleftharpoons H_2CO_3 + H_2O$
   (d) $HC_2H_3O_2 + H_2O \rightleftharpoons H_3O^+ + C_2H_3O_2^-$

35. Complete and balance the following equations:
   (a) $Mg(s) + HCl(aq) \longrightarrow$
   (b) $BaO(s) + HBr(aq) \longrightarrow$
   (c) $Al(s) + H_2SO_4(aq) \longrightarrow$
   (d) $Na_2CO_3(aq) + HCl(aq) \longrightarrow$
   (e) $Fe_2O_3(s) + HBr(aq) \longrightarrow$
   (f) $Ca(OH)_2(aq) + H_2CO_3(aq) \longrightarrow$

37. Which of the following compounds are electrolytes? Consider each substance to be mixed with water.
   (a) HCl
   (b) CO_2
   (c) $CaCl_2$
   (d) $C_{12}H_{22}O_{11}$ (sugar)
   (e) C_3H_7OH (rubbing alcohol)
   (f) CCl_4 (insoluble)

39. Calculate the molarity of the ions present in each of the following salt solutions. Assume each salt to be 100% dissociated:
   (a) 0.015 M NaCl
   (b) 4.25 M NaKSO$_4$
   (c) 0.20 M CaCl$_2$
   (d) 22.0 g KI in 500. mL of solution

41. In Exercise 39, how many grams of each ion would be present in 100. mL of each solution?

43. What is the molar concentration of all ions present in a solution prepared by mixing the following? Neglect the concentration of H^+ and OH^- from water. Also, assume volumes of solutions are additive.
   (a) 30.0 mL of 1.0 M NaCl and 40.0 mL of 1.0 M NaCl
   (b) 30.0 mL of 1.0 M HCl and 30.0 mL of 1.0 M NaOH
   *(c) 100.0 mL of 0.40 M KOH and 100.0 mL of 0.80 M HCl

34. Identify the conjugate acid–base pairs in the following equations:
   (a) $HC_2H_3O_2 + H_2SO_4 \rightleftharpoons H_2C_2H_3O_2^+ + HSO_4^-$
   (b) The two-step ionization of sulfuric acid,
   $$H_2SO_4 + H_2O \longrightarrow H_3O^+ + HSO_4^-$$
   $$HSO_4^- + H_2O \rightleftharpoons H_3O^+ + SO_4^{2-}$$
   (c) $HClO_4 + H_2O \longrightarrow H_3O^+ + ClO_4^-$
   (d) $CH_3O^- + H_3O^+ \longrightarrow CH_3OH + H_2O$

36. Complete and balance the following equations:
   (a) $NaOH(aq) + HBr(aq) \longrightarrow$
   (b) $KOH(aq) + HCl(aq) \longrightarrow$
   (c) $Ca(OH)_2(aq) + HI(aq) \longrightarrow$
   (d) $Al(OH)_3(s) + HBr(aq) \longrightarrow$
   (e) $Na_2O(s) + HClO_4(aq) \longrightarrow$
   (f) $LiOH(aq) + FeCl_3(aq) \longrightarrow$

38. Which of the following compounds are electrolytes? Consider each substance to be mixed with water.
   (a) $NaHCO_3$ (baking soda)
   (b) N_2 (insoluble gas)
   (c) $AgNO_3$
   (d) $HCOOH$ (formic acid)
   (e) $RbOH$
   (f) K_2CrO_4

40. Calculate the molarity of the ions present in each of the following salt solutions. Assume each salt to be 100% dissociated:
   (a) 0.75 M ZnBr$_2$
   (b) 1.65 M Al$_2$(SO$_4$)$_3$
   (c) 900. g (NH$_4$)$_2$SO$_4$ in 20.0 L of solution
   (d) 0.0120 g Mg(ClO$_3$)$_2$ in 1.00 mL of solution

42. In Exercise 40, how many grams of each ion would be present in 100. mL of each solution?

44. What is the molar concentration of all ions present in a solution prepared by mixing the following? Neglect the concentration of H^+ and OH^- from water. Also, assume volumes of solutions are additive.
   (a) 100.0 mL of 2.0 M KCl and 100.0 mL of 1.0 M CaCl$_2$
   (b) 35.0 mL of 0.20 M Ba(OH)$_2$ and 35.0 mL of 0.20 M H$_2$SO$_4$

45. Given the data for the following separate titrations, calculate the molarity of the HCl:

	mL HCl	Molarity HCl	mL NaOH	Molarity NaOH
(a)	40.13	M	37.70	0.728
(b)	19.00	M	33.66	0.306
(c)	27.25	M	18.00	0.555

46. Given the data for the following separate titrations, calculate the molarity of the NaOH:

	mL HCl	Molarity HCl	mL NaOH	Molarity NaOH
(a)	37.19	0.126	31.91	M
(b)	48.04	0.482	24.02	M
(c)	13.x13	1.425	39.39	M

47. Rewrite the following unbalanced equations, changing them into balanced net ionic equations. All reactions are in water solution:
(a) $K_2SO_4(aq) + Ba(NO_3)_2(aq) \longrightarrow$
$$KNO_3(aq) + BaSO_4(s)$$
(b) $CaCO_3(s) + HCl(aq) \longrightarrow$
$$CaCl_2(aq) + CO_2(g) + H_2O(l)$$
(c) $Mg(s) + HC_2H_3O_2(aq) \longrightarrow$
$$Mg(C_2H_3O_2)_2(aq) + H_2(g)$$

48. Rewrite the following unbalanced equations, changing them into balanced net ionic equations. All reactions are in water solution:
(a) $H_2S(g) + CdCl_2(aq) \longrightarrow CdS(s) + HCl(aq)$
(b) $Zn(s) + H_2SO_4(aq) \longrightarrow ZnSO_4(aq) + H_2(g)$
(c) $AlCl_3(aq) + Na_3PO_4(aq) \longrightarrow AlPO_4(s) + NaCl(aq)$

49. In each of the following pairs which solution is more acidic? All are water solutions. Explain your answer.
(a) 1 molar HCl or 1 molar H_2SO_4?
(b) 1 molar HCl or 1 molar $HC_2H_3O_2$?

50. In each of the following pairs which solution is more acidic? All are water solutions. Explain your answer.
(a) 1 molar HCl or 2 molar HCl?
(b) 1 normal H_2SO_4 or 1 molar H_2SO_4?

51. What volume (in milliliters) of 0.245 M HCl will neutralize 50.0 mL of 0.100 M $Ca(OH)_2$? The equation is
$2 HCl(aq) + Ca(OH)_2(aq) \longrightarrow CaCl_2(aq) + 2 H_2O(l)$.

52. What volume (in milliliters) of 0.245 M HCl will neutralize 10.0 g of $Al(OH)_3$? The equation is
$3 HCl(aq) + Al(OH)_3(s) \longrightarrow AlCl_3(aq) + 3 H_2O(l)$.

*53. A 0.200-g sample of impure NaOH requires 18.25 mL of 0.2406 M HCl for neutralization. What is the percent of NaOH in the sample?

*54. A batch of sodium hydroxide was found to contain sodium chloride as an impurity. To determine the amount of impurity, a 1.00-g sample was analyzed and found to require 49.90 mL of 0.466 M HCl for neutralization. What is the percent of NaCl in the sample?

*55. What volume of H_2 gas, measured at 27°C and 700. torr pressure, can be obtained by reacting 5.00 g of zinc metal with 100. mL of 0.350 M HCl? The equation is
$$Zn(s) + 2 HCl(aq) \longrightarrow ZnCl_2(aq) + H_2(g)$$

*56. What volume of H_2 gas, measured at 27°C and 700. torr, can be obtained by reacting 5.00 g of zinc metal with 200. mL of 0.350 M HCl? The equation is
$$Zn(s) + 2 HCl(aq) \longrightarrow ZnCl_2(aq) + H_2(g)$$

57. Calculate the pH of solutions having the following H^+ ion concentrations:
(a) 0.01 M
(b) 1.0 M
(c) $6.5 \times 10^{-9} M$

58. Calculate the pH of solutions having the following H^+ ion concentrations:
(a) $1 \times 10^{-7} M$
(b) 0.50 M
(c) 0.00010 M

59. Calculate the pH of the following:
(a) orange juice, $3.7 \times 10^{-4} M H^+$
(b) vinegar, $2.8 \times 10^{-3} M H^+$

60. Calculate the pH of the following:
(a) black coffee, $5.0 \times 10^{-5} M H^+$
(b) limewater, $3.4 \times 10^{-11} M H^+$

61. How many milliliters of 0.325 N HNO_3 are required to neutralize 32.8 mL of 0.225 N NaOH?

62. How many milliliters of 0.325 N H_2SO_4 are required to neutralize 32.8 mL of 0.225 N NaOH?

63. The equivalent masses of organic acids are often determined by titration with a standard solution of a base. Determine the equivalent mass of benzoic acid, if 0.305 g of it requires 25 mL of 0.10 N NaOH for neutralization.

*64. Determine the equivalent mass of succinic acid, if 0.738 g is required to neutralize 125 mL of 0.10 N base.

Additional Exercises

These exercises are not paired or labeled by topic and provide additional practice on concepts covered in this chapter.

65. What is the concentration of Ca^{2+} ions in a solution of CaI_2 having an I^- ion concentration of 0.520 M?

66. How many milliliters of 0.40 M HCl can be made by diluting 100. mL of 12 M HCl with water?

67. If 29.26 mL of 0.430 M HCl neutralizes 20.40 mL of $Ba(OH)_2$ solution, what is the molarity of the $Ba(OH)_2$ solution? The reaction is
$$Ba(OH)_2(aq) + 2\ HCl(aq) \longrightarrow BaCl_2(aq) + 2\ H_2O(l)$$

68. A 1 molal solution of acetic acid, $HC_2H_3O_2$, in water freezes at a lower temperature than a 1 molal solution of ethyl alcohol, C_2H_5OH, in water. Explain.

69. At the same cost per pound, which alcohol, CH_3OH or C_2H_5OH, would be more economical to purchase as an antifreeze for your car? Why?

70. How does a hydronium ion differ from a hydrogen ion?

71. Arrange, in decreasing order of freezing points, 1 molal aqueous solutions of HCl, $HC_2H_3O_2$, $C_{12}H_{22}O_{11}$ (sucrose), and $CaCl_2$. (List the one with the highest freezing point first.)

72. At 100°C the H^+ concentration in water is about 1×10^{-6} mol/L, about ten times that of water at 25°C. At which of these temperatures is
   (a) the pH of water the greater?
   (b) the hydrogen ion (hydronium ion) concentration the higher?
   (c) the water neutral?

73. What is the relative difference in H^+ concentration in solutions that differ by 1 pH unit?

74. A sample of pure sodium carbonate with a mass of 0.452 g was dissolved in water and neutralized with 42.4 mL of hydrochloric acid. Calculate the molarity of the acid:
$$Na_2CO_3(aq) + 2\ HCl(aq) \longrightarrow$$
$$2\ NaCl(aq) + CO_2(g) + H_2O(l)$$

75. What volume (mL) of 0.1234 M HCl is needed to neutralize 2.00 g $Ca(OH)_2$?

76. How many grams of KOH are required to neutralize 50.00 mL of 0.240 M HNO_3?

77. Two drops (0.1 mL) of 1.0 M HCl are added to water to make 1.0 L of solution. What is the pH of this solution if the HCl is 100% ionized?

78. What volume of concentrated (18.0 M) sulfuric acid must be used to prepare 50.0 L of 5.00 M solution?

79. Three (3.0) grams of NaOH are added to 500. mL of 0.10 M HCl. Will the resulting solution be acidic or basic? Show evidence for your answer.

80. A 10.00-mL sample of base solution requires 28.92 mL of 0.1240 N H_2SO_4 for neutralization. What is the normality of the base?

81. What is the normality and the molarity of a 25.00-mL sample of H_3PO_4 solution that requires 22.68 mL of 0.5000 N NaOH for complete neutralization?

82. How many milliliters of 0.10 N NaOH is required to neutralize 60.0 mL of 0.20 N H_2SO_4?

*83. A 25-mL sample of H_2SO_4 requires 40. mL of 0.20 N NaOH for neutralization.
   (a) What is the normality of the sulfuric acid solution?
   (b) How many grams of sulfuric acid are contained in the 25 mL sample?

*84. A solution of 40.0 mL of HCl is neutralized by 20.0 mL of NaOH solution. The resulting neutral solution is evaporated to dryness, and the residue is found to have a mass of 0.117 g. Calculate the normality of the HCl and NaOH solutions.

85. Compare and contrast a concentrated solution of a weak electrolyte with a dilute solution of a strong electrolyte.

86. Under what circumstances will a solution's normality be equal to its molarity?

87. How many milliliters of water must be added to 85.00 mL of 1.000 N H_3PO_4 to give a solution that is 0.6500 N H_3PO_4?

*88. If 380 mL of 0.35 M $Ba(OH)_2$ is added to 500.0 mL of 0.65 M HCl, will the mixture be acidic or basic? Find the pH of the resulting solution.

89. Fifty milliliters (50.00 mL) of 0.2000 M HCl is titrated with 0.2000 M NaOH. Find the pH of the solution after:
   (a) 0.000 mL of base have been added
   (b) 10.00 mL of base have been added
   (c) 25.00 mL of base have been added
   (d) 49.00 mL of base have been added
   (e) 49.90 mL of base have been added
   (f) 49.99 mL of base have been added
   (g) 50.00 mL of base have been added

   Plot your answers on a graph with pH on the y-axis and mL NaOH on the x-axis.

90. NaOH reacts with sulfuric acid:
   (a) Write a balanced equation for the reaction producing Na_2SO_4.
   (b) How many milliliters of 0.10 M NaOH are needed to react with 0.0050 mol H_2SO_4?
   (c) How many grams of Na_2SO_4 will also form?

*91. Lactic acid (found in sour milk) has an empirical formula of $HC_3H_5O_3$. A 1.0-g sample of lactic acid required 17.0 mL of 0.65 M NaOH to reach the end point of a titration. What is the molecular formula for lactic acid?

92. A 10.0-mL sample of HNO_3 was diluted to a volume of 100.00 mL. Then, 25 mL of that diluted solution was needed to neutralize 50.0 mL of 0.60 M KOH. What was the concentration of the original nitric acid?

93. The pH of a solution of a strong acid was determined to be 3. If water is then added to dilute this solution, would the pH change? Why or why not? Could enough water ever be added to raise the pH of an acid solution above 7?

94. Solution X has a pH of 2. Solution Y has a pH of 4. On the basis of this information, which of the following is true?
   (a) the $[H^+]$ of X is one half that of Y
   (b) the $[H^+]$ of X is twice that of Y
   (c) the $[H^+]$ of X is 100 times that of Y
   (d) the $[H^+]$ of X is 1/100 that of Y

95. A student is given three solutions: an acid, a base, and one that is neither acidic nor basic. The student performs tests on these solutions and records their properties. For each result, tell whether it is the property of an acid, a base, or whether you cannot decide.
   (a) the solution has $[H^+] = 1 \times 10^{-7} M$
   (b) the solution has $[OH^-] = 1 \times 10^{-2} M$
   (c) the solution turns litmus red
   (d) the solution is a good conductor of electricity

Answers to Practice Exercises

15.1 (a) HCO_3^-, (b) NO_2^-, (c) $C_2H_3O_2^-$
15.2 (a) H_2SO_4, (b) NH_4^+, (c) H_2O
15.3 (a) 0.050 M Mg^{2+}, 0.10 M Cl^-, (b) 0.070 M Al^{3+}, 0.21 M Cl^-

15.4 (a) 11.41, (b) 2.89, (c) 5.429
15.5 (c) 0.0620 M HCl
15.6 $3 H_2S(aq) + 2Bi^{3+}(aq) \longrightarrow Bi_2S_3(s) + 6 H^+(aq)$

1. Multiplication Multiplication is a process of adding any given number or quantity to itself a certain number of times. Thus, 4 times 2 means 4 added two times, or 2 added together four times, to give the product 8. Various ways of expressing multiplication are

$$ab \qquad a \times b \qquad a \cdot b \qquad a(b) \qquad (a)(b)$$

Each of these expressions means *a* times *b*, or *a* multiplied by *b*, or *b* times *a*.

When $a = 16$ and $b = 24$,   we have $16 \times 24 = 384$.

The expression $°F = (1.8 \times °C) + 32$ means that we are to multiply 1.8 times the Celsius degrees and add 32 to the product. When °C equals 50,

$$°F = (1.8 \times 50) + 32 = 90 + 32 = 122°F$$

The result of multiplying two or more numbers together is known as the *product.*

2. Division The word *division* has several meanings. As a mathematical expression, it is the process of finding how many times one number or quantity is contained in another. Various ways of expressing division are

$$a \div b \qquad \frac{a}{b} \qquad a/b$$

Each of these expressions means *a* divided by *b*.

When $a = 15$ and $b = 3$, $\dfrac{15}{3} = 5$.

The number above the line is called the *numerator*; the number below the line is the *denominator.* Both the horizontal and the slanted (/) division signs also mean "per." For example, in the expression for density, the mass per unit volume:

$$\text{density} = \text{mass/volume} = \frac{\text{mass}}{\text{volume}} = \text{g/mL}$$

The diagonal line still refers to a division of grams by the number of milliliters occupied by that mass. The result of dividing one number into another is called the *quotient.*

3. Fractions and Decimals A fraction is an expression of division, showing that the numerator is divided by the denominator. A *proper fraction* is one in which the numerator is smaller than the denominator. In an *improper fraction,* the numerator is the larger number. A decimal or a decimal fraction is a proper fraction in which the denominator is some power of 10. The decimal fraction is determined by carrying out the division of the proper fraction. Examples of proper fractions and their decimal fraction equivalents are shown in the accompanying table.

4. Addition of Numbers with Decimals To add numbers with decimals, we use the same procedure as that used when adding whole numbers, but we always line up the decimal points in the same column. For example, add $8.21 + 143.1 + 0.325$:

```
    8.21
+ 143.1
+   0.325
  151.635
```

When adding numbers that express units of measurement, we must be certain that the numbers added together all have the same units. For example, what is the total length of three pieces

Proper fraction	Decimal fraction	Proper fraction
$\dfrac{1}{8}$ =	0.125 =	$\dfrac{125}{1000}$
$\dfrac{1}{10}$ =	0.1 =	$\dfrac{1}{10}$
$\dfrac{3}{4}$ =	0.75 =	$\dfrac{75}{100}$
$\dfrac{1}{100}$ =	0.01 =	$\dfrac{1}{100}$
$\dfrac{1}{4}$ =	0.25 =	$\dfrac{25}{100}$

of glass tubing: 10.0 cm, 125 mm, and 8.4 cm? If we simply add the numbers, we obtain a value of 143.4, but we are not certain what the unit of measurement is. To add these lengths correctly, first change 125 mm to 12.5 cm. Now all the lengths are expressed in the same units and can be added:

$$
\begin{array}{r}
10.0 \text{ cm} \\
12.5 \text{ cm} \\
\underline{8.4 \text{ cm}} \\
30.9 \text{ cm}
\end{array}
$$

5. Subtraction of Numbers with Decimals   To subtract numbers containing decimals, we use the same procedure as for subtracting whole numbers, but we always line up the decimal points in the same column. For example, subtract 20.60 from 182.49:

$$
\begin{array}{r}
182.49 \\
- 20.60 \\
\hline
161.89
\end{array}
$$

6. Multiplication of Numbers with Decimals   To multiply two or more numbers together that contain decimals, we first multiply as if they were whole numbers. Then, to locate the decimal point in the product, we add together the number of digits to the right of the decimal in all the numbers multiplied together. The product should have this same number of digits to the right of the decimal point.

Multiply 2.05×2.05 (total of four digits to the right of the decimal):

$$
\begin{array}{r}
2.05 \\
\times 2.05 \\
\hline
1025 \\
4100 \\
\hline
4.2025
\end{array}
$$ (four digits to the right of the decimal)

Here are more examples:

$14.25 \times 6.01 \times 0.75 = 64.231875$ (six digits to the right of the decimal)
$39.26 \times 60 = 2355.60$ (two digits to the right of the decimal)

7. Division of Numbers with Decimals   To divide numbers containing decimals, we first relocate the decimal points of the numerator and denominator by moving them to the right as many places as needed to make the denominator a whole number. (Move the decimal of both the numerator and the denominator the same amount and in the same direction.) For example,

$$
\frac{136.94}{4.1} = \frac{1369.4}{41}
$$

The decimal point adjustment in this example is equivalent to multiplying both numerator and denominator by 10. Now we carry out the division normally, locating the decimal point immediately above its position in the dividend:

$$
\begin{array}{r}
33.4 \\
41\overline{)1369.4} \\
\underline{123} \\
139 \\
\underline{123} \\
164 \\
\underline{164}
\end{array}
\qquad
\frac{0.441}{26.25} = \frac{44.1}{2625} =
\begin{array}{r}
0.0168 \\
2625\overline{)44.1000} \\
\underline{2625} \\
17850 \\
\underline{15750} \\
21000 \\
\underline{21000}
\end{array}
$$

The foregoing examples are merely guides to the principles used in performing the various mathematical operations illustrated. There are, no doubt, shortcuts and other methods, and the student will discover these with experience. Every student of chemistry should learn to use a calculator for solving mathematical problems. The use of a calculator will save many hours of doing tedious longhand calculations. After solving a problem, the student should check for errors and evaluate the answer to see if it is logical and consistent with the data given.

8. Algebraic Equations   Many mathematical problems that are encountered in chemistry fall into the following algebraic forms. Solutions to these problems are simplified by first isolating the desired term on one side of the equation. This rearrangement is accomplished by treating both sides of the equation in an identical manner until the desired term is isolated.

(a) $a = \dfrac{b}{c}$

To solve for b, multiply both sides of the equation by c:

$$a \times c = \frac{b}{\not{c}} \times \not{c}$$

$$b = a \times c$$

To solve for c, multiply both sides of the equation by $\dfrac{c}{a}$:

$$\not{a} \times \frac{c}{\not{a}} = \frac{b}{\not{c}} \times \frac{\not{c}}{a}$$

$$c = \frac{b}{a}$$

(b) $\dfrac{a}{b} = \dfrac{c}{d}$

To solve for a, multiply both sides of the equation by b:

$$\frac{a}{\not{b}} \times \not{b} = \frac{c}{d} \times b$$

$$a = \frac{c \times b}{d}$$

To solve for b, multiply both sides of the equation by $\dfrac{b \times d}{c}$:

$$\frac{a}{\not{b}} \times \frac{\not{b} \times d}{c} = \frac{\not{c}}{d} \times \frac{b \times d}{\not{c}}$$

$$b = \frac{a \times d}{c}$$

(c) $a \times b = c \times d$

To solve for a, divide both sides of the equation by b:

$$\frac{a \times \not{b}}{\not{b}} = \frac{c \times d}{b}$$

$$a = \frac{c \times d}{b}$$

(d) $\dfrac{b - c}{a} = d$

To solve for b, first multiply both sides of the equation by a:

$$\dfrac{\cancel{a}(b - c)}{\cancel{a}} = d \times a$$

$$b - c = d \times a$$

Then add c to both sides of the equation:

$$b - \cancel{c} + \cancel{c} = d \times a + c$$

$$b = (d \times a) + c$$

When $a = 1.8$, $c = 32$, and $d = 35$,

$$b = (35 \times 1.8) + 32 = 63 + 32 = 95$$

9. Exponents, Powers of 10, Expression of Large and Small Numbers   In scientific measurements and calculations, we often encounter very large and very small numbers—for example, 0.00000384 and 602,000,000,000,000,000,000,000. These numbers are troublesome to write and awkward to work with, especially in calculations. A convenient method of expressing these large and small numbers in a simplified form is by means of exponents or powers of 10. This method of expressing numbers is known as **scientific or exponential notation**.

An **exponent** is a number written as a superscript following another number. Exponents are often called *powers* of numbers. The term *power* indicates how many times the number is used as a factor. In the number 10^2, 2 is the exponent, and the number means 10 squared, or 10 to the second power, or $10 \times 10 = 100$. Three other examples are

$$3^2 = 3 \times 3 = 9$$
$$3^4 = 3 \times 3 \times 3 \times 3 = 81$$
$$10^3 = 10 \times 10 \times 10 = 1000$$

For ease of handling, large and small numbers are expressed in powers of 10. Powers of 10 are used because multiplying or dividing by 10 coincides with moving the decimal point in a number by one place. Thus, a number multiplied by 10^1 would move the decimal point one place to the right; 10^2, two places to the right; 10^{-2}, two places to the left. To express a number in powers of 10, we move the decimal point in the original number to a new position, placing it so that the number is a value between 1 and 10. This new decimal number is multiplied by 10 raised to the proper power. For example, to write the number 42,389 in exponential form, the decimal point is placed between the 4 and the 2 (4.2389), and the number is multiplied by 10^4; thus, the number is 4.2389×10^4:

$$42{,}389 = 4.2389 \times 10^4$$
$$\overset{}{4\,3\,2\,1}$$

The exponent of 10 (4) tells us the number of places that the decimal point has been moved from its original position. If the decimal point is moved to the left, the exponent is a positive number; if it is moved to the right, the exponent is a negative number. To express the number 0.00248 in exponential notation (as a power of 10), the decimal point is moved three places to the right; the exponent of 10 is -3, and the number is 2.48×10^{-3}.

$$0.00248 = 2.48 \times 10^{-3}$$
$$\overset{}{1\,2\,3}$$

Study the following examples.

scientific or exponential notation

exponent

$$1237 = 1.237 \times 10^3$$
$$988 = 9.88 \times 10^2$$
$$147.2 = 1.472 \times 10^2$$
$$2,200,000 = 2.2 \times 10^6$$
$$0.0123 = 1.23 \times 10^{-2}$$
$$0.00005 = 5 \times 10^{-5}$$
$$0.000368 = 3.68 \times 10^{-4}$$

Exponents in multiplication and division   The use of powers of 10 in multiplication and division greatly simplifies locating the decimal point in the answer. In multiplication, first change all numbers to powers of 10, then multiply the numerical portion in the usual manner, and finally add the exponents of 10 algebraically, expressing them as a power of 10 in the product. In multiplication, the exponents (powers of 10) are added algebraically.

$$10^2 \times 10^3 = 10^{(2+3)} = 10^5$$
$$10^2 \times 10^2 \times 10^{-1} = 10^{(2+2-1)} = 10^3$$

Multiply: $40,000 \times 4200$

Change to powers of 10: $4 \times 10^4 \times 4.2 \times 10^3$

Rearrange: $4 \times 4.2 \times 10^4 \times 10^3$

$$16.8 \times 10^{(4+3)}$$
$$16.8 \times 10^7 \quad \text{or} \quad 1.68 \times 10^8 \quad \text{(Answer)}$$

Multiply: 380×0.00020
$$3.80 \times 10^2 \times 2.0 \times 10^{-4}$$
$$3.80 \times 2.0 \times 10^2 \times 10^{-4}$$
$$7.6 \times 10^{(2-4)}$$
$$7.6 \times 10^{-2} \quad \text{or} \quad 0.076 \quad \text{(Answer)}$$

Multiply: $125 \times 284 \times 0.150$
$$1.25 \times 10^2 \times 2.84 \times 10^2 \times 1.50 \times 10^{-1}$$
$$1.25 \times 2.84 \times 1.50 \times 10^2 \times 10^2 \times 10^{-1}$$
$$5.325 \times 10^{(2+2-1)}$$
$$5.33 \times 10^3 \quad \text{(Answer)}$$

In division, after changing the numbers to powers of 10, move the 10 and its exponent from the denominator to the numerator, changing the sign of the exponent. Carry out the division in the usual manner and evaluate the power of 10. The following is a proof of the equality of moving the power of 10 from the denominator to the numerator:

$$1 \times 10^{-2} = 0.01 = \frac{1}{100} = \frac{1}{10^2} = 1 \times 10^{-2}$$

In division, change the sign(s) of the exponent(s) of 10 in the denominator and move the 10 and its exponent(s) to the numerator. Then add all the exponents of 10 together. For example,

$$\frac{10^5}{10^3} = 10^5 \times 10^{-3} = 10^{(5-3)} = 10^2$$

$$\frac{10^3 \times 10^4}{10^{-2}} = 10^3 \times 10^4 \times 10^2 = 10^{(3+4+2)} = 10^9$$

10. Significant Figures in Calculations   The result of a calculation based on experimental measurements cannot be more precise than the measurement that has the greatest uncertainty. (See Section 2.5 for additional discussion.)

Addition and Subtraction   The result of an addition or subtraction should contain no more digits to the right of the decimal point than are contained in the quantity that has the least number of digits to the right of the decimal point.

Perform the operation indicated and then round off the number to the proper number of significant figures:

$$
\begin{array}{ll}
142.8 \ \ \text{g} & \\
18.843 \ \text{g} & 93.45 \ \text{mL} \\
\underline{36.42 \ \ \text{g}} & \underline{-18.0 \ \ \text{mL}} \\
198.063 \ \text{g} & 75.45 \ \text{mL} \\
198.1 \ \ \text{g (Answer)} & 75.5 \ \ \text{mL (Answer)}
\end{array}
$$

Multiplication and Division   In calculations involving multiplication or division, the answer should contain the same number of significant figures as the measurement that has the least number of significant figures. In multiplication or division the position of the decimal point has nothing to do with the number of significant figures in the answer. Study the following examples:

	Round off to
$2.05 \times 2.05 = 4.2025$	4.20
$18.48 \times 5.2 = 96.096$	96
$0.0126 \times 0.020 = 0.000252$ or	
$\quad 1.26 \times 10^{-2} \times 2.0 \times 10^{-2} = 2.520 \times 10^{-4}$	2.5×10^{-4}
$\dfrac{1369.4}{41} = 33.4$	33
$\dfrac{2268}{4.20} = 540.$	540.

11. Dimensional Analysis   Many problems of chemistry can be readily solved by dimensional analysis using the factor-label or conversion-factor method. Dimensional analysis involves the use of proper units of dimensions for all factors that are multiplied, divided, added, or subtracted in setting up and solving a problem. Dimensions are physical quantities such as length, mass, and time, which are expressed in such units as centimeters, grams, and seconds, respectively. In solving a problem, we treat these units mathematically just as though they were numbers, which gives us an answer that contains the correct dimensional units.

A measurement or quantity given in one kind of unit can be converted to any other kind of unit having the same dimension. To convert from one kind of unit to another, the original quantity or measurement is multiplied or divided by a conversion factor. The key to success lies in choosing the correct conversion factor. This general method of calculation is illustrated in the following examples.

Suppose we want to change 24 ft to inches. We need to multiply 24 ft by a conversion factor containing feet and inches. Two such conversion factors can be written relating inches to feet:

$$
\frac{12 \ \text{in.}}{1 \ \text{ft}} \qquad \text{or} \qquad \frac{1 \ \text{ft}}{12 \ \text{in.}}
$$

We choose the factor that will mathematically cancel feet and leave the answer in inches. Note that the units are treated in the same way we treat numbers, multiplying or dividing as required. Two possibilities then arise to change 24 ft to inches:

$$24 \text{ ft} \times \frac{12 \text{ in.}}{1 \text{ ft}} \quad \text{or} \quad 24 \text{ ft} \times \frac{1 \text{ ft}}{12 \text{ in.}}$$

In the first case (the correct method), feet in the numerator and the denominator cancel, giving us an answer of 288 in. In the second case, the units of the answer are ft²/in., the answer being 2.0 ft²/in. In the first case, the answer is reasonable since it is expressed in units having the proper dimensions. That is, the dimension of length expressed in feet has been converted to length in inches according to the mathematical expression

$$\text{ft} \times \frac{\text{in.}}{\text{ft}} = \text{in.}$$

In the second case, the answer is not reasonable since the units (ft²/in.) do not correspond to units of length. The answer is therefore incorrect. The units are the guiding factor for the proper conversion.

The reason we can multiply 24 ft times 12 in./ft and not change the value of the measurement is that the conversion factor is derived from two equivalent quantities. Therefore, the conversion factor 12 in./ft is equal to unity. When you multiply any factor by 1, it does not change the value:

$$12 \text{ in.} = 1 \text{ ft} \quad \text{and} \quad \frac{12 \text{ in.}}{1 \text{ ft}} = 1$$

Convert 16 kg to milligrams. In this problem it is best to proceed in this fashion:

$$\text{kg} \longrightarrow \text{g} \longrightarrow \text{mg}$$

The possible conversion factors are

$$\frac{1000 \text{ g}}{1 \text{ kg}} \quad \text{or} \quad \frac{1 \text{ kg}}{1000 \text{ g}} \qquad \frac{1000 \text{ mg}}{1 \text{ g}} \quad \text{or} \quad \frac{1 \text{ g}}{1000 \text{ mg}}$$

We use the conversion factor that leaves the proper unit at each step for the next conversion. The calculation is

$$16 \text{ kg} \times \frac{1000 \text{ g}}{1 \text{ kg}} \times \frac{1000 \text{ mg}}{1 \text{ g}} = 1.6 \times 10^7 \text{ mg}$$

Many problems can be solved by a sequence of steps involving unit conversion factors. This sound, basic approach to problem solving, together with neat and orderly setting up of data, will lead to correct answers having the right units, fewer errors, and considerable saving of time.

12. Graphical Representation of Data
A graph is often the most convenient way to present or display a set of data. Various kinds of graphs have been devised, but the most common type uses a set of horizontal and vertical coordinates to show the relationship of two variables. It is called an *x–y* graph because the data of one variable are represented on the horizontal or *x* axis (abscissa) and the data of the other variable are represented on the vertical or *y* axis (ordinate). See Figure I.1.

As a specific example of a simple graph, let us graph the relationship between Celsius and Fahrenheit temperature scales. Assume that initially we have only the information in the following table:

°C	°F
0	32
50	122
100	212

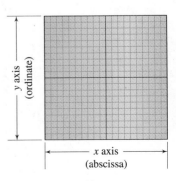

FIGURE I.1

On a set of horizontal and vertical coordinates (graph paper), scale off at least 100 Celsius degrees on the x axis and at least 212 Fahrenheit degrees on the y axis. Locate and mark the three points corresponding to the three temperatures given and draw a line connecting these points (see Figure I.2).

Here is how a point is located on the graph: Using the 50°C–122°F data, trace a vertical line up from 50°C on the x axis and a horizontal line across from 122°F on the y axis and mark the point where the two lines intersect. This process is called *plotting*. The other two points are plotted on the graph in the same way. [*Note*: The number of degrees per scale division was chosen to give a graph of convenient size. In this case, there are 5 Fahrenheit degrees per scale division and 2 Celsius degrees per scale division.]

The graph is Figure I.2 shows that the relationship between Celsius and Fahrenheit temperature is that of a straight line. The Fahrenheit temperature corresponding to any given Celsius temperature between 0 and 100° can be determined from the graph. For example, to find the Fahrenheit temperature corresponding to 40°C, trace a perpendicular line from 40°C on the x axis to the line plotted on the graph. Now trace a horizontal line from this point on the plotted line to the y axis and read the corresponding Fahrenheit temperature (104°F). See the dashed lines in Figure I.2. In turn, the Celsius temperature corresponding to any Fahrenheit temperature between 32 and 212° can be determined from the graph by tracing a horizontal line from the Fahrenheit temperature to the plotted line and reading the corresponding temperature on the Celsius scale directly below the point of intersection.

The mathematical relationship of Fahrenheit and Celsius temperatures is expressed by the equation °F = 1.8 × °C + 32. Figure I.2 is a graph of this equation. Because the graph is a straight line, it can be extended indefinitely at either end. Any desired Celsius temperature can be plotted against the corresponding Fahrenheit temperature by extending the scales along both axes as necessary.

FIGURE I.2 ▶

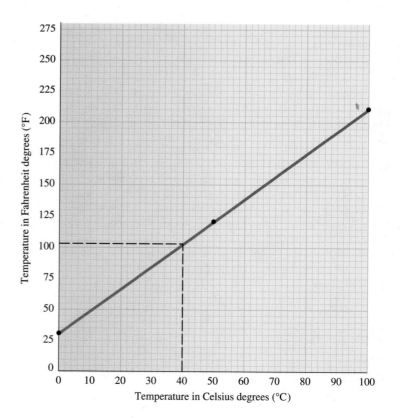

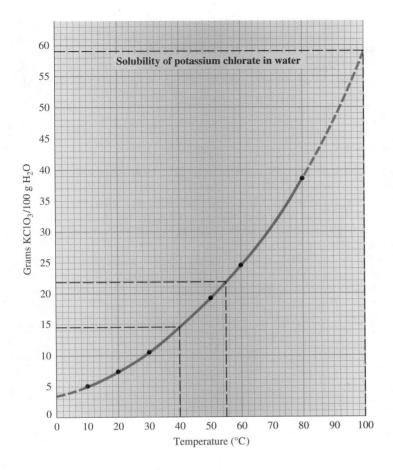

Temperature (°C)	Solubility (g KClO₃/100 g water)
10	5.0
20	7.4
30	10.5
50	19.3
60	24.5
80	38.5

◀ FIGURE I.3

Figure I.3 is a graph showing the solubility of potassium chlorate in water at various temperatures. The solubility curve on this graph was plotted from the data in the table next to the graph.

In contrast to the Celsius–Fahrenheit temperature relationship, there is no simple mathematical equation that describes the exact relationship between temperature and the solubility of potassium chlorate. The graph in Figure I.3 was constructed from experimentally determined solubilities at the six temperatures shown. These experimentally determined solubilities are all located on the smooth curve traced by the unbroken-line portion of the graph. We are therefore confident that the unbroken line represents a very good approximation of the solubility data for potassium chlorate over the temperature range from 10 to 80°C. All points on the plotted curve represent the composition of saturated solutions. Any point below the curve represents an unsaturated solution.

The dashed-line portions of the curve are *extrapolations*; that is, they extend the curve above and below the temperature range actually covered by the plotted solubility data. Curves such as this one are often extrapolated a short distance beyond the range of the known data, although the extrapolated portions may not be highly accurate. Extrapolation is justified only in the absence of more reliable information.

The graph in Figure I.3 can be used with confidence to obtain the solubility of KClO₃ at any temperature between 10 and 80°C, but the solubilities between 0 and 10°C and between 80 and 100°C are less reliable. For example, what is the solubility of $KClO_3$ at 55°C, at 40°C, and at 100°C?

First draw a perpendicular line from each temperature to the plotted solubility curve. Now trace a horizontal line to the solubility axis from each point on the curve and read the corresponding solubilities. The values that we read from the graph are

55°C	22.0 g $KClO_3$/100 g water
40°C	14.2 g $KClO_3$/100 g water
100°C	59 g $KClO_3$/100 g water

Of these solubilities, the one at 55°C is probably the most reliable because experimental points are plotted at 50°C and at 60°C. The 40°C solubility value is a bit less reliable because the nearest plotted points are at 30°C and 50°C. The 100°C solubility is the least reliable of the three values because it was taken from the extrapolated part of the curve, and the nearest plotted point is 80°C. Actual handbook solubility values are 14.0 and 57.0 g of $KClO_3$/100 g water at 40°C and 100°C, respectively.

The graph in Figure I.3 can also be used to determine whether a solution is saturated or unsaturated. For example, a solution contains 15 g of $KClO_3$/100 g of water and is at a temperature of 55°C. Is the solution saturated or unsaturated? *Answer*: The solution is unsaturated because the point corresponding to 15 g and 55°C on the graph is below the solubility curve; all points below the curve represent unsaturated solutions.

Temperature (°C)	Vapor pressure (torr)	Temperature (°C)	Vapor pressure (torr)
0	4.6	26	25.2
5	6.5	27	26.7
10	9.2	28	28.3
15	12.8	29	30.0
16	13.6	30	31.8
17	14.5	40	55.3
18	15.5	50	92.5
19	16.5	60	149.4
20	17.5	70	233.7
21	18.6	80	355.1
22	19.8	90	525.8
23	21.2	100	760.0
24	22.4	110	1074.6
25	23.8		

Physical Constants

Constant	Symbol	Value
Atomic mass unit	amu	1.6606×10^{-27} kg
Avogadro's number	N	6.022×10^{23}/mol
Gas constant	R (at STP)	0.08205 L atm/K mol
Mass of an electron	m_e	9.11×10^{-31} kg 5.486×10^{-4} amu
Mass of a neutron	m_n	1.675×10^{-27} kg 1.00866 amu
Mass of a proton	m_p	1.673×10^{-27} kg 1.00728 amu
Speed of light	c	2.997925×10^8 m/s

SI Units and Conversion Factors

Length	Mass
SI unit: meter (m)	SI unit: kilogram (kg)

Length

SI unit: meter (m)

1 meter	= 1.0936 yards
1 centimeter	= 0.3937 inch
1 inch	= 2.54 centimeters (exactly)
1 kilometer	= 0.62137 mile
1 mile	= 5280 feet
	= 1.609 kilometers
1 angstrom	= 10^{-10} meter

Mass

SI unit: kilogram (kg)

1 kilogram	= 1000 grams
	= 2.20 pounds
1 pound	= 453.59 grams
	= 0.45359 kilogram
	= 16 ounces
1 ton	= 2000 pounds
	= 907.185 kilograms
1 ounce	= 28.3 g
1 atomic mass unit	= 1.6606×10^{-27} kilogram

Volume

SI unit: cubic meter (m^3)

1 liter	= 10^{-3} m^3
	= 1 dm^3
	= 1.0567 quarts
1 gallon	= 4 quarts
	= 8 pints
	= 3.785 liters
1 quart	= 32 fluid ounces
	= 0.946 liter
1 fluid ounce	= 29.6 milliliters

Temperature

SI unit: kelvin (K)

$$0\ K = -273.15°C$$
$$= -459.67°F$$
$$K = °C + 273.15$$
$$°C = \frac{°F - 32}{1.8}$$
$$°F = 1.8(°C) + 32$$
$$°F = 1.8(°C + 40) - 40$$

Energy

SI unit: joule (J)

1 joule	= 1 kg m^2/s^2
	= 0.23901 calorie
1 calorie	= 4.184 joules

Pressure

SI unit: pascal (Pa)

1 pascal	= 1 kg/m s^2
1 atmosphere	= 101.325 kilopascals
	= 760 torr (mm Hg)
	= 14.70 pounds per square inch (psi)

Appendix IV Solubility Table

	F^-	Cl^-	Br^-	I^-	O^{2-}	S^{2-}	OH^-	NO_3^-	CO_3^{2-}	SO_4^{2-}	$C_2H_3O_2^-$
H^+	S	S	S	S	S	s	S	S	s	S	S
Na^+	S	S	S	S	S	S	S	S	S	S	S
K^+	S	S	S	S	S	S	S	S	S	S	S
NH_4^+	S	S	S	S	—	S	S	S	S	S	S
Ag^+	S	I	I	I	I	I	—	S	I	I	I
Mg^{2+}	I	S	S	S	I	d	I	S	I	S	S
Ca^{2+}	I	S	S	S	I	d	I	S	I	I	S
Ba^{2+}	I	S	S	S	s	d	s	S	I	I	S
Fe^{2+}	s	S	S	S	I	I	I	S	s	S	S
Fe^{3+}	I	S	S	—	I	I	I	S	I	S	I
Co^{2+}	S	S	S	S	I	I	I	S	I	S	S
Ni^{2+}	s	S	S	S	I	I	I	S	I	S	S
Cu^{2+}	s	S	S	—	I	I	I	S	I	S	S
Zn^{2+}	s	S	S	S	I	I	I	S	I	S	S
Hg^{2+}	d	S	I	I	I	I	I	S	I	d	S
Cd^{2+}	s	S	S	S	I	I	I	S	I	S	S
Sn^{2+}	S	S	S	s	I	I	I	S	I	S	S
Pb^{2+}	I	I	I	I	I	I	I	S	I	I	S
Mn^{2+}	s	S	S	S	I	I	I	S	I	S	S
Al^{3+}	I	S	S	S	I	d	I	S	—	S	S

Key: S = soluble in water
  s = slightly soluble in water
  I = insoluble in water (less than 1 g/100 g H_2O)
  d = decomposes in water

Chapter 2

2. 7.6 cm

4. The most dense (mercury) at the bottom and the least dense (glycerin) at the top. In the cylinder the solid magnesium would sink in the glycerin and float on the liquid mercury.

6. 0.789 g/mL < ice < 0.91 g/mL

8. $D = m/V$ specific gravity $= \dfrac{d_{\text{substance}}}{d_{\text{water}}}$

10. A lean person weighs more in water than an obese person and has a higher density.

12. **Rule 1.** When the first digit after those you want to retain is 4 or less, that digit and all others to its right are dropped. The last digit retained is not changed.

Rule 2. When the first digit after those you want to retain is 5 or greater, that digit and all others to the right of it are dropped and the last digit is increased by one.

14. The correct statements are: a, c, d, e, g, h, i, j, l, n, p, q.

16. (a) mg (d) nm
(b) kg (e) Å
(c) m (f) μL

18. (a) not significant (d) significant
(b) significant (e) significant
(c) not significant (f) significant

20. (a) 40.0 (3) (c) 129,042 (6)
(b) 0.081 (2) (d) 4.090×10^{-3} (4)

22. (a) 8.87 (c) 130. (1.30×10^2)
(b) 21.3 (d) 2.00×10^6

24. (a) 4.56×10^{-2} (c) 4.030×10^1
(b) 4.0822×10^3 (d) 1.2×10^7

26. (a) 28.1 (d) 2.010×10^3
(b) 58.5 (e) 2.49×10^{-4}
(c) 4.0×10^1 (f) 1.79×10^3

28. (a) $\frac{1}{4}$ (c) $1\frac{2}{3}$ or $\frac{5}{3}$
(b) $\frac{5}{8}$ (d) $\frac{8}{9}$

30. (a) 100
(b) 4.6 mL
(c) 22

32. (a) 4.5×10^8 Å (e) 6.5×10^5 mg
(b) 1.2×10^{-6} cm (f) 5.5×10^3 g
(c) 8.0×10^6 mm (g) 468 mL
(d) 0.164 g (h) 9.0×10^{-3} mL

34. (a) 117 ft (d) 4.3×10^4 g
(b) 10.3 mi (e) 75.7 L
(c) 7.4×10^4 mm^3 (f) 1.3×10^3 m^3

36. 50. ft/s

38. 0.102 km/s

40. 5.0×10^2 s

42. 3×10^4 mg

44. 3.0×10^3 hummingbirds to equal the mass of a condor

46. $2800

48. 57 L

50. 160 L

52. 4×10^5 m^2

54. 5 gal

56. 113°F Summer!

58. (a) 90.°F (c) 546 K
(b) -22.6°C (d) -300 K

60. -11.4°C $= 11.4$°F

62. 3.12 g/mL

64. 1.28 g/mL

66. 3.40×10^2 g

68. 7.0 lb

70. Yes, 116.5 L additional solution

72. -15°C > 4.5°F

74. B is 14 mL larger than A.

76. 76.9 g

78. 3.57×10^3 g

80. The container must hold at least 50 mL.

82. The gold bar is not pure gold.

84. 0.842 g/mL

Chapter 3

2. (a) Attractive forces among the ultimate particles of a solid (atoms, ions, or molecules) are strong enough to hold these particles in a fixed position within the solid and thus maintain the solid in a definite shape. Attractive forces among the ultimate particles of a liquid (usually molecules) are sufficiently strong to hold them together (preventing the liquid from rapidly becoming a gas) but are not strong enough to hold the particles in fixed positions (as in a solid).

(b) The ultimate particles in a liquid are quite closely packed (essentially in contact with each other) and thus the volume of the liquid is fixed at a given temperature. But, the ultimate particles in a gas are relatively far apart and essentially independent of each other. Consequently, the gas does not have a definite volume.

(c) In a gas the particles are relatively far apart and are easily compressed, but in a solid the particles are closely packed together and are virtually incompressible.

4. mercury and water

6. Three phases are present.

8. A system containing only one substance is not necessarily homogeneous.

10. 30 Si g/1 g H. There are more Si atoms than H atoms.

12. P Na
Al N
H Ni
K Ag
Mg Pu

14. sodium   silver
    potassium   tungsten
    iron   gold
    antimony   mercury
    tin   lead

16. In an element all atoms are alike, while a compound contains two or more elements which are chemically combined. Compounds may be decomposed into simpler substances while elements cannot.

18. 7 metals   1 metalloid   2 nonmetals

20. aurum

22. A *compound* is two or more elements chemically combined in a definite proportion by mass. Its properties differ from those of its components. A *mixture* is the physical combining of two or more substances (not necessarily elements). The composition may vary, the substances retain their properties and may be separated by physical means.

24. characteristic physical and chemical properties

26. Cations are positive, anions are negative.

28. Homogeneous are one phase, heterogeneous have two or more phases.

30. (a) H_2 (c) HCl   (e) NO

32. Sponge iron rusts easily. During this process hydrogen is generated and collected. The sponge iron can be regenerated from the rust formed and the process repeated indefinitely.

34. Charcoal—destructive distillation of wood.
    Bone black—destructive distillation of bones or waste.
    Carbon black—residue from burning natural gas.

36. Correct statements are: c, f, g, j, m, o, q, t, u, w, x, y.

38. (a) magnesium, bromine
    (b) carbon, chlorine
    (c) hydrogen, nitrogen, oxygen
    (d) barium, sulfur, oxygen
    (e) aluminum, phosphorus, oxygen

40. (a) $AlBr_3$ (c) $PbCrO_4$
    (b) CaF_2 (d) C_6H_6

42. (a) 1 atom Al, 3 atoms Br
    (b) 1 atom Ni, 2 atoms N, 6 atoms O
    (c) 12 atoms C, 22 atoms H, 11 atoms O

44. (a) 2 atoms   (d) 5 atoms
    (b) 2 atoms   (e) 17 atoms
    (c) 9 atoms

46. (a) 2 atoms H   (d) 4 atoms H
    (b) 6 atoms H   (e) 8 atoms H
    (c) 12 atoms H

48. (a) element   (c) element
    (b) compound   (d) mixture

50. (a) mixture   (c) mixture
    (b) element   (d) compound

52. (a) HO
    (b) C_2H_6O
    (c) $Na_2Cr_2O_7$

54. No. The only common liquid elements (at least at room temperature) are mercury and bromine.

56. 75% solids

58. 420 atoms

60. 40 atoms H

62. (a) magnesium, manganese, molybdenum, mendelevium, mercury
    (b) carbon, phosphorus, sulfur, selenium, iodine, astatine, boron
    (c) sodium, potassium, iron, silver, tin, antimony

64. (a) As temperature decreases, density increases.
    (b) 1.28 g/L, 1.17 g/L, 1.08 g/L

Chapter 4

2. solid

4. Water disappears. Gas appears above each electrode and as bubbles in solution.

6. A new substance is always formed during a chemical change, but never formed during physical changes.

8. The hot pack contains a solution of sodium acetate or sodium thiosulfate. A small crystal is added by squeezing a corner of the bag or bending a small metal activator. The solution crystallizes and heat is released to the surroundings. To reuse it, the pack is heated in boiling water until the crystals dissolve. Then, it is slowly cooled and stored until needed.

10. Potential energy is the energy of position. Kinetic energy is the energy matter possesses due to its motion.

12. The correct statements are: a, f, h, i.

14. (a) chemical   (d) chemical
    (b) physical   (e) chemical
    (c) physical   (f) physical

16. The copper wire, like the platinum wire, changed to a glowing red color when heated. Upon cooling, a new substance, black copper(II) oxide, had appeared.

18. Reactant: water
    Products: hydrogen, oxygen

20. the transformation of kinetic energy to thermal energy

22. (a) +   (d) +
    (b) −   (e) −
    (c) −

24. 2.2×10^3 J

26. 5.03×10^{-2} J/g °C

28. 5 °C

30. 29.1 °C

32. 45.7 g coal

34. 656 °C

36. 16.7 °C

38. 6:06 and 54 s

40. at the same rate

42. The mercury and sulfur react to form a compound since the properties of the product are different from the properties of either reactant.

Chapter 5

2. A neutron is about 1840 times heavier than an electron.

4. An atom is electrically neutral. An ion has a charge.

6. a Wintergreen Lifesaver

8. SIRA technique is stable isotope ratio analysis.

10. The correct statements are: a, b, c, f, g.

12. The correct statements are: b, d.

14. (a) The nucleus of the atom contains most of the mass.
(b) The nucleus of the atom is positively charged.
(c) The atom is mostly empty space.

16. The nucleus of an atom contains nearly all of its mass.

18. Electrons: Dalton—Electrons are not part of his model.
    Thomson—Electrons are scattered throughout the positive mass of matter in the atom.
    Rutherford—Electrons are located out in space away from the central positive mass.

Positive matter: Dalton—No positive matter in his model.
    Thomson—Positive matter is distributed throughout the atom.
    Rutherford—Positive matter is concentrated in a small central nucleus.

20. The isotope of C with a mass of 12 is an exact number.

22. Three isotopes of hydrogen have the same number of protons and electrons but differ in the number of neutrons.

24. (a) 80 protons; $+80$
(b) Hg

26. 40

28. (a) 27 protons, 32 neutrons
(b) 15 protons, 16 neutrons
(c) 74 protons, 110 neutrons
(d) 92 protons, 143 neutrons

30. 24.31 amu

32. 6.716 amu

34. $1.0 \times 10^{15}:1$

36. (a) isotopes
(b) adjacent to each other on the periodic table

38. (a) The $(+)$ particles are much lighter mass particles.
(b) negative

40. the number of protons and electrons

42. $^{60}Q = 50\%$
$^{63}Q = 50\%$

44. 6.03×10^{24} atoms

46.

Atomic number	Mass number	Symbol	Protons	Neutrons
(a) 8	16	O	8	8
(b) 28	58	Ni	28	30
(c) 80	199	Hg	80	119

Chapter 6

2. Charges on their ions must be equal and opposite in sign.

4. Seaborg is still alive. Elements can only be named for a person after he is dead according to IUPAC rules.

6. The correct statements are: a, b, c, e, f, h, i, k, n, p, q.

8. (a) BaO (d) $BeBr_2$
(b) H_2S (e) Li_4Si
(c) $AlCl_3$ (f) Mg_3P_2

10. Cl^- HSO_4^-
Br^- HSO_3^-
F^- CrO_4^{2-}
I^- CO_3^{2-}
CN^- HCO_3^-
O^{2-} $C_2H_3O_2^-$
OH^- ClO_3^-
S^{2-} MnO_4^-
SO_4^{2-} $C_2O_4^{2-}$

12. $(NH_4)_2SO_4$ NH_4Cl $(NH_4)_3AsO_4$ $NH_4C_2H_3O_2$ $(NH_4)_2CrO_4$
$CaSO_4$ $CaCl_2$ $Ca_3(AsO_4)_2$ $Ca(C_2H_3O_2)_2$ $CaCrO_4$
$Fe_2(SO_4)_3$ $FeCl_3$ $FeAsO_4$ $Fe(C_2H_3O_2)_3$ $Fe_2(CrO_4)_3$
Ag_2SO_4 $AgCl$ Ag_3AsO_4 $AgC_2H_3O_2$ Ag_2CrO_4
$CuSO_4$ $CuCl_2$ $Cu_3(AsO_4)_2$ $Cu(C_2H_3O_2)_2$ $CuCrO_4$

14. (a) carbon dioxide
(b) dinitrogen oxide
(c) phosphorus pentachloride
(d) carbon tetrachloride
(e) sulfur dioxide
(f) dinitrogen tetroxide
(g) diphosphorus pentoxide
(h) oxygen difluoride
(i) nitrogen trifluoride
(j) carbon disulfide

16. (a) potassium oxide
(b) ammonium bromide
(c) calcium iodide
(d) barium carbonate
(e) sodium phosphate
(f) aluminum oxide
(g) zinc nitrate
(h) silver sulfate

18. (a) $SnBr_4$ (d) $Hg(NO_2)_2$
(b) Cu_2SO_4 (e) TiS_2
(c) $Fe_2(CO_3)_3$ (f) $Fe(C_2H_3O_2)_2$

20. (a) $HC_2H_3O_2$ (d) H_3BO_3
(b) HF (e) HNO_2
(c) $HClO$ (f) H_2S

22. (a) phosphoric acid   (e) hypochlorous acid
(b) carbonic acid   (f) nitric acid
(c) iodic acid   (g) hydroiodic acid
(d) hydrochloric acid   (h) perchloric acid

24. (a) Na_2CrO_4 (h) $Co(HCO_3)_2$
(b) MgH_2 (i) $NaClO$
(c) $Ni(C_2H_3O_2)_2$ (j) $As_2(CO_3)_5$
(d) $Ca(ClO_3)_2$ (k) $Cr_2(SO_3)_3$
(e) $Pb(NO_3)_2$ (l) $Sb_2(SO_4)_3$
(f) KH_2PO_4 (m) $Na_2C_2O_4$
(g) $Mn(OH)_2$ (n) $KSCN$

26. (a) calcium hydrogen sulfate
    (b) arsenic(III) sulfite
    (c) tin(II) nitrite
    (d) iron(III) bromide
    (e) potassium hydrogen carbonate
    (f) bismuth(III) arsenate
    (g) iron(II) bromate
    (h) ammonium monohydrogen phosphate
    (i) sodium hypochlorite
    (j) potassium permanganate

28. (a) FeS_2 (e) $Mg(OH)_2$
    (b) $NaNO_3$ (f) $Na_2CO_3 \cdot 10\ H_2O$
    (c) $CaCO_3$ (g) C_2H_5OH
    (d) $C_{12}H_{22}O_{11}$

30. *-ide:*   Suffix is used to indicate a binary compound except for hydroxides, cyanides, and ammonium compounds.
    -ous: Used in acids to indicate that the polyatomic anion continued the *-ite* suffix; also used for the lower ionic charge of a multivalent metal.
    hypo: Used as a prefix in acids or salts when the polyatomic ion contains less oxygen than that of *-ous* acid or the *-ite* salt.
    per: Used as a prefix in acids or salts when the polyatomic ion contains more oxygen than that of the *-ic* acid or the *-ate* salt.
    -ite: The suffix of a salt derived from an *-ous* acid.
    -ate: The suffix of a salt derived from an *-ic* acid.
    Roman numerals indicate the charge on the metal cation.

32. (a) $50e^-$, 50p
    (b) $48e^-$, 50p
    (c) $46e^-$, 50p

34. $Li_3Fe(CN)_6$
    $AlFe(CN)_6$
    $Zn_3[Fe(CN)_6]_2$

36. ammonium oxide   zinc chloride
    ammonium carbonate   zinc acetate
    ammonium chloride   carbonic acid
    ammonium acetate   acetic acid
    zinc oxide   hydrochloric acid
    zinc carbonate

Chapter 7

2. A mole of gold has a higher mass than a mole of potassium.
4. A mole of gold atoms contains more electrons than a mole of potassium atoms.
6. 6.022×10^{23}
8. (a) 6.022×10^{23} atoms   (d) 16.00 g
   (b) 6.022×10^{23} molecules   (e) 32.00 g
   (c) 1.204×10^{24} atoms
10. Choosing 100 g of a compound allows us to simply drop the % sign and use grams for each percent.
12. calculation based on body mass in mg additive/kg mass/day

14. An empirical formula gives the smallest whole number ratio of the atoms present in a compound. The molecular formula represents the actual number of atoms of each element in a molecule of the compound. It may be the same as the empirical formula or may be a multiple of the empirical formula.

16. The correct answers are: a, d, f, g, j, k, m.

18. (a) 40.00   (f) 122.1
    (b) 275.8   (g) 180.2
    (c) 152.0   (h) 368.4
    (d) 96.09   (i) 244.2
    (e) 146.3

20. (a) 0.625 mol NaOH
    (b) 0.275 mol Br_2
    (c) 7.18×10^{-3} mol $MgCl_2$
    (d) 0.462 mol CH_3OH
    (e) 2.03×10^{-2} mol Na_2SO_4
    (f) 5.97 mol ZnI_2

22. (a) 0.0417 g H_2SO_4 (c) 0.122 g Ti
    (b) 11 g CCl_4 (d) 8.0×10^{-7} g S

24. (a) 1.05×10^{24} molecules Cl_2
    (b) 1.6×10^{23} molecules C_2H_6
    (c) 1.64×10^{23} molecules CO_2
    (d) 3.75×10^{24} molecules CH_4

26. (a) 3.271×10^{-22} g Au
    (b) 3.952×10^{-22} g U
    (c) 2.828×10^{-23} g NH_3
    (d) 1.795×10^{-22} g $C_6H_4(NH_2)_2$

28. (a) 0.886 mol S
    (b) 42.8 mol NaCl
    (c) 1.05×10^{24} atoms Mg
    (d) 9.47 mol Br_2

30. (a) 6.022×10^{23} molecules NH_3
    (b) 6.022×10^{23} N atoms
    (c) 1.807×10^{24} H atoms
    (d) 2.41×10^{24} atoms

32. (a) 6.0×10^{24} atoms O
    (b) 5.46×10^{24} atoms O
    (c) 5.0×10^{18} atoms O

34. (a) 1.27 g Cl
    (b) 9.07 g H
    (c) 23.0 g I

36. (a) 47.97% Zn   (d) 21.21% N
         52.02% Cl   6.104% H
    (b) 18.17% N   24.27% S
         9.153% H   48.45% O
         31.16% C   (e) 23.09% Fe
         41.51% O   17.37% N
    (c) 12.26% Mg   59.53% O
         31.24% P   (f) 54.39% I
         56.48% O   45.61% Cl

38. (a) 47.55% Cl   (c) 83.46% Cl
    (b) 34.05% Cl   (d) 83.63% Cl

40. 24.2% C
4.04% H
71.72% Cl

42. (a) $KClO_3$
(b) $KHSO_4$
(c) Na_2CrO_4

44. (a) CuCl   (d) K_3PO_4
(b) $CuCl_2$ (e) $BaCr_2O_7$
(c) Cr_2S_3 (f) PBr_8Cl_3

46. V_2O_5

48. The empirical formula is CH_2O. The molecular formula is $C_6H_{12}O_6$.

50. 5.88 g Na

52. 5.54×10^{19} m

54. (a) 8×10^{16} drops
(b) 8×10^6 mi^3

56. 10.3 mol H_2SO_4

58. (a) H_2O
(b) CH_3OH

60. 8.66 g Li

62. There is not sufficient S present.

64. 4.77 g O

66. (a) CCl_4
(b) C_2Cl_6
(c) C_6Cl_6
(d) C_3Cl_8

68. 2.4×10^{22} atoms Cu

70. 8×10^{-15} mol people

72. 32 g Mg

74. carbon

Chapter 8

2. the number of moles of each of the chemical species in the reaction

4. The correct statements are: a, d, e, f, h, i, j, l, n, o, p, q.

6. (a) $H_2 + Br_2 \longrightarrow 2\ HBr$

(b) $4\ Al + 3\ C \xrightarrow{\Delta} Al_4C_3$

(c) $Ba(ClO_3)_2 \xrightarrow{\Delta} BaCl_2 + 3\ O_2$

(d) $CrCl_3 + 3\ AgNO_3 \longrightarrow Cr(NO_3)_3 + 3\ AgCl$

(e) $2\ H_2O_2 \longrightarrow 2\ H_2O + O_2$

8. (a) combination
(b) combination
(c) decomposition
(d) double displacement
(e) decomposition

10. (a) $2\ MnO_2 + CO \longrightarrow Mn_2O_3 + CO_2$
(b) $Mg_3N_2 + 6\ H_2O \longrightarrow 3\ Mg(OH)_2 + 2\ NH_3$
(c) $4\ C_3H_5(NO_3)_3 \longrightarrow 12\ CO_2 + 10\ H_2O + 6\ N_2 + O_2$
(d) $4\ FeS + 7\ O_2 \longrightarrow 2\ Fe_2O_3 + 4\ SO_2$
(e) $2\ Cu(NO_3)_2 \longrightarrow 2\ CuO + 4\ NO_2 + O_2$

(f) $3\ NO_2 + H_2O \longrightarrow 2\ HNO_3 + NO$
(g) $2\ Al + 3\ H_2SO_4 \longrightarrow Al_2(SO_4)_3 + 3\ H_2$
(h) $4\ HCN + 5\ O_2 \longrightarrow 2\ N_2 + 4\ CO_2 + 2\ H_2O$
(i) $2\ B_5H_9 + 12\ O_2 \longrightarrow 5\ B_2O_3 + 9\ H_2O$

12. (a) $2\ H_2O \longrightarrow 2\ H_2 + O_2$
(b) $HC_2H_3O_2 + KOH \longrightarrow KC_2H_3O_2 + H_2O$
(c) $2\ P + 3\ I_2 \longrightarrow 2\ PI_3$
(d) $2\ Al + 3\ CuSO_4 \longrightarrow 3\ Cu + Al_2(SO_4)_3$
(e) $(NH_4)_2SO_4 + BaCl_2 \longrightarrow 2\ NH_4Cl + BaSO_4$
(f) $SF_4 + 2\ H_2O \longrightarrow SO_2 + 4\ HF$

(g) $Cr_2(CO_3)_3 \xrightarrow{\Delta} Cr_2O_3 + 3\ CO_2$

14. (a) $Cu + FeCl_3(aq) \longrightarrow$ no reaction

(b) $H_2 + Al_2O_3(s) \xrightarrow{\Delta}$ no reaction

(c) $2\ Al + 6\ HBr(aq) \longrightarrow 3\ H_2(g) + 2\ AlBr_3(aq)$
(d) $I_2 + HCl(aq) \longrightarrow$ no reaction

16. (a) $SO_2 + H_2O \longrightarrow H_2SO_3$
(b) $SO_3 + H_2O \longrightarrow H_2SO_4$
(c) $Ca + 2\ H_2O \longrightarrow Ca(OH)_2 + H_2$
(d) $2\ Bi(NO_3)_3 + 3\ H_2S \longrightarrow Bi_2S_3 + 6\ HNO_3$

18. (a) $C + O_2 \xrightarrow{\Delta} CO_2$

(b) $2\ Al(ClO_3)_3 \xrightarrow{\Delta} 9\ O_2 + 2\ AlCl_3$
(c) $CuBr_2 + Cl_2 \longrightarrow CuCl_2 + Br_2$
(d) $2\ SbCl_3 + 3\ (NH_4)_2S \longrightarrow Sb_2S_3 + 6\ NH_4Cl$

(e) $2\ NaNO_3 \xrightarrow{\Delta} 2\ NaNO_2 + O_2$

20. (a) exothermic
(b) endothermic

22. (a) $2\ Al + 3\ I_2 \xrightarrow{H_2O} 2\ AlI_3 + $ heat
(b) $4\ CuO + CH_4 + $ heat $\longrightarrow 4\ Cu + CO_2 + 2\ H_2O$
(c) $Fe_2O_3 + 2\ Al \longrightarrow 2\ Fe + Al_2O_3 + $ heat

24. 58 O on each side

26. A balanced equation tells us
(a) the type of atoms/molecules involved in the reaction.
(b) the relationship between quantities of the substances in the reaction.
A balanced equation gives no information about
(a) the time required for the reaction.
(b) odor or colors which may result.

28. Mg above Zn on the activity series

30. (a) $4\ K + O_2 \longrightarrow 2\ K_2O$
(b) $2\ Al + 3\ Cl_2 \longrightarrow 2\ AlCl_3$
(c) $CO_2 + H_2O \longrightarrow H_2CO_3$
(d) $CaO + H_2O \longrightarrow Ca(OH)_2$

32. (a) $Zn + H_2SO_4 \longrightarrow H_2 + ZnSO_4$
(b) $2\ AlI_3 + 3\ Cl_2 \longrightarrow 2\ AlCl_3 + 3\ I_2$
(c) $Mg + 2\ AgNO_3 \longrightarrow Mg(NO_3)_2 + 2\ Ag$
(d) $2\ Al + 3\ CoSO_4 \longrightarrow Al_2(SO_4)_3 + 3\ Co$

34. (a) $AgNO_3(aq) + KCl(aq) \longrightarrow AgCl(s) + KNO_3(aq)$
  (b) $Ba(NO_3)_2(aq) + MgSO_4(aq) \longrightarrow$
$$Mg(NO_3)_2(aq) + BaSO_4(s)$$
  (c) $H_2SO_4(aq) + Mg(OH)_2(aq) \longrightarrow$
$$2 H_2O(l) + MgSO_4(aq)$$
  (d) $MgO(s) + H_2SO_4(aq) \longrightarrow H_2O(l) + MgSO_4(aq)$
  (e) $Na_2CO_3(aq) + NH_4Cl(aq) \longrightarrow$ no reaction

36. chlorophyll   photosynthesis
  carotenoids   absorb high energy oxygen and
                 release it appropriately
  anthocyanins   diversity in leaf color

38. Leaves containing all pigments:

40. CO_2, CH_4, H_2O They act to trap the heat near the surface of the earth.

42. One-half the CO_2 released remains in the air.

Chapter 9

2. (a) correct   (d) incorrect
  (b) incorrect   (e) correct
  (c) correct   (f) incorrect

4. (a) 25.0 mol $NaHCO_3$
  (b) 3.85×10^{-3} mol $ZnCl_2$
  (c) 16 mol CO_2
  (d) 4.3 mol C_2H_5OH

6. (a) 1.31 g $NiSO_4$
  (b) 3.60 g $HC_2H_3O_2$
  (c) 373 g Bi_2S_3
  (d) 1.35 g $C_6H_{12}O_6$
  (e) 18 g K_2CrO_4

8. HCl

10. (a) $\dfrac{3 \text{ mol } CaCl_2}{1 \text{ mol } Ca_3(PO_4)_2}$ (d) $\dfrac{1 \text{ mol } Ca_3(PO_4)_2}{2 \text{ mol } H_3PO_4}$

  (b) $\dfrac{6 \text{ mol } HCl}{2 \text{ mol } H_3PO_4}$ (e) $\dfrac{6 \text{ mol } HCl}{1 \text{ mol } Ca_3(PO_4)_2}$

  (c) $\dfrac{3 \text{ mol } CaCl_2}{2 \text{ mol } H_3PO_4}$ (f) $\dfrac{2 \text{ mol } H_3PO_4}{6 \text{ mol } HCl}$

12. 15.5 mol CO_2
14. 4.20 mol HCl
16. 19.7 g $Zn_3(PO_4)_2$
18. 117 g H_2O
  271 g Fe
20. (a) 0.500 mol Fe_2O_3
  (b) 12.4 mol O_2
  (c) 6.20 mol SO_2
  (d) 65.6 g SO_2
  (e) 0.871 mol O_2
  (f) 332 g FeS_2
22. (a) H_2S is the limiting reactant and $Bi(NO_3)_3$ is in excess.
  (b) H_2O is the limiting reactant and Fe is in excess.
24. (a) 16.5 g CO_2
  (b) 59.9 g CO_2
  (c) 6.0 mol CO_2, 8.0 mol H_2O, and 4.0 mol O_2
26. sulfur
28. 95.0% yield
30. 77.8% CaC_2
32. A subscript is used to indicate the number of atoms in a formula. It cannot be changed without changing the identity of the substance. Coefficients are used only to balance atoms in chemical equations. They may be changed as needed to achieve a balanced equation.
34. (a) 380 g C_2H_5OH
      370 g CO_2
  (b) 480 mL C_2H_5OH
36. 65 g O_2
38. 1.0 g Ag_2S
40. (a) 2.0 mol Cu, 2.0 mol $FeSO_4$, and 1.0 mol $CuSO_4$
  (b) 15.9 g Cu, 38.1 g $FeSO_4$, 6.0 g Fe, and no $CuSO_4$
42. (a) 3.2×10^2 g C_2H_5OH
  (b) 1.10×10^3 g $C_6H_{12}O_6$
44. 3.7×10^2 kg Li_2O
46. 13 tablets
48. The mask protects the parts of the machine that are not to be etched away.
50. Micromachines could be used as smart pills, drug reservoirs, or mini computers.

Chapter 10

2. A second electron may enter an orbital already occupied if its spin is opposite the electron already in the orbital, and if all other orbitals of the same sublevel contain an electron.
4. Both $1s$ and $2s$ orbitals are spherical in shape and located symmetrically around the nucleus. The radius of $2s$ is larger than the $1s$.
6. $1s$, $2s$, $2p$, $3s$, $3p$, $4s$, $3d$, $4p$
8. The Bohr orbit has an electron traveling a specific path while an orbital is a region of space where the electron is most probably found.

10. *s* orbital

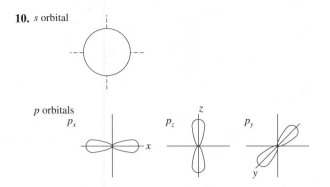

p orbitals

p_x p_z p_y

12. Transition elements are found in the center of the periodic table. The last electrons for these elements are found in the *d* or *f* orbitals. Representative elements are located on either side of the periodic table. The valence electrons for these elements are found in the *s* and/or *p* orbitals.

14.

Atomic number	Symbol
8	O
16	S
34	Se
52	Te
84	Po

All these elements have an outermost electron structure of s^2p^4.

16. 32; the 6th period has this number of elements.

18. Ar and K; Co and Ni; Te and I; Th and Pa; U and Np; Es and Fm; Md and No.

20. The correct statements are: a, c, e, f, g, i, and n.

22. (a) F   9 protons
       (b) Ag   47 protons
       (c) Br   35 protons
       (d) Sb   51 protons

24. (a) Cl $1s^2 2s^2 2p^6 3s^2 3p^5$
       (b) Ag $1s^2 2s^2 2p^6 3s^2 3p^6 4s^2 3d^{10} 4p^6 5s^1 4d^{10}$
       (c) Li $1s^2 2s^1$
       (d) Fe $1s^2 2s^2 2p^6 3s^2 3p^6 4s^2 3d^6$
       (e) I $1s^2 2s^2 2p^6 3s^2 3p^6 4s^2 3d^{10} 4p^6 5s^2 4d^{10} 5p^5$

26. Each line corresponds to a change from one orbit to another.

28. 32 electrons

30. (a) (14p / 14n) $2e^- 8e^- 4e^-$ $^{28}_{14}Si$

       (b) (16p / 16n) $2e^- 8e^- 6e^-$ $^{32}_{16}S$

       (c) (18p / 22n) $2e^- 8e^- 8e^-$ $^{40}_{18}Ar$

       (d) (23p / 28n) $2e^- 8e^- 11e^- 2e^-$ $^{51}_{23}V$

       (e) (15p / 16n) $2e^- 8e^- 5e^-$ $^{31}_{15}P$

32. (a) Sc
       (b) Zr
       (c) Sn
       (d) Cs

34.

Atomic number		Electron structure
(a)	9	$[He]2s^2 2p^5$
(b)	26	$[Ar]4s^2 3d^6$
(c)	31	$[Ar]4s^2 3d^{10} 4p^1$
(d)	39	$[Kr]5s^2 4d^1$
(e)	52	$[Kr]5s^2 4d^{10} 5p^4$
(f)	10	$[He]2s^2 2p^6$

36. (a) (13p / 14n) $2e^- 8e^- 3e^-$ $^{27}_{13}Al$

       (b) (22p / 26n) $2e^- 8e^- 10e^- 2e^-$ $^{48}_{22}Ti$

38. It is in the fourth energy level because the 4*s* orbital is at a lower energy level than the 3*d* orbital.

40. Noble gases each have filled *s* and *p* orbitals in the outer energy level.

42. All the elements in a Group have the same number of outer-shell electrons.

44. All of these elements have an $s^2 d^{10}$ electron configuration in their outermost energy levels.

46. (a) and (f)
       (e) and (h)

48. 7, 33 since they are in the same group.

50. (a) nonmetal
       (b) metal
       (c) metal
       (d) metalloid

52. Period 6, lanthinide series

54. Group VIIA contains 7 valence electrons.
       Group VIIB contains 2 electrons in the outermost level and 5 electrons in an inner *d* orbital.

56. Nitrogen has more valence electrons on more energy levels. More varied electron jumps are possible.

58. (a) 100%   (d) 24%
       (b) 100%   (e) 20%
       (c) 19%

60. $\dfrac{1.5 \times 10^8}{1}$

62. In the scanning-probe microscope, a probe is placed near the surface of a sample and a parameter such as voltage is measured. The signals are translated electronically into a topographic image of the object. In a light microscope the image is formed by light reflecting from the object.

64. In the atomic force microscope, the probe measures the electric forces between electrons in the molecule, while in scanning-tunneling the actual movement of the electrons is measured.

66. Transition elements are found in Groups IB–VIIB and VIII.

68. 8, 16, 34, 52, 84. All have 6 electrons in their outer shell.

70. 35 would be in VIIA and 37 would be in IA.

72. (a) $[Rn]7s^2 6d^{10} 5f^{14} 7p^5$
    (b) 7 valence electrons
    (c) F, Cl, Br, I, At
    (d) halogen family, Group VIIA

74. Most gases are located in the upper right part of the periodic table (H is the exception). They are nonmetals. Liquids show no pattern. Neither do solids, except that the vast majority of solids are metals.

76. Oil can be traced by added labeled fullerenes.

Chapter 11

2. More energy is required for neon because it has a very stable outer electron structure consisting of an octet of electrons in filled orbitals (noble gas electron structure).

4. The first ionization energy decreases from top to bottom because, in successive alkali metals, the outermost electron is farther away from the nucleus and is shielded from the positive nucleus by additional electron shells.

6. Barium has a lower ionization energy than beryllium.

8. (a) K > Na   (d) I > Br
   (b) Na > Mg   (e) Zr > Ti
   (c) O > F

10. Atomic size increases down the column since each successive element has an additional energy level.

12. Cs·   Ba: Tl: ·Pb: ·Po: ·At: :Rn:
    Each of these is a representative element and has the same number of electrons in its outer shell as its periodic group number.

14. Valence electrons are the electrons found in the outermost energy level of an atom.

16. An aluminum ion has a +3 charge because it has lost 3 electrons in acquiring a noble gas electron structure.

18. The correct statements are: b, c, g, h, j, k, n, and o.

20. A bromine atom is smaller since it has one less electron than the bromine ion in its outer shell.

22.
	+	−
(a)	H	Cl
(b)	Li	H
(c)	C	Cl
(d)	I	Br
(e)	Mg	H
(f)	O	F

24. (a) covalent
    (b) ionic
    (c) covalent
    (d) ionic

26. (a) $F + 1e^- \rightarrow F^-$ (b) $Ca \rightarrow Ca^{2+} + 2e^-$

28. (a) Ca: + :O: ⟶ CaO

    (b) Na· + ·Br: ⟶ NaBr

30. Si (4)   N (5)   P (5)   O (6)   Cl (7)

32. (a) Chloride ion, none
    (b) Nitrogen atom, gain $3e^-$ or lose $5e^-$
    (c) Potassium atom, lose $1e^-$

34. (a) A magnesium ion, Mg^{2+}, is larger than an aluminum ion, Al^{3+}.
    (b) The Fe^{2+} ion is larger than the Fe^{3+} ion.

36. (a) SbH_3, Sb_2O_3
    (b) H_2Se, SeO_3
    (c) HCl, Cl_2O_7
    (d) CCl_4, CO_2

38. BeBr, beryllium bromide
    $MgBr_2$, magnesium bromide
    $RaBr_2$, radium bromide

40. (a) Ga: (b) $[Ga]^{3+}$ (c) $[Ca]^{2+}$

42. (a) covalent   (c) covalent
    (b) ionic   (d) covalent

44. (a) covalent
    (b) covalent
    (c) ionic

46. (a) :O::O: (b) :Br:Br: (c) :I:I:

48. (a) :S:H
        H
    (b) :S::C::S:
    (c) H:N:H
          H
    (d) [H:N:H]⁺ [:Cl:]⁻
          H

50. (a) [:I:]⁻
    (b) [:S:]²⁻
    (c) [:O:C::O:]²⁻
          :O:
    (d) [:O:Cl:O:]⁻
          :O:
    (e) [:O:N::O:]⁻
          :O:

52. (a) nonpolar
    (b) nonpolar
    (c) polar

54. (a) 2 electron pairs, linear
    (b) 4 electron pairs, tetrahedral
    (c) 4 electron pairs, tetrahedral

56. (a) tetrahedral
    (b) pyramidal
    (c) tetrahedral

58. (a) tetrahedral
   (b) bent
   (c) bent
60. potassium
62. (a) Zn
   (b) Be
   (c) Ne
64. Lithium has a +1 charge after the first electron is removed. It takes more energy to overcome that charge than to remove an e^- from helium
66. $SnBr_2$, $GeBr_2$
68. A covalent bond results from the sharing of a pair of electrons between two atoms, while an ionic bond involves the transfer of one electron or more from one atom to another.
70. N, O, F, Cl
72. It is possible for a molecule to be nonpolar even though it contains polar bonds.
74. (a) 105°
   (b) 107°
   (c) 109.5°
   (d) 109.5°
76. 77 K is the boiling point of liquid N_2. If a substance superconducts at this temperature, liquid N_2 can be used to cool the material.
78. Current material used in superconductors is brittle, nonmalleable, and does not carry a high current per cross-sectional area.
80. Normally, the LCD acts as a mirror reflecting light. It has a series of layers, however. When the molecules at the top align with the lines etched on the first layer of glass and those at the bottom align with the grooves on the bottom layer of glass, the molecules in between form a twisted spiral (trying to align with those near them). If a current is applied to specific segments of the etched glass, the plates become charged and the spirals of molecules are attached to the charged plate destroying the arrangement. The pattern of reflected light is changed and a number appears.
82. (a) Both use the p orbitals for bonding. B uses one s and $2p$ while N uses $3p$ orbitals.
   (b) BF_3 is trigonal planar; NF_3 is pyramidal.
   (c) BF_3 has no lone pairs; NF_3 has one lone pair.
   (d) BF_3 has 3 very covalent bonds; NF_3 has 3 covalent bonds.
84. Each element in a particular group has the same number of valence electrons.
86. C_2Cl_4

Chapter 12

2. The air pressure inside the balloon is greater than the air pressure outside the balloon.
4. 1 torr = 1 mm Hg.
6. 1 atm corresponds to 4 L.
8. The piston would move downward.
10. O_2, H_2S, HCl, F_2, CO_2
12. Rn, F_2, N_2, CH_4, He, H_2, decreasing molar mass

14. (a) pressure   (c) temperature
   (b) volume   (d) number of moles
16. A gas is least likely to behave ideally at low temperatures.
18. Equal volumes of H_2 and O_2 at the same T and P:
   (a) have equal numbers of molecules (Avogadro's law)
   (b) mass O_2 = 16 × mass H_2
   (c) moles O_2 = moles H_2
   (d) average kinetic energies are the same (T same)
   (e) rate H_2 = 4 (rate O_2) (Graham's law of effusion)
   (f) density O_2 = 16 (density H_2)
20. $N_2(g) + O_2(g) \longrightarrow 2NO(g)$
   1 vol + 1 vol $\longrightarrow$ 2 vol
22. Conversion of oxygen to ozone is an endothermic reaction.
24. Heating a mole of N_2 gas at constant pressure has the following effects:
   (a) Density will decrease.
   (b) Mass does not change.
   (c) Average kinetic energy of the molecules increases.
   (d) Average velocity of the molecules will increase.
   (e) Number of N_2 molecules remains unchanged.
26. The correct statements are: b, d, f1, f4, h, i, n, o, p, q, t.
28. (a) 715 torr
   (b) 953 mbar
   (c) 95.3 kPa
30. (a) 0.082 atm
   (b) 55.92 atm
   (c) 0.296 atm
   (d) 0.0066 atm
32. (a) 132 mL
   (b) 615 mL
34. 711 mm Hg
36. (a) 6.17 L
   (b) 8.35 L
38. 7.8×10^2 mL
40. 33.4 L
42. 681 torr
44. 1.45×10^3 torr
46. 1.19 L C_3H_8
48. 28.0 L N_2
50. 1.33 g NH_3
52. (a) 11 mol H_2S
   (b) 14.7 L H_2S
   (c) 31.8 L H_2S
54. 2.69×10^{22} molecules CH_4
56. (a) 0.179 g/L He
   (b) 2.50 g/L C_4H_8
58. (a) 3.17 g/L Cl_2
   (b) 1.46 g/L Cl_2
60. 19 L Kr
62. 72.8 L
64. (a) 5.6 mol NH_3
   (b) 0.640 L NO
   (c) 1.1×10^2 g O_2

66. 153 L SO_2

68. The can will explode.

70. (a) 5 L Cl_2
(b) 5.0 L NH_3
(c) 5.9 L SO_3

72. (a) CH_3
(b) C_2H_6

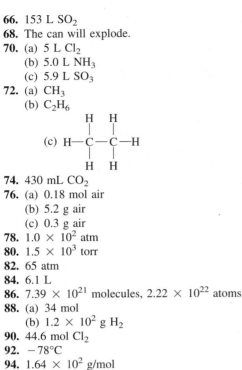

(c)

74. 430 mL CO_2

76. (a) 0.18 mol air
(b) 5.2 g air
(c) 0.3 g air

78. 1.0×10^2 atm

80. 1.5×10^3 torr

82. 65 atm

84. 6.1 L

86. 7.39×10^{21} molecules, 2.22×10^{22} atoms

88. (a) 34 mol
(b) 1.2×10^2 g H_2

90. 44.6 mol Cl_2

92. $-78°C$

94. 1.64×10^2 g/mol

96. 0.13 mol N_2

98. 9.0 atm

100. (a) Helium effuses twice as fast as CH_4.
(b) The gases meet 66.7 cm from the helium end.

102. (a) 10.0 mol CO_2; 3.0 mol O_2; no CO_2
(b) 29 atm

104. Some ammonia gas dissolves in the water squirted into the flask, lowering the pressure inside the flask. The atmospheric pressure outside is greater than the pressure inside the flask and thus pushes water from the beaker up the tube and into the flask.

106. (a) 1.1×10^2 torr CO_2; 13 torr H_2
(b) 120 torr

108. Air enters the room.

Chapter 13

2. H_2S, H_2Se, and H_2Te are gases

4.

6. Prefixes preceding the word *hydrate* are used to indicate the number of molecules of water present in the formulas.

8. about 70°C

10. case (b)

12. remain unchanged

14. (a) 88°C
(b) 78°C
(c) 16°C

16. melting point, 0°C; boiling point, 100°C (at 1 atm pressure); colorless; odorless; tasteless; heat of fusion, 335 J/g

(80 cal/g); heat of vaporization, 2.26 kJ/g (540 cal/g); density = 1.0 g/mL (at 4°C); specific heat = 4.184 J/g °C

18. If you apply heat to an ice-water mixture, the heat energy is absorbed to melt the ice, rather than to warm the water, so the temperature remains constant until all the ice has melted.

20. Ice floats in water because it is less dense than water. Ice sinks in ethyl alcohol because it is more dense than the alcohol.

22. Ethyl alcohol exhibits hydrogen bonding; ethyl ether does not.

24. Ammonia exhibits hydrogen bonding; methane does not.

26. $H_2NCH_2CH_2NH_2$

28. (a) mercury, acetic acid, water, toluene, benzene, carbon tetrachloride, methyl alcohol, bromine
(b) Highest is mercury; lowest is bromine.

30. In a pressure cooker, the temperature at which water boils increases above its normal boiling point, because the water vapor (steam) formed by boiling cannot escape. This results in an increased pressure over the water, and consequently, an increased boiling temperature.

32. As temperature increases, molecular velocities increase.

34. Ammonia

36. HF has a higher boiling point because of the strong H-bonding in HF.

38. 34.6°C

40. The expected temperature would be 4°C at the bottom of the lake.

42. endothermic

44. The correct statements are: a, b, c, f, h, l, m, o, p, s, t, w

46. $[HClO_4, Cl_2O_7]$ $[H_2CO_3, CO_2]$ $[H_3PO_4, P_2O_5]$

48. $[Ca(OH)_2CaO]$ $[KOH, K_2O]$ $[Ba(OH)_2, BaO]$

50. (a) $Li_2O + H_2O \longrightarrow 2\ LiOH$
(b) $2\ KOH \overset{\Delta}{\longrightarrow} K_2O + H_2O$
(c) $Ba + 2\ H_2O \longrightarrow Ba(OH)_2 + H_2$
(d) $Cl_2 + H_2O \longrightarrow HCl + HClO$
(e) $SO_3 + H_2O \longrightarrow H_2SO_4$
(f) $H_2SO_3 + 2\ KOH \longrightarrow K_2SO_3 + 2\ H_2O$

52. (a) magnesium ammonium phosphate hexahydrate
(b) iron(II) sulfate heptahydrate
(c) tin(IV) chloride pentahydrate

54. (a) Distilled water has been vaporized by boiling and recondensed.
(b) Natural waters are generally not pure, but contain dissolved minerals and suspended matter, and can even contain harmful bacteria.

56. 0.262 mol $FeI_2 \cdot 4\ H_2O$

58. 1.05 mol H_2O

60. 48.7% H_2O

62. $FePO_4 \cdot 4\ H_2O$

64. 5.5×10^4 J

66. 42 g

68. The system will be at 0°C. It will be a mixture of ice and water.

70. (a) 0.784 g H_2O
(b) 0.447 g H_2O
(c) 0.167 g H_2O

72. Eventually the water will lose enough energy to change from a liquid to a solid (freeze).

74. (a) From 0°C to 40°C solid X warms until at 40°C it begins to melt. The temperature remains at 40°C until all of X is melted. After that, liquid X will warm steadily to 65°C where it will begin to boil and remain at 65°C until all the liquid becomes vapor. Beyond 65°C the vapor will warm steadily until 100°C.

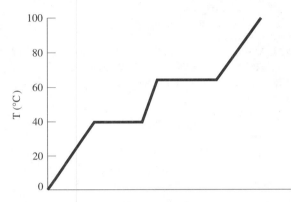

(b) 37,000 J

76. 75°C at 270 torr

78. $MgSO_4 \cdot 7 H_2O$ $Na_2HPO_4 \cdot 12 H_2O$

80. chlorine

82. When organic pollutants in water are oxidized by dissolved oxygen, there may not be sufficient dissolved oxygen to sustain marine life.

84. Na_2 zeolite(s) + $Mg^{2+}(aq) \longrightarrow$
$$Mg \text{ zeolite}(s) + 2Na^+(aq)$$

86. Humectants are polar compounds while emollients are nonpolar compounds.

88. The triangle theory of sweetness states that "sweet" molecules contain three specific sites that produce the proper structure to attach to the taste buds and trigger the "sweet" response.

90. 1.6×10^5 cal

92. 3.1 kcal

94. 2.30×10^6 J

96. 40.2 g H_2O

98. 6.97×10^{18} molecules/s

100. (a) 40.0 mL O_2
(b) 20.0 mL O_2 unreacted

Chapter 14

2. 4.5 g NaF

4. For lithium and sodium halides, solubility is
$$F^- < Cl^- < Br^- < I^-$$
For potassium halides, solubility is
$$Cl^- < Br^- < F^- < I^-$$

6. KNO_3

8. 6×10^2 cm^2

10. The dissolving process involves solvent molecules attaching to the solute ions or molecules. This rate decreases

as more of the solvent molecules are already attached to solute molecules. As the solution becomes saturated, the number of unused solvent molecules decreases. Also, the rate of recrystallization increases as the concentration of dissolved solute increases.

12. The solution level in the thistle tube will fall.

14. It is not always apparent which component in a solution is the solute.

16. yes

18. Hexane and benzene are both nonpolar molecules.

20. Air is considered to be a solution because it is a homogeneous mixture of several gaseous substances and does not have a fixed composition.

22. The solubility of gases in liquids is greatly affected by the pressure of a gas above the liquid; little effect for solids in liquids.

24. In a saturated solution, the net rate of dissolution is zero.

26. 16 moles HNO_3/L of solution

28. The champagne would spray out of the bottle.

30. Water molecules can pass through in both directions.

32. A lettuce leaf immersed in salad dressing containing salt and vinegar will become limp and wilted as a result of osmosis.

34. The correct statements are: a, b, f, h, j, k, l, n, p, r, s, t, v, x, y, z

36. Since there are 2 equivalents per mole of H_2SO_4, 18 M is equivalent to 36 N.

38. (a) 1 M NaOH
(b) 0.6 M Ba(OH)$_2$
(c) 2 M KOH
(d) 1.5 M Ca(OH)$_2$

40. The vapor pressure of the solution equals the vapor pressure of the pure solvent at the freezing point. See Figure 14.8.

42. The presence of the methanol lowers the freezing point of the water.

44. The molarity of a 5 molal solution is less than 5 M.

46. When the paper on a scratch and sniff label is scratched or pulled open the fragrance is released into the air.

48. Three types of microencapsulation systems are:
(a) water diffuses through capsule forming a solution which diffuses out again
(b) mechanical
(c) thermal

50. Reasonably soluble: (c) $CaCl_2$ (d) $Fe(NCl_3)_3$
Insoluble: (a) PbI_2 (b) $MgCO_3$ (e) $BaSO_4$

52. (a) 7.41% $Mg(NO_3)_2$
(b) 6.53% $NaNO_3$

54. 544 g solution

56. (a) 8.815% NaOH
(b) 15% $C_6H_{12}O_6$

58. 37.5 g K_2CrO_4

60. 33.6% NaCl

62. 22% C_6H_{14}

64. (a) 2.5 M HCl
(b) 0.59 M BaCl$_2 \cdot 2H_2O$
(c) 2.19×10^{-3} M Al$_2(SO_4)_3$
(d) 0.172 M Ca(NO$_3$)$_2$

66. (a) 3.5×10^{-4} mol NaOH
(b) 16 mol $CoCl_2$

68. (a) 13 g HCl
(b) 8.58 g $Na_2C_2O_4$

70. (a) 3.91×10^4 mL
(b) 7.82×10^3 mL

72. (a) 0.250 L
(b) 1.81 M NaCl

74. (a) 20. mL 15 M NH_3
(b) 69 mL 18 M H_2SO_4

76. (a) 1.83 M H_2SO_4
(b) 0.30 M H_2SO_4

78. (a) 33.3 mL of 0.250 M Na_3PO_4
(b) 1.10 g $Mg_3(PO_4)_2$

80. (a) 0.13 mol Cl_2
(b) 16 mol HCl
(c) 1.3×10^2 mL of 6 M HCl
(d) 3.2 L Cl_2

82. (a) $\left(\dfrac{98.08 \text{ g } H_2SO_4}{\text{eq}}\right)$

$\left(\dfrac{56.11 \text{ g KOH}}{\text{eq}}\right)$

(b) $\left(\dfrac{49.00 \text{ g } H_3PO_4}{\text{eq}}\right)$

$\left(\dfrac{23.95 \text{ g LiOH}}{\text{eq}}\right)$

(c) $\left(\dfrac{63.02 \text{ g } HNO_3}{\text{eq}}\right)$

$\left(\dfrac{40.00 \text{ g NaOH}}{\text{eq}}\right)$

84. (a) 5.55 N H_3PO_4
(b) 0.250 N $HC_2H_3O_2$
(c) 1.25 N NaOH

86. (a) 11.94 mL NaOH
(b) 10.09 mL NaOH

88. (a) 5.5 m $C_6H_{12}O_6$
(b) 0.25 m I_2

90. (a) 0.544 m
(b) 2.7°C
(c) 81.5°C

92. 163 g/mol

94. 97 g 10% NaOH solution

96. $C_8H_4N_2$

98. 0.14 M

100. (a) 4.5 g NaCl
(b) 450. mL H_2O must be evaporated.

102. 210. mL solution

104. 6.72 M HNO_3

106. 540. mL water to be added

108. To make 250. mL of 0.625 M KOH, take 31.3 mL of 5.00 M KOH and dilute with water to a volume of 250. mL.

110. 2.08 M HCl

112. 12.0 g $Mg(OH)_2$

114. 6.2 m H_2SO_4
5.0 M H_2SO_4

116. (a) 2.9 m
(b) 101.5°C

118. (a) 8.04×10^3 g $C_2H_6O_2$
(b) 7.24×10^3 mL $C_2H_6O_2$
(c) -4.0°F

120. 0.46 L HCl

122. (a) 7.7 L H_2O must be added
(b) 0.0178 mol
(c) 0.0015 mol

124. Mix together 667 mL 3.00 M HNO_3 and 333 mL 12.0 M HNO_3 to get 1000. mL of 6.00 M HNO_3

126. 2.84 g $Ba(OH)_2$ is formed.

128. (a) 0.011 mol
(b) 14 g
(c) 3.2×10^2 mL solution
(d) 1.5%

130. 8.9×10^{-3} M

Chapter 15

2. An electrolyte must be present in the solution for the bulb to glow.

4. First the orientation of the polar water molecules about the Na^+ and Cl^- ions is different. Second more water molecules will fit around Cl^-, since it is larger than the Na^+ ion.

6. tomato juice

8. Arrhenius: $HCl + NaOH \longrightarrow NaCl + H_2O$
Brønsted-Lowry: $HCl + KCN \longrightarrow HCN + KCl$
Lewis: $AlCl_3 + NaCl \longrightarrow AlCl_4^- + Na^+$

10. acids, bases, salts

12. Hydrogen chloride dissolved in water conducts an electric current. HCl does not ionize in benzene.

14. CH_3OH is a nonelectrolyte; NaOH is an electrolyte. This indicates that the OH group in CH_3OH must be covalently bonded to the CH_3 group.

16. The dissolving of NaCl is a dissociation process, while the dissolving of HCl in water is an ionization process.

18. Ions are hydrated in solution because there is an electrical attraction between the charged ions and the polar water molecules.

20. (a) $[H^+] = [OH^-]$
(b) $[H^+] > [OH^-]$
(c) $[OH^-] > [H^+]$

22. HCl is much more soluble in the polar solvent, water, than the nonpolar solvent, benzene.

24. The correct statements are: a, c, f, i, j, k, l, n, p, r, t, u

26. Colloids are prepared by two methods:
(a) dispersion
(b) condensation

28. *Adsorption* refers to the adhesion of particles to a surface while *absorption* refers to the taking in of one material by another.

30. Dialysis is the process of removing dissolved solutes from colloidal dispersion by the use of a dialyzing membrane.

32. Acidic shampoo breaks hydrogen bonds and salt bridges in the hair leaving only disulfide bonds. In a depilatory, a basic solution is used which breaks all types of bonds (H, salt bridges, disulfide) and the hair dissolves.

34. (a) $H_2SO_4 - HSO_4^-$; $H_2C_2H_3O_2^+ - HC_2H_3O_2$

    (b) Step 1: $H_2SO_4 - HSO_4^-$; $H_3O^+ - H_2O$

        Step 2: $HSO_4^- - SO_4^{2-}$; $H_3O^+ - H_2O$

    (c) $HClO_4 - ClO_4^-$; $H_3O^+ - H_2O$

    (d) $H_3O^+ - H_2O$; $CH_3OH - CH_3O^-$

36. (a) $NaOH(aq) + HBr(aq) \longrightarrow NaBr(aq) + H_2O(l)$

    (b) $KOH(aq) + HCl(aq) \longrightarrow KCl(aq) + H_2O(l)$

    (c) $Ca(OH)_2(aq) + 2 HI(aq) \longrightarrow CaI_2(aq) + 2 H_2O(l)$

    (d) $Al(OH)_3(s) + 3 HBr(aq) \longrightarrow AlBr_3(aq) + 3 H_2O(l)$

    (e) $Na_2O(s) + 2 HClO_4(aq) \longrightarrow 2 NaClO_4(aq) + H_2O(l)$

    (f) $3 LiOH(aq) + FeCl_3(aq) \longrightarrow Fe(OH)_3(s) + 3 LiCl(aq)$

38. (a) $NaHCO_3$—salt

    (c) $AgNO_3$—salt

    (d) $HCOOH$—acid

    (e) $RbOH$—base

    (f) K_2CrO_4—salt

40. (a) $0.75\ M\ Zn^{2+}$, $1.5\ M\ Br^-$

    (b) $4.95\ M\ SO_4^{2-}$, $3.30\ M\ Al^{3+}$

    (c) $0.682\ M\ NH_4^+$, $0.341\ M\ SO_4^{2-}$

    (d) $0.0628\ M\ Mg^{2+}$, $0.126\ M\ ClO_3^-$

42. (a) $4.9\ g\ Zn^{2+}$, $12\ g\ Br^-$

    (b) $8.90\ g\ Al^{3+}$, $47.5\ g\ SO_4^{2-}$

    (c) $1.23\ g\ NH_4^+$, $3.28\ g\ SO_4^{2-}$

    (d) $0.153\ g\ Mg^{2+}$, $1.05\ g\ ClO_3^-$

44. (a) $[K^+] = 1.0\ M$, $[Ca^{2+}] = 0.5\ M$, $[Cl^-] = 2.0\ M$

    (b) No ions are present in the solution.

    (c) $0.67\ M\ Na^+$, $0.67\ M\ NO_3^-$

46. (a) $0.147\ M\ NaOH$

    (b) $0.964\ M\ NaOH$

    (c) $0.4750\ M\ NaOH$

48. (a) $H_2S(g) + Cd^{2+}(aq) \longrightarrow CdS(s) + 2 H^+(aq)$

    (b) $Zn(s) + 2H^+(aq) \longrightarrow Zn^{2+}(aq) + H_2(g)$

    (c) $Al^{3+}(aq) + PO_4^{3-}(aq) \longrightarrow AlPO_4(s)$

50. (a) $2\ M\ HCl$

    (b) $1\ M\ H_2SO_4$

52. 1.57×10^3 mL of $0.245\ M$ HCl

54. 7.0% NaCl in sample

56. $0.936\ L\ H_2$

58. (a) 7.0

    (b) 0.30

    (c) 4.00

60. (a) 4.30

    (b) 10.47

62. 22.7 mL of $0.325\ N\ H_2SO_4$

64. 57 g/eq

66. 3.0×10^3 mL

68. The ionization of the acetic acid solution increases the particle concentration above that of the alcohol solution.

70. A hydronium ion is a hydrated hydrogen ion.

72. (a) $100°C$ $pH = 6.0$

       $25°C$ $pH = 7.0$

    (b) H^+ concentration is higher at $100°C$.

    (c) The water is neutral at both temperatures.

74. $0.201\ M$ HCl

76. 0.673 g KOH

78. 13.9 L of $18.0\ M\ H_2SO_4$

80. $0.3586\ N$ NaOH

82. 1.2×10^2 mL

84. $0.100\ N$ NaOH, $0.0500\ N$ HCl

86. The molarity and normality of an acid solution will be the same when the acid has one ionizable hydrogen. The molarity and the normality of a base solution will be the same when the base has one ionizable hydroxide. The normality and molarity of a salt (ionic compound) solution will be the same when the salt contains one ionizable cation with a charge of $+1$.

88. 1.1, acidic

90. (a) $2\ NaOH(aq) + H_2SO_4(aq) \longrightarrow$
$$Na_2SO_4(aq) + 2\ H_2O(l)$$

    (b) 1.0×10^2 mL NaOH

    (c) 0.71 g Na_2SO_4

92. $12\ M\ HNO_3$

94. (c)

absolute zero $-273°C$, the zero point on the Kelvin (absolute) temperature scale. *See also* Kelvin scale. [2.11, 12.6]

acid (1) A substance that produces H^+ (H_3O^+) when dissolved in water. (2) A proton donor. (3) An electron-pair acceptor. A substance that bonds to an electron pair. [15.1]

acid anhydride A nonmetal oxide that reacts with water to form an acid. [13.12]

activation energy The amount of energy needed to start a chemical reaction. [8.5, 16.15]

actual yield The amount of product actually produced in a chemical reaction (as compared to the theoretical yield). [9.6]

alkali metal An element (except H) from Group IA of the periodic table. [10.4]

alkaline earth metal An element from Group IIA of the periodic table. [10.4]

allotrope A substance existing in two or more molecular or crystalline forms (example: graphite and diamond are two allotropic forms of carbon). [12.17]

amorphous A solid without shape or form. [3.2]

amphoteric (substance) A substance having properties of both an acid and a base. [15.3]

anion A negatively charged ion. *See also* ion. [3.9, 5.5, 6.2]

1 atmosphere The standard atmospheric pressure; that is, the pressure exerted by a column of mercury 760 mm high at a temperature of 0°C. *See also* atmospheric pressure. [12.3]

atmospheric pressure The pressure experienced by objects on the earth as a result of the layer of air surrounding our planet. A pressure of 1 atmosphere (1 atm) is the pressure that will support a column of mercury 760 mm high at 0°C. [12.3]

atom The smallest particle of an element that can enter into a chemical reaction. [3.4]

atomic mass The average relative mass of the isotopes of an element referred to the atomic mass of carbon-12. [5.11]

atomic mass unit (amu) A unit of mass equal to one-twelfth the mass of a carbon-12 atom. [5.11]

atomic number The number of protons in the nucleus of an atom of a given element. *See also* isotopic notation. [5.9]

atomic theory The theory that substances are composed of atoms, and that chemical reactions are explained by the properties and the interactions of these atoms. [5.2, Ch. 10]

Avogadro's law Equal volumes of different gases at the same temperature and pressure contain equal numbers of molecules. [12.11]

Avogadro's number 6.022×10^{23}; the number of formula units in 1 mole. [7.1, 9.1]

balanced equation A chemical equation having the same number of each kind of atom and the same electrical charge on each side of the equation. [8.2]

barometer A device used to measure atmospheric pressure. [12.3]

base A substance whose properties are due to the liberation of hydroxide (OH^-) ions into a water solution. [15.1]

basic anhydride A metal oxide that reacts with water to form a base. [13.12]

binary compound A compound composed of two different elements. [6.4]

boiling point The temperature at which the vapor pressure of a liquid is equal to the pressure above the liquid. [13.5]

bond length The distance between two nuclei that are joined by a chemical bond. [13.10]

Boyle's law At constant temperature, the volume of a fixed mass of gas is inversely proportional to the pressure (PV = constant). [12.5]

Brownian movement The random motion of colloidal particles. [15.15]

calorie (cal) A commonly used unit of heat energy; one calorie is a quantity of heat energy that will raise the temperature of 1 g of water 1°C (from 14.5 to 15.5°C). Also, 4.184 joules = 1 calorie exactly. *See also* joule. [4.6]

capillary action The spontaneous rising of a liquid in a narrow tube, which results from the cohesive forces within the liquid and the adhesive forces between the liquid and the walls of the container. [13.4]

cation A positively charged ion. *See also* ion. [3.9, 5.5]

Celsius temperature scale (°C) The temperature scale on which water freezes at 0°C and boils at 100°C at 1 atm pressure. [2.11]

Charles' law At constant pressure, the volume of a fixed mass of any gas is directly proportional to the absolute temperature (V/T = constant). [12.6]

chemical bond The attractive force that holds atoms together in a compound. [Ch.11]

chemical change A change producing products that differ in composition from the original substances. [4.3]

chemical equation A shorthand expression showing the reactants and the products of a chemical change (for example, $2 H_2O = 2 H_2 + O_2$). [4.3, 8.1]

chemical family *See* groups or families of elements.

chemical formula A shorthand method for showing the composition of a compound using symbols of the elements. [3.11]

chemical properties The ability of a substance to form new substances either by reaction with other substances or by decomposition. [4.1]

chemistry The science of the composition, structure, properties, and reactions of matter, especially of atomic and molecular systems. [1.2]

chlorofluorocarbons A group of compounds made of carbon, chlorine, and fluorine. [CIA p. 62]

colligative properties Properties of a solution that depend on the number of solute particles in solution and not on the nature of the solute (examples: vapor pressure lowering, freezing point lowering, boiling point elevation). [14.7]

colloid A dispersion in which the dispersed particles are larger than the solute ions or molecules of a true solution and smaller than the particles of a mechanical suspension. [15.13]

combination reaction A direct union or combination of two substances to produce one new substance. [8.4]

combustion A chemical reaction in which heat and light are given off; generally, the process of burning or uniting a substance with oxygen. [8.5]

common names Arbitrary names that are not based on the chemical compostion of compounds (examples: quicksilver for mercury, laughing gas for nitrous oxide). [6.1]

compound A distinct substance composed of two or more elements combined in a definite proportion by mass. [3.9]

concentrated solution A solution containing a relatively large amount of dissolved solute. [14.6]

concentration of a solution A quantitative expression of the amount of dissolved solute in a certain quantity of solvent. [14.2]

condensation The process by which molecules in the gaseous state return to the liquid state. [13.3]

conjugate acid-base Two molecules or ions whose formulas differ by one H^+. (The acid is the species with the H^+, and the base is the species without the H^+.) [15.1]

covalent bond A chemical bond formed between two atoms by sharing a pair of electrons. [11.5]

Dalton's atomic theory The first modern atomic theory to state that elements are composed of minute individual particles called *atoms*. [5.2]

Dalton's law of partial pressures The total pressure of a mixture of gases is the sum of the partial pressures exerted by each of the gases in the mixture. [12.10]

decomposition reaction A breaking down, or decomposition, of one substance into two or more different substances. [8.5]

deliquescence The absorption of water from the atmosphere by a substance until it forms a solution. [13.14]

density The mass of an object divided by its volume. [2.12]

dialysis The process by which a parchment membrane allows the passage of true solutions but prevents the passage of colloidal dispersions. [15.17]

diatomic molecules The molecules of elements that always contain two atoms. Seven of the elements occur as diatomic molecules: H_2, N_2, O_2, F_2, Cl_2, Br_2, and I_2. [3.10]

diffusion The property by which gases and liquids mix spontaneously because of the random motion of their particles. [12.2]

dilute solution A solution containing a relatively small amount of dissolved solute. [14.6]

dipole A molecule that is electrically asymmetrical, causing it to be oppositely charged at two points. [11.6]

dissociation The process by which a salt separates into individual ions when dissolved in water. [15.6]

double bond A covalent bond in which two pairs of electrons are shared. [11.5]

double-displacement reaction A reaction of two compounds to produce two different compounds by exchanging the components of the reacting compounds. [8.4]

effusion The process by which gas molecules pass through a tiny orifice from a region of high pressure to a region of lower pressure. [12.2]

electrolyte A substance whose aqueous solution conducts electricity. [15.5]

electron A subatomic particle that exists outside the nucleus and carries a negative electrical charge. [5.6]

electron configuration The orbital arrangement of electrons in an atom. [10.5]

electron-dot structure *See* Lewis structure.

electronegativity The relative attraction that an atom has for a pair of shared electrons in a covalent bond. [11.6]

electron shell *See* principal energy levels of electrons.

element A basic building block of matter that cannot be broken down into simpler substances by ordinary chemical changes; in 1994 there were 111 known elements. [3.4]

empirical formula A chemical formula that gives the smallest whole-number ratio of atoms in a compound, that is, the relative number of atoms of each element in the compound; also known as the simplest formula. [7.4]

endothermic reaction A chemical reaction that absorbs heat. [8.5]

energy The capacity of matter to do work. [4.5]

energy levels of electrons Areas in which electrons are located at various distances from the nucleus. [10.4]

energy sublevels The *s, p, d,* and *f* orbitals within a principal energy level occupied by electrons in an atom. [10.4]

equivalent mass That mass of a substance that will react with, combine with, contain, replace, or in any other way be equivalent to 1 mole of hydrogen atoms or hydrogen ions. [14.6]

evaporation The escape of molecules from the liquid state to the gas or vapor state. [13.2]

exothermic reaction A chemical reaction in which heat is released as a product. [8.5]

Fahrenheit temperature scale (°F) The temperature scale on which water freezes at 32°F and boils at 212°F at 1 atm pressure. [2.11]

formula unit The atom or molecule indicated by the formula of the substance under consideration (examples: Mg, O_2, H_2O). [7.1]

freezing or melting point The temperature at which the solid and liquid states of a substance are in equilibrium. [13.6]

frequency speed A measurement of the number of waves that pass a particular point per second. [10.2]

gas A state of matter that has no shape or definite volume so that the substance completely fills its container. [3.2]

Gay-Lussac's law At constant volume, the pressure of a fixed mass of gas is directly proportional to the absolute temperature (P/T = constant). [12.7]

Gay-Lussac's law of combining volumes (of gases) When measured at the same temperature and pressure, the ratios of the volumes of reacting gases are small whole numbers. [12.11]

Graham's law of effusion The rates of effusion of two gases at the same temperature and pressure are inversely proportional to the square roots of their densities or molar masses. [12.2]

ground state The lowest available energy level within an atom. [10.3]

groups or families (of elements) Vertical groups of elements in the periodic table (IA, IIA, and so on). Families of elements that have similar outer-orbital electron structures. [10.6]

halogens Group VIIA of the periodic table; consists of the elements fluorine, chlorine. bromine, iodine, and astatine. [10.4]

heat A form of energy associated with the motion of small particles of matter. [2.11]

heat of fusion The energy required to change 1 gram of a solid into a liquid at its melting point. [13.7]

heat of reaction The quantity of heat produced by a chemical reaction. [8.5]

heat of vaporization The amount of heat required to change 1 gram of a liquid to a vapor at its normal boiling point. [13.7]

heterogeneous Matter without uniform composition—having two or more components or phases. [3.3]

homogeneous Matter that has uniform properties throughout. [3.3]

hydrate A solid that contains water molecules as a part of its crystalline structure. [13.13]

hydrocarbon A compound composed entirely of carbon and hydrogen. [CIA p. 62, 8.5]

hydrogen bond The intermolecular force acting between molecules that contain hydrogen covalently bonded to the highly electronegative elements, F, O, and N. [13.11]

hydronium ion The result of a proton combining with a polar water molecule to form a hydrated hydrogen ion (H_3O^+). [15.1]

hygroscopic substance A substance that readily absorbs and retains water vapor. [13.14]

hypothesis A tentative explanation of certain facts to provide a basis for further experimentation. [1.4]

ideal gas A gas that behaves precisely according to the Kinetic Molecular Theory; also called a perfect gas. [12.2]

ideal gas equation $PV = nRT$; that is, the volume of a gas varies *directly* with the number of gas molecules and the absolute temperature and *inversely* with the pressure. [12.14]

immiscible Incapable of mixing; immiscible liquids do not form a solution with one another. [14.2]

ion A positively or negatively charged atom or group of atoms. *See also* cation, anion. [3.9, 5.5]

ionic bond The chemical bond between a positively charged ion and a negatively charged ion. [11.3]

ionic compound A compound that will conduct electricity when dissolved in water. [6.3]

ionization The formation of ions, which occurs as the result of a chemical reaction of certain substances with water. [15.6]

ionization energy The energy required to remove an electron from an atom, an ion, or a molecule. [11.1]

isotope An atom of an element that has the same atomic number but a different atomic mass. Since their atomic numbers are identical, isotopes vary only in the number of neutrons in the nucleus. [5.10]

isotopic notation Notation for an isotope of an element where the subscript is the atomic number, the superscript is the mass number, and they are attached on the left of the symbol for the element. (For example, hydrogen-1 is notated as ^1_1H.) *See also* atomic number, mass number. [5.10]

IUPAC International Union of Pure and Applied Chemistry, which devised (in 1921) and continually upgrades the system of nomenclature for inorganic and organic compounds. [6.1]

joule (J) The SI unit of energy. *See also* calorie. [4.6]

Kelvin temperature scale (K) Absolute temperature scale starting at absolute zero, the lowest temperature possible. Freezing and boiling points of water on this scale are 273 K and 373 K, respectively, at 1 atm pressure. *See also* absolute zero. [2.11, 12.6]

kilocalorie (kcal) 1000 cal; the kilocalorie is also known as the nutritional or large Calorie, used for measuring the energy produced by food. [4.6]

kilogram (kg) The standard unit of mass in the metric system; 1 kilogram equals 2.205 pounds. [2.9]

kilojoule (kJ) 1000J. [4.6]

kinetic energy (KE) The energy that matter possesses due to its motion; $KE = 1/2\ mv^2$. [4.5]

Kinetic-Molecular Theory (KMT) A group of assumptions used to explain the behavior and properties of gases. [12.2]

law A statement of the occurrence of natural phenomena that occur with unvarying uniformity under the same conditions. [1.4]

Law of Conservation of Energy Energy can neither be created nor destroyed, but it may be transformed from one form to another. [4.8]

Law of Conservation of Mass No change is observed in the total mass of the substances involved in a chemical reaction; that is, the mass of the products equals the mass of the reactants. [4.4]

Law of Definite Composition A compound always contains two or more elements in a definite proportion by mass. [5.3]

Law of Multiple Proportions Atoms of two or more elements may combine in different ratios to produce more than one compound. [5.3]

Lewis structure A method of indicating the covalent bonds between atoms in a molecule or an ion such that a pair of electrons (:) represents the valence electrons forming the covalent bond. [11.2]

limiting reactant A reactant that limits the amount of product formed because it is present in insufficient amount compared to the other reactants. [9.6]

line spectrum Colored lines generated when light emitted by a gas is passed through a spectroscope. Each element possesses a unique set of these lines. [10.3]

linear structure In the VSEPR model, an arrangement where the pairs of electrons are arranged 180° apart for maximum separation. [11.11]

liquid A state of matter in which the particles move about freely while the substance retains a definite volume; thus, liquids flow and take the shape of their containers. [3.2]

liter (L) A unit of volume commonly used in chemistry; 1 L = 1000 mL; the volume of a kilogram of water at 4°C. [2.10]

logarithm (log) The power to which 10 must be raised to give a certain number. The log of 100 is 2. [15.9]

mass The quantity or amount of matter that an object possesses. [2.1]

mass number The sum of the protons and neutrons in the nucleus of a given isotope of an atom. *See also* isotopic notation. [5.10]

mass percent solution The grams of solute in 100 g of a solution. [14.6]

matter Anything that has mass and occupies space. [3.1]

melting or freezing point *See* freezing or melting point.

meniscus The concave shape of the surface of a liquid when placed in a glass cylinder. [13.4]

metal An element that is solid at room temperature and whose properties include luster, ductility, malleability, and good conductivity of heat and electricity; metals tend to lose their valence electrons and become positive ions. [3.8]

metalloid An element having properties that are intermediate between those of metals and nonmetals (for example, silicon); these elements are useful in electronics. [3.8]

meter (m) The standard unit of length in the SI and metric systems; 1 meter equals 39.37 inches. [2.7]

metric system A decimal system of measurements. *See also* SI. [2.6]

miscible Capable of mixing and forming a solution. [14.2]

mixture Matter containing two or more substances, which can be present in variable amounts; mixtures can be homogeneous (sugar water) or heterogeneous (sand and water). [3.3, 3.12]

molality (m) An expression of the number of moles of solute dissolved in 1000 grams of solvent. [14.7]

molarity (M) The number of moles of solute per liter of solution. [14.6]

molar mass The mass of Avogadro's number of atoms or molecules. [7.1, 9.1]

molar solution A solution containing 1 mole of solute per liter of solution. [14.6]

molar volume (of a gas) The volume of 1 mol of a gas at STP equals 22.4 L/mol. [12.12]

mole The amount of a substance containing the same number of formula units (6.022×10^{23}) as there are in exactly 12 g of C-12. One mole is equal to the molar mass in grams of any substance. [7.1]

mole ratio A ratio between the number of moles of any two species involved in a chemical reaction; the mole ratio is used as a conversion factor in stoichiometric calculations. [9.2]

molecular equation *See* un-ionized equation.

molecular formula The total number of atoms of each element present in one molecule of a compound; also known as the true formula. *See also* empirical formula. [7.4]

molecule The smallest uncharged individual unit of a compound formed by the union of two or more atoms. [3.9]

net ionic equation A chemical equation that includes only those molecules and ions that have changed in the chemical reaction. [15.12]

neutralization The reaction of an acid and a base to form a salt plus water. [15.10]

neutron A subatomic particle that is electrically neutral and is found in the nucleus of an atom. [5.6]

noble gases A family of elements in the periodic table—helium, neon, argon, krypton, xenon, and radon—that contain a particularly stable electron structure. [10.4]

nonelectrolyte A substance whose aqueous solutions do not conduct electricity. [15.5]

nonmetal An element that has properties the opposite of metals: lack of luster, relatively low melting point and density, and generally poor conduction of heat and electricity. Non-metals may or may not be solid at room temperature (examples: carbon, bromine, nitrogen); many are gases. They are located mainly in the upper right-hand corner of the periodic table. [3.8]

nonpolar covalent bond A covalent bond between two atoms with the same electronegativity value; thus the electrons are shared equally between the two atoms. [11.6]

normal boiling point The temperature at which the vapor pressure of a liquid equals 1 atm or 760 torr pressure. [13.5]

normality The number of equivalent masses of solute per liter of solution. [14.6]

nucleus The central part of an atom that contains all its protons and neutrons. The nucleus is very dense and has a positive electrical charge. [5.8]

one atmosphere The standard atmospheric pressure; that is, the pressure exerted by a column of mercury 760 mm high at a temperature of 0°C. [12.3]

orbital A cloudlike region around the nucleus where electrons are located. Orbitals are considered to be energy sublevels (*s, p, d, f*) within the principal energy levels. *See also* principal energy levels. [10.3, 10.4]

organic diagram A way of showing the arrangement of electrons in an atom where orbitals are represented by boxes grouped by sublevel with small arrows indicating the electrons. [10.5]

osmosis The diffusion of water, either pure or from a dilute solution, through a semipermeable membrane into a solution of higher concentration. [14.8]

partial pressure The pressure exerted independently by each gas in a mixture of gases. [12.10]

parts per million (ppm) A measurement of the concentration of dilute solutions now commonly used by chemists in place of mass percent. [14.6]

Pauli exclusion principle An atomic orbital can hold a maximum of two electrons, which must have opposite spins. [10.4]

percent composition of a compound The mass percent represented by each element in a compound. [7.3]

percent yield The ratio of the actual yield to the theoretical yield multiplied by 100. [10.6]

perfect gas A gas that behaves precisely according to theory; also called an ideal gas [12.2]

period of elements The horizontal groupings of elements in the periodic table. [10.6]

periodic table An arrangement of the elements according to their atomic numbers. The table consists of horizontal rows or periods and vertical columns or families of elements. Each period ends with a noble gas. [10.6]

pH A method of expressing the H^+ concentration (acidity) of a solution; pH $= -\log [H^+]$. pH $= 7$ is a neutral solution, pH < 7 is acidic, and pH > 7 is basic. [15.9]

phase A homogeneous part of a system separated from other parts by a physical boundary. [3.3]

photon Theoretically, a tiny packet of energy that streams with others of its kind to produce a beam of light. [10.2]

photosynthesis The process by which green plants utilize light energy to synthesize carbohydrates. [CIA p. 154]

physical change A change in form (such as size, shape, physical state) without a change in composition. [4.2]

physical properties Inherent physical characteristics of a substance that can be determined without altering its composition: color, taste, odor, state of matter, density, melting point, boiling point. [4.1]

physical states of matter Solids, liquids, and gases. [3.2]

physiological saline solution A solution of 0.90% sodium chloride that is isotonic (has the same osmotic pressure) with blood plasma. [14.8]

polar covalent bond A covalent bond between two atoms with differing electronegativity values, resulting in unequal sharing of bonding electrons. [11.5]

polyatomic ion An ion composed of more than one atom. [6.5]

potential energy Stored energy, or the energy of an object due to its relative position. [4.5]

pressure Force per unit area; expressed in many units, such as mm Hg, atm, $lb/in.^2$, torr, pascal. [12.3]

principal energy levels of electrons Existing within the atom, these energy levels contain orbitals within which electrons are found. *See also* orbital, electron. [10.4]

product A chemical substance produced from reactants by a chemical change. [4.3]

properties The characteristics, or traits, of substances that give them their unique identities. Properties are classified as physical or chemical. [4.1]

proton A subatomic particle found in the nucleus of the atom that carries a positive electrical charge and a mass of about 1 amu. An H^+ ion is a proton. [5.6]

quanta Small discrete increments of energy. From the theory proposed by physicist Max Planck that energy is emitted in energy *quanta* rather than a continuous stream. [10.3]

quantum mechanics or wave mechanics The modern theory of atomic structure based on the wave properties of matter. [10.3]

reactant A chemical substance entering into a reaction. [4.3]

representative element An element in one of the A groups in the periodic table. [10.6]

resonance structure A molecule or ion that has multiple Lewis structures. *See also* Lewis structure. [11.8]

rounding off numbers The process by which the value of the last digit retained is determined after dropping nonsignificant digits. [2.3]

salts Ionic compounds of cations and anions. [Ch. 6, 15.4]

saturated solution A solution containing dissolved solute in equilibrium with undissolved solute. [14.6]

scientific laws Simple statements of natural phenomena to which no exceptions are known under the given conditions. [1.4]

scientific method A method of solving problems by observation; recording and evaluating data of an experiment; formulating hypotheses and theories to explain the behavior of nature; and devising additional experiments to test the hypotheses and theories to see if they are correct. [1.4]

scientific notation Writing a number as a power of 10; to do this, move the decimal point in the original number so that it is located after the first nonzero digit, follow the new number by a multiplication sign and 10 with an exponent (called its *power*) that is the number of places the decimal point was moved. Example: $2468 = 2.468 \times 10^3$. [2.4]

semipermeable membrane A membrane that allows the passage of water (solvent) molecules through it in either direction but prevents the passage of larger solute molecules or ions. [14.8]

SI An agreed-upon standard system of measurements used by scientists around the world (*Système Internationale*). *See also* metric system. [2.6]

significant figures The number of digits that are known plus one estimated digit are considered significant in a measured quantity; also called significant digits. [2.2, Appendix I]

simplest formula *See* empirical formula.

single bond A covalent bond in which one pair of electrons is shared between two atoms. [11.5]

single-displacement reaction A reaction of an element and a compound that yields a different element and a different compound. [8.4]

sol A colloidal dispersion in a liquid. [15.14]

solid A state of matter having a definite shape and a definite volume, whose particles cohere rigidly to one another, so that a solid can be independent of its container. [3.2]

solubility An amount of solute that will dissolve in a specific amount of solvent under stated conditions. [14.2]

solute The substance that is dissolved—or the least abundant component—in a solution. [14.1]

solution A system in which one or more substances are homogeneously mixed or dissolved in another substance. [14.1]

solvent The dissolving agent or the most abundant component in a solution. [14.1]

specific gravity The ratio of the density of one substance to the density of another substance taken as a standard. Water is usually the standard for liquids and solids; air, for gases. [2.12]

specific heat The quantity of heat required to change the temperature of 1 gram of any substance by 1°C. [4.6]

spectator ion An ion in solution that does not undergo chemical change during a chemical reaction. [15.10]

speed (of a wave) A measurement of how fast a wave travels through space. [10.2]

spin A property of an electron that describes its appearance of spinning on an axis like a globe; the electron can only spin in two directions and, to occupy the same orbital, two electrons must spin in opposite directions. *See also* orbital. [10.4]

standard boiling point *See* normal boiling point.

standard conditions *See* STP.

standard temperature and pressure *See* STP.

Stock (nomenclature) System A system that uses Roman numerals to name elements that form more than one type of cation. (For example: Fe^{2+}, iron (II); Fe^{3+}, iron (III).) [6.4]

stoichiometry The area of chemistry that deals with the quantitative relationships among reactants and products in a chemical reaction. [9.2]

STP (standard temperature and pressure) 0°C (273 K) and 1 atm (760 torr); also known as standard conditions. [12.8]

strong electrolyte An electrolyte that is essentially 100% ionized in aqueous solution. [15.7]

subatomic particles Particles found within the atom, mainly protons, neutrons, and electrons. [5.6]

sublimation The process of going directly from the solid state to the vapor state without becoming a liquid. [13.2]

subscript Number that appears partially below the line and to the right of a symbol of an element (example: H_2SO_4). [3.11]

substance Matter that is homogeneous and has a definite, fixed composition; substances occur in two forms—as elements and as compounds. [3.3]

supersaturated solution A solution containing more solute than needed for a saturated solution at a particular temperature. Supersaturated solutions tend to be unstable; jarring the container or dropping in a "seed" crystal will cause crystallization of the excess solute. [14.6]

surface tension The resistance of a liquid to an increase in its surface area. [13.4]

symbol In chemistry, an abbreviation for the name of an element. [3.7]

system A body of matter under consideration. [3.3]

temperature A measure of the intensity of heat, or of how hot or cold a system is; the SI unit is the Kelvin (K). [2.11]

tetrahedral structure An arrangement of the VSEPR model where four pairs of electrons are placed 109.5 degrees apart to form a tetrahedron. [11.11]

theoretical yield The maximum amount of product that can be produced according to a balanced equation. [9.6]

theory An explanation of the general principles of certain phenomena with considerable evidence to support it; a well-established hypothesis. [1.4]

Thomson model of the atom Thomson asserted that atoms are not indivisible but are composed of smaller parts; they contain both positively and negatively charged particles—protons as well as electrons. [5.6]

titration The process of measuring the volume of one reagent required to react with a measured mass or volume of another reagent. [15.10]

torr A unit of pressure (1 torr = 1 mm Hg). [12.3]

total ionic equation An equation that shows compounds in the form in which they actually exist. Strong electrolytes are written as ions in solution, whereas nonelectrolytes, weak electrolytes, precipitates, and gases are written in the un-ionized form. [15.12]

transition elements The metallic elements characterized by increasing numbers of d and f electrons in an inner shell. These elements are located in Groups IB through VIIB and in Group VIII of the periodic table. [10.6]

trigonal planar An arrangement of atoms in the VSEPR model where the three pairs of electrons are placed 120° apart on a flat plane. [11.11]

triple bond A covalent bond in which three pairs of electrons are shared between two atoms. [11.5]

Tyndall effect An intense beam of light passed through a colloidal dispersion is clearly visible but is not visible when passed through a true solution. [15.15]

un-ionized equation A chemical equation in which all the reactants and products are written in their molecular, or normal, formula expression; also called a molecular equation. [15.12]

unsaturated solution A solution containing less solute per unit volume than its corresponding saturated solution. [14.6]

valence electron An electron in the outermost energy level of an atom; these electrons are the ones involved in bonding atoms together to form compounds. [10.5]

vaporization *See* evaporation. [13.2]

vapor pressure The pressure exerted by a vapor in equilibrium with its liquid. [13.3]

vapor pressure curve A graph generated by plotting the temperature of a liquid on the x axis and its vapor pressure on the y axis. Any point on the curve represents an equilibrium between the vapor and liquid. [13.5]

volatile (substance) A substance that evaporates readily; a liquid with a high vapor pressure and a low boiling point. [13.3]

volume The amount of space occupied by matter; measured in SI units by cubic meters (m^3), but also commonly in liters and milliliters. [2.10]

volume percent (solution) The volume of solute in 100 mL of solution. [14.6]

VSEPR Valence shell electron pair repulsion; a simple model for predicting the shapes of molecules. [11.11]

water of crystallization Water molecules that are part of a crystalline structure, as in a hydrate; also called water of hydration. [13.13]

water of hydration *See* water of crystallization.

wavelength The distance between consecutive peaks and troughs in a wave; symbolized by the Greek letter lambda. [10.2]

weak electrolyte A substance that is ionized to a small extent in aqueous solution. [15.7]

weight A measure of the earth's gravitational attraction for a body (object). [2.1]

word equation A statement in words, in equation form, of the substances involved in a chemical reaction. [8.2]

Front Matter

p. i, Phototake/The Creative Link; p. ii, Tony Stone Images/Ian Murphy; p. vi, Tony Stone Images/Steve Leonard; p. xi, Courtesy of IBM/Tom Way; xii, Courtesy of Merck and Company (Black Star/Michael Sheil); p. xiv, Peter Arnold/David Cavagnaro.

Chapter 1

p. 1, The Image Bank/Murray Alcosser; p. 3, Both photos courtesy of Merck and Company (left photo Black Star/Michael Sheil); p. 4, Lori Ritzenthaler; p. 5, Courtesy of William Buehler; p. 8, Tony Stone Images/Steve Leonard; p. 9, Phototake/Richard Nowitz; p. 10, The Image Bank/Stephen Marks.

Chapter 2

p. 11, The Image Bank/Murray Alcosser; p. 13, The Image Bank/Andy Caulfield; p. 14, Courtesy of NASA; p. 18, Left: The Image Bank/Joseph Drivas; Right: Photo Researchers/ CNRI/Science Photo Library; p. 21, Lori Ritzenthaler; p. 28, Rita Amaya (4 photos); p. 34, Lori Ritzenthaler; p. 37, Photo Researchers/Will and Deni McIntyre.

Chapter 3

p. 46, Peter Arnold/Michel Viard; p. 49, Phototake/Edward S. Ross; p. 53, Courtesy of IBM/ Tom Way; p. 54, The Granger Collection; p. 55, Courtesy of International Academy of Science; pp. 56-57, Rita Amaya; p. 61, Courtesy of GE Superabrasives; p. 62, Rita Amaya (3 photos).

Chapter 4

p. 68, Photo Researchers/Fritz Henle; p. 71, PeterArnold/Leonard Lessin (2 photos); p. 72, Phototake/Yoav Levy; p. 75, Tony Stone Images/Chris McCooey; p. 76, Consumer Reports (3 photos); p. 78, Photo Researchers/Lee F. Snyder; p. 79, Lori Ritzenthaler.

Chapter 5

p. 83 Photo Researchers/Kent Wood; p. 85, The Granger Collection; p. 86, The Granger Collection; p. 87 Julie Kranhold; p. 88, Courtesy of IBM Almaden Research Center; p. 89, The Granger Collection; p. 90, Photo Researchers/Peter Fowler/Science Photo Library; p. 94, Lori Ritzenthaler.

Chapter 6

p. 100, The Image Bank/Antonio Rosario; p. 103, San Jose Mercury News/Jason Grow; p. 106, Fundamental Photographs/Richard Megna; p. 108, Courtesy of Sargent-Welch, a VWR Corporation; p. 119, Lori Ritzenthaler.

Chapter 7

p. 124, The Image Bank/Michael Skott; p. 126, The Granger Collection; p. 130, Rita Amaya; p. 135, Julie Kranhold.

Chapter 8

p. 147, Fundamental Photographs/Richard Megna; p. 150, Phototake/Yoav Levy; p. 153, Lori Ritzenthaler; p. 154, Peter Arnold/David Cavagnaro; p. 157, Phototake/Yoav Levy; p. 159, Fundamental Photographs/Richard Megna; p. 160, Fundamental Photographs/Michael Dalton; p. 161, Rita Amaya; p. 162, Tony Stone Images/Andy Sacks; p. 163, Rita Amaya (2 photos); p. 164, Tony Stone Images/Dave Bjorn.

Chapter 9

p. 171, Courtesy of Merck and Company/Michael Taufic; p. 177, Courtesy of NASA; p. 181, Peter Arnold/Manfred Kage; p. 183, Gazelle Technologies/© 1991 Preferred Stock.

Chapter 10

p. 193, Tony Stone Images/Tim Brown; p. 195, Courtesy of The National Institute of Standards and Technology; p. 196, The Granger Collection; p. 197, The Granger Collection; p. 198, Courtesy of IBM Almaden Research Center; p. 199, Tony Stone Images/Craig Wells; p. 201, Left: Phototake/Carolina Biological Supply; Middle: Phototake; Right: Photo Researchers/ Lawrence Livermore Laboratory/Science Photo Library; p. 203, Courtesy of IBM Almaden Research Center.

Chapter 11

p. 214, Imagebank/Andrea Pistolesi; p. 219, Courtesy of Bancroft Library/University of California, Berkeley; p. 224, Peter Arnold/IBM Research; p. 231, AP/Wide World Photos; p. 233, Left: Courtesy of Bharat Bhushan/Ohio State University; p. 233, Right: Courtesy of IBM Almaden Research Center.

Chapter 12

p. 251, Phototake/Ray Nelson; p. 258, The Granger Collection; p. 263, Rita Amaya (3 photos); p. 265, The Granger Collection; p. 270, Peter Arnold/John Allison; p. 277, Tony Stone Images/ Stephen Frink; p. 285, Peter Arnold/Peggy and Yoram Kahana; p. 286, Courtesy of NASA.

Chapter 13

p. 294, Courtesy of NASA; p. 298, Lori Ritzenthaler; p. 299, Phototake/Yoav Levy; p. 302, Courtesy of Matthew Senay and David Jewitt/University of Hawaii; p. 304, Rita Amaya; p. 308, Lori Ritzenthaler; p. 312, Fundamental Photographs/Paul Silverman; p. 313, Sara Hotchkiss/SKA; p. 317, Courtesy of Southern California Edison/Mitch Kaufman.

Chapter 14

p. 324, Tony Stone Images/Warren Bolster; p. 326, Fundamental Photographs/Richard Megna; p. 327, Kip Peticolas/Fundamental Photographs; p. 330, Lori Ritzenthaler; p. 335, Fundamental Photographs/Richard Megna; p. 341, Courtesy of Sargent Welch/A VWR Corporation; pp. 346 and 348, Lori Ritzenthaler; p. 349, Peter Arnold/Craig Newbauer; p. 352, Phototake/ Dennis Kunkel (3 photos)

Chapter 15

p. 361, Tony Stone Images/G. Brad Lewis; p. 369, The Image Bank/Charles C. Place; pp. 370 and 377, Rita Amaya; p. 378, Lori Ritzenthaler; p. 379, Left: Rita Amaya; Right: Peter Arnold/Leonard Lessin; p. 382, The Image Bank/Tim Bieber; p. 388, Rita Amaya; p. 390, Photo Researchers.

INDEX

Entries in boldface appear in the Glossary.

NAMES, FORMULAS AND CHARGES OF COMMON IONS

Positive Ions (Cations)		
1+	Ammonium	NH_4^+
	Copper(I)	Cu^+
	(Cuprous)	
	Hydrogen	H^+
	Potassium	K^+
	Silver	Ag^+
	Sodium	Na^+
2+	Barium	Ba^{2+}
	Cadmium	Cd^{2+}
	Calcium	Ca^{2+}
	Cobalt(II)	Co^{2+}
	Copper(II)	Cu^{2+}
	(Cupric)	
	Iron(II)	Fe^{2+}
	(Ferrous)	
	Lead(II)	Pb^{2+}
	Magnesium	Mg^{2+}
	Manganese(II)	Mn^{2+}
	Mercury(II)	Hg^{2+}
	(Mercuric)	
	Nickel(II)	Ni^{2+}
	Tin(II)	Sn^{2+}
	(Stannous)	
	Zinc	Zn^{2+}
3+	Aluminum	Al^{3+}
	Antimony(III)	Sb^{3+}
	Arsenic(III)	As^{3+}
	Bismuth(III)	Bi^{3+}
	Chromium(III)	Cr^{3+}
	Iron(III)	Fe^{3+}
	(Ferric)	
	Titanium(III)	Ti^{3+}
	(Titanous)	
4+	Manganese(IV)	Mn^{4+}
	Tin(IV)	Sn^{4+}
	(Stannic)	
	Titanium(IV)	Ti^{4+}
	(Titanic)	
5+	Antimony(V)	Sb^{5+}
	Arsenic(V)	As^{5+}

Negative Ions (Anions)		
1−	Acetate	$C_2H_3O_2^-$
	Bromate	BrO_3^-
	Bromide	Br^-
	Chlorate	ClO_3^-
	Chloride	Cl^-
	Chlorite	ClO_2^-
	Cyanide	CN^-
	Fluoride	F^-
	Hydride	H^-
	Hydrogen carbonate	HCO_3^-
	(Bicarbonate)	
	Hydrogen sulfate	HSO_4^-
	(Bisulfate)	
	Hydrogen sulfite	HSO_3^-
	(Bisulfite)	
	Hydroxide	OH^-
	Hypochlorite	ClO^-
	Iodate	IO_3^-
	Iodide	I^-
	Nitrate	NO_3^-
	Nitrite	NO_2^-
	Perchlorate	ClO_4^-
	Permanganate	MnO_4^-
	Thiocyanate	SCN^-
2−	Carbonate	CO_3^{2-}
	Chromate	CrO_4^{2-}
	Dichromate	$Cr_2O_7^{2-}$
	Oxalate	$C_2O_4^{2-}$
	Oxide	O^{2-}
	Peroxide	O_2^{2-}
	Silicate	SiO_3^{2-}
	Sulfate	SO_4^{2-}
	Sulfide	S^{2-}
	Sulfite	SO_3^{2-}
3−	Arsenate	AsO_4^{3-}
	Borate	BO_3^{3-}
	Phosphate	PO_4^{3-}
	Phosphide	P^{3-}
	Phosphite	PO_3^{3-}